Basics of Mechanical Engineering

Surinder Kumar

Associate Professor
Mechanical Engineering Department at
Sri Ram College of Engineering and Management
(SRCEM), Palwal

Ane Books Pvt. Ltd.

New Delhi ♦ Chennai

Basics of Mechanical Engineering

Surinder Kumar

Published by

Ane Books Pvt. Ltd.

4821, Parwana Bhawan, 1st Floor, 24 Ansari Road, Darya Ganj,
New Delhi - 110 002, India
Tel.: +91(011) 23276843-44, Fax: +91(011) 23276863
e-mail: kapoor@anebooks.com, Website: www.anebooks.com

Branch Office:

- Avantika Niwas, 1st Floor, 19 Doraiswamy Road, T. Nagar, **Chennai** - 600 017, Tel.: +91(044) 28141554, 28141209
e-mail: anebooks_tn@airtelmail.in

ISBN : 978-93-8061-807-4

Printed at: R.K. Offset Process, New Delhi.

Basics of Mechanical Engineering

Preface

The subject "Basics of Mechanical Engineering" is a course which is taught to the first year students of all the disciplines of Engineering in all technical institutions of Haryana which are affiliated to the M.D. University, Rohtak.

The course titled Basics of Mechanical Engineering covers the following topics:

- Introduction to Machine Tool commonly used in workshop
- Basic Concept of Thermodynamics
- Properties of Steam and Steam Generators
- Refrigeration and Air Conditioning
- Hydraulic Turbines and Pumps
- Power Transmission Methods and Devices
- Stress and Strains
- Manufacturing System

The book has been written in simple and lucid language. Text and diagrams have been presented in a conecrete and comprehensive manner. It covers a broad spectrum of critical engineering topics and helps the readers understand the fundamentals. Each chapter is well-supported with necessary illustrations, solved examples and unsolved problems. Hightlights of the chapter and Review questions make the text more interesting. University examination questions have been added for the benefit of students.

I wish to express my profound thanks to all those who helped in making this book a reality. I am indebted to Mr. Sunil Saxena and Mr. J.R. Kapoor for their considerable cooperation and encouragement through the publication of this book. The never-ending family support and the presence of our lovely children Krish and Arshvi who infused vigour in me while writing this book.

My effort in writing this book will be amply rewarded if the readers send their constructive suggestions with a view to enhance the usefulness of the book.

S. Kumar

Acknowledgement

I would like to express my sincere gratitude to Sh. Narender Gupta, Hon'ble Chairman, Shri Ram College of Engineering and Management (SRCEM), Palwal (Haryana) for their encouragement and kind cooperation.

I will also sincerely acknowledge Dr. S.K. Gupta, Director-Principal, faculty members and staff of SRCEM for their constant support and suggestions.

I extend my heartiest gratitude to all those who have been associated with this book directly and indirectly. I also pay my gratitude to my students whose inquistive queries motivated me to write this book and I also express my thanks to all my colleages, friends and well-wishers who have cooperated me to complete this book successfully.

I am extremely grateful to Mr. Sunil Saxena of Ane Books Pvt. Ltd., New Delhi, and his team for their considerable cooperation and encouragement. The never-ending family support and the presence of our lovely children Krish and Arshvi who infused vigour in me while writing this book.

S. Kumar

Syllabus for

Maharishi Dayanand University, Rohtak

ME-101F: Basics of Mechanical Engineering

Section A

Introduction to Machine Tool to Commonly used Machine Tools in a Workshop: Lathe, Shaper, Planer, Milling, Drilling, Slotter. Introduction to Metal Cutting.

Basic Concept of Thermodynamics: Introduction, States, Work, Heat, Temperature, Zeroth, 1st, 2nd and 3rd law of thermodynamics, Concept of internal energy, enthalpy and entropy. Problems

Properties of Steam and Steam Generator: Formation of steam at constant pressure, Thermodynamic properties of Steam, Use of steam tables, Measurement of dryness fraction by throttling calorimeter.

Section B

Refrigeration and Airconditioning: Introduction to refrigeration and air-conditioning, Rating of refrigeration machines, Coefficient of performance, Simple refrigeration vapour compression cycle, Psychrometric charts and its use, Human comforts.

Hydraulic Turbines and Pumps: Introduction, Classification, Construction details and working of Pelton, Francis and Kaplan turbines, Specific speed and selection of turbines, Classification of water pumps and their working.

Section C

Power Transmission Methods and Devices: Introduction to Power transmission, Belt, Rope, Chain and Gear drive. Types and functioning of clutches.

Stresses and Strains: Introduction, Concept and types of Stresses and strains, Poisson's ratio, stresses and strains in simple and compound bars under axial, flexure and torsional loading, Stress-strain diagrams, Hook's law, Elastic constants and their relationships.

Section D

Introduction to Manufacturing Systems, Fundamentals of Numerical Control (NC), Advantage of NC systems, Classifications of NC, Comparison of NC and CNC.

Contents

1

Introduction to Machine Tool

1.1 MACHINE TOOL

A machine tool is a power operated metal cutting machine which is used for shaping, sizing or processing a product to a desired accuracy by removing the excess material in the form of metal chips known as machine tool.

Some common examples of machine tools are : lathe, shaper, planer, milling, drilling, slotter etc.

1.2 CLASSIFICATION OF MACHINE TOOLS

Machine tools may be classified as follows:

1. Classification Based on the Purpose of the Machine Tool:

(*i*) Single purpose machine tool

(*ii*) Multipurpose machine tool

(*iii*) Special purpose machine tool

(*iv*) Transfer machine tool

(*v*) Tape control machine tool

2. Classification Based on the Types of Surface Produced:

(*a*) *Machine tool producing mainly flat surfaces, e.g.*

(*i*) Slotting machine

(*ii*) Shaping machine

(*iii*) Planing machine

(*iv*) Milling machine

(*v*) Broaching machine etc.

(*b*) *Machine tool producing mainly cylindrical work:*

(*i*) Lathe

(*ii*) Drilling machine

(*iii*) Boring machine.

3. Classification Based Upon the Size of Chip Removal:

(*a*) Machine tool using cutting tools and producing comparatively larger sized chips.

(*b*) Machine tool using abrasives and producing chips of very small size.

1.3 FUNCTIONS OF A MACHINE TOOL

In general, machine tools are required to produce shapes (like flat, cylindrical, splines etc.) on the raw material by performing some metal removal operation in the form of chips from the work piece.

Functions of machine tools are as follows:

(*i*) To hold, support and guide the workpiece to be machined.

(*ii*) To hold, support and guide the cutting tools.

(*iii*) To regulate the cutting speed and also the feeding movement between the tool and workpiece.

(*iv*) To provide requisite, motion to the workpiece, tool or both.

(*v*) To perform various operations by using different attachments.

1.4 LATHE

Lathe machine is a tool which removes the material in the form of chips by rotating a workpiece against a single point cutting tool known as Lathe.

The main purpose of the lathe machine is to produce flat, tapered, circular, hollow and internal, external threads on nut, bolts etc. Some common examples of lathe work are as follows:

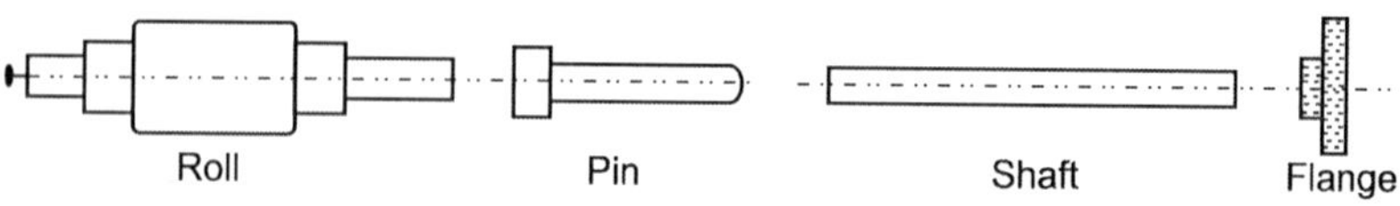

(*i*) Simply Turned Parts

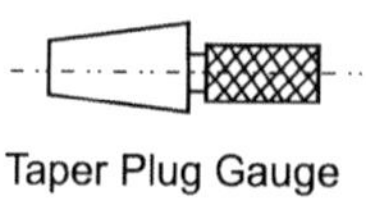

(*ii*) Taper Turned Parts

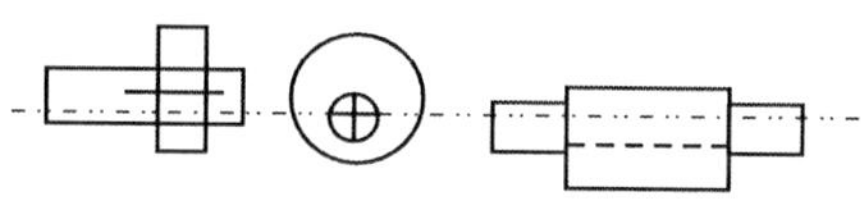

(*iii*) Eccentric Turned Parts

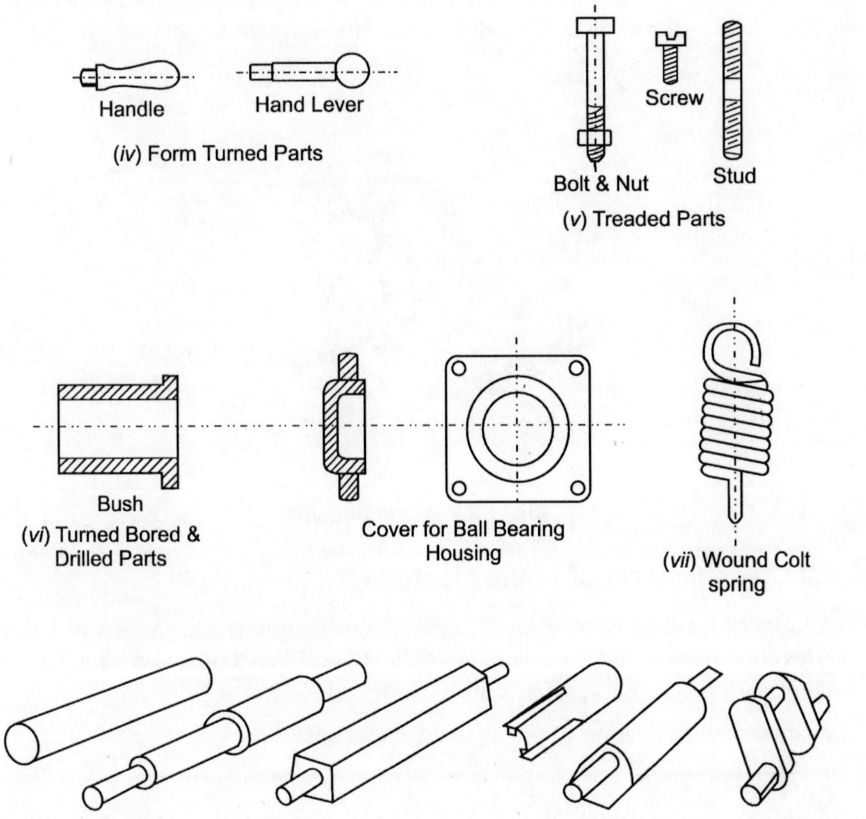

Fig. 1.1 Products

1.4.1 Function of Lathe

The main function of the lathe is metal cutting in a desired shape and size. This is accomplished by holding the workpiece rigidly between the centers. The workpiece rotated against a single point cutting tool to remove the material in the form of chips. The tool material should be hard, then the workpiece material to perform the proper cutting action.

1.4.2 Working Principle of Lathe

Lathe is a machine tool which removes material by rotating the work piece against a single point cutting tool to give it required shape and size when the work piece is held between centers, supported by Jaws of a Chuck, face plate and tail stock center of a lathe. The cutting tool is rigidly held and supported in a tool post securely and rigidly on a machine and is fed against revolving work to remove metal from the work in the form of Chips. A lathe is basically used to produce cylindrical as well as plane surfaces at right angles to the axis of rotation.

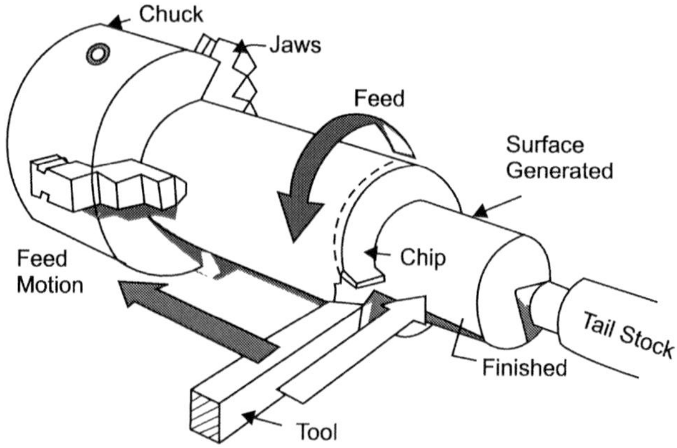

Fig. 1.2 Working of Lathe

1.4.3 Classification of Lathe Machine

The lathe of various designs and construction have been developed to suit the various conditions of metal machining. But all type of lathes have some fundamental principle of operation to perform the same function.

The general classification of the lathe are as follows:

1. Engine lathe
   (*a*) step cone pullely
   (*b*) gear head lathe
   (*c*) variable speed lathe
2. Speed lathe
   (*a*) wood working lathe
   (*b*) polishing lathe
   (*c*) centering lathe
   (*d*) spinning lathe
3. Bench lathe
4. Tool Room lathe
5. Turret lathe
6. Capstan lathe
7. Gap bed lathe
8. Hollow spindle lathe
9. Automotive lathe
   (*a*) crank shaft lathe
   (*b*) duplicating lathe
   (*c*) crank shaft lathe
   (*d*) facing lathe
10. Special purpose lathe.

1. **Engine Lathe:** It is the most important member of the lathe family. It was most popular in olden days. Its name is driven by separate engine with over head shaft and belt. The stepped cone pulley geared head are often used to vary the speed of lathe spindle. A tail stock is provided to facilitate

holding the work between centers and permit cutting tool both in cross and longitudinal direction with refrence to the lathe axis. It is also capable in performing a wide variety of operations with the help of various attachments. It is available in sizes to handle the job ranges 1m to 4m in length and upto 1m in diameters.

2. **Speed Lathe:** As it is indicated by its name speed lathe means high speed of the head stock spindle. It is simple in construction and operation having head stock, a tail stock and tool post. It has no gear feed box, lead screw and carriage tools are usually hand operated to provide the speed variation in spindle speed a stepped cone pulley is used. Such machines are used in wood turning, metal spinning and polishing operations.
3. **Bench Lathe:** It is a small lathe and is mounted on a separate prepared bench or cabinet. It is generally used for small and precision work with high accuracy. It is usually provided with all the attachment as a lower lathe and is capable in performing almost all operations as a engine lathe.
4. **Tool Room Lathe:** It is the modern engine lathe which is equipped with all necessary accessories and attachments for accurate tool room work. It is geared head with variable spindle speed and feeds. It is mainly used for precision work on tools, dies and gauges in machining work, where accuracy in needed. It is costlier than engine lathe.
5. **Turret Lathe:** This machine is used to perform large no. of operations *used in mass production*. The distinguishing feature of this type of lathe the *tail stock* of an engine lathe is *replaced by* a *hexagonal turret*, on the face of which multiple tools may be fitted and fed in to the work in a proper sequence and parting off tools with minimum wastage of time. Operations like drilling, counter, sinking, reaming, tapping etc. can be performed simultaneously.
6. **Capston Lathe:** It is similar to the turret lathe and incorporate capstan slide which moves on an auxiliary slide and can be clamped in any position. It is best suited for less production of small parts necessary in light weight and short stroke of capaston slide.
7. **Gap Bed Lathe:** A gap is provided on the bed near the head stock with a view to handle jobs having flanges and some other projected parts. Generally a removable portion is provided in the bed so that when not required, it can be inserted.
8. **Hollow Spindle Lathe:** These lathes are provided with spindles having large through bores in order to facilitate turning the ends of long tubler work pieces. The long jobs are supported on some outboard support.
9. **Automotive Lathe:** It is a high speed, heavy duty, mass production lathe with complete automatic control. Once tools are set and machine starts, it performs automatically all the operations to finish the job. The changing of tools, speeds and feeds are also done automatically. These machines may have single or multiple spindle.

10. **Special Purpose Lathe:** As the name implies, these are used for special purpose and for jobs which cannot be accommodated on a standard lathe.

    The "wheel lathe" is made for finishing the journals and turning the threads, rail road, car and locomotive as well as "T lathe" is used for jet engines. The axis of the bed is at right angles to the axis of the head stock spindle in the form of T. Mechanical, air and hydraulic devices are used to coordinate the movements of the tool to reproduce the accurate shape of the template. It is used for turning very long parts such as turbine, engine parts and crank shafts etc.

1.4.4 Principal Parts of a Lathe

The Lathe carries the following parts, as illustrated by a block diagram in Fig. 1.3. Detailed mechanical features of the lathe are shown in Fig. 1.4.

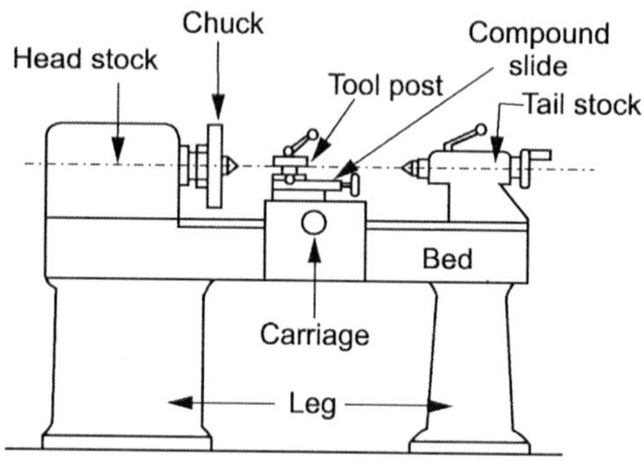

Fig. 1.3. Block diagram of a Lathe

The main parts of a lathe are:

1. Bed
2. Headstock
3. Tailstock
4. Carriage
    - (*a*) Saddle
    - (*b*) Cross slide
    - (*c*) Compound rest
    - (*d*) Tool post
    - (*e*) Apron

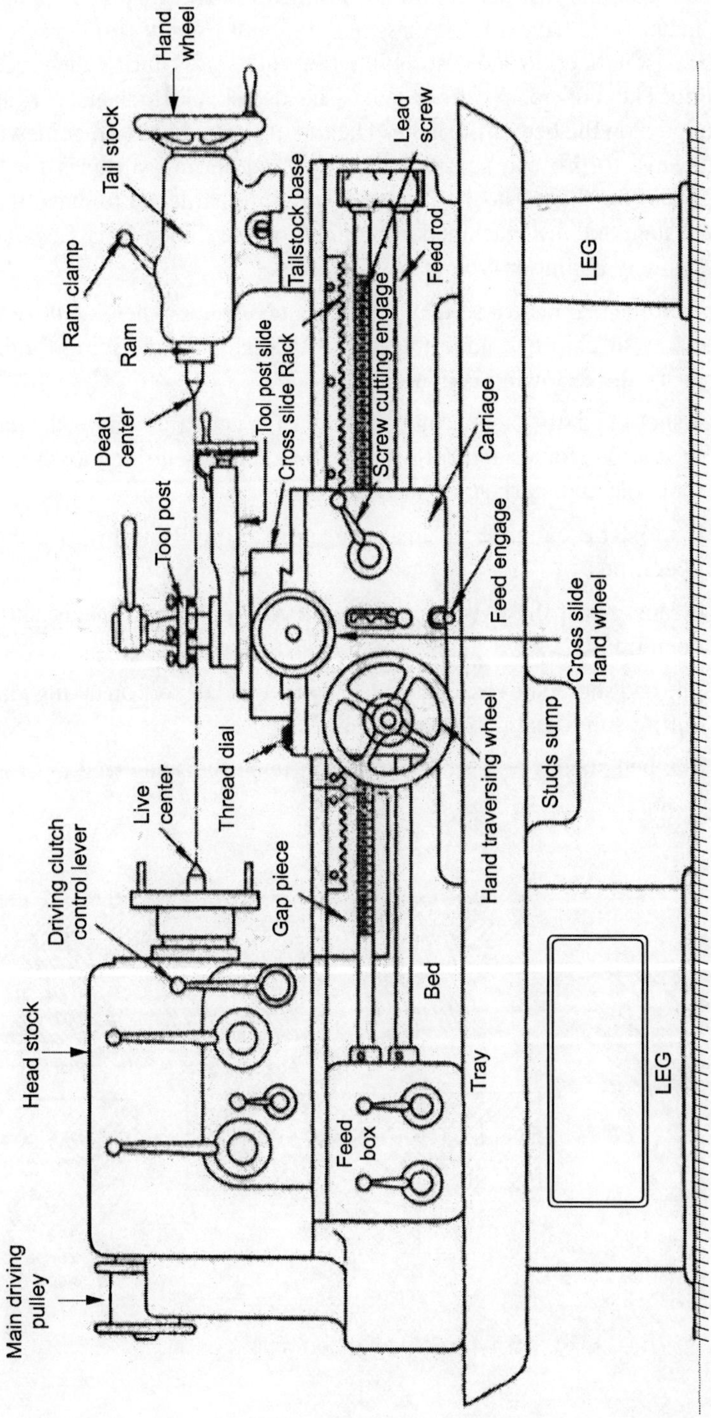

Fig. 1.4. Lathe Parts

1. **Bed:** The bed is the base of the lathe and it supports all the major components of the lathe. Beds have a large mass and are built usually from grey Cast Iron to resist deflection and absorb vibrations generated during the process of cutting. The three main parts of lathe—headstock, tailstock and carriage, are mounted on the bed of the lathe. The top of the bed has two guideways or slideways to provide support (bearing) and sliding surfaces for the carriage and for the tailstock. The guideways are hardened to make them wear resistant and are machined accurately to give dimensional accuracy. The ways may be flat or type in cross-section.

   Legs under the bed raise it so that the top is at convenient height above the floor. A tray or chip pan under the bed collcts chips and cutting fluid. It must satisfy the following conditions:

   (*i*) It should be sufficiently rigid and robust in construction to withstand the various forces and deflection exerted on the cutting tool through tool post and carriage on the lathe bed.

   (*ii*) It most be capable in absorbing vibrations (damping) during operations.

   (*iii*) It must resist the twisting stresses developed during metal cutting operations.

   (*iv*) The bed should be seasoned naturally to avoid distortion taking place during solidification of casting.

   (*v*) The bed should be tested for level in length-wise as well as cross-wise.

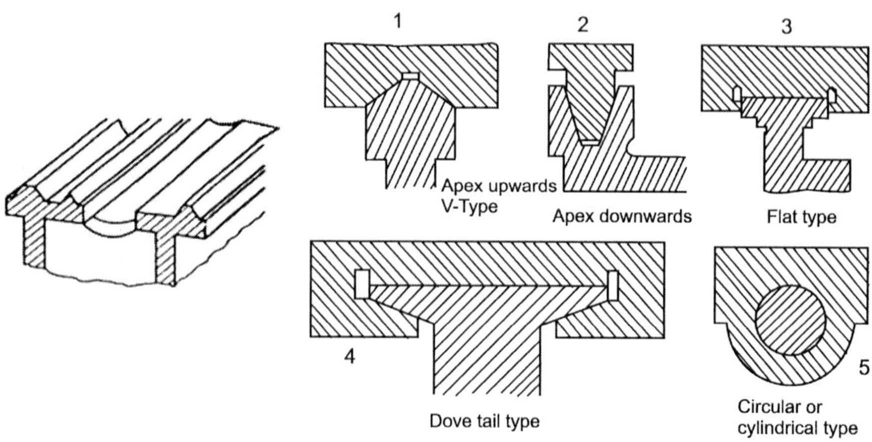

Fig. 1.5 Types of Lathe bedways

2. **Headstock:** The headstock is permanently mounted at the left hand end of the lathe bed. All lathes receive their power through the headstock, which may be equipped with step cone pulley drive or a geared head drive.

   Headstock supports the spindle and is equipped with the power driving mechanism for the spindle. The spindle speed can be set through speed selector knobs. The spindle is hollow to facilitate holding of long workpiece. The work holding devices such as chucks, centres and collects are attached to the spindle.

   (*i*) Cone pulley

   (*ii*) Back gears and back gear lever

   (*iii*) Main spindle and head stock spindle

   (*iv*) Live centre

   (*v*) Feed reverse lever.

   The spindle of head stock is made of carbon or nickel chrome steel.

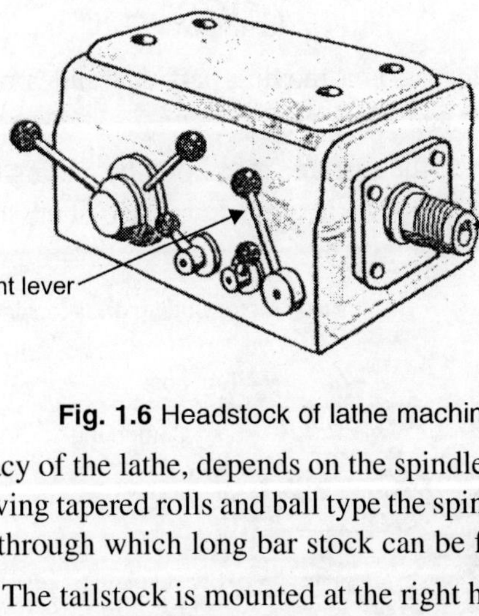

Fig. 1.6 Headstock of lathe machine.

   The accuracy of the lathe, depends on the spindle which is mounted on the bearing having tapered rolls and ball type the spindle had a hole throughout its length, through which long bar stock can be fed for machining.

3. **Tailstock:** The tailstock is mounted at the right hand end of the bed. It can be moved along the guideways and can be clamped in any position on the bed. The main function of the tailstock is to hold the dead centre and to support the long workpieces during machining and to hold twist drills, reamers and other tools that may be used to perform operations on jobs.

   The tailstock spindle is a hollow shaft. It is mounted in the tailstock body. The left end of its hole is tapered to hold the dead centre, or to hold a twist drill etc. Tailstock of a lathe is shown in Fig. 1.7

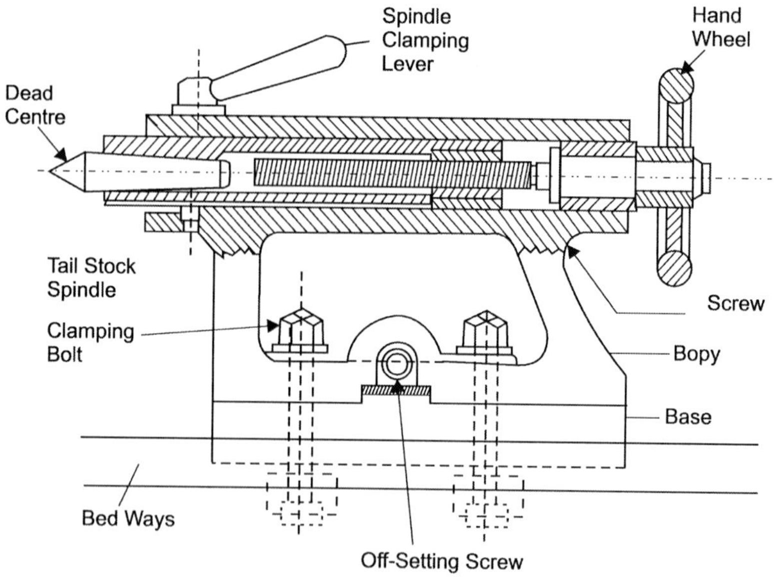

Fig. 1.7 Tail Stock

4. **Carriage:** The carriage is a moving part that slides over the guideways between headstock and tailstock.

   The main function of the carriage is to hold the cutting tool and move it to give longitudinal and or cross feed to it. The main parts of the carriage are shown in Fig. 1.8.

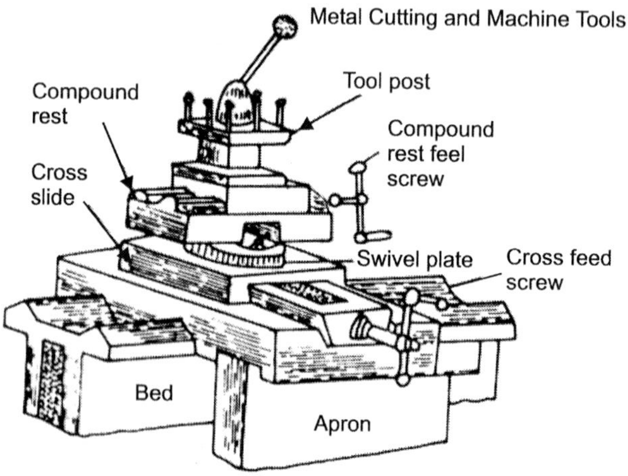

Fig. 1.8 Lathe carriage

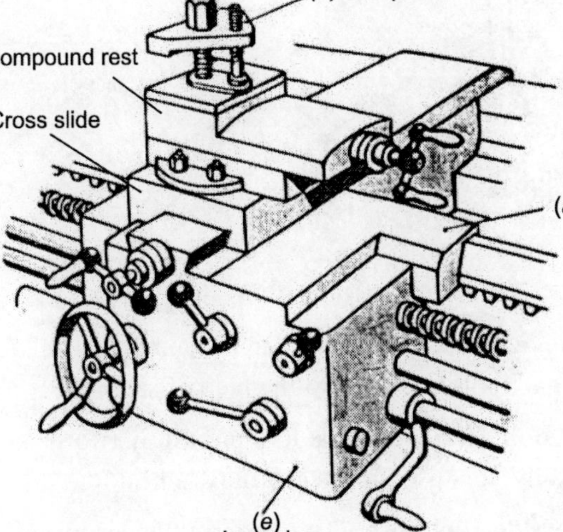

Fig. 1.9 Carriage of the machine (*a*) saddle, (*b*) cross slide, (*c*) compound rest, (*d*) tool post, (*e*) apron box

The slides should move in the guide ways freely without any play.

(*a*) *Saddle*: The Saddle is an H shaped casting that fits over the bed and slide along the ways. It carries the cross slide and tool post.

(*b*) *The cross slide*: Cross slide is mounted on the saddle and carries compound rest and tool post on the upper face of the cross slide. It can be moved by means of feed screw that is operated by a small hand wheel having graduated dial on its rim. It is used for facing and cut off directions.

(*c*) *The compound rest:* The compound rest or compound slide is mounted on the top of the cross-slide and has a circular base graduated in degrees. It is used for angular cuts and short tapers. The slide may be swivelled to any angle within a circle. A required depth of cut can be given on the work piece by hand operated.

(*d*) *Tool post*: It is located on the top of the compound rest to hold the tool for convenient working position. Common type of tool post are as follows:

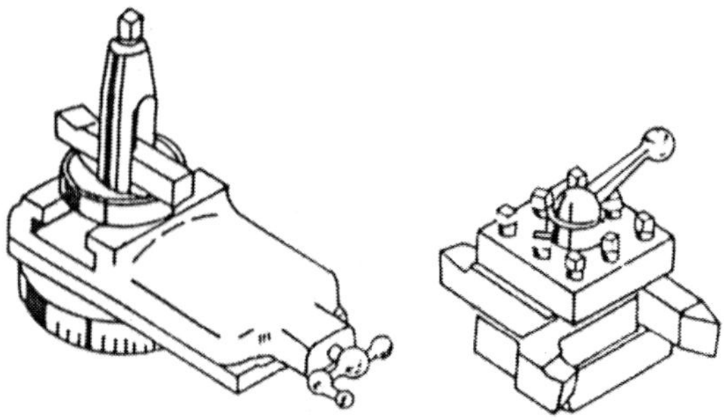

Fig. 1.10 (*a*) Single, (*b*) Four tool post

(*i*) **Single Screw Tool Post:** It can accommodate any a single tool at a time and tool is fixed with the help of set screw.

(*ii*) **Four Bolt Tool Post:** The tool is held by two straps and four bolts generally. It is used for heavy duty lathes.

(*iii*) **Open Side Tool Post:** The tool is held independently by fixing bolt and clamped in position by two set screws. The arrangement ensures quick replacement of the tool.

(*iv*) **Four Way Tool Post:** In this tool four post slides are open to accommodate four tools at a time. The tools are held in position by seprate screw and a locking bolt mounted at the centre.

(*e*) *The apron*: The portion of the carriage that extends in front of the lathe is called an apron. It consists gears, clutches and levers for operating the carriage by hand and power feed.

(*f*) *The lead screw:* The lead screw is a long threaded shaft which is slightly below and parallel to the bed ways extending from head stock to tail stock. The lead screw is used generally for cuting threads.

The lead screw is a feed rod that transmit power from quick change box to drive the apron mechanism for cross and longitudinal feed.

5. **Feed Mechanism:** The movement of the tool relative to the work is known as feed. Feeds of the lathe tool are of three types:

(*i*) *Longitudinal feed*: When the tool moves parallel to lathe axis; and is effected by the movement of the carriage.

(*ii*) *Cross feed*: When the tool moves at right angle to the lathe axis with the help of cross-slide.

(*iii*) *Angular feed*: When the tool moves with the help of compound slide which is swivelled at an angle to the lathe axis.

Cross and longitudinal feeds are both hand and power operated, but angular feed is hand operated only.

The feed mechanism of a lathe is obtained by using a train of gears (series of gears in mesh) that transmit motion from the headstock to the main spindle and lead screw. From the feed rod, the motion is transmitted to various gears in the apron. The feed gears are controlled by friction through small knobs located in front of the apron.

6. **Thread Cutting Mechanism:** Any mechanism used for transmitting motion from the main spindle to the lead screw for thread cutting is known as thread cutting mechanism (shown in Fig. 1.11).

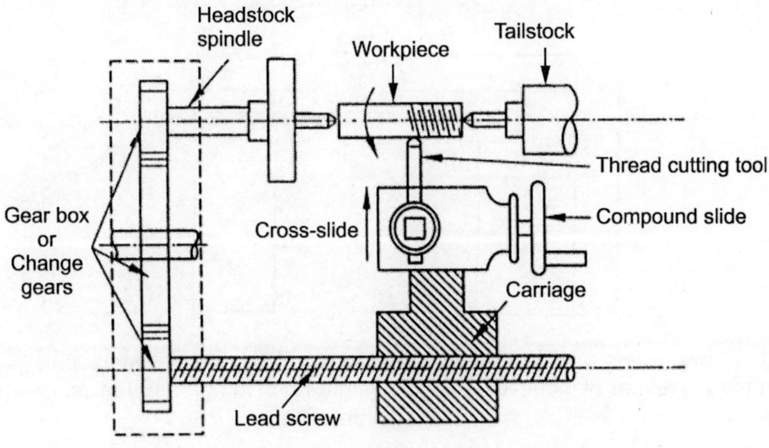

Fig. 1.11 Thread Cutting Mechanism

The rotation of the lead screw is used to travel the tool along the work to produce screw threads. The lead screw has a definite pitch. A split nut in the apron is used to engage the lead screw with the carriage.

1.4.5 Size and Specification of a Lathe Machine

The size of the lathe is expressed completely by the following considerations:

1. The maximum diameter that can be held.
2. The swing over bed.
3. The height of the centres measured from the lathe bed.
4. The length of the bed.
5. Type of bed (*i*) straight, (*ii*) semigap type.
6. Width of the bed.

7. The maximum length of the work that can be turned between centres.
8. Range of spindle speeds.
9. The range of thread that can be cut-metric or english.
10. Range of feeds.
11. Type of spindle nose.
12. Size of spindle bore.
13. The maximum bar diameter.
14. Cross slide travel (*a*) top slide travel, (*b*) tool section.

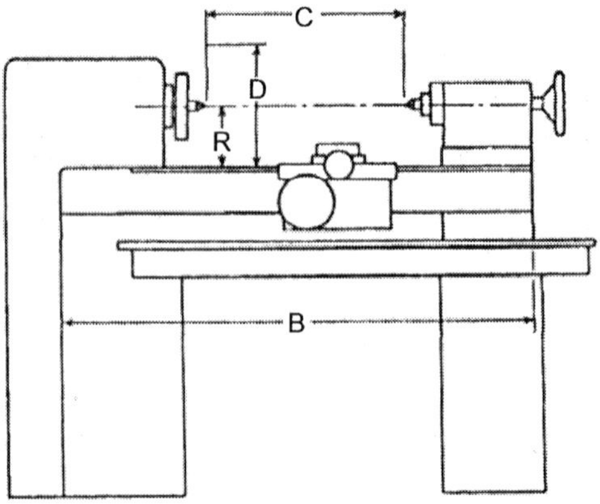

Fig. 1.12 Measuring a lathe for size. (C), maximum distance between centers: (D), maximum diameter of workpiece over ways (swing of lathe); (R), radius, one-half swing; (B), length of bed.

15. Tail stock sleeve travel.
16. Floor area required.
17. Horse power and RPM of Electric Motor.
18. Shipping dimensions – length × width × height × weight

1.4.6 Lathe Accessories and Attachments

Accessories

Accessories are the devices employed for holding and supporting the work and the tool on the lathe are called accessories. They include the devices like centers, chucks, catch plate and carriers, dogs, tool holder and tool posts, collets, mandrels jigs and fixtures etc.

Attachments

Attachments are the additional devices used for specific purposes like:

1. Taper turning
2. Milling
3. Grinding
4. Gear cutting
5. Turret
6. Cutter and crank pin turning attachments.

Accessories

1. **Chucks:** Chuck is an important device for holding and rotating the piece of work in a lathe. There are different types of the chucks used in a lathe like:
   (*a*) Three jaw universal chuck or self-centric chuck.
   (*b*) Four jaw independent chuck
   (*c*) Six jaw universal chuck
   (*d*) Collect Chuck (*i*) Draw in type, (*ii*) Push out type
   (*e*) Air or hydraulic chuck
   (*f*) Drill chuck
   (*g*) Magnetic chuck

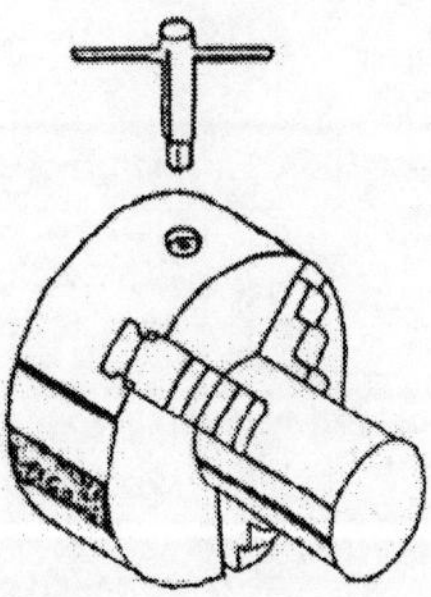
(*a*) Three jaw chuck

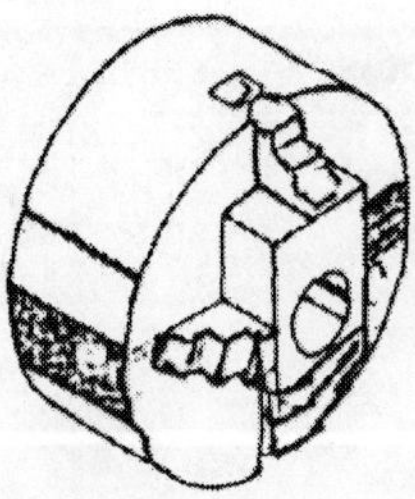
(*b*) Four jaw chuck

(*c*) Six jaw universal chuck

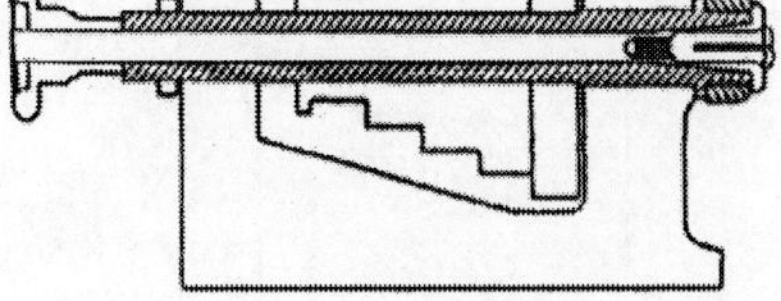
(*d*) (*i*) Draw in collect chuck

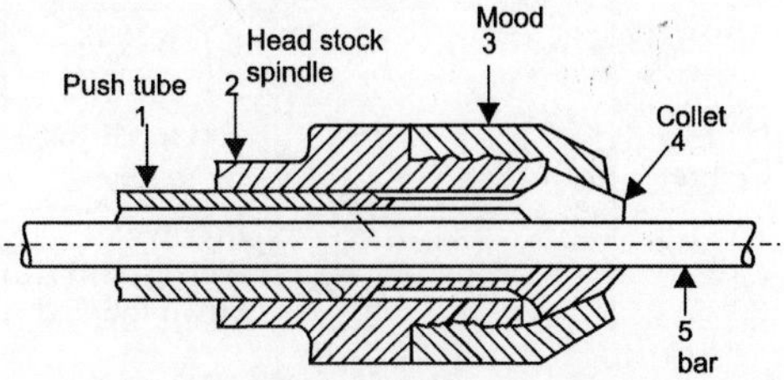

(*d*) (*ii*) Draw in push out type chuck

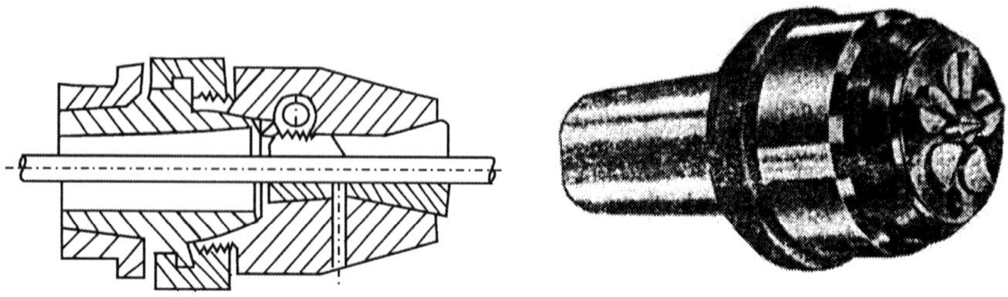

(*e*) (*f*) Drill chuck

Fig. 1.13 Accessories

2. **Lathe Centres:** These are the common accessories/devices for holding the work between the live and dead centers to prevent the thrust produced in metal cutting. These are of very hard mate to resist deflection and wear.

Angles:

For general purpose work	—	60°
For heavy work	—	75°
For morse taper	—	4° - 6°
For metric taper	—	(4° - 6°)

Table 1.1 Centres

Types	Sketch	Application	Types	Sketch	Application
Ordinary Centre	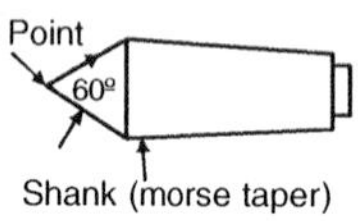	For general purpose	Tipped Centre	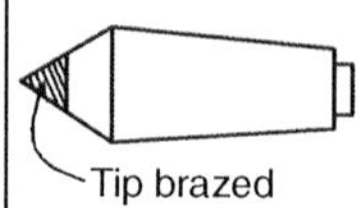	Less wear and strain than inserted bit type centre and expensive. May be used in production work.
Half Centre	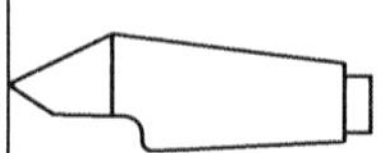	It is used at the tailstock end to allow more space for the tool when turning down the end.	Inserted Bit Type Centre	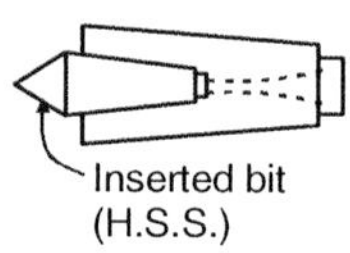	It minimises the running cost of the centre as bit made of H.S.S. will reduce wear.

Contd...

Types	Sketch	Application	Types	Sketch	Application
Pipe Centre	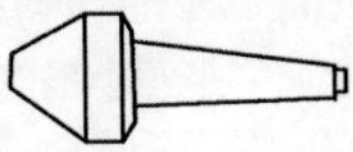	It is used for holding the jobs such as shell, pipes etc.	Ball Centre		It minimises wear and strain during taper turning by tailstock setover method.
Friction less Centre	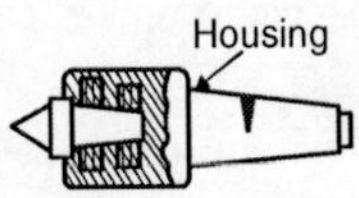	It is used in tailstock end for supporting heavy job rotating in high speed.			

3. **Face Plate:** It consists of a circular disc bored out and threaded to fit the nose of the lathe spindle. It has radially placed slots used for clamping the work by means of a bolt.

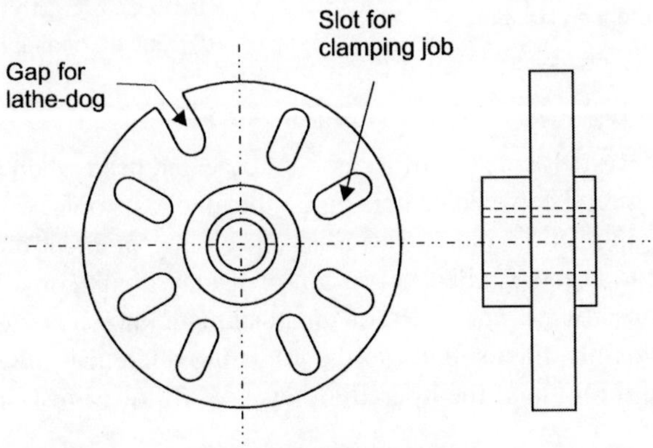

Fig. 1.14 Face plate

4. **Angle Plate:** It is a cast iron plate having its two faces machined to make them absolutely at right angles to each other slots are provided in angle plate to hold the work like flanges of a pipe elbow and ecentric jobs etc with the help of nut and bolt.

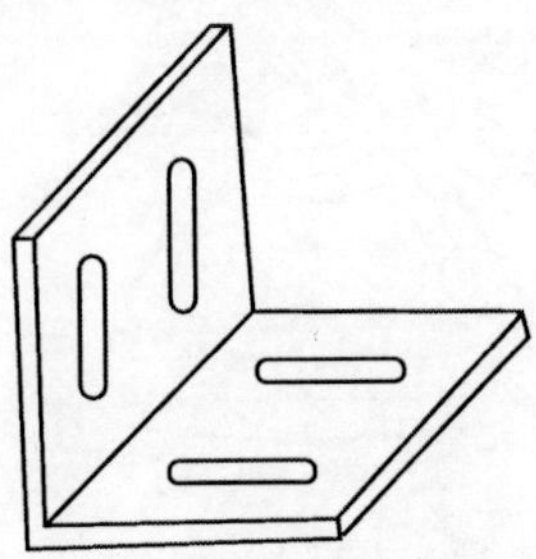

Fig. 1.15 Angle Plates

5. **Mandrels:** A mandrel is a device used for holding and rotating a hollow piece of work that is already drilled and bored. It is mounted between the live and dead centres for machining the outer surface of the work. To avoid distortion and wear it is made of high carbon steel. Different types of mandrels are employed according to specific requirements.

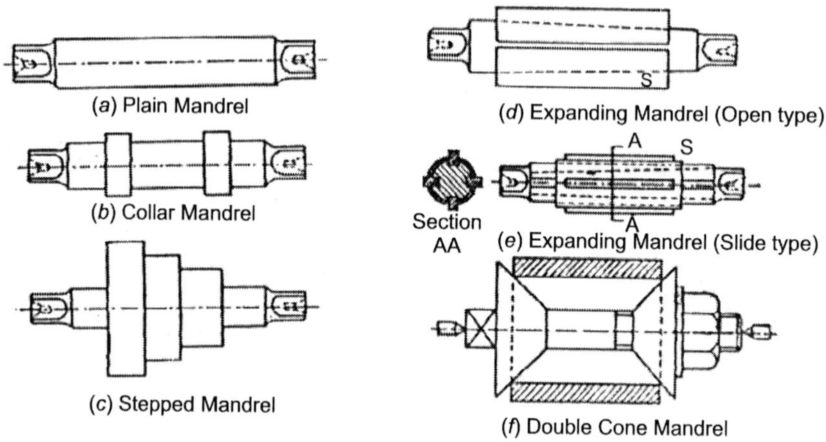

Fig. 1.16 Collar mandrel

6. **Steady Rest:** It consists of a cast iron base and used when a very long job is to be turned between centers on a lathe, due to its own weight, it provides a springing action and carries a lot of bending moment due to which the turning tool gets spoiled very soon or it may break some time. To avoid these jobs always, supported on an attachment known as steady rest. If the job is flexible, then support to the job is provided near the cutting edge of the tool throughout the operation hence a "follower rest" is used for this purpose.

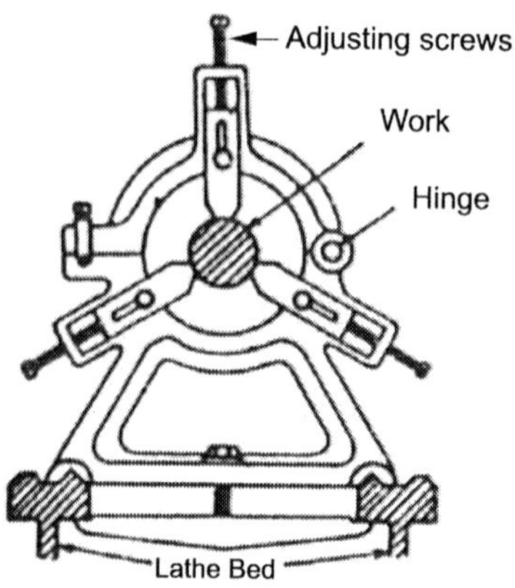

Fig. 1.17 Steady rest

7. **Lathe Carriers and Dogs:** These are used in conjunction with the driving plate. The work is inserted in V shaped hole and the position of work is set by a screw provided in the dog or carrier. Straight and bend tail type of dogs are generally used.

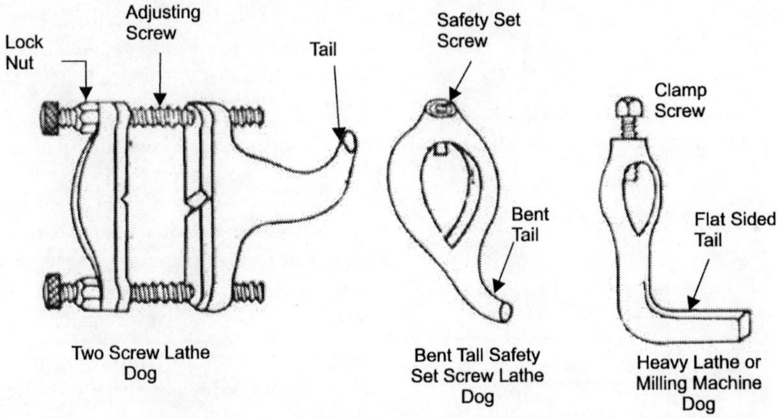

Fig. 1.18 Lathe carriers and dogs

1.4.7 Lathe Operations

The operations perform on a lathe machine are shown in Table 1.2.

Table 1.2 Lathe Operations

Operation	Sketch	Function
1. Facing	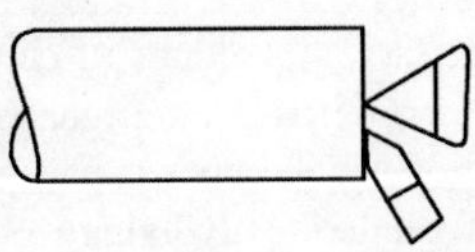	Facing is the operation to make the end of flat a workpiece smooth and finish of the required length. It is also known as squaring operation. It is accomplished by causing the work to revolve while the tool being fed perpendicular to the work piece and pells off chip from end face of the work. A side or facing tool or a regular turning tool may be used.
2. Turning	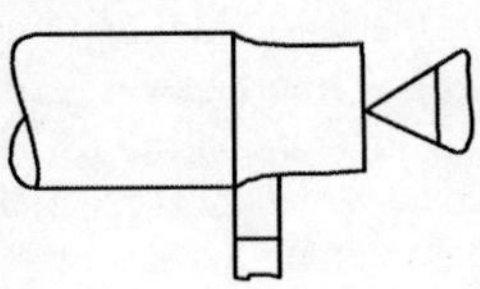	Turning is the operation of producing perfectly cylindrical external surface of the work while the work is held between centers or in a chuck to accomplish the work revolve and tool is fed longitudinally, peels off a chip from the external surface of the work.
3. Shoulder Turning	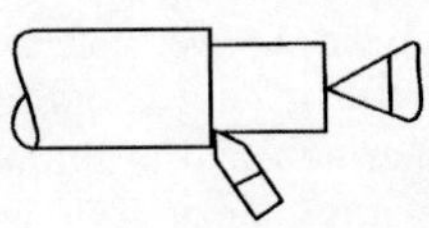	It is used to finish the side or shoulders of a stepped work known as shoulder turning. It is also similar to the operation of facing.

Contd...

Operation	Sketch	Function
4. Taper Turning	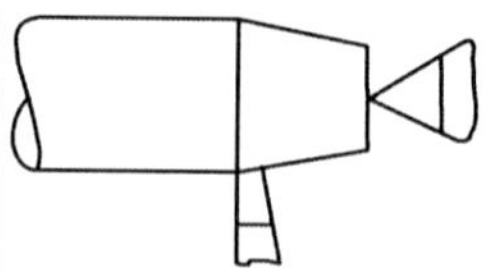	It is a operation of finishing or producing an external conical surface on a work. A chamfering tool is used for small tapers while the long tapers can be produced by swivelling the compound rest at the required angle off setting the tail stock or with the help of a taper turning attachment.
5. Chamfering	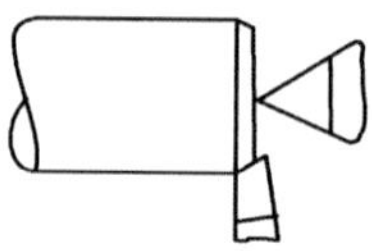	The lathe operation of bevelling the external edge of a work pieces with the help of a chamfering tool known as chamfering. Chamfering is an essential operation to avoid injury to the hand, for thread cutting operations to pass the nut freely as well as stress concentration point of view in parts.
6. Forming	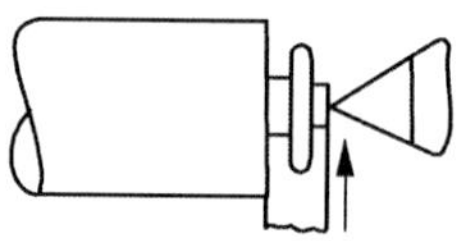	The lathe operation of turning a convex, concave or any irregular shape is known as forming. It is generally accomplished by: (1) Using a form tool. (2) Using simultaneous cross and longitudinal feed of the tool. (3) Tracing or copying a template.
7. Grooving	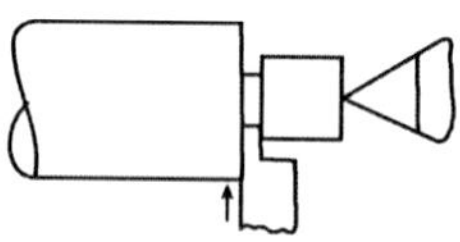	The lathe operation of reducing the diameter of a work piece over a very narrow surface known as grooving. The grooving tool is fed into the revolving work upto desired depth at right angle to the centre line of the work piece.
8. Parting Off	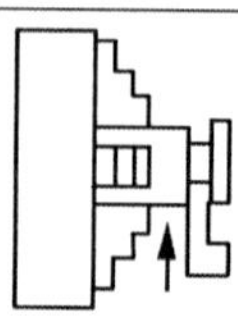	The lathe operations of cutting a work piece from a long bar after machined to the desired size. Shape known as parting off and the tool is fed into the revolving work at right angle to the centre line of the work piece.
9. Knurlings	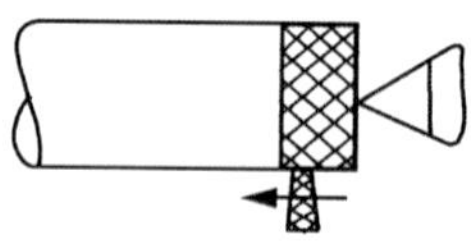	The lathe operation used to make depression into the work surface by rolling known as knurling. Knurling wheels are used for making cut or impressions on the face of work piece by hard wheels. It is generally used to provide the grip on the surface of a nut or screw etc.
10. Spinning	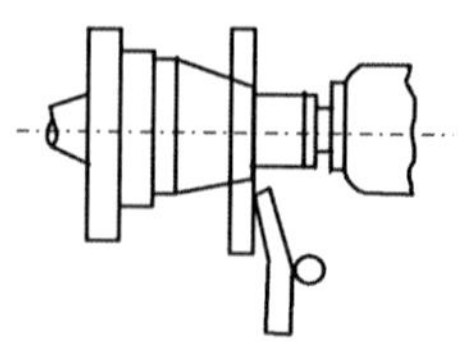	It is used to forming a thin sheet of metal by revolving the work piece at high speed and pressing it with a spinning tool against a former attached to the head stock spindle known as spinning. This operation is generally done on spinning lathe.

Contd...

Operation	Sketch	Function
11. Filling	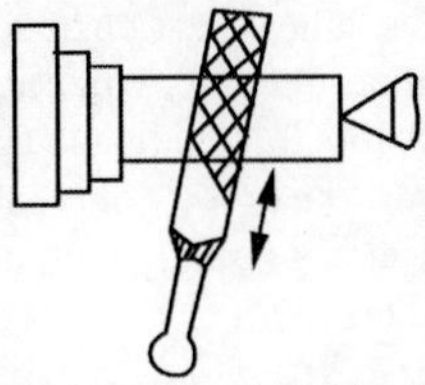	The operation of finishing an external surface generally with the help of a flat single butt file by revolving the work at high speed known as filling. Filling is done to remove burs, sharp corners and fed marks on the work piece.
12. Polishing	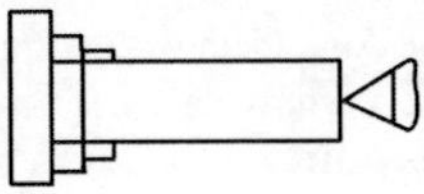	It is used to improve surface quality of a revolving work piece against successively finer grade of emery cloth known as polishing. It is made to perform after filling to improve surface quality of the work.
13. External Thread Cutting	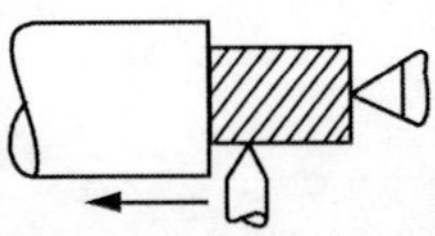	The lathe operation of cutting a helical groove on an external cylindrical or conical surface by feeding the tool longitudinally to the revolving work known as external thread cutting. In this operation the longitudinal feed is equal to pitch of the thread.
14. Drilling	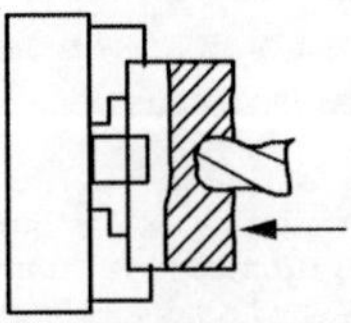	It is an operation of producing a hole or enlarging an already existing cylindrical hole with the help of a drilling tool (called drill) is known as drilling. In this operation the drill is held in a Chuck and feed against a revolving job to produce the hole.
15. Boring	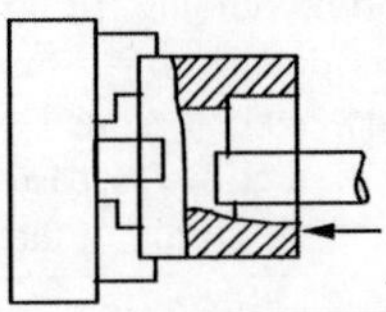	The operation of enlarging an already existing cylindrical hole with the help of a boring tool known as boring.
16. Counter Boring	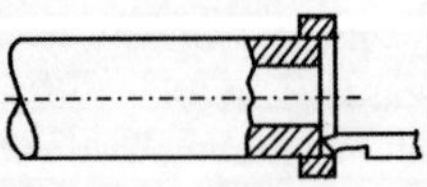	The operation of enlarging a small end portion with a boring tool of an already existing hole in a work piece known as counter boring.
17. Taper Boring	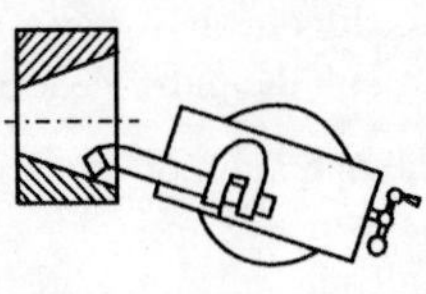	The operation of producing a taper hole from an already existing or drilled hole, with the help of a boring tool known as taper boring. Taper boring can be done by swivelling the compound slide or by using a taper turning attachment.
18. Reaming	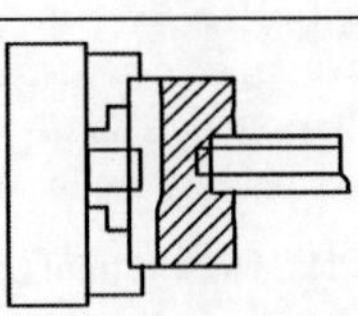	The operation of fine finishing and sizing an already drilled or bored hole with a reaming tool (*i.e.*, Reamer) call reaming.

Contd...

Operation	Sketch	Function
19. Internal Threading	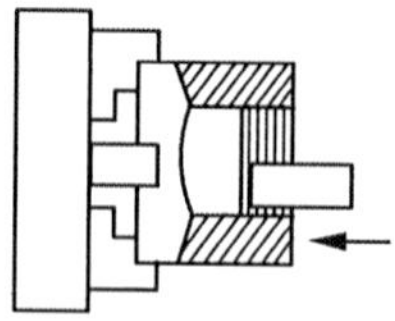	The lathe operation of cutting a helical groove on an internal cylindrical surface by feeding the tool longitudinally against to the revolving work known as internal threading. The tool used is similar to the boring tool with cutting edges.
20. Eccentric Turning	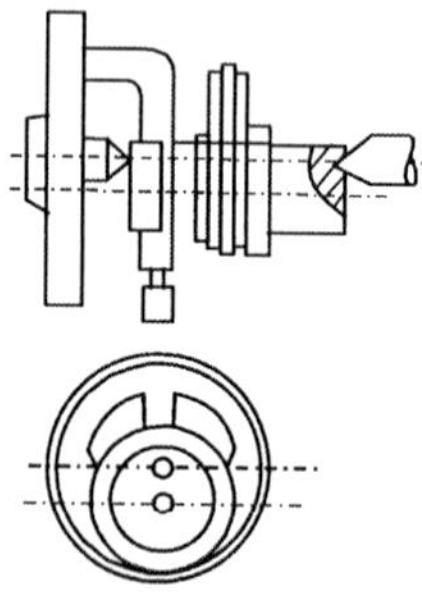	The operation of turning a cylindrical surface of a work piece with its axis parallel to but not coincident with the normal axis of the work piece known as eccentric turning. To turn the eccentric surface it is necessary to provide centres off set from the normal axis by an amount equal to one half the eccentricity desired.
21. Under Cutting	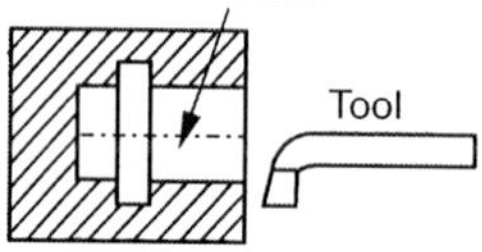	The operation of enlarging the diameter of a small surface within the length of a drilled or bored hole known as under cutting. The tool used is similar to the boring tool.
22. Spring Winding	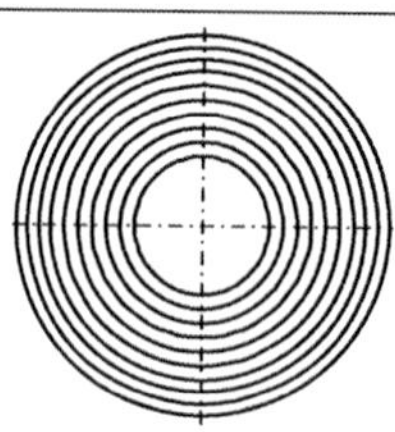	The operation of making a coiled spring from a spring wire on a lathe machine is known as spring winding. In this operation the wire is passed around a mandrel which revolve between centers and a small hole is provided in a steel bar to pass the wire through it to wind the spring of the required pitch.
23. Grinding	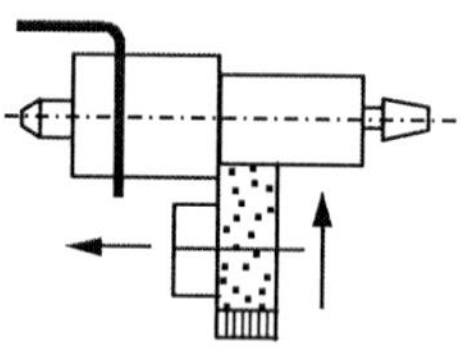	The operation of removing metal in the form of minute chips by rotating abrasive wheel against a work, known as grinding. Both external and internal cylindrical grinding can be carried out on a lathe by using a grinding attachment and depth of cut is controlled through the movement of the cross slide.
24. Milling	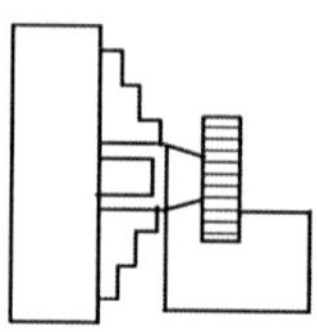	The operation of removing metal by feeding the work against a rotating cutter having multiple cutting edges known as milling, generally it is done by the two methods: (I) work piece held in a special attachment on the cross-slide and is fed against a revolving cutter held by a chuck. (II) The work is held between centres and kept stationary while the revolving cutter, held in an attachment on the carriage.

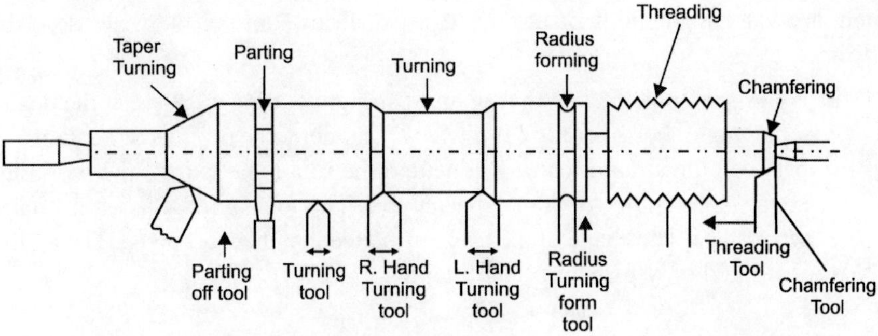

Fig. 1.19 Schematic diagram of perform on a lathe machine.

1.4.8 Methods for Taper Turning

Taper Turning Methods

Taper Turning: It is a operation of finishing or producing a external or internal conical surface on a work by using chamfering tool, small tapers and large tapers can be produced by swivelling the compound rest at the required angle.

Tapers are turned bodies with uniform reduction. Workpieces with tapers and taper holes are used for various purposes, like fastening and sealing.

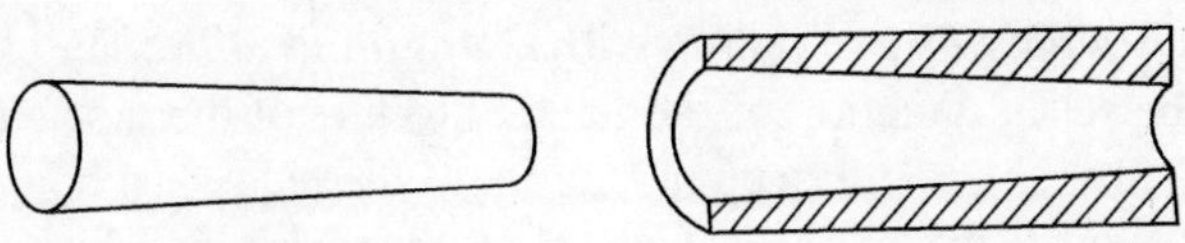

(*i*) Outside taper (*ii*) Inside taper

Fig. 1.20 Taper

Some examples of the tapered parts are as follows :

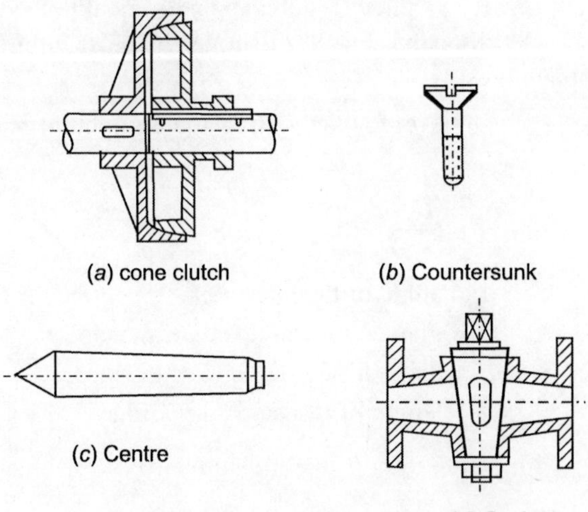

(*a*) cone clutch (*b*) Countersunk (*c*) Centre (*d*) Cock

Fig. 1.21 Tapered parts.

There are various methods of turning taper surfaces some of them are described below:

1. **By Form Tool:** The form tool has a straight cutting edge set at the desired taper angle. Form tool is a replica of the taper to be produced, *i.e.* the angle between the straight cutting edge and the rotational axis of the workpiece equal taper angle α (half taper angle). The tool is fed against a rotating workpiece. The shape of the tool reproduced on the workpiece. The method is limited only for short length tapers.

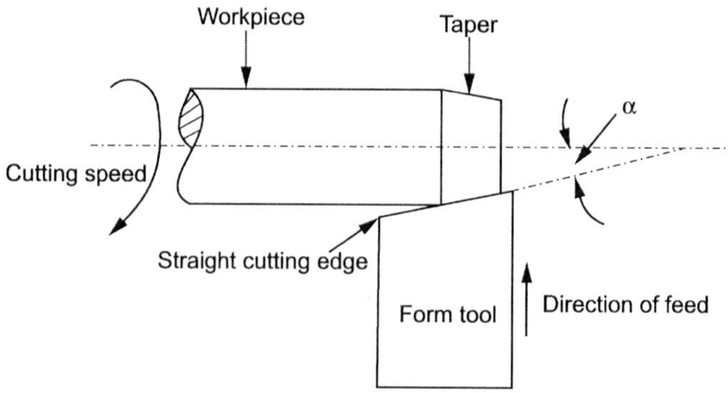

Fig. 1.22 Taper by form tool.

2. **Taper Turning by Swivelling with Compound Slide:** The compound slide must be set in direction of the lateral arc line of the taper. The method is suitable for slight and steep tapers. Since the feed must be manipulated by hand, an unclean surface of the workpiece may result. The operation length of the compound slide is limited, therefore, as a rule, only short tapers can be turned. The compound rest has a circular base graduated in degress, which can be swivelled at any angle. While turning a taper, the base of compound rest is swivelled through an angle equal to the taper angle. The tool is then fed by hand. Fig. 1.24 shows the taper turning by swivelling the compound rest.

   Half the taper angle α for swivelling the compound slide is given the formula:

$$\tan \alpha = \frac{D-d}{2l}$$

where
α = angle of taper,
D = diameter of large end of the taper
d = diamter of small end of the taper
l = length of the tapered portion

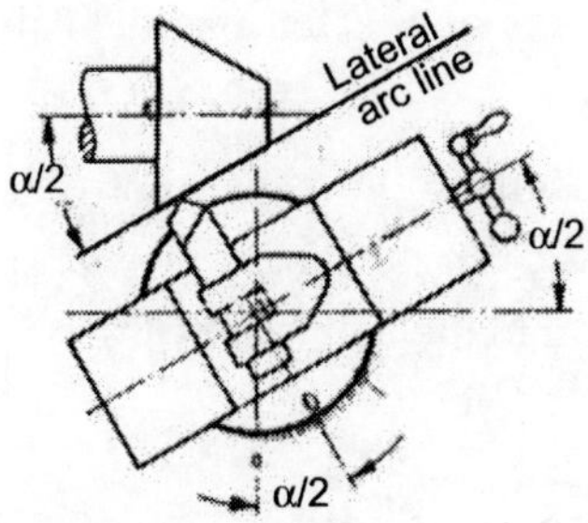

Fig. 1.23 Taper turning with the compound slide.

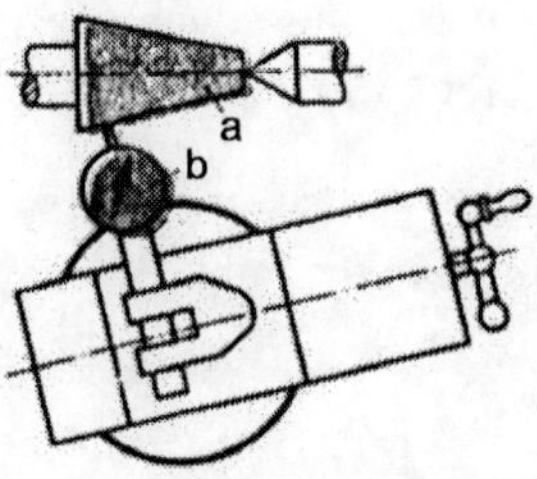

Fig. 1.24 Setting according to master specimen (*a*) Master, (*b*) Dial indicator.

Setting of the Compound Slide by Means of the Graduation: The compound slide is traversed from the zero position by the setting angle and tightened with the screws.

Setting of the Compound Slide in Accordance with the Master specimen: A taper plug gauge can, for instance, surve as a master specimen. A dial indicator which touches the master with its feeler, is clamped in the tool holder. While moving the relocated compound slide, along the lateral arc line, the dial indicator must not show any divergence.

3. **Taper Turning with the Taper Turning Attachment:** With the taper turning attachment, which is installed on some lathes, outside and inside tapers can be turned with a setting angle upto 10° with automatic feed. This method utilises the principle of guiding the tool in a straight path set at an angle to the axis of rotation of the workpiece. While the workpiece is revolving between the centres or by a chuck aligned to the lathe axis.

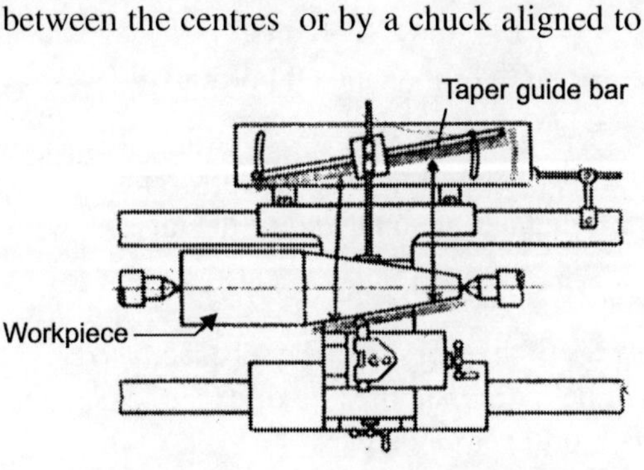

Fig. 1.25 Taper turning with the taper turning atachment.

4. **Taper Turning with Relocated Tailstock Centre:** If the tailstock centre is shifted off centre, a taper shape resutls when the tool carriage is moved in longitudinal direction. The amount of shift should not be greater than 1/50 of the length of the workpice, otherwise the lathe centres have a poor contact. Therefore, the method is only used for turning long and slight

tapers. It has the advantage that the automatic longitudinal feed can be engaged.

This method uses the principle of shifting the axis rotation of the workpiece at an angle to the lathe axis Fig. 1.26. The angle at which the axis f rotation of the workpiece is shifted is equal to the half angle of the taper.

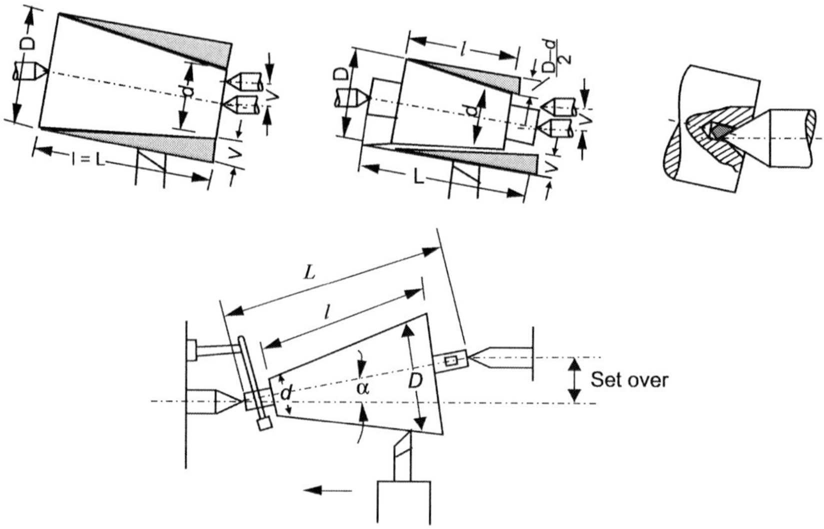

Fig. 1.26 Taper turning by setting over tailsotock centre

$$\text{Set over} = L \times \frac{D-d}{2l}$$

where l = length of the tapered portion

L = length of workpiece

D = diameter of large end of the taper

d = diamter of small end of the taper

Rules of Taper Turning

1. The cutting edge of the tool should be set exactly to centre height, otherwise the taper will be inaccurate in spite of exact adjustment of the compound slide, tailstock or taper guide bar.
2. When turning a taper with the compound slide between centres, the latter must be in proper alignment, otherwise an inaccurate taper will result even with exact adjustment of the compound slide.
3. If several equal tapered workpieces are to be turned with relocated tailstock centre, the length of the workpieces and depth of the centre holes must be the same.

4. Care must be taken to lubricate gliding parts, while using the taper turning attachment.

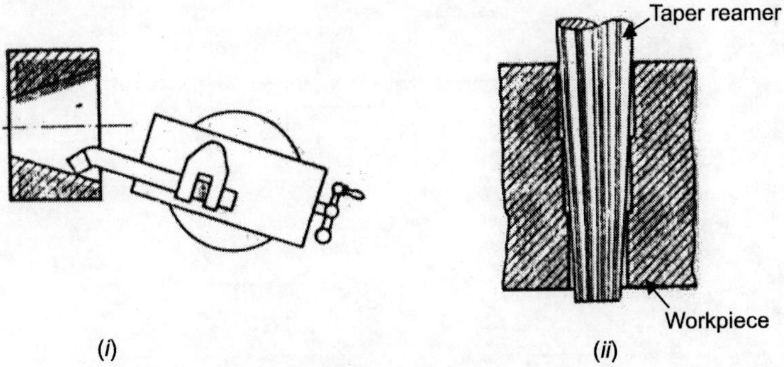

Fig. 1.27 (*i*) Turning of an inside taper (*ii*) Stepped rough drilling of bigger taper holes.

1.4.9 Cutting Speed, Feed and Depth of Cut

Cutting Speed: The cutting speed of a tool is the speed at which the metal is removed by the tool from the work piece in one pass and its is expressed in meter per minute.

$$\text{Cutting speed} = \frac{\pi DN}{1000} \text{ m/min}$$

where D = diameter of job in mm

N = RPM of the job or spindle.

Feed: It is the tool advancement per revolution of job parallel to the surface being machined. It is expressed in mm per revolution and in the British system it is expressed in inches per revolution.

Depth of Cut: It is the perpendicular distance measured from the machined surface to the in cut surface of the work piece. It is expressed in mm.

$$\text{Depth of } (t) \text{ cut} = \frac{d_1 - d_2}{2}$$

where: d_1 = Diameter of the unmachined surface

d_2 = Diameter of the machined surface

(t) = Depth of cut in mm

generally, depth of cut varies from 1 mm to 5 mm oughing operations and 0.2 mm to 1 mm finishing operations.

The ratio of depth of cut to the feed varies from 10:1.

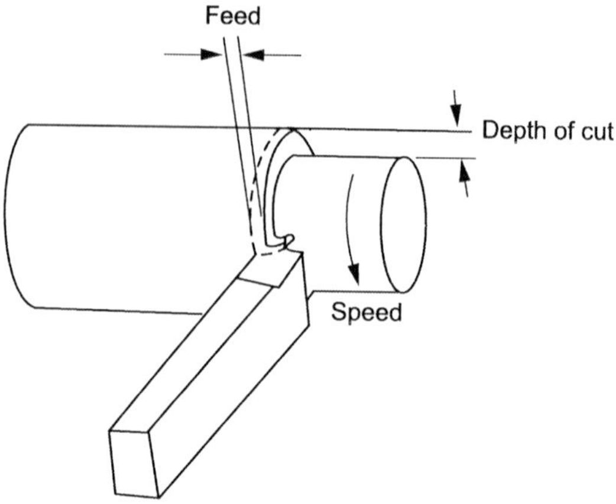

Fig. 1.28 Concept of cutting speed, feed and depth of cut.

1.4.10 Safety Precautions Associated with Lathe

1. Before operating machine, known all the operation controls and how to stop it.
2. Revolving chuck should never be stopped by hand.
3. Use manual power (Hand) when holding and removing the job from the machine chuck.
4. Correct size centers should be used.
5. Do not shift gears while the machine is in operation.
6. Use brush to clean the chips.
7. While working on a lathe always wear appron shoes and goggles to protect from accident.
8. Chuck key should be removed from chuck after fixing or tightening the job in chuck.
9. All gaurds should be in their position.
10. Tool and instruments should not placed over bed lathe.
11. Before starting the lathe spindle by power. Its runout should be checked by moving machine chuck with hand to avoid eccentricity of the job.
12. Moving and sliding parts of the lathe should be cleaned and lubricated properly.

1.5 SHAPER

Shaper is a versatile machine tool which is used for producing flat surfaces with a reciprocating tool that takes a straight line cut by using single point cutting tool. The machine can produce:

(*a*) Horizontal, vertical and angular surfaces.

(*b*) It can cut external and internal slots grooves and key ways with the help of a single point tool driven by quick return mechanism.

(*c*) It can also produce contours of concave, convex and irregular surfaces also.

The main significance of this machine is its greater flexibility in work holding and quick adjustment of tools due to simple design generally it is used in tool room, die making and for general repair work. Some examples of shaper work are as follows:

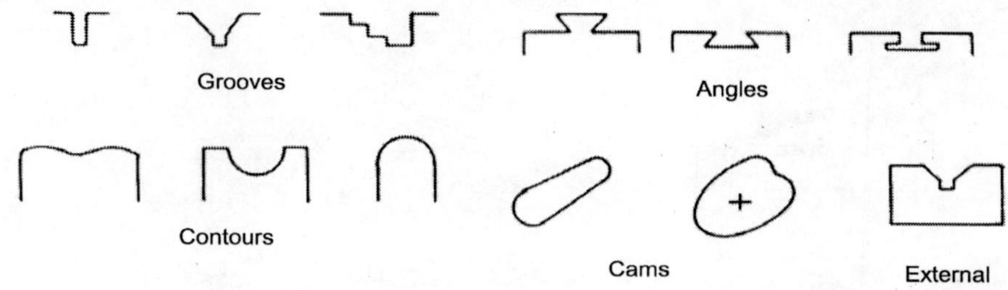

Fig. 1.29 Surface produced on Shaper

1.5.1 Working Principle of a Shaper

In a shaper, a single point cutting tool reciprocates over the stationary work piece which is rigidly held in a machine table. The tool is held in a tool post mounted on the ram of the machine. This ram reciprocates in forward and backward direction. The material is cut when the tool is moving in forward direction *i.e.*, is known as forward stroke. No cut of material takes place during the return stroke of the ram hence it is termed as Backward or ideal stroke. These strokes are obtained by quick return mechanism generally return stroke take the half time as compared to forward stroke.

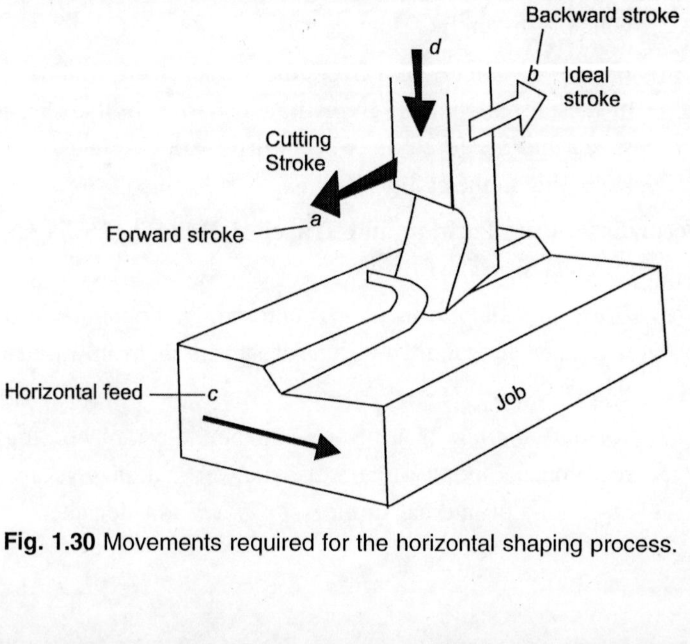

Fig. 1.30 Movements required for the horizontal shaping process.

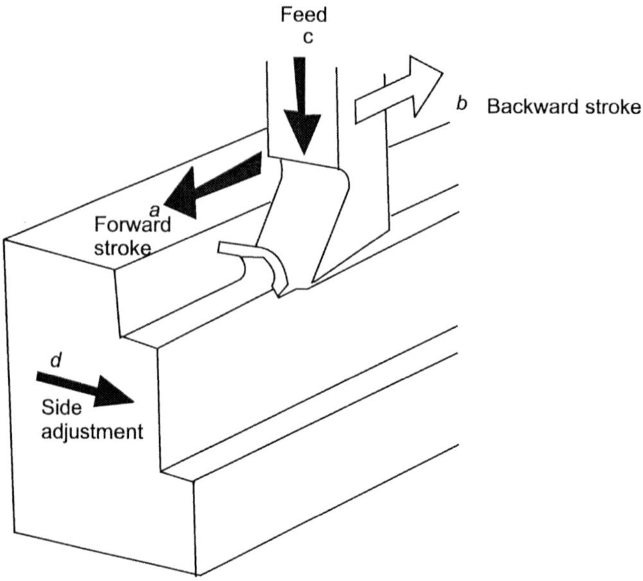

Fig. 1.31 Movemens required for the shaping process (vertical)

1.5.2 Classification of Shaper

1. **According to the Ram Driving Mechanism:** According to this basis, the shapers are classified as follows:

   (*a*) *Crank shaper*: In a crank shaper, a crank and a slotted lever quick return motion mechanism are used to give the reciprocating motion to the ram.

   (*b*) *Geared shaper*: In a geared shaper, the ram carries a rack below it, which is driven by a spur gear.

   (*c*) *Hydraulic shaper*: In a hydraulic shaper, reciprocating movement of the ram is obtained by hydraulic power. In this shaper, cutting speed and the force of the ram drive are constant from the very beginning to the end of cut.

2. **According to the Position and Travel of Ram:** According to this, the shapers are classified as follows:

   (*a*) *Horizontal shaper*: In a horizontal shaper, the ram reciprocates in a horizontal direction. These type of shapers are mainly used to produce flat surfaces.

   (*b*) *Vertical shaper*: In a vertical shaper, the ram holding the tool reciprocates along a vertical axis. Vertical shapers are used for machining of internal surfaces, keyways and slots etc.

(*c*) *Travelling head shaper*: In this shaper, the ram carrying the tool while it reciprocates moves crosswise to give the required feed so if heavy jobs are to be shaped, it is difficult to provide the feed by the table. In that case this shaper is used because feed is also supplied by the ram.

3. **According to the Direction of Cutting Stroke:** According to the direction of cutting stroke, the shapers are classified as follows:
   (*a*) *Push cut shaper*: In this the cutting stoke is forward stroke while an idle stroke is return stroke.
   (*b*) *Draw cut shaper*: In this shaper, the return stroke is cutting stroke while forward stroke is an idle stroke.
4. **According to the Design of the Table:** According to this basis, the shapers are classified as follows:
   (*a*) *Standard or plain shaper*: In this shaper, table has only two movements *i.e.* horizontal and vertical, to give the feed, it cannot be swivelled or tilted.
   (*b*) *Universal shaper*: In this shaper, the table can be moved horizontal, vertical and can also be swivelled and tilted. This is mostly used in tool room work.

1.5.3 Principal Parts of Shaper

Principal parts of a shaper are as follows:

1. Base 2. Column 3. Cross rail 4. Table 5. Ram
6. Tool head 7. Vice.

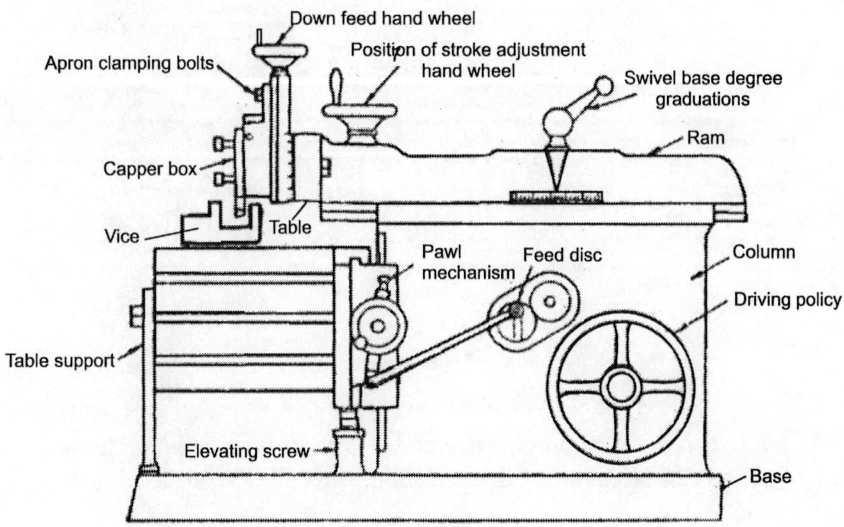

Fig. 1.32 Parts of standard shaper

The principal parts of a shaper, as shown in Fig. 1.33 on prev. page.

1. **Base:** It is a heavy and robust cast iron body which acts as a support for all other parts of the machine which are mounted over it.
2. **Column:** It is a box type iron body mounted upon the base. It acts as a housing for the operating mechanism of the machine, electrical, cross rail and ram. On the top it is having two guideways on which the ram reciprocates.
3. **Cross Rail:** It is a heavy cast iron construction, attached to the column at its front on the vertical guideways. It carries two mechanism, one for elevating the table and the other for cross travel of the table.
4. **Table:** It is made of cast iron and used for holding the workpiece. T slots are provided on its top and sides for securing the work on to it. It slides along the cross rail to provide feed to the work.
5. **Ram:** It reciprocates on the guideways provided above the column. It carries the tool head and mechanism for adjusting the stroke length.
6. **Tool Head:** It is attached to the front portion of the ram and is used to hold the tool rigidly. It also provides the vertical and angular movement to the tool for cutting.

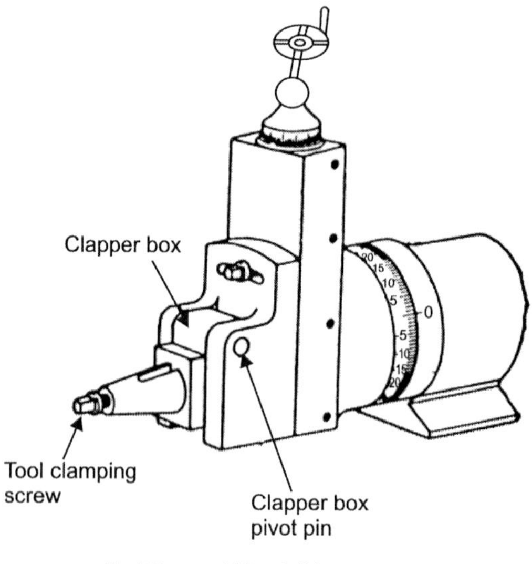

Tool Box and Headslide

Fig. 1.33 Tool head and its details

7. **Vice:** It is a job holding device and is mounted on the table. It holds and supports the work during the operation. Directly job can be clamped on the table.

1.5.4 Size and Specification of a Shaper

The size of shaper is determined by the length of stroke of ram.

1. Maximum length of stroke - 300 mm
2. Size of the table *i.e.* length, width and depth of the table
3. Maximum vertical travel of tool.
4. Power in input - HP of motor - 1.5 HP to 2 HP
5. Maximum no. of stroke per minute
6. Cutting to return stroke ratio.
7. No. and amount of feed
8. Type of quick return mechanism
9. Floor area required for installation
10. Weight of machine – 850 kg
11. Maxi vice opening
12. Maxi distance from table to ram - 450 mm
13. Size of table - ϕ - 275 mm
14. Dimension of base plate - 700 mm × 490 mm
15. Bed dimensions = Length × Weight × Height

    = 1375 mm × 412 mm × 575 mm

1.5.5 Principle of Quick Return Motion (Crank and Slotted Link) Mechanism

The principle of quick return mechanism can be understand from the figure in shaper metal cutting takes place while the tool is moving in the forward stroke and no cutting takes place along the return stroke *i.e.* is termed as ideal stroke. The mechanism used for this purpose is called quick return mechanism which helps to increase the rate of production. The mechanism consisting a slotted link called rocker arm. The bottom end of rocker arm is pivoted on a block and its upper end is connected through a link which is attached to hand lever. At the back of rocker arm a bull gear is provided which is mounted to the machine frame. The rocker arm is provided with a slider block. The bull gear receives power from the pinion and rotate. The driving pin also rotate with the bull gear. Hence, the oscillating motion of the slotted link is converted into reciprocating motion of the ram.

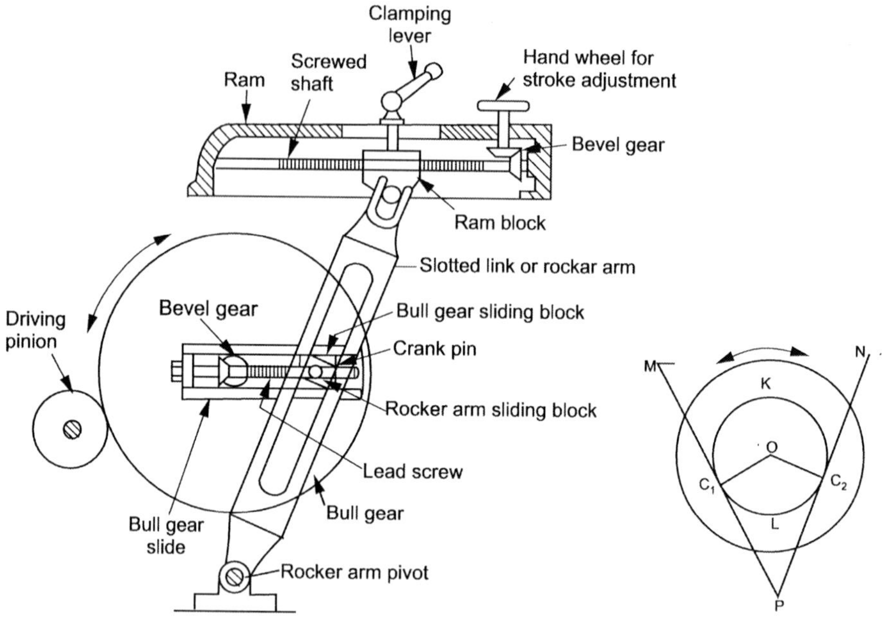

Fig. 1.34 Frank and slotted link mechine

When the link is in the position P.M. the ram will be at the extreme backward position of its stroke and when it is at PN the ram will be at extreme forward position. Now PM and PN are tangent to the crank pin circle. The forward cutting stroke will takes place when crank will rotate through an angle $C_1\ k\ C_2$ and during retrun-stroke it will rotate on angle $C_2\ L\ C_1$. Hence, the angle made by cutting stroke in greater than return stroke.

The ratio between cutting times and return time can be determined by the relation.

$$\frac{\text{Cutting time}}{\text{Return time}} = \frac{C_1kC_2}{C_2LC_1}$$

Cutting time to return time ratio usually varies between 2 : 1 and practical limit is 3 : 2. The cutting and return speed is not constant throughout the stroke.

1.5.6 Feed Mechanism

In a shaper speed and feed two basic movement of a shaper are mutually perpendicular in direction for each forward stroke of the ram, the distance moved by the work side ways towards the cutting tool is called feed. The distance moved by the work is achieved during the ideal stroke of ram to avoid the damage of tool. The cross feed screw is connected with the table. The connecting rod connects the pawl/arm and crank wheel. As the crank wheel rotates. The pawl arm oscillates to and fro around the cross feed screw. Pawl is pressed on the ratchet wheel by the spring. When the cross feed screw rotates. The table moves automatically during each return stroke of the ram. During material feed, pawl is pulled up and the pin is placed in disengaged position. Manual feed is given by rotating cross feed screw rod with a handle during the return stroke of the ram.

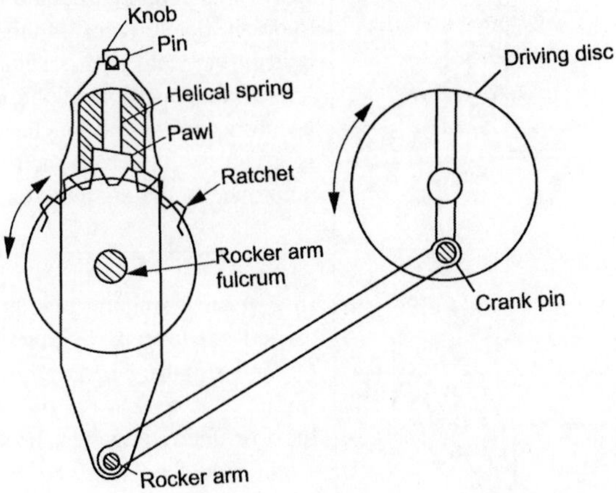

Fig. 1.35 Automatic fed mechanism of a shaper

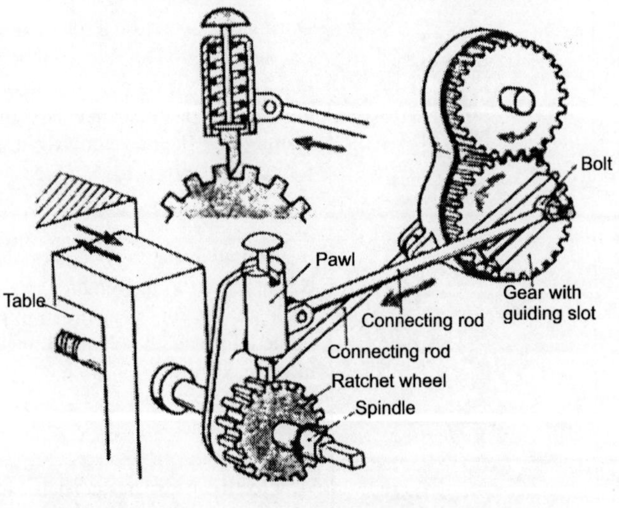

Fig. 1.36 Assembly of the automatic fed mechanism

1.5.7 Shaper Operations

1. Horizontal shaping (Facing at top surface)
2. Vertical shaping (Facing at side)
3. Groove shaping (Keyways etc.)
4. Angular shaping (Dovetail cutting)
5. Cutting of gear teeth
6. Irregular cutting

Operation	Sketch	Description
1. Horizontal Shaping	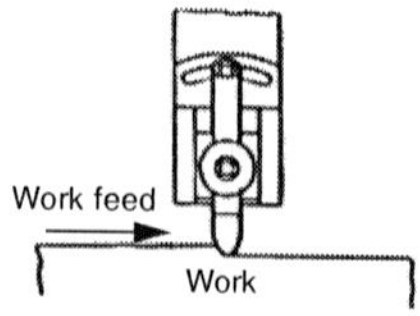	Horizontal shaping operation is done generally to obtain flat surfaces. In this operation work is held in vice and tool is held in tool head. The length and position of the stroke is adjusted according to the job length. Initially cross feel is given by hand and then the power feed is employed to cut the material over the job.
2. Vertical shaping	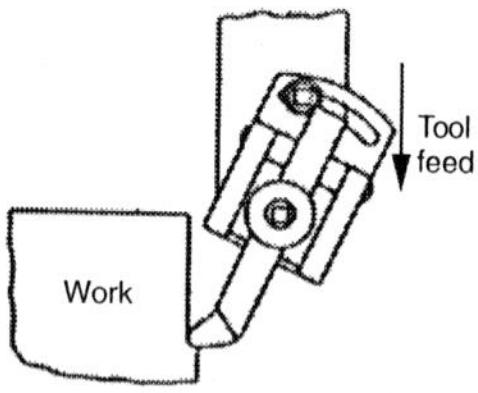	In a vertical shaping operation, the cutting is carried out to make grooves, keyways parting off and squaring up the ends and shoulders. Initially the tool is fed downward by rotating the downfeed screw by hand at the end of return stroke and feed is given about 0.25 mm for cutting purpose.
3. Groove shaping	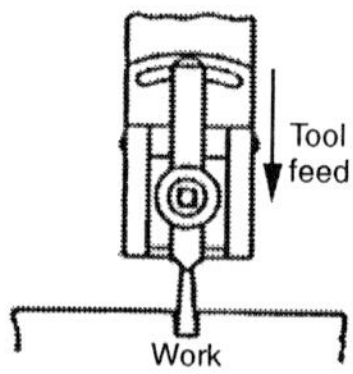	During this operation a square nose tool is used for shaping grooves, slots and keyways ensure that tool (Keyways etc.) should be fixed rigidly and no more clearance is required during operation. A flat nose tool is genrally employed for cutting keyways.
4. Angular (Dovetail cutting)	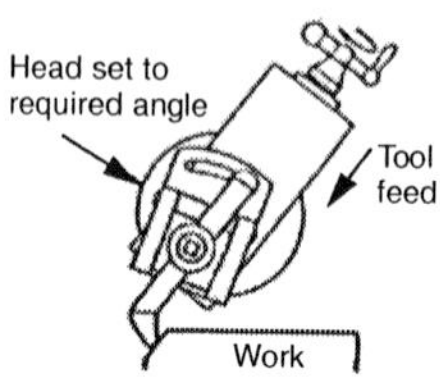	In an angular shaping operation the swivel head is shaping set to the required angle. The proper setting at the tool is again an important factor. Sometime it can be obtained by setting the job in inclined surface position.
5. Cutting of gear teeth	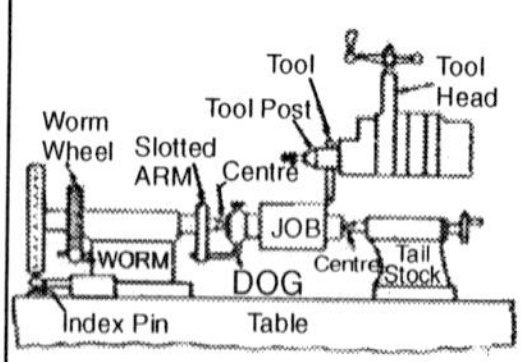	Gear teeth can also be cut on the shaper machine by using indexing head arrangement on the table. The periphery of a gear blank is devided equally by using index plate having proper hole circles on it and a form tool is used for this purpose.
6. Irregular cutting	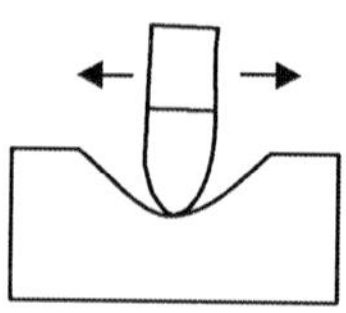	A shaper can produce contoured, convex and cancave surfaces. To produce small contoured surface a firm tool is used and for big surfaces a round nose tool is selected for machining irregular surfaces.

1.5.8 Shaper Tools

The shape of shaper and planer tools are generally similar that of those of the lathe tools but differ a lot in the various angles. There are large variety of tools in use but they are broadly classified as follows:

1. **According to shape:** straight, crank, goose, neck etc.
2. **According to direction of cutting:** left hand and right hand.
3. **According to the operations:** down cutting, side recessing, parting off, squaring etc.
4. **According to degree of finish:** roughing and finishing.
5. **According to the shape and cutting edge:** round nose or square.

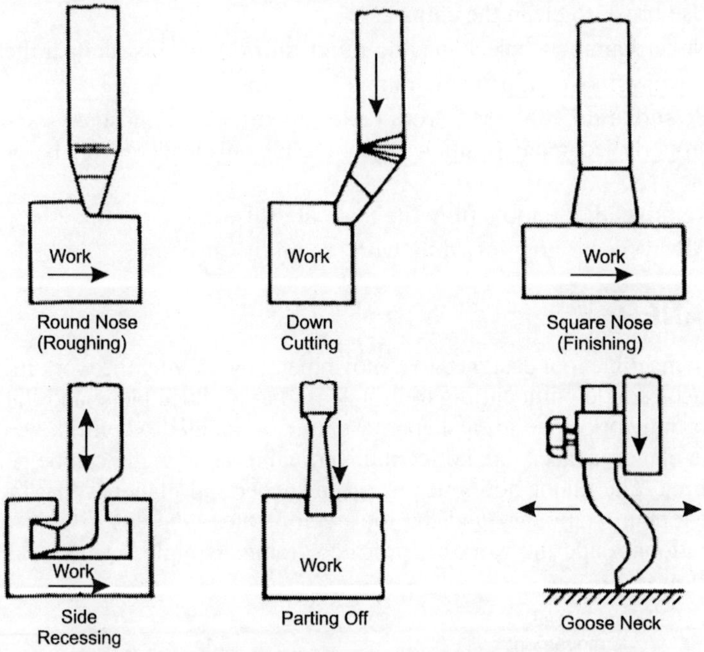

Fig. 1.37 Types of shaper tools

1.5.9 Cutting Speed, Feed and Depth of Cut of a Shaper

Cutting speed: The distance travelled by the shaping tool in metres per minute across the job during the working stroke is known as cutting speed for shaping.

The speed in metres per minute of the tool during the non-cutting stroke is known as return speed.

The return stroke in a shaping machine is faster than cutting stroke. Usually, the ratio of return stroke time to cutting stroke time is kept as 2:3.

Feed: The distance travelled in millimetre per working stroke by the job across the tool (in horizontal shaping) or by the tool across the job in vertical or angular shaping is known as feed of the shaping operation.

Depth of cut: The depth of penetration in millimetres of the tool into the job during a shaping cut is known as the depth of cut for the shaping operation.

1.5.10 Safety Precautions Associated with Shaper

1. Check up all the controls of the machine before starting the operation.
2. Always maintain the parallel rear.
3. Always select a proper tool, proper tool material, proper tool angle, proper speed and feed.
4. Place a metal shield over the tool to catch the chips.
5. Use brush to clean the chips
6. Never hammer a piece, having rough surface, against the parallel.
7. Never remove chips while ram is in motion.
8. Be sure, ram, tool head, work table support, clamping screws and vice are properly secured in position and tool head, tool clear the work before starting the shaper.
9. Remove all the tools from the table after use.
10. Always wear goggles while working on the machine.

1.6 PLANER

Planer is a machine tool designed to remove metal by moving the work in a straight line against a single point cutting tool. It is used to produce plane and flat surfaces similar to the work done on a shaper, a planer is adopted when the work size is large. The plane surfaces like horizontal, vertical or at an angle can be perform on this machine. The major difference between shaper and planer is that in a shaper the work remain stationary and the tool reciprocates whereas in planer the tool remain stationary and the work reciprocates. Some examples of the planing work are as follows:

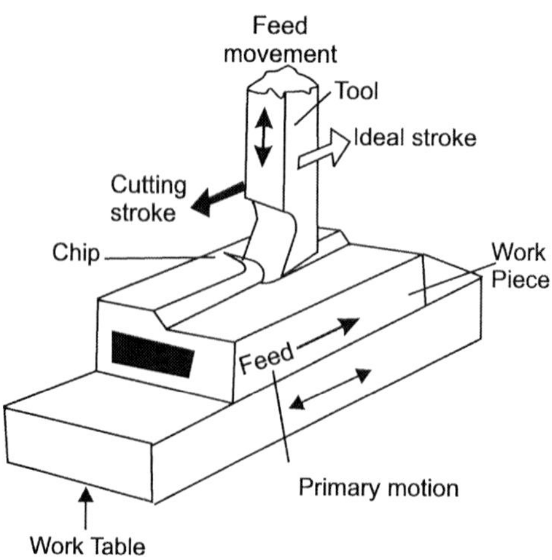

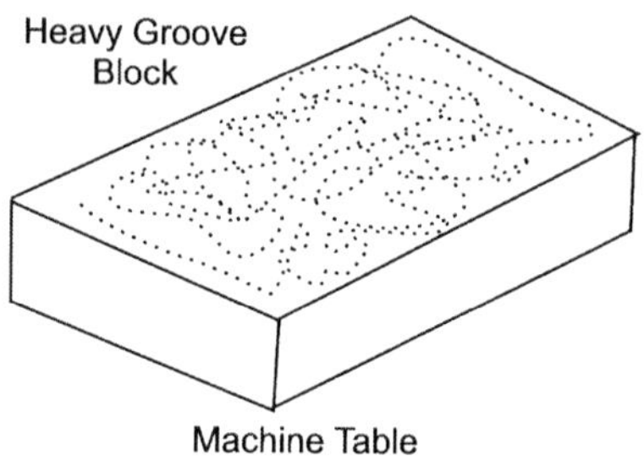

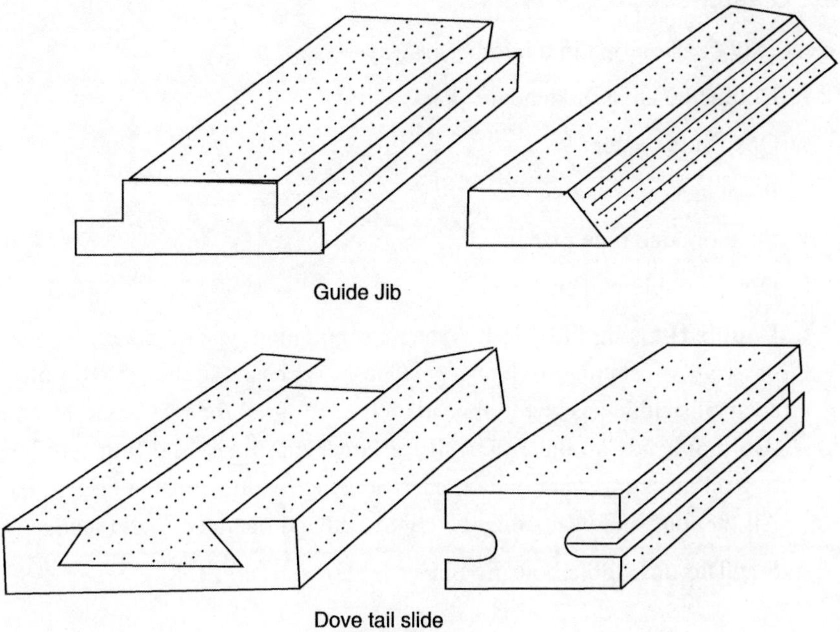

Fig. 1.38 Planer

1.6.1 Working Principle of a Planer

In the working of a planer, the work is held rigidly on the worktable (platen) while tool is held in the tool head of the machine.

The work table reciprocates and thus moves the work in a straight line path of stationary tool. In the reciprocation of the table working and idle strokes take place alternately. The tool performs metal removal operation during the working stroke. The idle stroke is made faster than the working stroke to reduce the idle time in the operation.

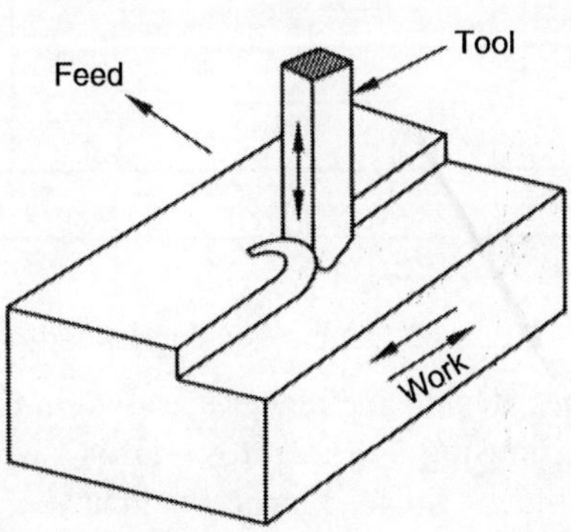

Fig. 1.39 Working principle of the planer.

1.6.2 Classification of Planers

Planers may be classified in a number of ways:

1. Double housing or standard planer
2. Open side planer
3. Pit planer
4. Plate or edge type planer
5. Divided table planer.

1. **Double Housing Planer:** It is the most common type of planer. It consists of two vertical column (housing). One on each side of the table (or platen) such a machine has one tool head on the cross-rail for very small machine, two tool heads on the cross-rail for larger machines; and four tool heads two on cross-rails and one on each housing for some of the larger planer. All the four tool heads may be controlled and operated individually.

   Standard or Double housing planer is shown in Fig. 1.40.

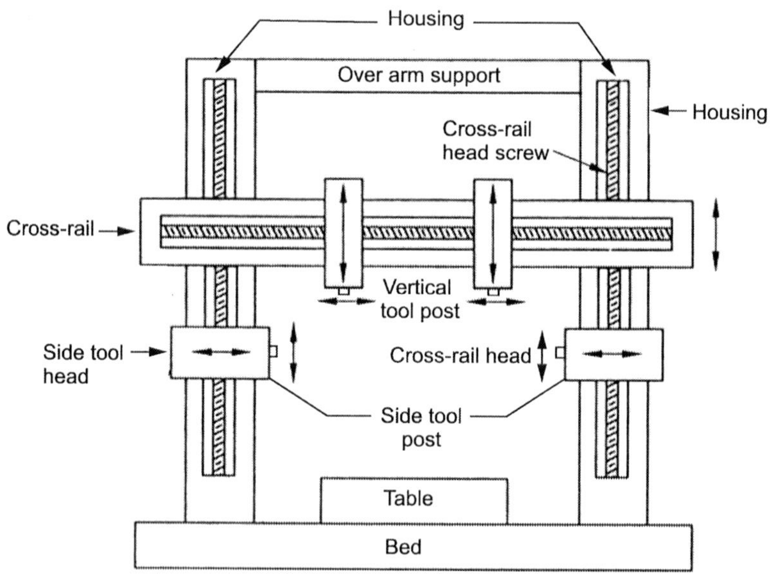

Fig. 1.40 Double housing planer

2. **Open Side Planer:** Open side planer is shown in Fig. 1.41. In this type of planer, only one housing and the cross-rail is suspended from the single housing as a cantilever. Such a planer would have a maximum of three tool heads and all other essential features of a standard planer.

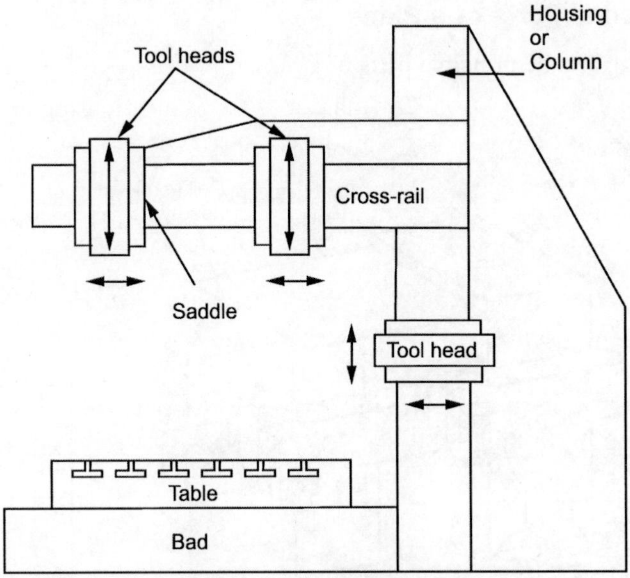

Fig. 1.41 Open side planer.

As one side of the planer being opened, large and wide jobs to be clamped on the table and reciprocate without being interfered by the housing.

3. **Pit Planer:** Its working principle is similar to a shaper and not to a conventional planer. This machine is intended for machining work larger than those taken by conventional planers.

   The conventional planer, to handle such a work, would require bed twice as long as the ways of this machine and become a space-taking as well as-a heavy machine. The work is held on a stationary table in the centre or the machine, or levelled up directly on the shop floor and the tools are passed over it. The cross-rail of this machine is very long and housings column rather short. The tool heads, rail and housings move along the ways past the work. It is simpler to move lighter weight of the machine than to move the work.

4. **Plate Planer:** It is a single purpose machine whose main function is to plane the edges of heavy steel plates such as armour plates, plates of ships and pressure vessels. The edges may be machined square or beveled.

   This machine is completely different from the conventional planer. The work is held stationary, and the tool and operator move to and fro along the work for the feed and the cut. The machine is relatively long, so the operator can stand on a plateform extending from the carriage to observe the cut.

5. **Divided Table Planer:** This type of planer has two tables on the bed which may be reciprocated separately or together.

   This type of design reduced much of idle time while setting the work.

1.6.3 Principal Parts of a Planer

The following are the principle parts of a planer.

1. Bed
2. Table
3. Housing or column
4. Cross rail
5. Tool heads

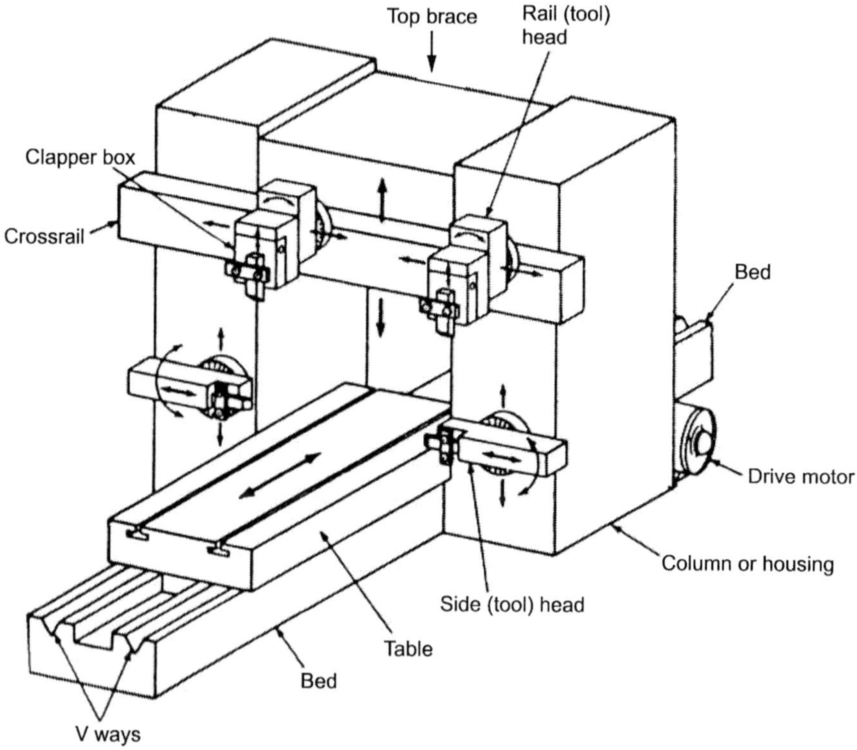

Fig. 1.42 A Double housing planer

1. **Bed:** Bed is made of heavy casting structure having a V type guide ways on which the working table slides.
2. **Table:** It is a box type structure made up of cast iron reciprocates in a V type guide ways. At the top it carries longitudinal T-slots and holes for clamping purpose.
3. **Housing or Column:** The housing are the rigid castings placed on each side of the bed. It carries cross rail elevating screws, vertical feed shaft and cross feed bar to transmit the power to the upper parts of the machine.
4. **Cross Rail:** It is mounted on the vertical guide ways of the two housing. It can be raised or lowered. The ways are provided at the front of the cross rail for two vertical tool heads.
5. **Tool Heads:** Two tool heads are provided on the cross-rail and the other two on the vertical columns each column carries one side tool head. All the four tool heads work independently. They can also work separately as desired.

1.6.4 Size and Specifications of a Planer

1. Maximum width and height of a job that can pass through housing
2. Maximum width of table
3. Maximum length of table
4. Maximum stroke of table in mm
5. Length of bed
6. No. of speed and feed available
7. Floor space required
8. Net weight of machine
9. Type of drive
10. H.P. of the motor
11. Power input.

1.6.5 Planer Operations

Commonly operations performed on a planer are as follows:

1. Machining horizontal flat surfaces.
2. Machining vertical flat surfaces.
3. Machining angular surface like dove tail.
4. Machining different types of slots and grooves.
5. Machining contours and curved surfaces.

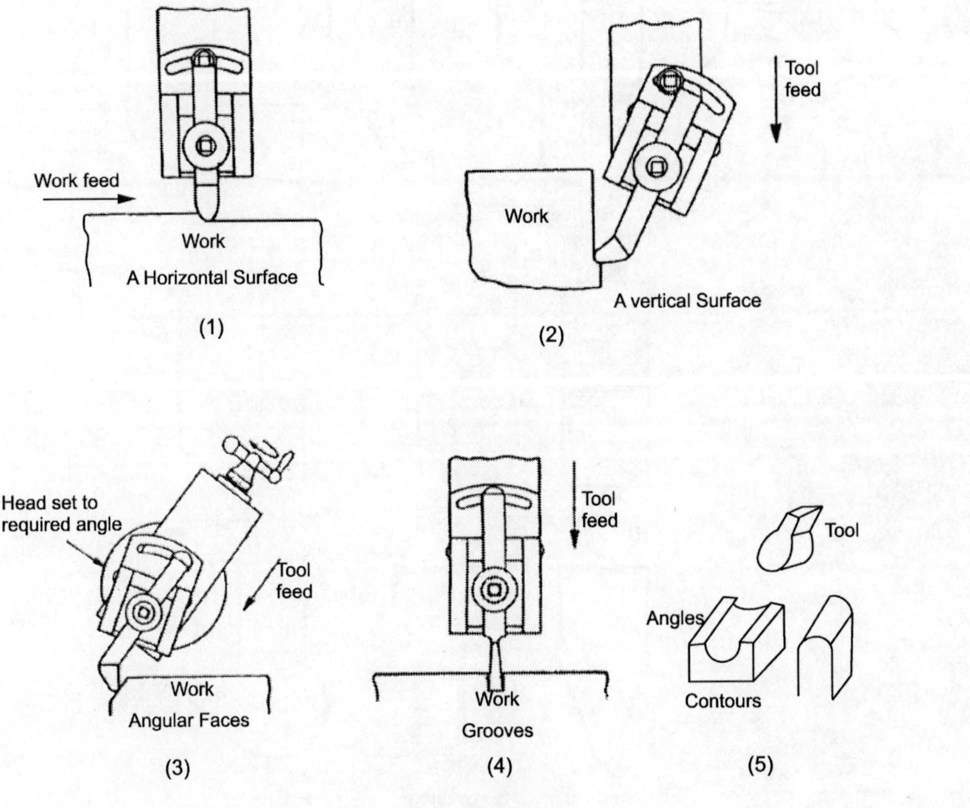

Fig.1.43 Planer operations

1.6.6 Planer Tools

Planer tools are similar to shaper tools but in case of planer depth of cut and cutting strokes are large so the tools are also large in size and made of high speed steel as well cemented carbide tips are also used for the production work generally two types of tools used in planer.

(*i*) Forged solid type.

(*ii*) Tool bit type held in holder.

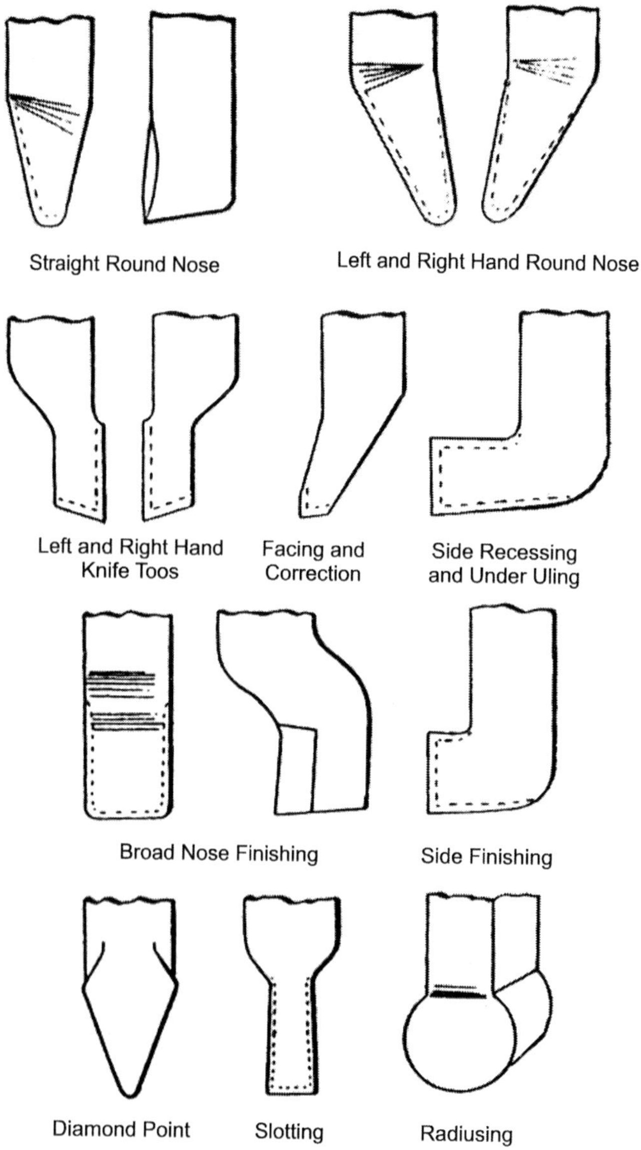

Fig. 1.44 Two types of planer tools

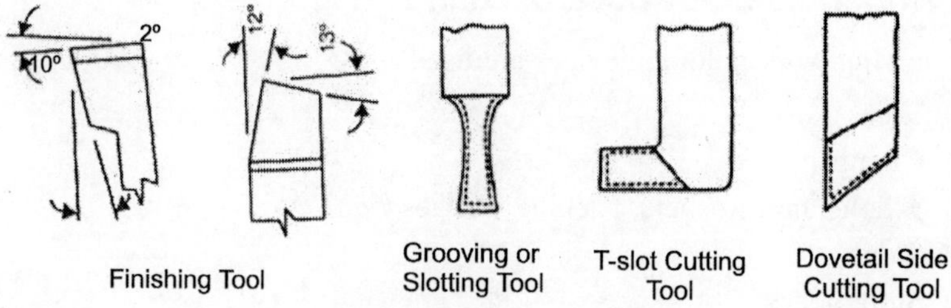

Fig. 1.45 Planer tools

1.6.7 Comparison between a Planer and a Shaper

Planer	Shaper
1. In a planer work is moved against a stationary tool.	1. In shaper tool moves while the work remain stationary.
2. More than one cutting tool can be used at a time	2. Only one cutting tool is used at a time.
3. It is used for large work.	3. It is used for small work.
4. It is heavier, rigid and costlier machine.	4. It is comparatively light and cheaper machine.
5. It requires more floor area.	5. It requires less floor area.
6. Tool used on a planer are larger, heavier and stronger.	6. The tool used on a shaper is small in size as compared to planer.
7. Heavy feeds are used.	7. Lighter feeds are used.
8. It can take deep cuts.	8. It can not take deep cuts.
9. Work setting requires more skill and time	9. Work may be clamped easily and quickly.
10. Indexed feed is given to the tool during ideal stroke of the work table.	10. Indexed feed is given to the work during the ideal stroke of the ram.
11. Tool is feed into the work.	11. Work is feed usually across the tool.

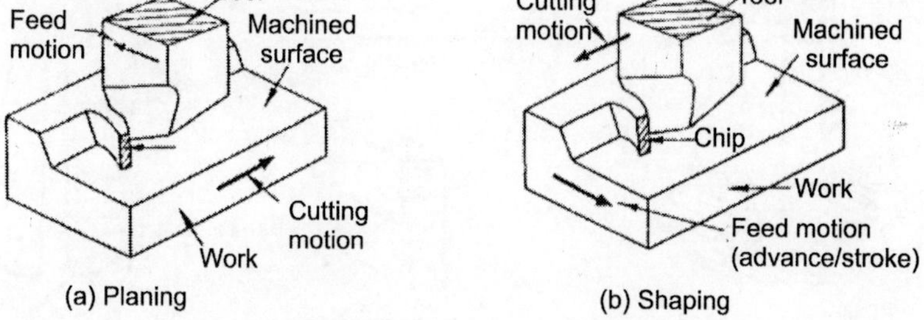

Fig. 1.46 Planer and Shaper

1.6.8 Work Holding Devices Used in a Planer

The following work holding devices are used:

(*a*) Vices
(*b*) Clamps
(*c*) T bolts, nut, washers, packing wedges etc.
(*d*) Stops
(*e*) Stop blocks
(*f*) Planer Jacks
(*g*) V. Blocks etc.
(*h*) Angle plate
(*i*) Holding down dogs
(*j*) Parallel strips etc.

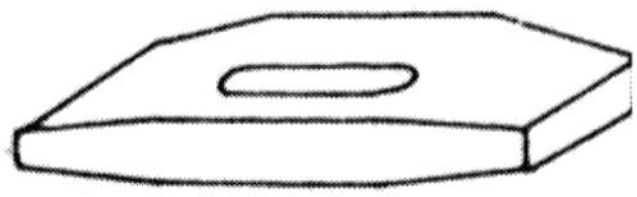
Plain clamps

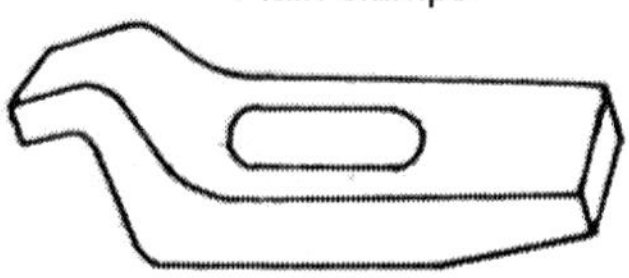
Offset clamps

U-clamp boith pin end

T-shot bolt

T-bolt

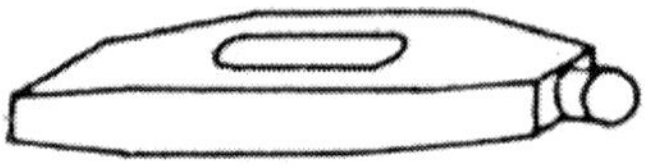
Pin-clamp

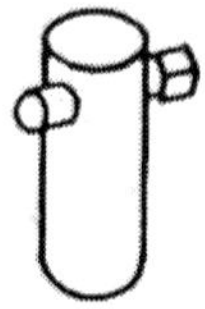
Stop pin

T slot removable nut

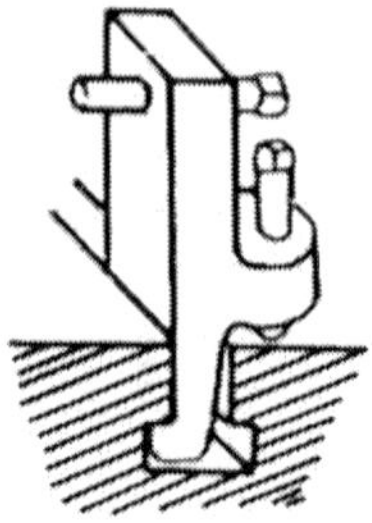
T slot stop bracket

Chisel point

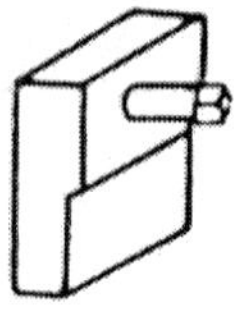
T slot shop blocks

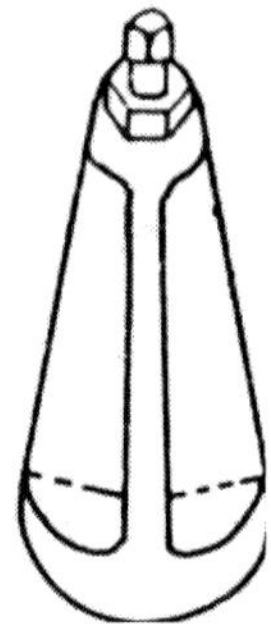
Jack screw

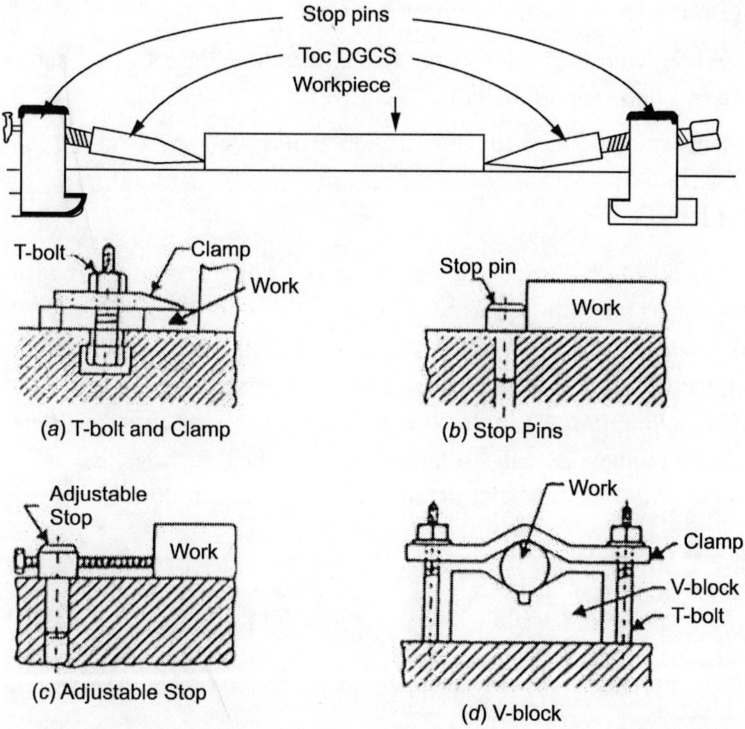

Fig. 1.47 Common accessories for setting up the work and for clamping purposes.

1.6.9 Cutting Speed, Feed and Depth of Cut

Cutting Speed: The rate at which the metal is removed from the work piece during the cutting stroke. It is expressed in m/min.

Feed: It is the distance in which the cutting tool travels at right angle to the direction of motion of the workable at the beginning of each cutting stroke. It is also expressed in mm.

Depth of Cut: It is the distance by which the tool digs into the work piece during the cutting stroks depth of cut is expressed in mm.

1.6.10 Safety Precuations in Planer

The following are the safety precautions in planer :

1. Ensure that work holding devices are at proper place and properly clamped.
2. Keep planer ideal while adjusting length of bed stroke and speed of machine to suit work.
3. After the work is fastened, cross rail should be clear.
4. Check the feed rod and its attachments whether they are in proper location or not.

5. Do not leave the tool between the ways.
6. While removing or loosening tool holder, hold the tool with hand or place a wooden support under it.
7. Never ride on the bed while the job is moving.

1.7 MILLING

Milling machine is a most versatile machine tool which is used to produce flat and curved surfaces on both side as well out side the work piece. It is a economical method of producing a wide variety of machining operations and creating accurate parts for interchanagable manufacture due to its high accuracy and higher production rate this machine is most preferable as compare to shaper and slotter. Milling machine can produce external, internal threads helical surfaces and various cross sections etc. Some examples of the milling work are as follows:

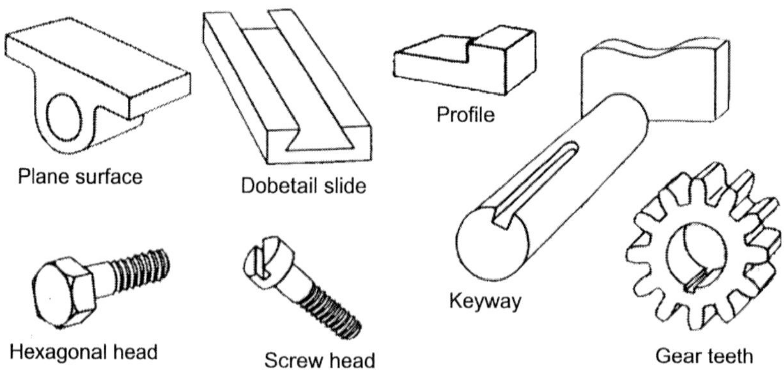

Fig. 1.48 Milling

1.7.1 Working Principle of Milling Machine

Milling machine is a power driven tool in which metal is removed by means of a multipoint revolving cutter. As the work moves against cutting edge the metal is removed from the work piece in the form of chips. The work may be fed against the cutter longitudinally, transversely or vertically This machine is very much suitable for tool room work due to the variety of operations, accuracy and better surface finish.

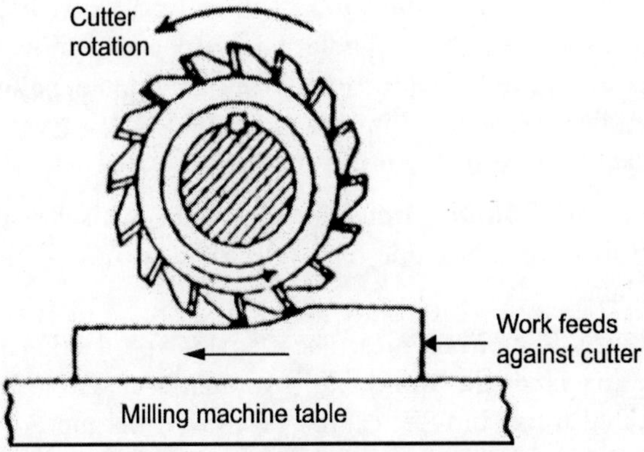

Fig. 1.49 Working Principle of Milling

1.7.2 Classification of Milling Machine

Milling machines are classified according to the design and operations performed on them. Classification is as follows:

1. **According to Column and knee:**
   (*a*) Horizontal milling machine (*b*) Vertical milling machine
   (*c*) Universal milling machine (*d*) Ram type milling machine
2. **According to Bed type:**
   (*a*) simple milling machine (*b*) Duplex milling machine
   (*c*) Triplex milling machine
3. **According to Special purpose type:**
   (*a*) Rotary table milling machine (*b*) Drum milling machine
   (*c*) Planetry milling machine (*d*) Profit milling machine
   (*e*) Pantagraph milling machine (*f*) Offset milling machine

1. **Column and Knee Type Milling Machines:** These milling machines are low production machines. In these machines, worktable is supported on a knee like casting which can slide in a vertical direction along a vertical column. The table can travel longitudinally and crosswise. It can also be moved up and down.

   These machines can further be sub-divided into three types as follows:

   (*a*) *Plain Milling Machines or Horizontal Milling Machines*: The cutter revolves about the horizontal axis of the machine spindle and the worktable can have movement in longitudinal, cross, and vertical directions manually or by automatic feeding worktable cannot be swivelled in a horizontal plane.

(*b*) *Vertical Milling Machines*: The cutter revolves about the vertical axis of the machine spindle. The work can be moved in longitudinal, cross and vertical directions. Worktable cannot be swivelled. In some modern vertical milling machines the spindle head (or milling head) can be swivelled in a vertical plane.

(*c*) *Universal Milling Machines*: This machine is very much similar to a plain milling machine but its table can also be swivelled in horizontal plane.

2. **Fixed-Bed Milling Machines:** These are high production milling machines. In these machines the worktable is mounted on a fixed bed and it has a longitudinal travel only. It cannot be moved up and down or crosswise. The adjustable spindle or spindle head (carrier) attached to a vertical column car : move along vertical ways on the column to adjust tool to the work.

   These machines can further be sub-divided in three types as follows:

   (*a*) *Simplex Milling Machine*: It has single adjustable, horizontal spindle head.

   (*b*) *Duplex Milling Machine*: It has two adjustable, horizontal spindle heads, each attached to a separate vertical column, one in each side of the fixed bed.

   (*c*) *Triphx Milling Machine*: It has two adjustable horizontal spindle heads like a duplex machine attached to two different vertical columns, and one vertical spindle head on a cross-rail.

3. **Planer Milling Machines:** These milling machines are high production machines for heavy jobs. It is very much similar to a double housing planer. In these machines, spindle carriers are mounted in place of planer tool posts which are operated by individual motors. In addition, two vertical milling heads are fitted on a cross-rail and can travel along it.

4. **Special Purpose Milling Machines:** These are special purpose machines. Some of these are as follows: .

   (*a*) *Rotary Table Milling Machine*: It is a mass production machine with a heavy, rotating, circular worktable mounted on the base of the machine.

   (*b*) *Drum Milling Machine*: It is a production type machine. It has a drum fixture upon whose *D* periphery workpieces are mounted. The drum fixture rotates slowly carrying work against rotating cutters.

   (*c*) *Planetary Milling Machine*: The workpiece is held stationary and cutter performs machining operation by revolving about its own axis as well as around the work (*i.e.* like a planet) shown in Fig.1.50.

   A planetary milling machine is used for (*i.e.* milling of internal) and external threads on all kinds of tapered surfaces, bearing surfaces, rear axle end holes etc.

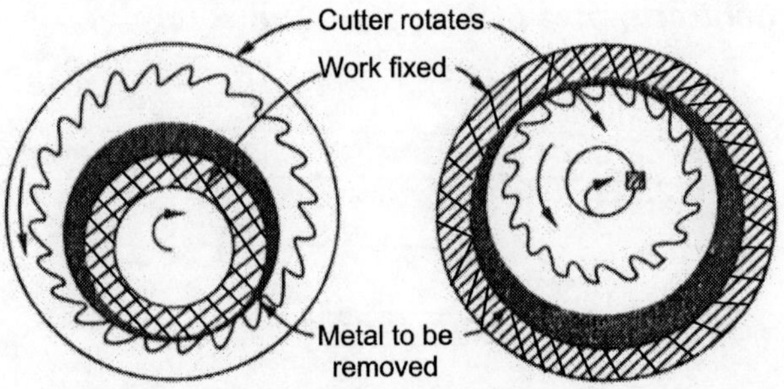

Fig. 1.50 Planetary Milling Machine

(*d*) *Profiling Machine*: In this machine, the cutter movement is guided by the path of a tracer which travels along the outside surface on a template, thus reproducing the same profile on the work as that of the template.

(*e*) *Pantograph Milling Machine*: This machine makes use of the pantograph linkage to guide the direction of the feed of the cutter to reproduce form of a template at either an enlarged or a reproduce scale.

(*f*) *Offset Milling Machine*: In this machine the cutter is mounted on a vertical spindle which extends into a fixture. The fixture is mounted eccentrically with cutter, so that, as it revolves, the work is milled as it passes by.

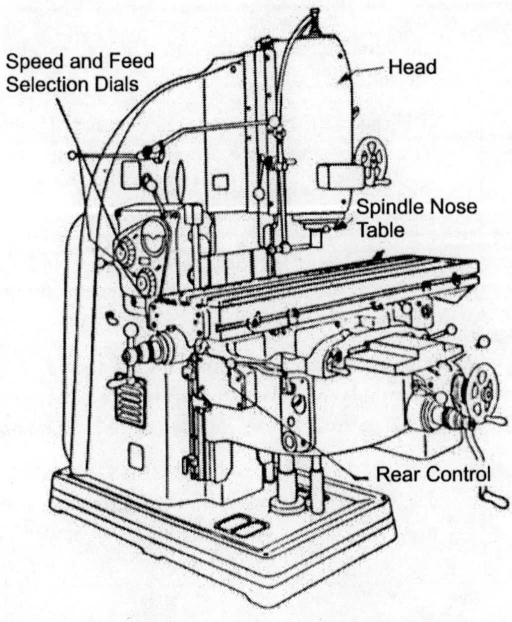

Fig. 1.51 Plain Vertical milling machine

1.7.3 Principal Parts of a Milling Machine

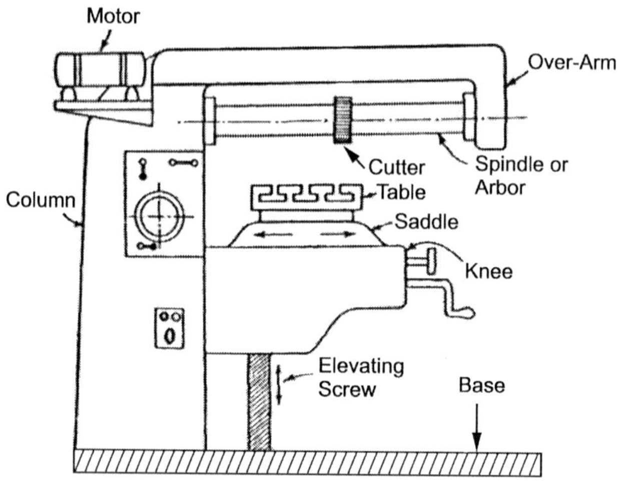

Fig. 1.52 Schematic digram of a milling machine

1. Base 2. Column 3. Knee 4. Saddle
5. Table 6. Over arm 7. Spindle 8. Arbor

Part Name	Description
1. Base	It is the base of the machine and all the other parts which rest upon it. It is generally made of grey cast iron to absorb shock and vibration. Base of the machine is hollow which serves as a reservoir for cutting fluid.
2. Column	It is the main supporting frame mounted vertically on the base. The motor and other driving mechanism are fixed on in it. It supports and guides the knee in its vertical travel it carries the jack for elevating the knee.
3. Knee	It is a structural member attached to the column and which moves vertically on the column which moves the table vertically for adjusting the depth of cut.
4. Saddle	The saddle supports and carries the table and is adjustable traversely on ways on top of the knee. It is provided with graduations for exact movement and can be operated by hand or power.
5. Table	The table rests on ways on the saddle and travel is longitudinally in a horizontal plane. It supports the work piece, fixtures etc.

Contd...

Part Name	Description
6. **Over arm**	The over arm is mounted on and guided by the top of the column it is adjusted in an out by hand to the position of maximum support for the arbor and then clamped.
7. **Spindle**	The spindle is mounted on the upper part of the column it receives power from the motor through belt, gears, clutches etc. and can be rotated at different speeds by using cone pulley and trasmits it to arbor.
8. **Arbor**	It is the extension of the spindle on which various cutters are mounted. It is tapered at one end to fit the spindle nose and has two slots to fit the nose keys for locating and driving it.

1.7.4 Size and Specifications of Milling Machine

1. The milling machine is specified by working surface of the table and its maximum length of longitudinal, cross and vertical travel of the table,
2. Knee movement in degrees,
3. Range and no. of spindle speeds,
4. Spindle nose tapes,
5. Power available,
6. Net weight
7. Floor space required etc.

Table length × width = 1100 mm × 3100 mm

Power traverse = longitudinal × cross × vertical = 650 mm × 235 mm × 420 mm

1.7.5 Milling Methods

There are two methods generally used for milling which are as follows:

Up Milling: It is also known as conventional milling, in which the cutter rotates against the direction in which the work is being fed.

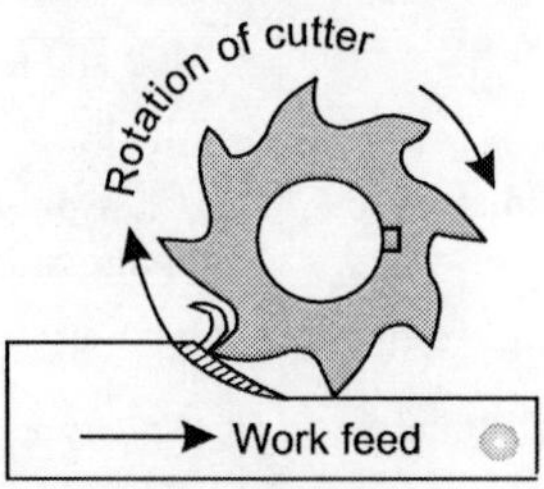

Fig. 1.53 Up milling.

Down Milling: Down milling which is also known as climb milling. The cutter rotates in the same direction in which the work is being fed.

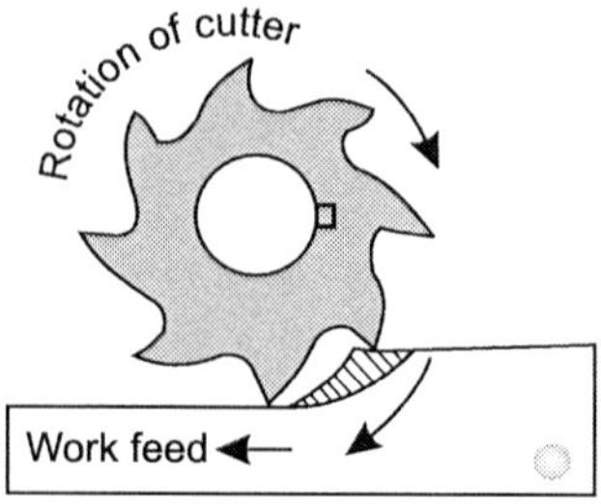

Fig. 1.54 Down milling

1.7.6 Comparison between Up Milling and Down Milling.

Up Milling	Down Milling
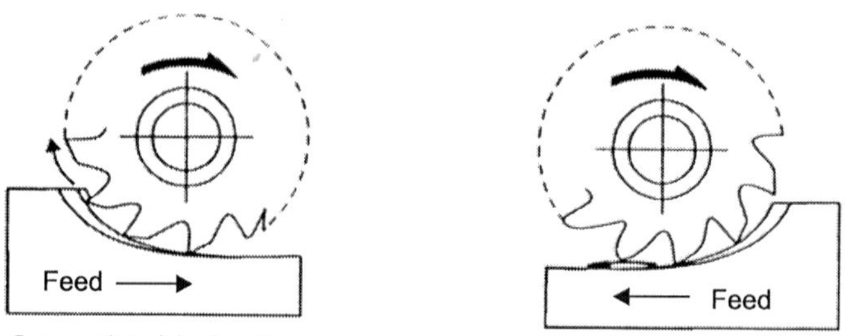 Conventional (up) milling	Climb (down milling)
Fig. 1.55 Up milling and down milling	
1. The cutter rotates against the direction in which work is being fed.	1. The cutter rotates in the same direction in which the work is being fed.
2. It is also known as conventional milling.	2. It is also known as climb milling.
3. Surface finish is better.	3. Surface finish is better and free from defects like backless error.
4. Cutting forces vary from zero to maximum	4. Cutting forces vary from maximum to zero
5. Job and tool motion is in the opposite direction.	5. Job and tool motion is in the same direction.
6. Chip thickness varies from minimum to maximum.	6. Chips thickness varies from maximum to minimum.
7. Use of cutting fluid is difficult.	7. Use of cutting fluid is easy.
8. Cutting forces tendency is to lift the job so clamping and fixing the job is necessary.	8. Forces are sufficient on the job so less pressure for clamping is required.
9. It is practicable.	9. It is impracticable.

1.7.7 Elements of a Plain Milling Cutter

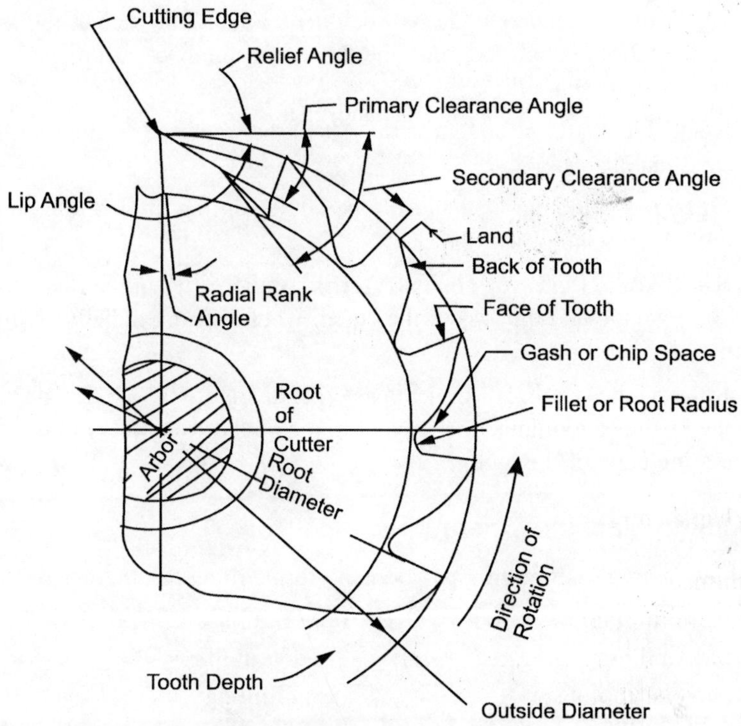

Fig. 1.56 Elements of a plain milling cutter

1. **Arbor:** The shaft on which the milling cutter is mounted and driven.
2. **Body:** That part of the cutter left after exclusion of the teeth and the portions to which teeth are attached.
3. **Clearance angles:** Angles formed by primary or secondary clearance and the tangent to the periphery of the cutter at the cutting edge. They are called the primary and secondary clearance angles, respectively.
4. **Cutting edge:** Edge formed by the intersection of the face and the circular land or the surface left by the provision of primary clearance.
5. **Face:** That portion of the gash, adjacent to the cutting edge on which chip impinges as it is cut from the work.
6. **Fillet:** Curved surface at the bottom of gash which joins the face of one tooth to the back of the tooth, immediately ahead.
7. **Gash:** Chip space between the back of one tooth and face of the next tooth.
8. **Land:** This is part of the back of the tooth adjacent to the cutting edge, which is relieved to avoid interference between the surface being machined and the cutter.

9. **Lead:** Axial advance of the helix of the cutting edge in one complete revolution of the cutter.
10. **Lip Angle:** Included angle between the land and the face of the tooth, or alternately the angle between the tangent to the back at the cutting edge and face of the tooth
11. **Neck:** Undercut, of cutters with a shank between the cutting portion and the shank.
12. **Outside Diameter:** Diameter of the circle passing through the peripheral cutting edge.
13. **Rake Angle (Axis for Helical Teeth):** Angle between the line peripheral cutting edge and the axis of the cutter, when looking radially, at the point of intersection.
14. **Rake Angle (Radial):** Angle, in a perpendicular to the axis, angle between the land of a tooth and the tangent to the outside diameter of cutter, at the cutting edge of the tooth.

1.7.8 Milling Machine Operations

The following are the different operations performed in a milling machine:

1. Plain milling
2. Face milling
3. Side milling
4. Straddle milling
5. Angular milling
6. Gang milling
7. Form milling
8. Profile milling
9. End milling
10. Saw milling
11. Milling key ways grooves and sloats
12. Gear cutting
13. Helical milling
14. Cam milling
15. Thread milling

Operation	Sketch	Description
1. Plain milling	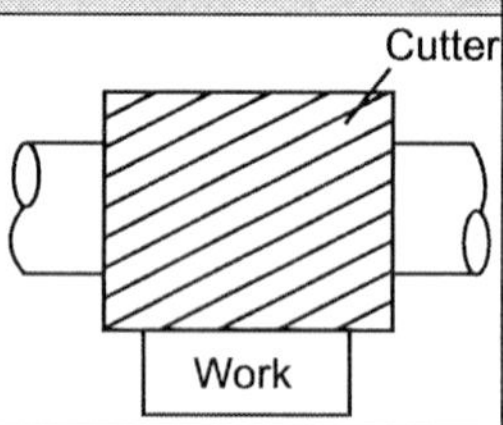	This operation is used to produce, a plain flat horizontal surface parallel to the axis of rotation of a plain milling cutter.
2. Face milling	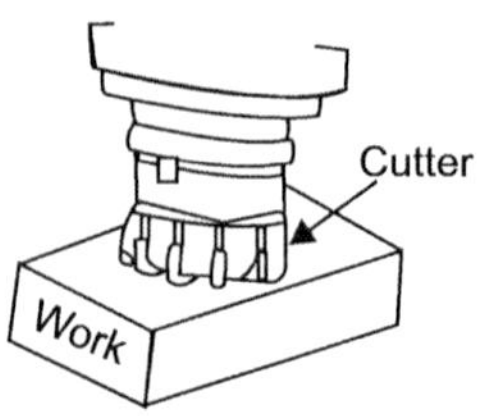	Face milling cutter is used to machine a surface at right angles to the spindle axis and parallel to the face of the cutter.

Contd...

Operation	Sketch	Description
3. Side milling	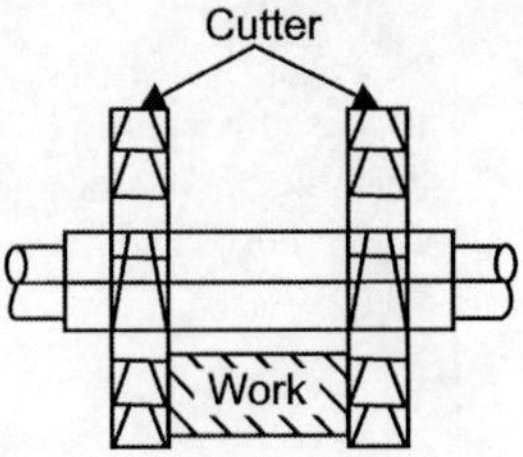	It is used to produce a flat vertical surface on the side of the work pieces.
4. Straddle milling	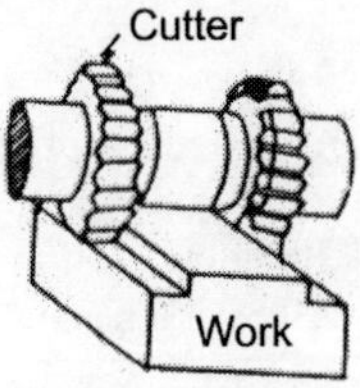	It is used to produce flat vertical surface on both sides of a work piece by using two side milling cutters mounted on the same arbor.
5. Angular milling	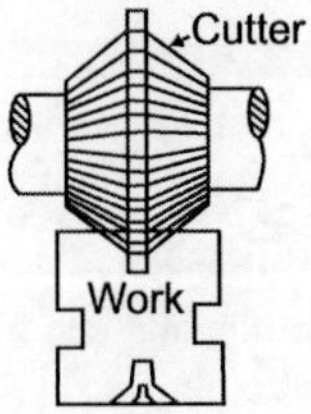	It is used to produce an angular surface on a work piece other than right angles to the axis of the milling machine spindle angular surfaces like bevels, compass and tapers etc.
6. Gang milling	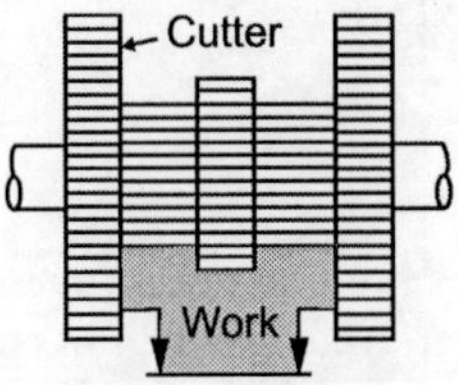	Gang milling is used for mounting several cutters on the arbor at time and make it possible to machine several surfaces in one pass.
7. Form milling	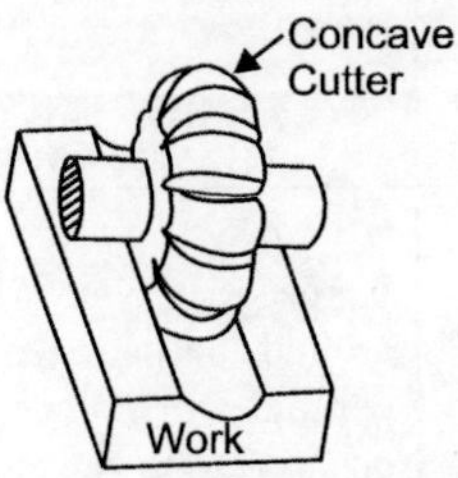	In this milling operation, the cutting teeth of the form cutter will be the same as that of the profile of the surface to be produced it may be concave, convex or only other irregular shape.
8. Profile milling	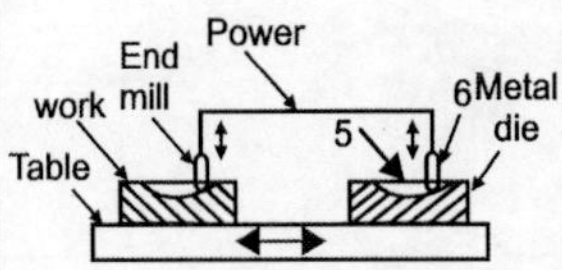	It is used for reproduction of an outline of a template or complex shape of a master die on a work piece. Different cutters may be used for profile milling.

Contd...

Operation	Sketch	Description
9. End milling	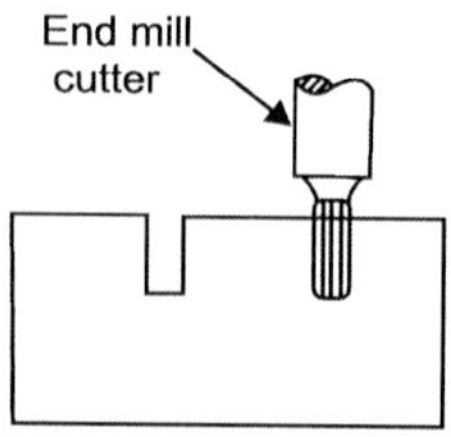	It is used generally on the vertical milling machine to cut slots, grooves or key ways on the horizontal vertical or an angle.
10. Saw milling	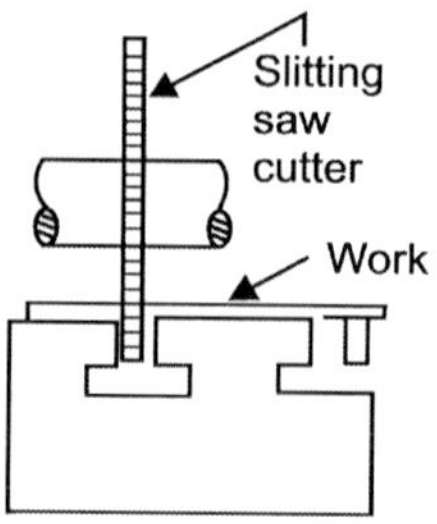	It is used to produce narrow slots or grooves on a work piece by using a saw milling cutter. It is also used for complete parting off operation.
11. Milling key ways grooves and slots	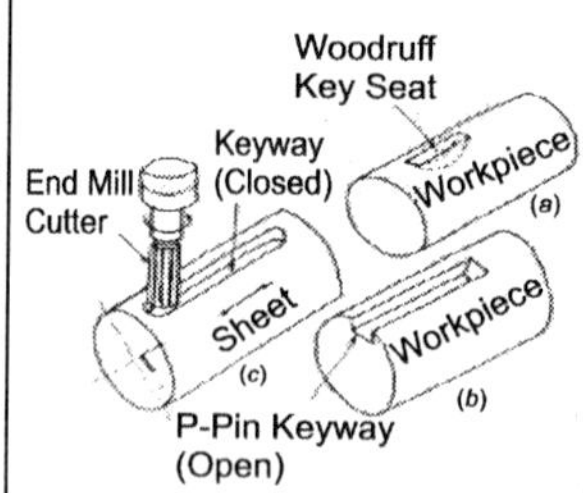	Key way may be out by using a plain milling cutter by a slitting saw for slot cutting and end mill cutter or side milling cutter are used to cut the T slots.
12. Gear cutting	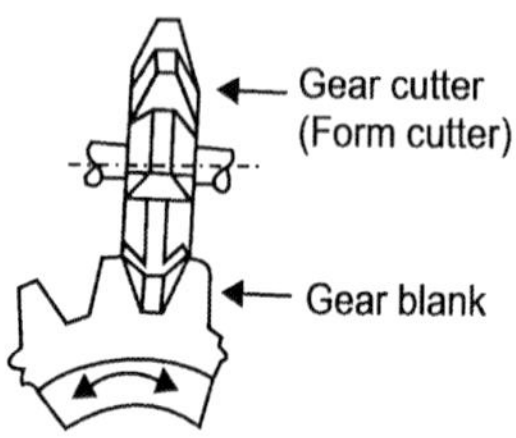	It is used in a milling machine by using a form relieved cutter. The cutter may be cylindrical type or end mill type which is used for gear cutting.
13. Helical milling	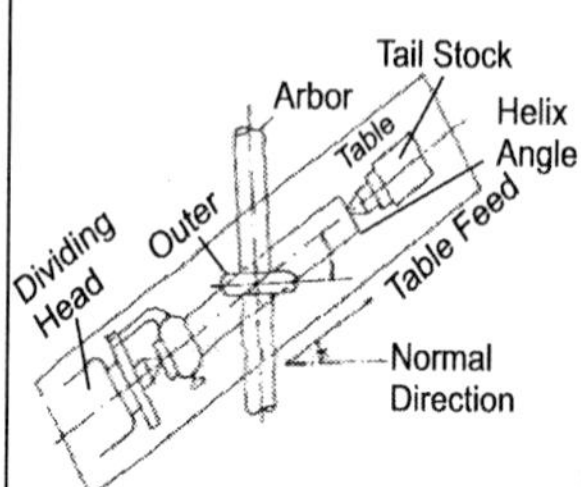	It is used to produce a helical flute or groove around the periphery of a cylindrical or coninal work piece or jobs like helical gears, cutting helical grooves or flutes on a drill blank or reamer.

Contd...

Operation	Sketch	Description
14. Cam milling	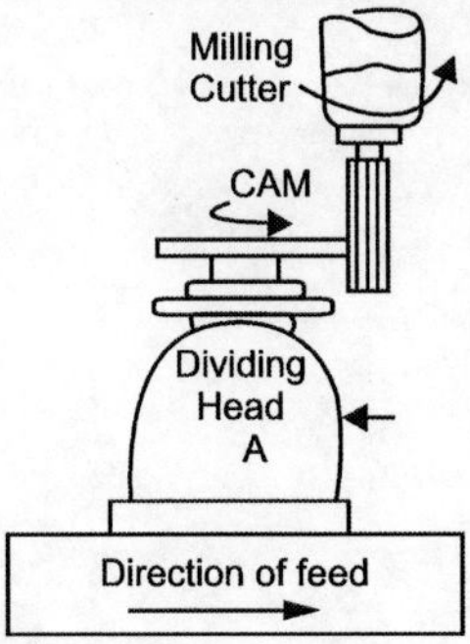	It is used to produce cam in a milling machine by the use of a universal dividing head and a vertical milling attachment.
15. Thread milling	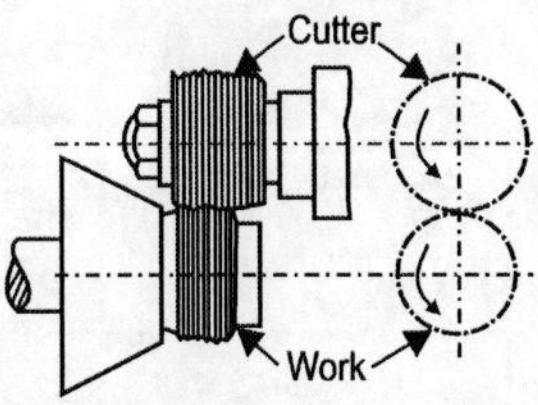	It is used to produce of threads by using a single or multiple thread milling cutter thread milling operation requires three dividing motions in a machine. One for cutter, one for work and third for logitudinal movement of the cutter.

1.7.9 Milling Cutters

Milling cutters are the revolving bodies with cutting teeth on the pheriphery. They are suitable in different types and sizes. These cutters are usually classified according to their shapes. Cutters are generally made of high carbon steel (HCS), high speed steel (HSS), satellite and sintred carbide tips.Generally there are five categories of plain milling cutters which are as follows:

1. Plain milling cutter
2. Face milling cutter
3. Side milling cutter
4. Staggered tooth side milling cutter
5. Face end mills

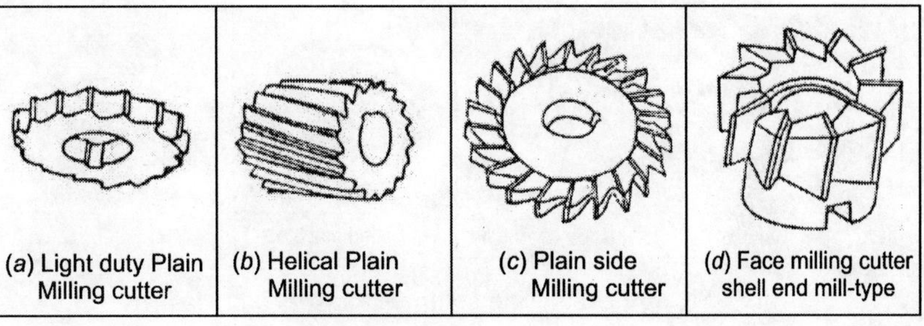

(*a*) Light duty Plain Milling cutter (*b*) Helical Plain Milling cutter (*c*) Plain side Milling cutter (*d*) Face milling cutter shell end mill-type

Fig. 1.57 Milling cutters

1. Plain milling cutters
   - (*a*) Straight teeth
   - (*b*) Helical plain milling
   - (*c*) Helical milling heavy duty
   - (*d*) Alternate gash plain milling
   - (*e*) Helical milling shank type
   - (*f*) Helical milling arbor type
   - (*g*) Sprial mill with inserted blades
   - (*h*) Carbide tipped with plain milling with cast body.

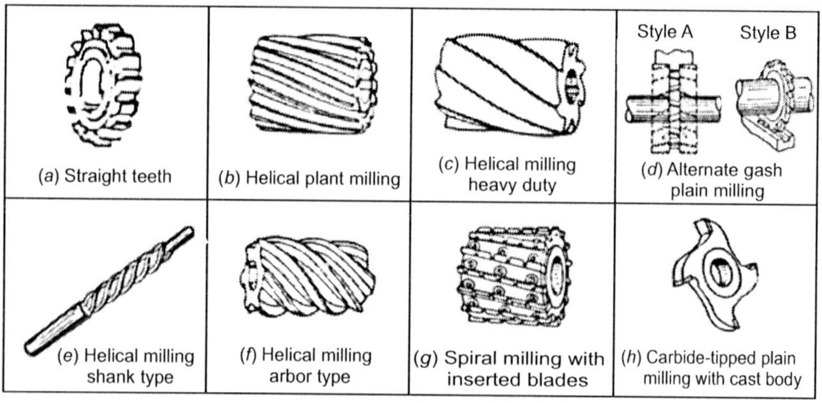

Fig. 1.58 Plain milling cutters.

2. Face milling cutter with inserted blades
   - (*a*) Inside cone type
   - (*b*) Outside cone type
   - (*c*) Cylindrical type
   - (*d*) Shear clear cutter

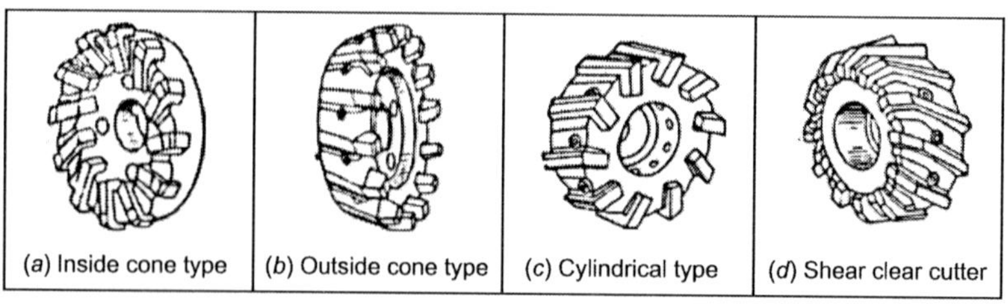

Fig. 1.59 Face milling cutter with inserted blades

3. Side Milling Cutters

(*a*) Integral tooth side mills (*b*) Inserted blade

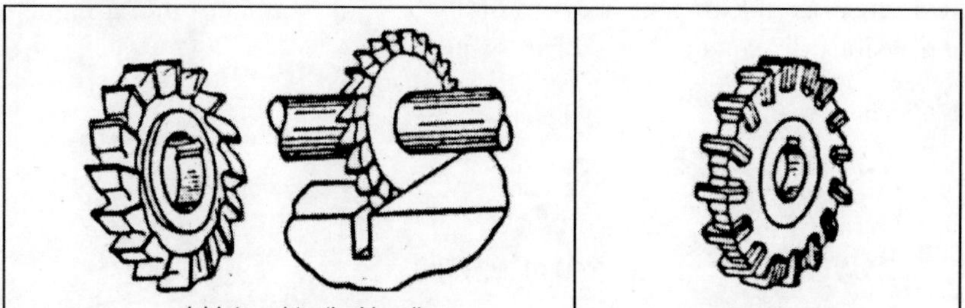

(*a*) Integral-tooth side mils (*b*) Inserted blade

Fig. 1.60 Side milling cutters

4. Staggered-tooth side milling cutters

(*a*) Channelling cutter (*b*) Integral tooth

(*c*) Herring bone type (*d*) Interlocking side mill

(*e*) Inserted blade

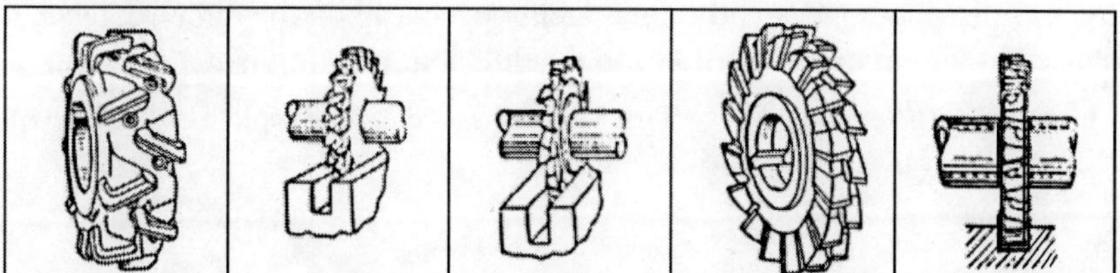

(*a*) Channeling cutter (*b*) Integral-tooth (*c*) Herring bone type (*d*) Interlocking side mill (*e*) Inserted blade

Fig. 1.61 Staggered-tooth side milling cutters

5. Face end milling cutter

(*a*) Half side mills set up for straddle milling

(*b*) Solid face mill with brazed carbide tips

(*c*) Inserted blade combination face and end mill.

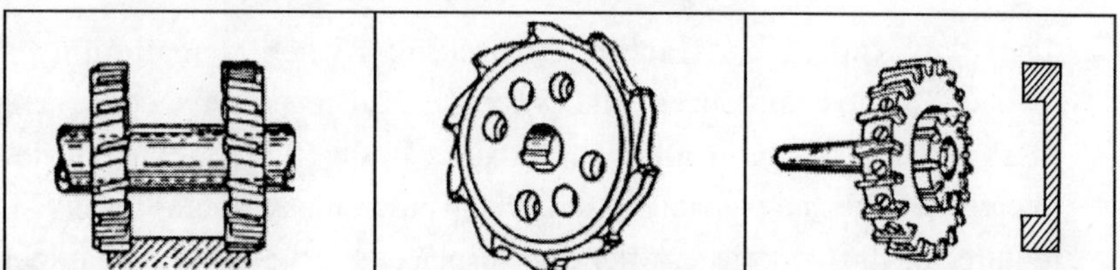

(*a*) Half side mills set up for straddle milling (*b*) Solid side mills with brazed carbide type (*c*) Inserted blade Combination face and end mills

Fig. 1.62 Face and end mills

1.7.10 Indexing

The operation of rotating the job through a required angle between two successive cuts is termed as indexing. It helps to divide the job periphery into a number of equal divisions *i.e.* square, hexagonal, octagonal etc.

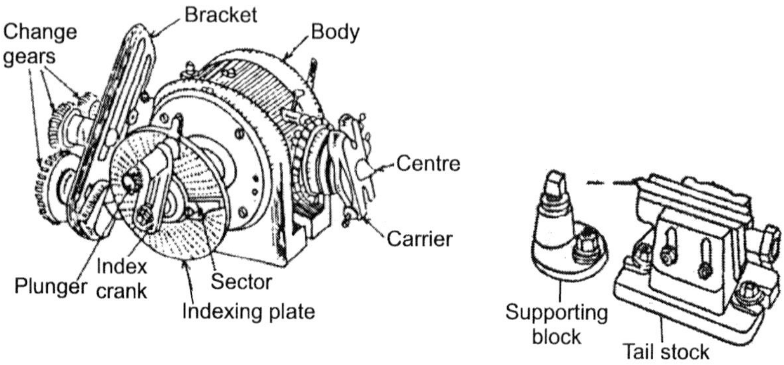

Fig. 1.63 Indexing head arrangement

1.7.11 Dividing with Indexing Attachments

In order to distribute faces and milled areas accurately on the circumference of a workpiece, dividing heads are used. In this method, marking is not necessary.

1. **Direct Indexing:** The dividing head has an index plate which is fitted directly on the spindle.

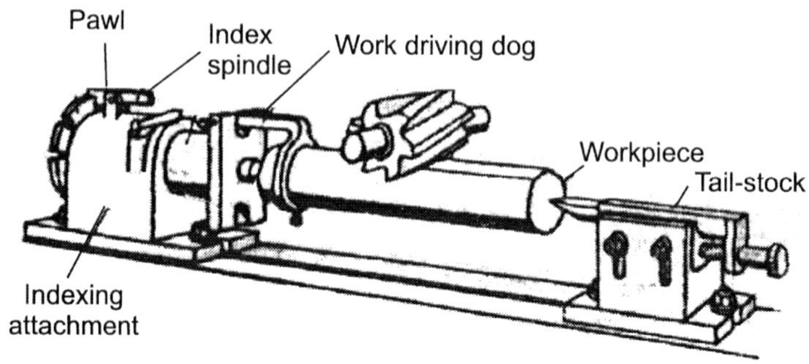

Fig. 1.64 Cylindrical indexing attachment

2. **The Plain Indexing Attachment:** It suffices when a small number of divisions has to be produced. The workpiece is clamped between the centres of the indexing head and the tail-stock. On the index spindle an interchangeable index plate is fixed, which has as many slots as the divisions required on the workpiece. After each displacement the index plate is locked in its position by a pawl. This method of indexing is called direct indexing.

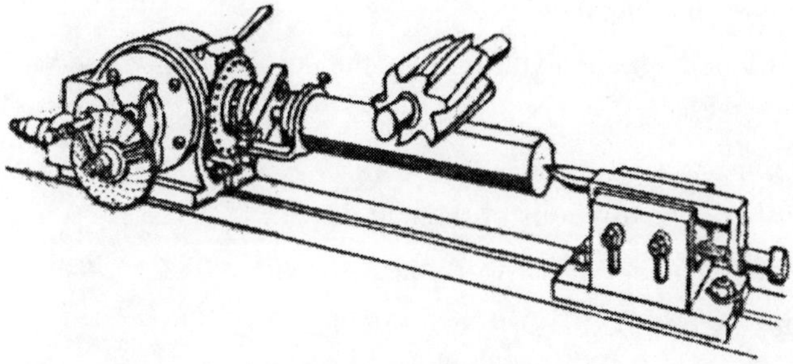

Fig.1.65 Dividing head

3. **Dividing Head:** The dividing based can be used for any number of divisions. In the housing, a worm gear with a ratio of 40:1 is installed. The worm wheel is fitted tightly on the index spindle. The interchangeable index plate is fixed and connected with the housing by a spring bolt. A dividing head has three index plates, the hole circles of which have different number of holes. The handle for turning the worm is adjustable towards the centre. It has an index pin, by means of which the division on the index plate is set. The sector arms save the counting of holes while indexing. Since the workpiece is shifted by means of the work gear, the procedure is known as compound indexing.

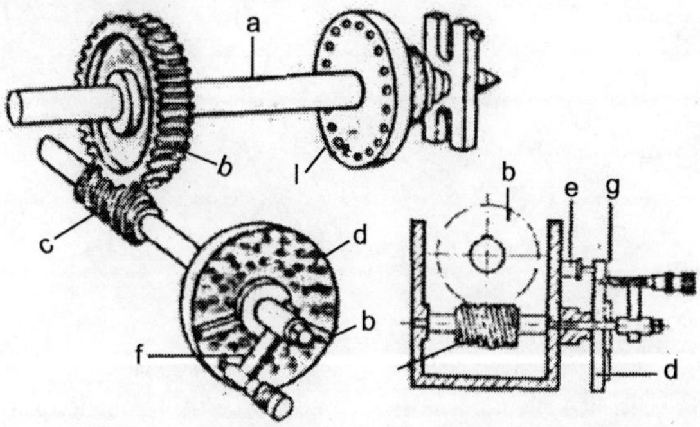

Fig.1.66 Main parts of the dividing head (*a*) index spindle, (*b*) worm gear, (*c*) worm (*d*) index plate, (*e*) spring bolt, (*f*) handle, (*g*) index pin, (*h*) sector arm, (*i*) index plate for direct indexing

Customary numbers of holes in the hole circles of the indexing plates

I	15	16	17	18	19	20
II	21	23	27	29	31	33
III	37	39	41	43	47	49

Brown and shape

Indexing with the Dividing head

To be able to divide with the dividing head, the number of revolutions of the handle must be ascertained.

Terms used:

nk = number of revolutions of the handle.

z = number of teeth of the worm gear (usually 40).

t = indexing number (*e.g.*, 4, 6, 8, 10, 12 divisions)

The numbers of revolutions of the handle result when the number of teeth of the worm, gear are divided through the index number.

Example: A hexagon has to be milled. How many revolutions have to be made so that the workpiece turns 1/6 after each milling of a face:

Result: Numbers of revolution of the handle $nk = \frac{z}{t}$; $nk = \frac{40}{6} = 6\frac{4}{6}$ revolution;

i.e., the handle must be turned around $6\frac{4}{6} = 6\frac{2}{3}$ times.

Performance: A hole circle which can be divided by 3, is selected, *e.g.*, a hole circle with 15 holes.

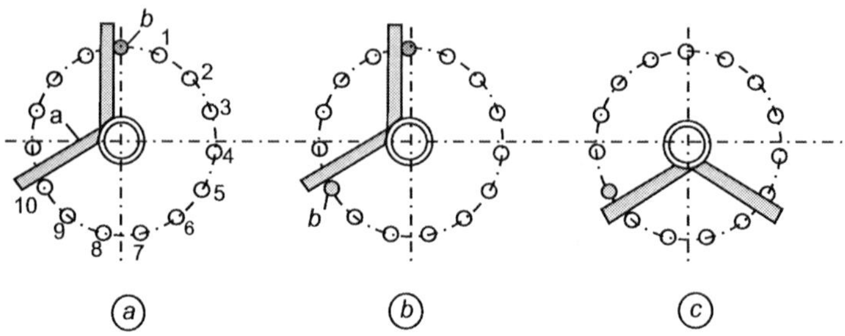

Fig. 1.67 Dividing with the dividing head (*a*) sector arm, (*b*) index pin on the handle.

(*a*) On the hole circle $\frac{2}{3} \times 15 = 10$ holes are counted and width of brace is set.

(*b*) After milling of the first face of the hexagon, the handle is turned 6 times around and 10 holes in addition.

(*c*) Sector arms are immediately set to new position.

Differential Indexing: Divisions which cannot be made by direct of compound indexing are made possible with the help of change gears by differential indexing.

During differential indexing the index plate is loose. It receives an advancing or retarding movement from the index spindle via the change gears. The number of teeth of the changes gears are ascertained by calculation.

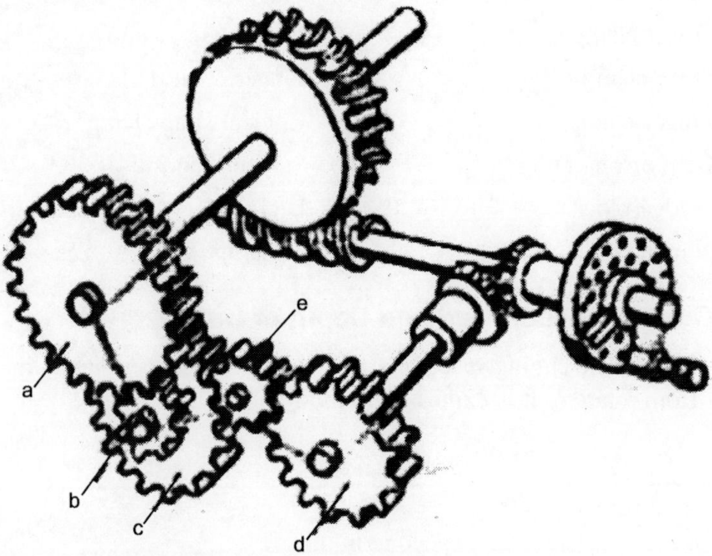

Fig. 1.68 Differential indexing and change gears

Parkinson: Number of teeth of the change gear.

I	24	24	28	32	36	40	44
II	48	56	64	72	86	100	

1.7.12 Work Holding and Clamping Devices Used as Milling Machine

For workholding purpose T-slots are provided on milling machine tables so the work piece can be clamped directly to the table for machining operations. There are some devices used for clamping are as follows:

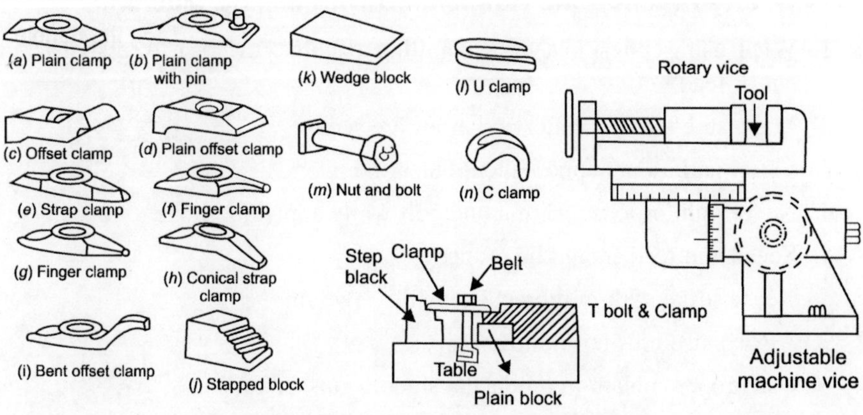

Fig. 1.69 Devices used for clamping.

(*a*) Plain clamp
(*b*) Plain clamp with pin
(*c*) Offset clamp
(*d*) Plain offset clamp
(*e*) Strap clamp
(*f*) Finger conical clamp
(*g*) Finger clamp
(*h*) Conical strap clamp
(*i*) Bent offset clamp
(*j*) Stapped block
(*k*) Wedge block
(*l*) U clamp
(*m*) Nut and bolt
(*n*) C clamp

1.7.13 Cutting Speed, Feed and Depth of Cut

Cutting Speed: The cutting speed of a milling cutter is its peripheral linear speed resulting from rotation. It is expressed in m/min.

$$\text{Cutting speed} \quad (V) = \frac{\pi DN}{1000}$$

where V = cutting speed in m/min

D = diameter of the cutter

N = speed in rpm

Feed: Feed is defined as the rate with which the work piece advances under the cutter feed is expressed in milling machine by different method:

1. Feed/tooth
2. Feed/rev
3. Feed/min.

Depth of Cut: The depth of cut in milling is the thickness of the material removed in one pass of the work under the cutter. It is expressed in mm.

1.7.14 Safety Precautions Associated with Milling Machine

1. When using cutters in a vertical milling machine never use excessive heavy cut or feed.
2. Make sure that the cutter and arbor are secure.
3. Cutter and arbor support should clear the work.
4. Never tight or loose the nut and bolt while applying power to machine.
5. Speed and feed should be proper.
6. Never strike with milling cutter while revolving.
7. Always use brush to remove chips.
8. Use proper coolant to produce a smooth surface finish.

1.8 DRILLING

Drilling machine is one of the simplest, moderate and accurate machine tool used in production shop and tool room. It consists of a spindle which provides rotary motion to the drilling tool. Drilling machine is designed for drilling holes in metal to take up, rivets, screws, bolts, and shafts etc. Some examples of the drilling work are as follows:

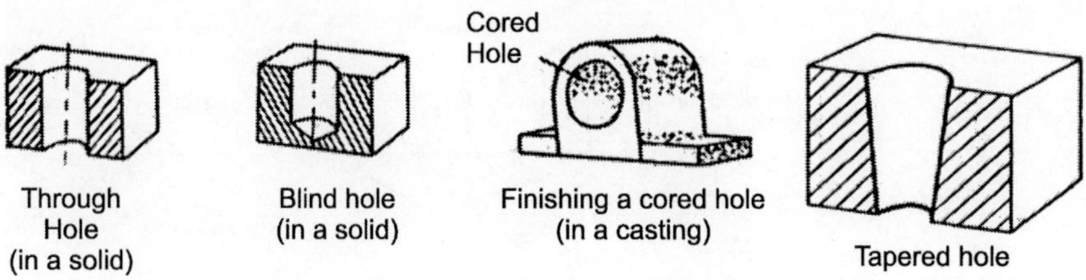

Fig. 1.70 Drilling holes in metal

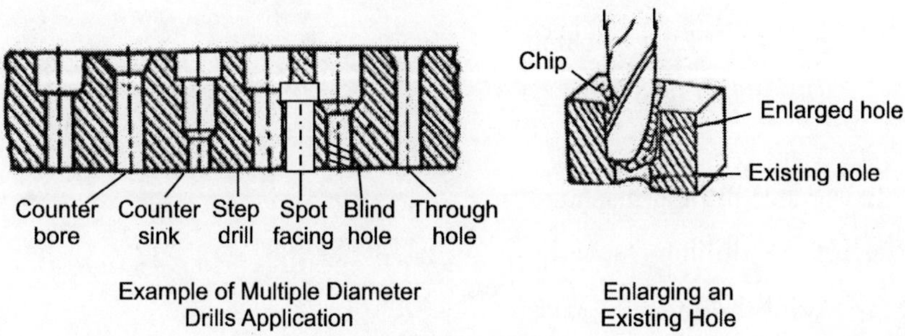

Fig. 1.71 Drilling

1.8.1 Working Principle of Drilling Machine

Drilling is a process of making hole or enlarging a drilled hole in an object by forcing a rotating tool called drill. When drilling is done work piece is usually kept stationary and the motion is given to the cutting tool called drill. It is generally done by the two ways.

(*i*) Rotary motion of the tool.

(*ii*) Feed motion in a straight line towards the work piece.

During operation, the drill rotating about its axis, is led in a straight line towards the fixed workpiece, the twist drill generally have the two cutting edges. The simultannous rotations and feed motion of the drill helps in cutting the chips.

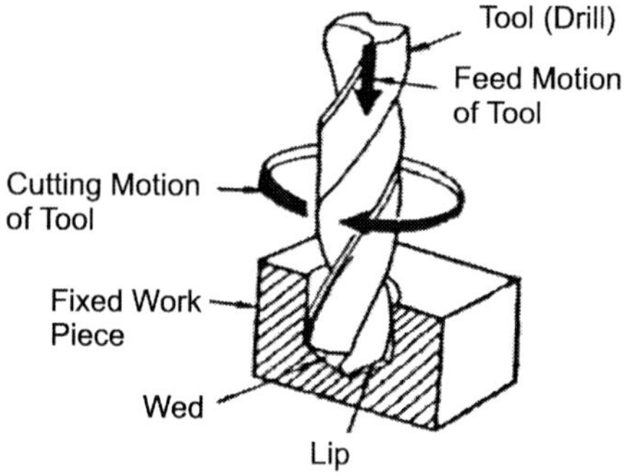

Fig. 1.72 Working Principle of Drilling Machine

1.8.2 Classification of Drilling Machines

1. Sensitive drilling machine
   (*a*) Bench type
   (*b*) Floor type
2. Portable drilling machine
3. Upright drilling machine
   (*a*) Round column type
   (*b*) Box column type
4. Radial drilling machine
   (*a*) Plain type
   (*b*) Semi universal type
   (*c*) Universal type
5. Gang drilling machine
6. Multispindle drilling machine
   (*a*) Single unit type
   (*b*) Way type

7. Deep hole drilling machine
    (*a*) Horizontal type
    (*b*) Vertical type

1. **Sensitive Bench Drilling Machine:** In a sensitive drill machine the feed lever is operated by hand which allows the operator to feel how the drill is cutting. If the drill gets into trouble, the pressure on it can be released. This design is used for light work using small drills upto 12.5 mm diameter either in a chuck or directly mounted in the taper nose of the spindle. It is used for tool room and shops for small work.

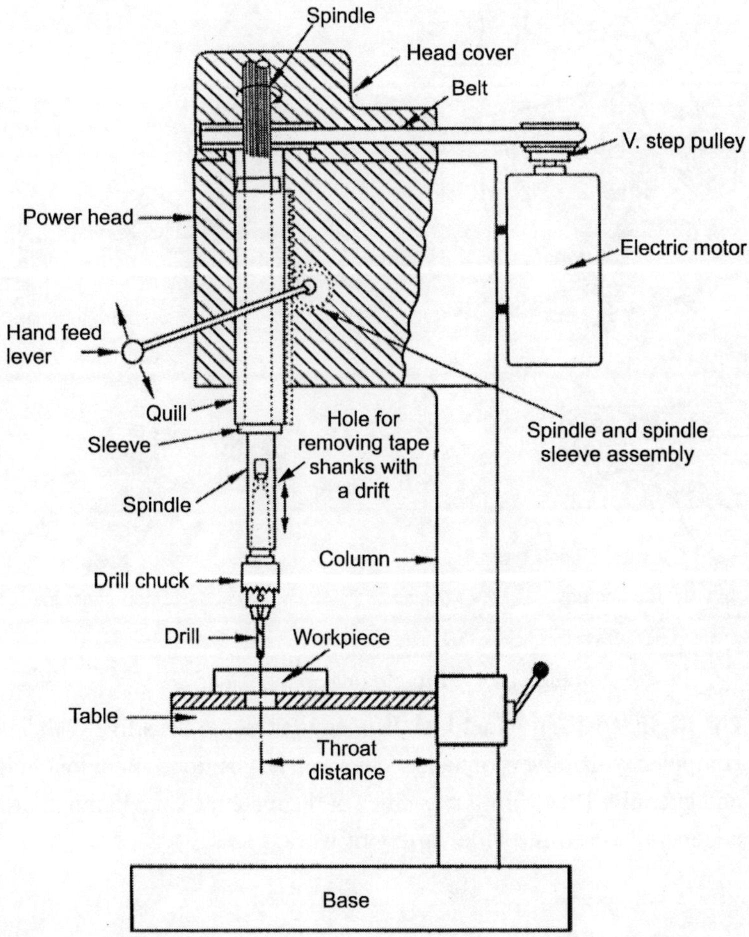

Fig. 1.73 Sensitive bench drilling machine of sensitive bench from finer + join

2. **Portable Drilling Machine:** A portable drilling machine is a small compact drilling machine and is used for drilling operation on big jobs which cannot be brought under a regular drilling machine or transported to the shop.

   A portable drilling machine may be used for drilling small diameter holes in large castings, or weldments upto 12 mm diameter at that place itself where they are lying.

   A hand-operated drill is the simplest type of portable drilling machine. hand dril, breast drill, electric drill and pneumatic drill are shown in Fig. 1.74.

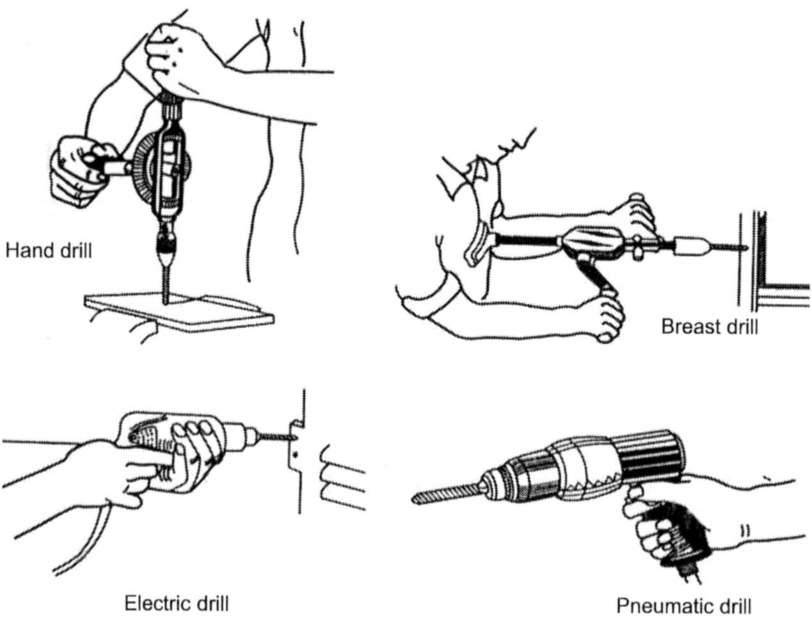

Fig. 1.74 Portable drilling machines

3. **Up Right Drilling Machine:** It is similar to the sensitive drill machine equippled with univers of table. Which can be positioned both longitudinally and laterally. The drilling capacities of the machine is ... 75 mm diameter.It is generally used for wide variety of work.

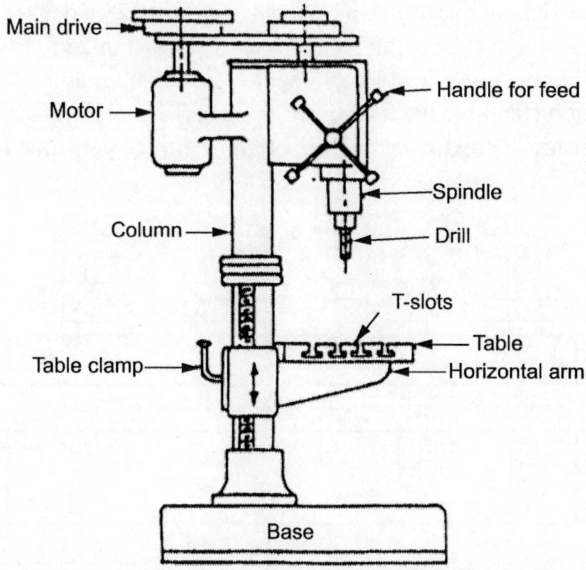

Fig. 1.75 Upright drilling machine

4. **Radial Drilling Machine:** It is used when a drilling operation is to be performed on heavy or bulk workpiece. Also used where the workpiece cannot be adjusted to locate the point of drilling.

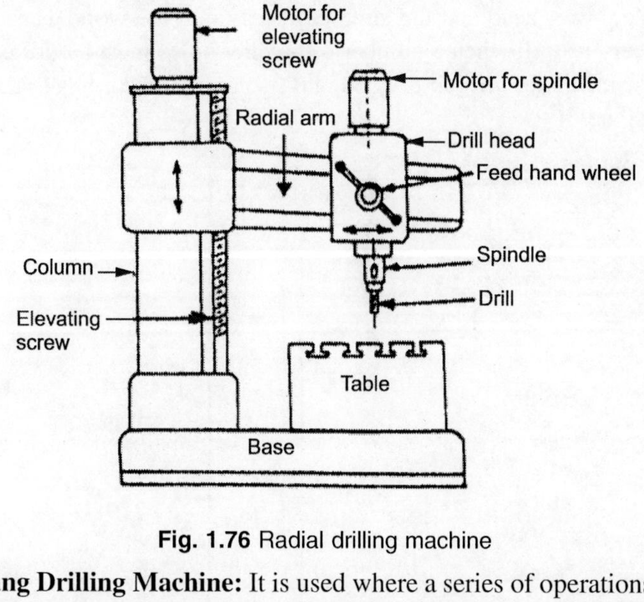

Fig. 1.76 Radial drilling machine

5. **Gang Drilling Machine:** It is used where a series of operations have to be performd like drilling, boring, reaming, tapping etc. A gang drilling machine is equivalent to 2,3 or 4 or more (upto 8 to 10) upright drill machines in a row with a single common long table or base (Fig. 1.77) This machine is

very convenient because of its compactness and is used in mass production. The machine can be setup so that work can be transferred from spindle to spindle to undergo two or more operations which may be similar (drilling holes of different diameters) or different (drilling, reaming, counter-boring etc.). In another set up, the same operations may be performed at all spindles.

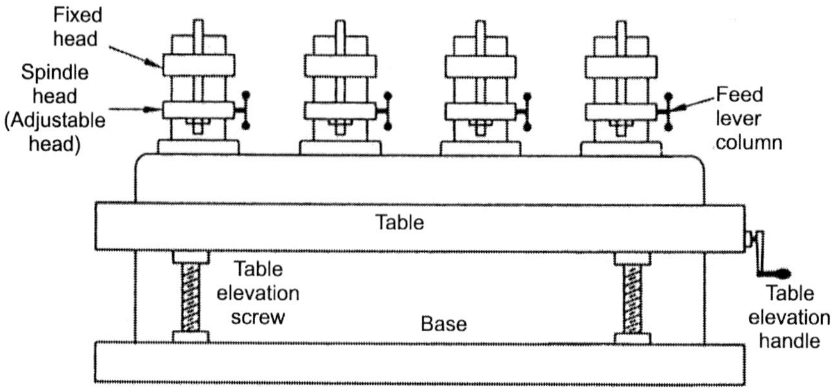

Fig. 1.77 Gang drilling machine

6. **Multispindle Drilling Machines:** It is used wherever a number of holes are to be drilled on a workpiece. These machines are mass production machines with the several spindles. (from 4 to 48 or more) driven by a single power head and fed simultaneously into the work. These machines are very usefull when a number of parallel holes must be drilled in a part. These machines are sub-divided into two group (according to the design to the spindle).

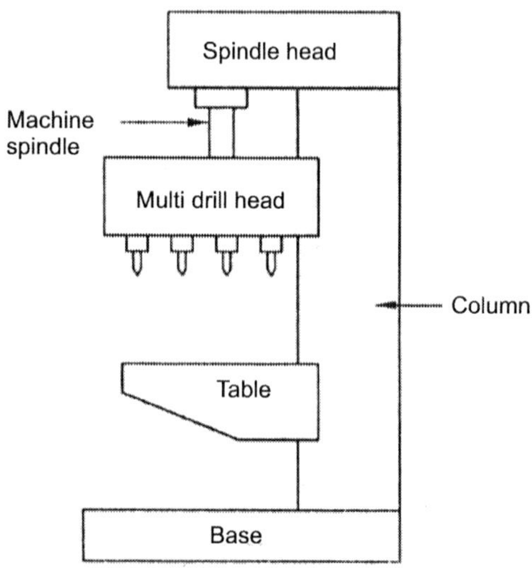

Fig. 1.78 Multiple-spindle drilling machine

7. **Deep Hole Drilling Machine:** When hole with a length to diameter ratio over 5 are called deep holes. (Fig.1.79). For example, hole through machine tool spindles, hollow shafts etc. The drilling of deep holes on many components like as rifle barrels, long spindles cannot be carried out on ordinary drilling machines, because it is impossible to obtain accurately straight hole. The straightness of the hole depends upon the rigidity of the twist drill and the guiding action of the margins, adjacent to the flutes, which slide on the surface of the drilled portion of the hole. So, deep hole drilling machines are used for such purposes.

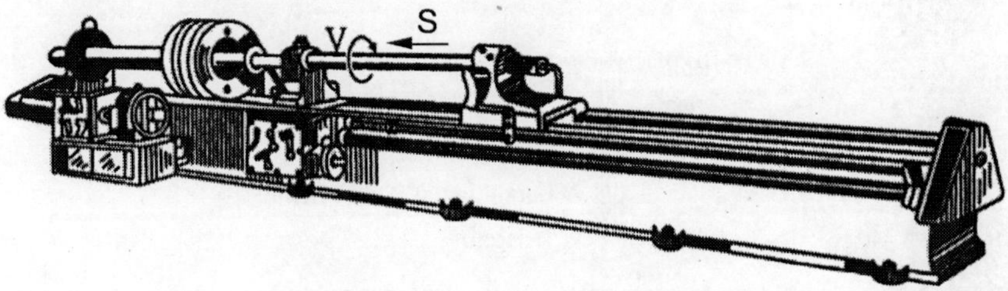

Fig. 1.79 Deep hole drilling machine

1.8.3 Principal Parts of a Drilling Machine

Different parts of a drilling machine are as follows:

1. Base
2. Column
3. Table
4. Power head
5. Spindle and spindle sleeve assembly
6. Drill chuck.

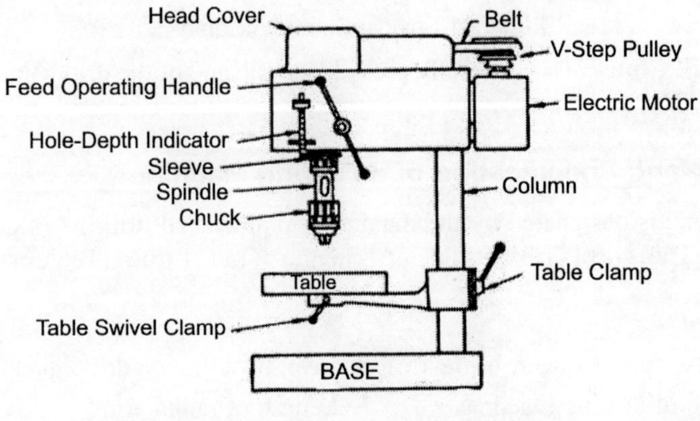

Fig. 1.80 Sensitive Bench Drilling Machine

1. **Base:** It is a heavy casting and it supports the drill machine structure. The base supports the column, table and head etc.
2. **Column:** It is a round vertical member which rests on the base. It supports the power head and table of the drilling machine.

3. **Table:** It is used for supporting the work piece and work piece holding device. It can swing around out of the way so that work piece can be supported directly on the base.
4. **Power Head:** The head containing electric motor, V pulleys and V belt which trasmit rotary motion to the drill spindle at a number of speeds.
5. **Spindle and Spindle Sleeve Assembly:** Spindle is the rotating part which hold the drill. It rotate and move up and down in a sleeve slides on bearings. The spindle speed or the drill speed is changed with the help of V-belt and V-stepped pulley. The lower end of a spindle has a morse tapered bore for drill chuck.

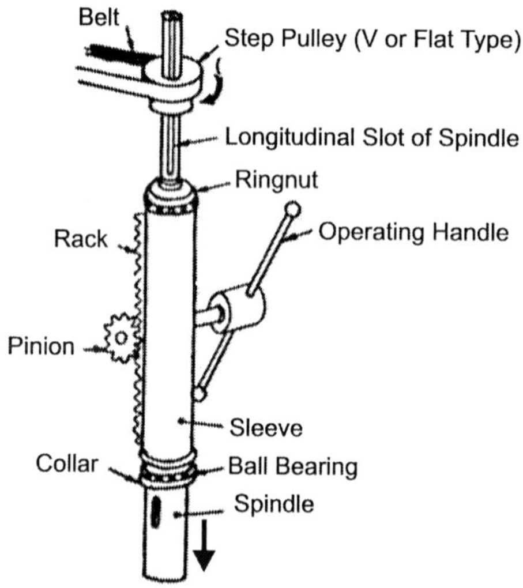

Fig. 1.81 Spindle-Sleeve Assembly

6. **Drill Chuck:** It is generally provided at the end of the drill spindle to hold the drill bit.

1.8.4 Size and Specification of a Drilling Machine

Drill machine is designated by the maximum diameter of drill that it can hold in drill chuck. The size of drilling machine is length of radial arm and column diameter.

Specifications

1. Maximum diameter of the drill that can be held in a drill chuck.
2. Size of drilling machine
3. Length of radial arm
4. Diameter of the column
5. Working surface of table.
6. Range of spindle speed.
7. Drilling capacity.
8. Traverse of spindle.
9. Range of powerfeed/rev.
10. Distance between spindle and column.
11. Motor speed.
12. H.P. of Motor power.

1.8.5 Nomenclature of a Twist Drill

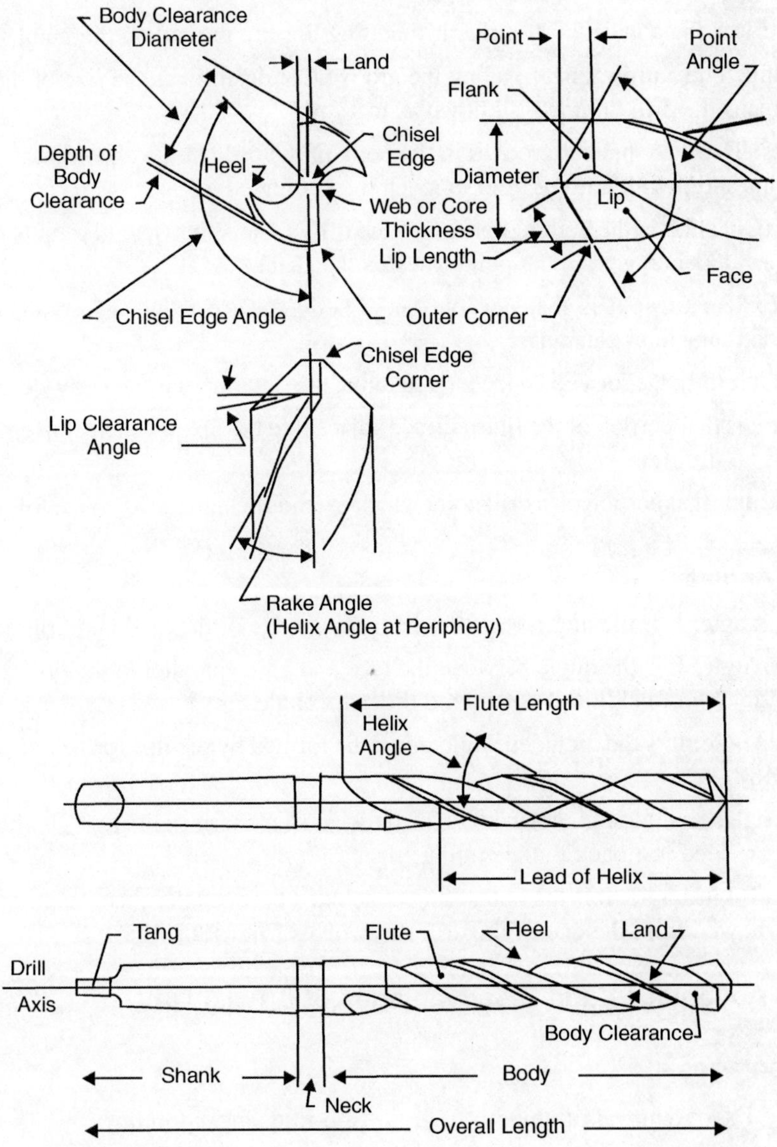

Fig. 1.82 Twist Drill Nomenclature

Body: The fluted portion of a drill.

Neck: The diametrically reduced portion between body and shank of drill.

Shank: The part of the drill by which it is held and driven.

Tang: The flattend end of a taper shank known as tang.

Drill Axis: The longitudinal centre line of the drill.

Lips: The cutting edge formed by intersection of face and flank both lips should have equal length, correct clearance and same angle of inclination.

Point: It is the conical part of the drill cutting lips are ground on the point.

Margin: The narrow surface along the groove that determines the size of the drill and keeps the drill alligned is known as margin.

Flutes: These are helical grooves in the body of the drill which allow the removal of chips and permit cutting fluid to reach the cutting edge.

Web (Core): It is the central portion of the drill situated between the roots of the flute and extending from the point towards the shank.

Body Clearance: It is that portion of the body surface reduced in diameter to provide diameteral clearance.

Heel: It is the edge formed by the interection of the flute surface and body clearance.

Face: It is that portion of the flute surface adjacent to the lip on which chip emerges as it is cut from the work.

Flank: It is that portion on a drill point which extends behind the lip to the following flute.

Drill Angles

Helix Angle: It is the angle between the leading edge of the land and drill axis.

Rake Angle: It is the angle between the face and a line parallel to the drill axis at the periphery of the drill it is equal to the helix angle.

Point Angle: It is the included angle of a cone formed by the lips for general work it is 118°.

Lip Angle: It is the angle formed by the flank and a plane at right angles to the drill axis or relived just back to the cutting edge.

Chisel Edge Angle: It is the obtuse angle included between the chisel edge and the lip as viewed from the end of the drill. It revolves from 120°-135°.

1.8.6 Advantages and Disadvantages of a Twist Drill

Advantages

1. Easy Removal of chips
2. Non-frequent regrindings
3. Saving of time
4. Lesser a power requirement.

Disadvantages

1. Digging tentency in soft metals.
2. Tentancy to hook into thin stock.
3. Unsuitable for drilling deep holes.

1.8.7 Types of Drills

Drill having two flutes for starting and making a hole whereas 3 or more flutes are used for enlarging or finishing drilled holes. Some illustrations of the different drills used for different purpose are as given below.

Following types of drill are in use:

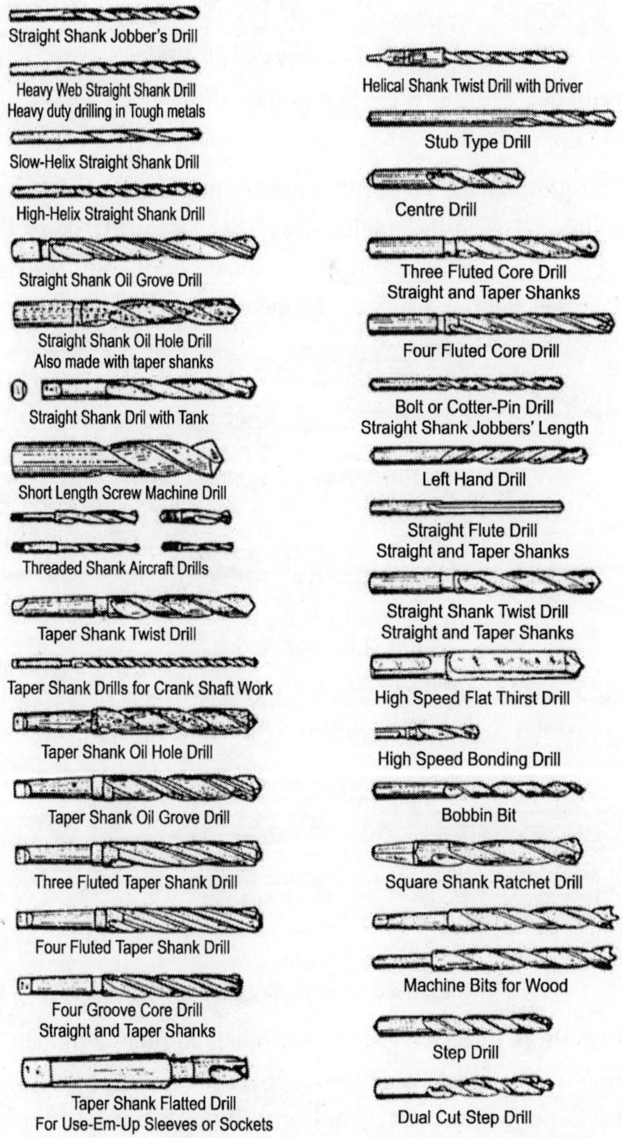

Fig. 1.83 Types of Drills

1. Flat drill or spade drill
2. Straight flute drill
3. Twist drill
   (*a*) Parallel shank drill
   (*b*) Taper shank drill
   (*c*) Jobber's drill
   (*d*) Stub drill
4. Center drill
5. Multiple flute core drills
6. Oil hole drills
7. Step drill
8. Subland drill
9. Special drills
10. Gun drill, etc.
11. Square drill
12. Carbide tipped drill.

1. **Flat or Spade Drill:** Flat drill is used sometimes when the twist drill is not available. The cutting angle varies from 90° to 120° and relief clearance at the cutting edge -3°–8°. It can be made easily by forging a piece of round carbon steel of a suitable diameter rod.

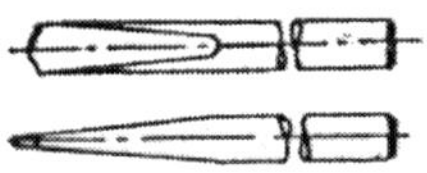

Fig. 1.84 A Flat drill

Fig. 1.85 Spade drill

2. **Straight Fluted Drill:** It has grooves or flutes running parallel to the drill axis. It has zero rake angles. It is used for softer material like brass, copper *etc*.

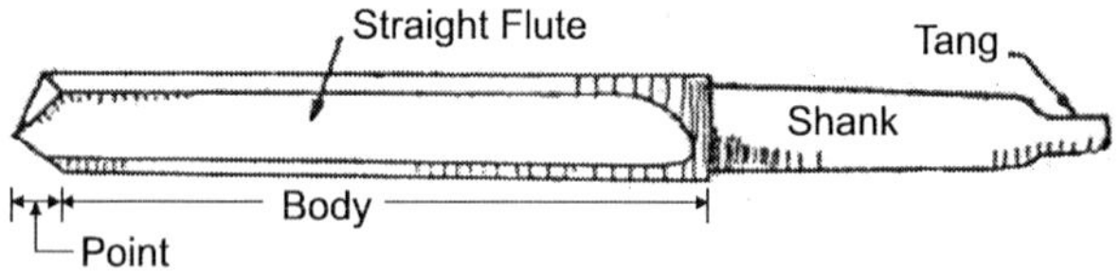

Fig. 1.86 Straight fluted drill.

3. **Twist Drill:** Twist drill consists of two helical flutes symmetrically cut on the cylindrically body. The helical grooves in the body appears to be twisted hence it is known as twist drill. Generally, it is made of high speed steel (HSS) and from high carbon steel. These are made in various sizes to suit the work, like tapered shank or parallel shank.

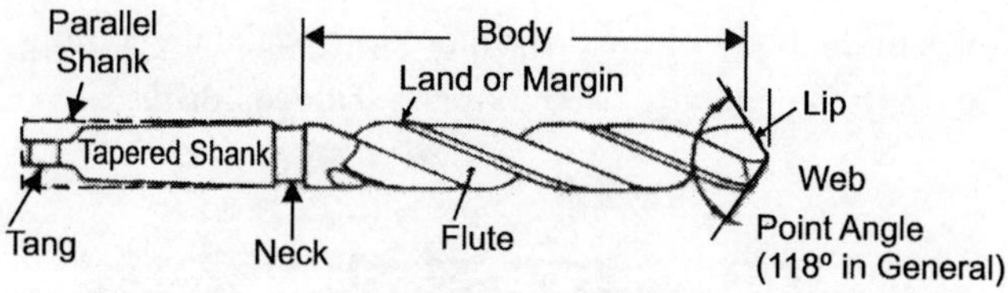

Fig. 1.87 Twist drill.

The flutes of a twist drill perform the following main four functions:

(*a*) To form correct and efficient lips at the point.

(*b*) To curl the chip in small ball shape which occupies mini. space.

(*c*) To form grooves to allow the removal of chips.

(*d*) To allow cutting fluid to reach the cutting edges.

Small helix angle is used for materials that break into very fine bits or powder like drilling hard rubber backalite and plastic laminates.

4. **Centre Drills:** The centre drills are straight shank, two fluted twist drills used when centre holes are drilled on the ends of a shaft.

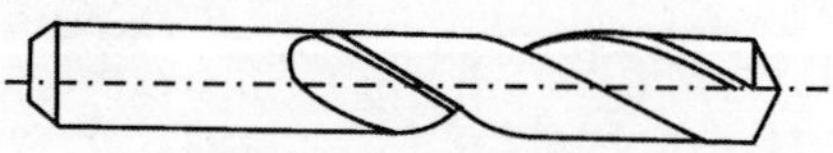

Fig.1.88 Centre Drills.

5. **Multiple Fluted Core Drill:** Multiple diameter drills are used to cut down on secondary operations which are time consuming and less accurate. It may be step type or subland type.

Where-ever a step drill provided the various diameters needed for operation and a Sub-land type drill provide flutes which may be maximum four in number and this type of drill never shapened on the diameter. That provide good concetricity as well life of the drill.

Fig.1.89 Core Drill (Three/Four Flue Drill)

Multiple Diameter Drill (Step Type)

Fig.1.90 Multiple Diameter Drill (sub-Land Type)

6. **Oil Hole Drill:** Oil hole drill means of flowing the coolent (oil) through the holes in the body of the drill in order to maintain cooling action at the cutting tip. It is generally used to make the deep drills.

Fig. 1.91 Oil hole drill

7. **Special Drills:**

   (*a*) **Gun Drill :**A gun drill is generally made with a single flute gun drill or a Tublar in shape. A single flute gun drill require the coolant during cutting and is also helps in flush out the chips. Wherever two flute gun drill provides balance cutting action and helps in size depth and tolerance control generally it is used in gun barrels, cam shafts crank shaft, connecting rod and machine tool spindles etc..

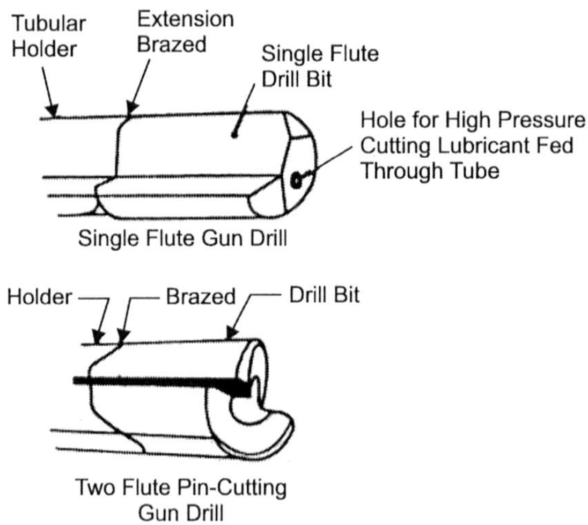

Fig. 1.92 Gun Drill.

(*b*) **Square Drills:** Any symmetrical shape such as square, hexagonal, octagonal or triagular as well blind holes may be drilled by special drill a square drill is a three sided drill for drilling a square hole. It requires floating chuck a guide plate and special drill which cuts on the end like a conventional drill.

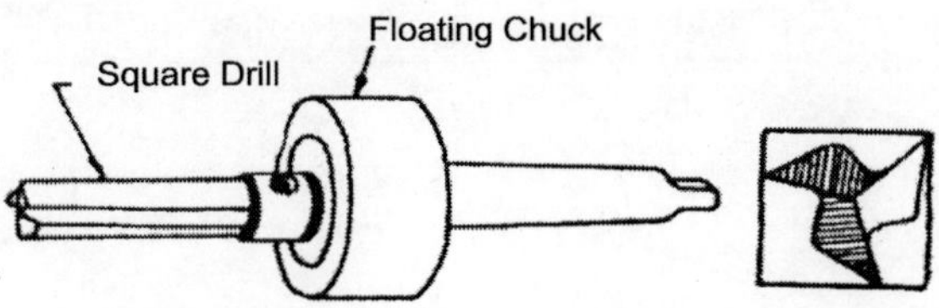

Fig. 1.93 Square drill

(*c*) **Carbide Tipped Drills:** The drill body is made of alloys and carbon tool steel. Having a cemented carbide tips which are brazed into the lips of a drill with copper or brass brazing fillet material. They have longer life high accuracy and high efficiency at high cutting speeds. These are used for drilling soft and hardended steel, cast iron, plastice, glass and other materials.

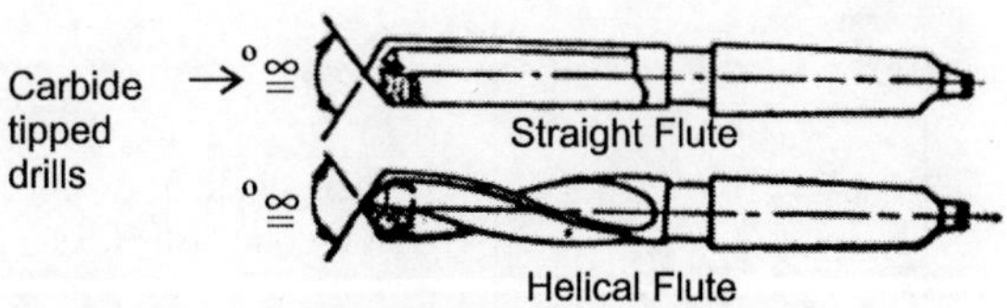

Fig. 1.94 Carbide tipped drill

1.8.8 Operations Performed on a Drilling Machine

The operations performed on a drilling machies are as follows:

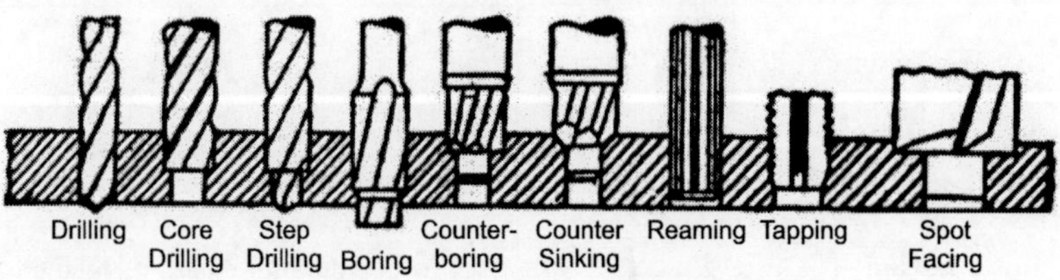

Fig. 1.95 Drilling operations

1. Drilling
2. Core drilling
3. Step drilling
4. Boring
5. Counter boring
6. Spot facing
7. Counter sinking
8. Reaming
9. Trapanning or circle cutting
10. Tapping
11. Surface grinding
12. Buffing
13. Hole milling

Operation	Sketch	Descriptions
1. Drilling	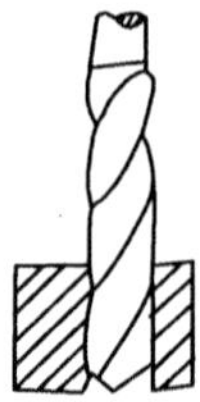	It is a operation of producing a cylindrical hole by removing metal.
2. Core drilling	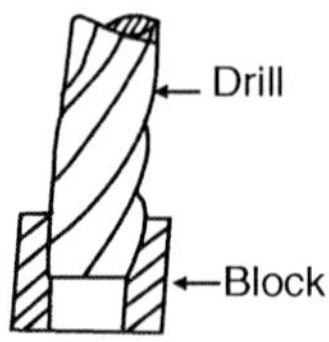	It is the operations of enlarging an already existing hole with the help of a drilling tool. It is generally used for making holes in casting.
3. Step drilling	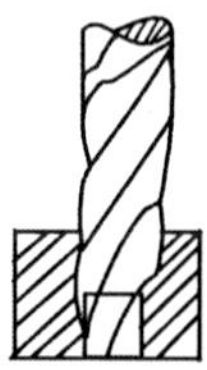	When a hole of two or more diameters is cut by one drill is called step drilling.
4. Boring	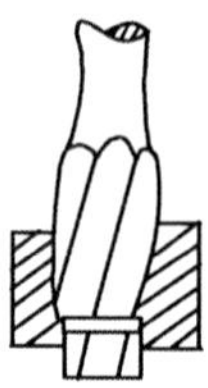	It is enlarging a hole with the aim of producing more accurate hole called boring.
5. Counter boring	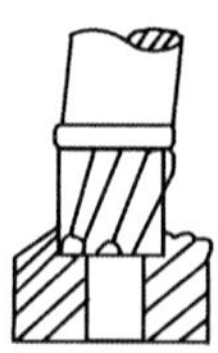	It is operation of enlarging a hole for a limited depth.
6. Spot facing	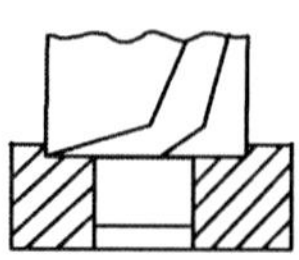	It is the operation of removing the metal at the top of a hole for nut, washer or cap of a screw may be brought flat.

Contd...

Operation	Sketch	Descriptions
7. Counter sinking	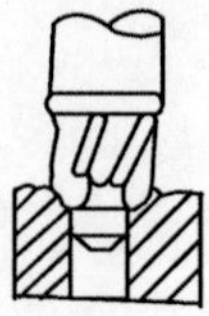	The operation of making a cone shaped enlargement of the end of a hole to produce a recess for a flat head screw, bolt or nut.
8. Reaming	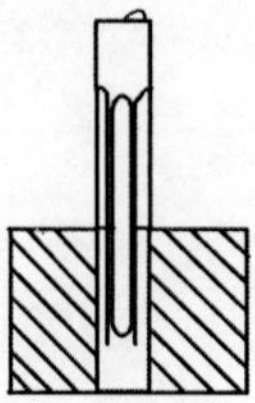	It is the operation of sizing and finishing the inside of a drilled hole using a reamer for exact size.
9. Trepanning or circle cutting	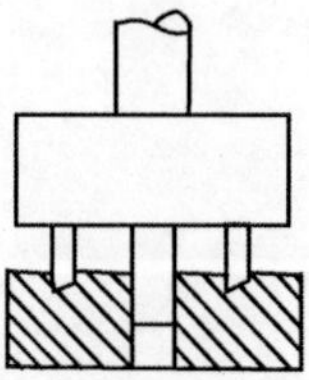	The operation of cutting large hole in a shaft or plate with a trepanning cutter.
10. Tapping	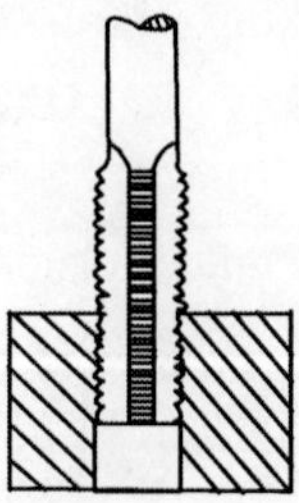	It is the operation of cutting an internal thread in a drilled hole with an externally threaded tool called a tap.
11. Surface grinding	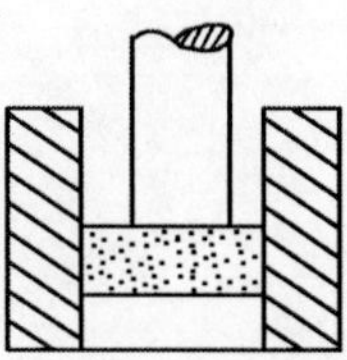	It is the operation of removing the metal in a very small amount from the metal surface. It is a finishing operation. The grinding wheel is mounted in the spindle of a drill for serve the purpose.

Contd...

Operation	Sketch	Descriptions
12. Buffing	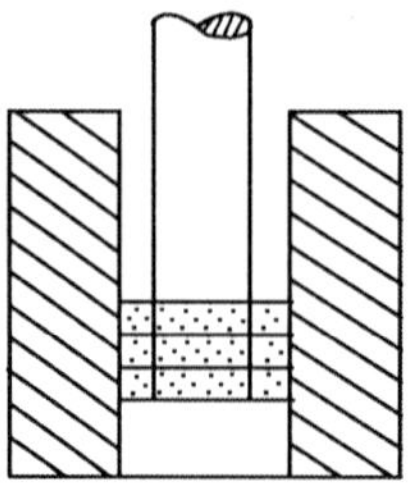	Buffing wheel is mounted in a drill chuck in the spindle of the drilling machine and holding the work accurately on the machine table for buffing or polishing purpose.
13. Hole milling		Hole milling is the operation in while the end mill is hold in the spindle and work is held on table to serve the purpose.

1.8.9 Work Holding Devices

A work held in a drilling machine for drilling purpose may cause accident and poor result, so various devices are used for holding the work in a drilling machine are:

(*a*) Vice
(*b*) Parallel clamp
(*c*) T-bolt and U-clamp
(*d*) Clamps
(*e*) Step blocks
(*f*) Jack screw
(*g*) Angle plate
(*h*) V-block
(*i*) Drill jig
(*j*) Straps

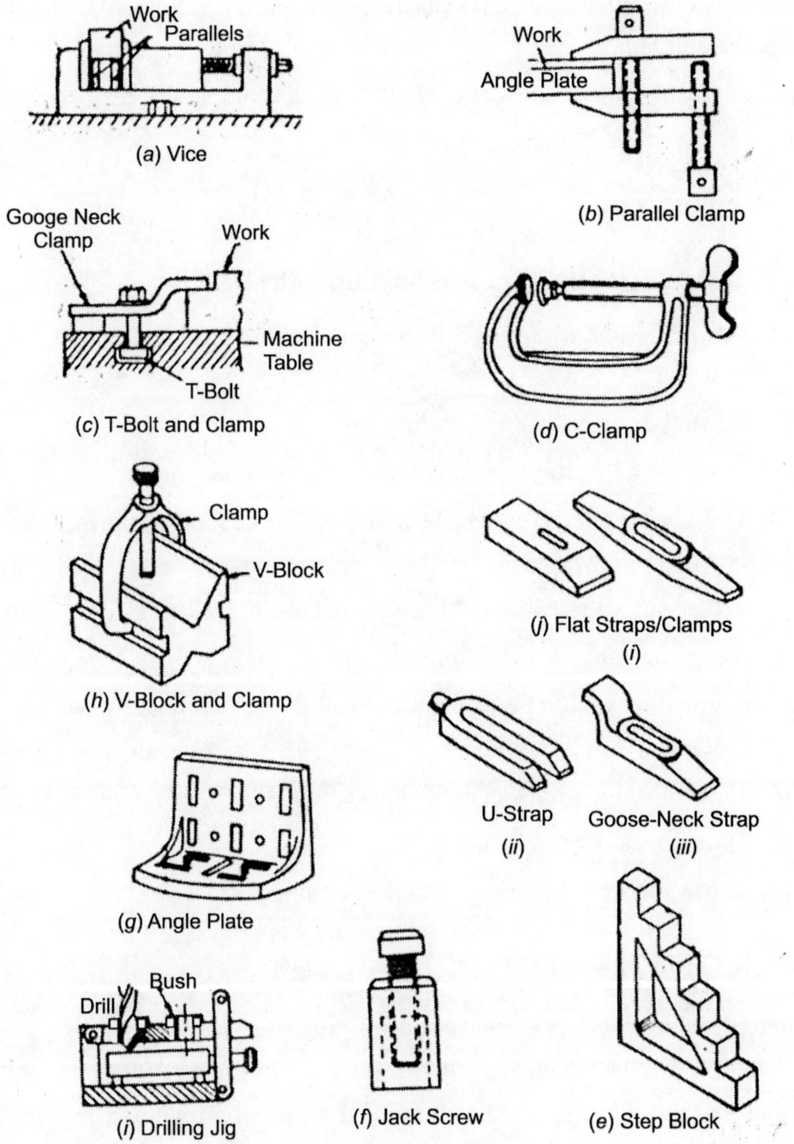

Fig. 1.96 Work holding devices

1.8.10 Cutting Speed, Feed and Depth of Cut

Cutting speed: Cutting speed in drilling is the peripheral speed of the drill relative to the drill. It is usually expressed in m/min.

$$\text{Cutting speed } V = \frac{\pi DN}{1000} = \text{m/min}, \quad \text{where } D = \text{Diameter of the drill}$$

$$N = \text{Spindle speed in rpm}$$

Feed: It is the distance by which tool advances in each revolution of the spindle. It is expressed in mm.

Depth of Cut: Depth of cut is equal to the one-half of the drill diameter.

$$t = \frac{d}{2} \text{mm}$$

1.8.11 Safety Precuations Associated with Drilling

1. Always select a proper drill material to suit the hardness of the work material.
2. Always use a correct size drill.
3. Maintain the cutting speed within the permisible range.
4. Use correct speed, correct feed and proper coolant to prevent excessive heating.
5. Proper size of sleeve should be selected to suit the taper shank of the drill.
6. Keep your fingers away from the drill.
7. Chips flow should be continous out of the drilled hole to avoid the drill clogging.
8. Ensure that the work piece is clamped properly before starting the operation.
9. Do not wear loose clothing.
10. Always wear safety goggles to protect the eyes.

1.9 SLOTTER – SLOTTING MACHINE

A slotting machine tool is designed for finishing slots or other enclosed parts not capable of doing machining by a shaper or a planer machine known as slotting.

Slotting machine is considered as a vertical shaper. In a slotting machine only vertical movement of the ram takes place for, downward cutting action, slotting machine may be operated by hand or power. The ram of the machine cannot be swivelled more than 5° to the vertical. The slotter is used for cutting grooves, Keyways and slots of various shapes for making regular and irregular surface both internal and external or internal and external gears can also be cut conveniently on this machine.

Some examples of the slotting machine work are as follows:

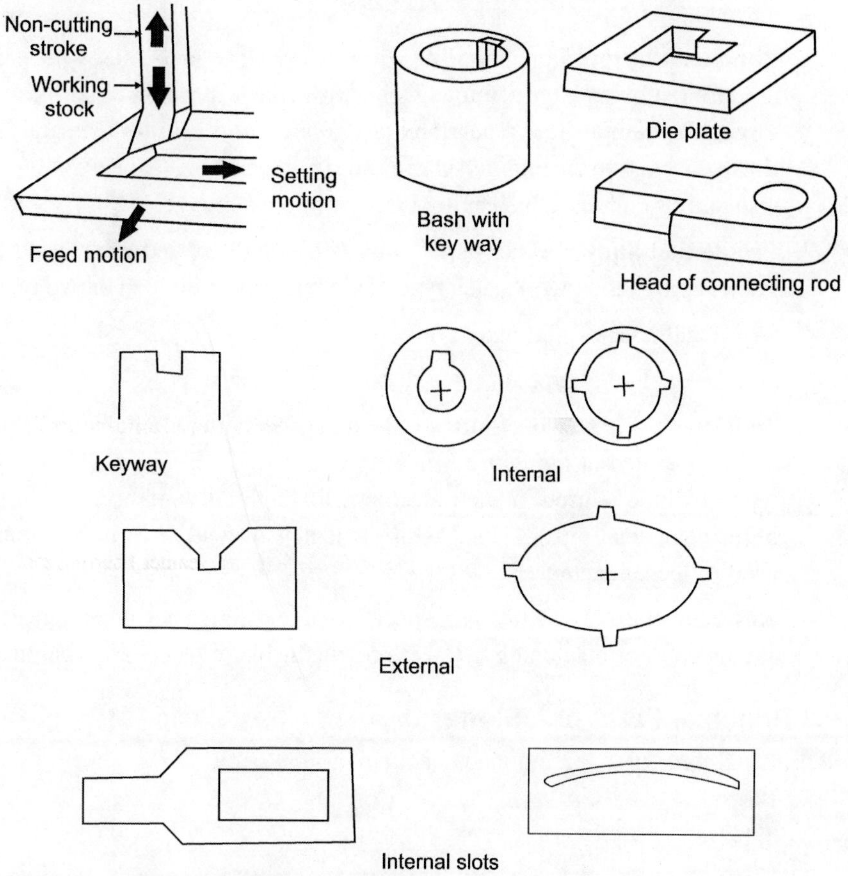

Fig. 1.97 Slotting machine work.

1.9.1 Working Principle of a Slotter

The cutting tool moves vertically in a cutting and return stroke. The tool performs the cutting operations during its downward stroke (working stroke). Its upward stroke (idle stroke) is faster than the downward stroke. The work is fixed on a horizontal rotary table and can be fed in cross or longitudinal directions or be given rotary feed in a horizontal plane about the axis of the rotary table. Figure 1.98 illustrates the working principle of a slotting machine.

1.9.2 Classification of Slotting Machines

There are mainly following types of the machines.

1. Puncher slotter.
2. Production slotter.

3. Tool room slotter (precision type)
4. Key slotter.

1. **Puncher Slotter:** Puncher slotter are heavy duty machines which are used for removing large amount of metal from the heavy work. Like sawing, forging or stamping etc. These are generally used to cut surplus metal and finish the work to the required shape and size. According to the requirement a square or circular table can be fitted on the machine.
2. **Production Slotter:** It consist of a heavy casting base and vastly used for general production work, such type of slotters can be divided in two types.
   1. Stationary -solid body type.
   2. Tilted body type for internal and external tapers.
3. **Tool Room or Precision Slotter :** The precision slotter is a lighter machine and is operated at high speed for very accurate work. Generally, a tilted type of frame is used in such machines for high cut with accurate finish by using special jigs. A quick return mechanism type drive is commonly used in these machines.
4. **Key Slotter :** Key slotter is a special type of slotter designed and used exclusively for machining key ways on the inside of wheel and gear hubs.

1.9.3 Principal Parts of a Slotter

The principal parts of a slotting machine are as follows :

1. Base
2. Column
3. Table
4. Ram
5. Cross slide

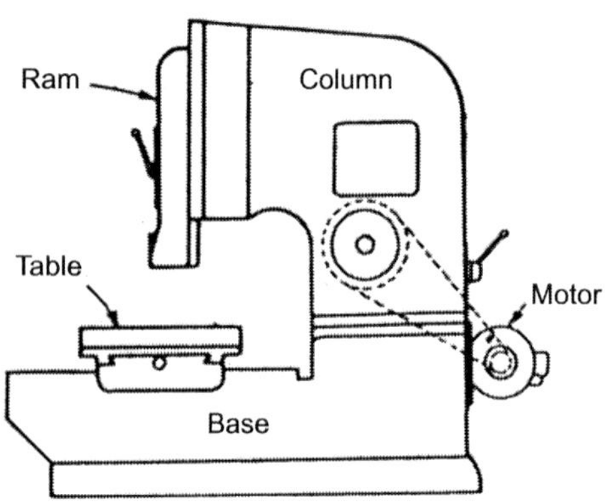

Fig. 1.98 Slotting machine

1. **Base:** Base is made up of cast iron also known as bed. It acts as a support for column, table, ram, driving mechanism etc. The top of the bed carries horizontal ways along with the work table for traverse.
2. **Column:** It is a vertical member with heavy cast iron body and acts as a housing for driving mechanism. It carries vertical ways along which ram moves up and down.
3. **Table :** The table of a slotter hold the work piece and is adjustable in longitudnal and cross-wise direction. The table can be rotated about its center. The top of the table provide T-slots to clamp the work and fixtures etc.
4. **Ram :** It reciprocates vertical up and down between the vertical guide ways provided in front of the column. The cutting action takes place during the down ward movement of the ram.
5. **Cross-slide:** It can move parallel to the face of the column. The circular work table is mounted on the top of the cross-slide.

1.9.4 Size and Specification of a Slotter

The size of slotting machine is expressed generally by maximum length of stroke and also given by the manufacturer which are as follows:

Specifications

Maximum stroke length		–	300 mm
Minimum stroke length		–	0
Height between table and head		–	450 mm
Maximum dia machined		–	900 mm
Diameter of table		–	500 mm
Traverse of table	Longitudnal	–	450 mm
	Traverse	–	350 mm
Face of head		–	575 mm × 250 mm
Over all height		–	2000 mm
Length, width and height of bed		–	1375 mm × 412 mm × 575 mm
Height of head		–	1275 mm
HP of motor		–	2 HP

1.9.5 Slotter Operations

The operations performed in a slotter machine are as follows:

1. Machining cylindrical surfaces.
2. Machinery irregular surface and cam machining.
3. Maching flat sunfaces.
4. Maching slots, grooves, keys ways and recesses.

S.N.	Operation	Description
1.	**Machining Cylindrical Surface**	The external and internal cylindrial surface can also be machined in slotting machine. The work is placed centrally on the top of the table and clamped by clamping devices. Now the tool is set radially on the work for operation. Saddle is clamped in position and feed is provided by the rotary table which rotates the table through a small arc at the beginning of each cutting stroke.
2.	**Machining Irregular Surfaces or Cams**	The work is set on the table and necessary adjustments are made with the tool. Longitudinal and rotary feed movements are provided to the table for machining purpose.
3.	**Machining Flat Surfaces**	The external and internal flat surfaces may be generated on the work piece easily in a slotting machine. The work is then clamped properly on the table and the position of length of stroke. is adjusted. A clearance of 20-25 mm is provided in the beginning of the cutting stroke, so that the feeding movement may takes place during ideal stroke. The cross feed is supplied at the beginning of each cutting stroke and the work is completed by using a rough and finishing tool.
4.	**Maching Slots, Grooves, Key Ways and Recesses**	Internal and external slots, grooves, key ways and recesses can be cut conveniently on a slotting machine. Slotter is specially intended for cutting internal grooves which are difficult to produce on other machines. External or internal gear teeth can also be machined by slotter keeping equally spaced on the pheriphery of the work. Indexing attachment can be used for accurate divisions.

1.9.6 Slotter Cutting Tools

The cutting tools used by the slotter are differ from the shaper. These are basically of two types:

1. Forged form solid tool steel bar
2. Tool bit type held in tool holder.

A slotter tool cuts parallel with its shank rather then cutting on a line perpendicular to its shank as in shaper or planer.

The slotter tools have front rake and side clearance but no side rake. The clearance angle should be $4° - 5°$ as well rake angle should be $10 - 12°$.

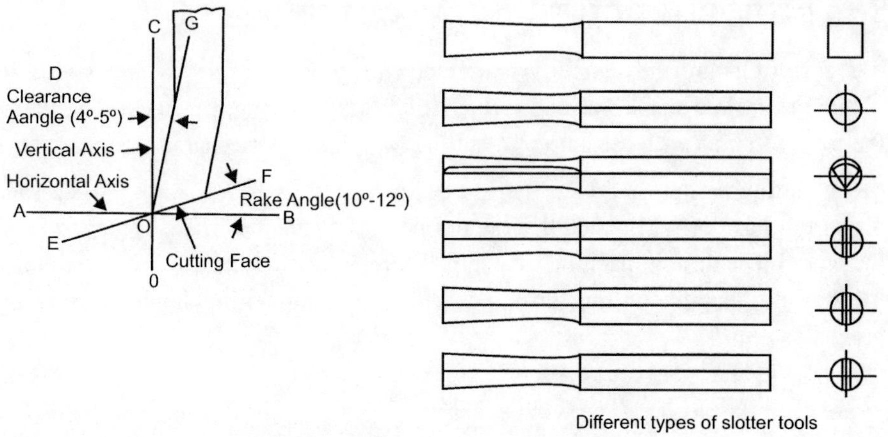

Fig. 1.99 Types of cutting tool used for slotter

1.9.7 Difference between Shaper and Slotter

S.No.	Name	Shaper	Slotter
1.	Movement of cutting tool	The cutting tool moves horizontally to and fro while performing the cutting operation.	The cutting tool moves vertically to and fro while performing the cutting operation.
2.	Job holding	Generally, rectangular job in shape is held on a fixed bed.	Generally, circular job in shape is held on a fixed bed.
3.	Feed motion	In shaper no rotary feed of the table.	In slotter longitudinal, cross or rotary feed of the table.
4.	Types of job	The shaper is generally used to produce flat surface.	The slotter is used for cutting grooves, keyways and slots on internal and external surfaces.
5.	Cutting	Cutting stroke is horizontal and slower than idle stroke.	Cutting stroke is vertical and slower than idle stroke.

1.9.8 Cutting Speed, Feed and Depth of Cut

Cutting Speed

The rate at which the metal is removed during the downward cutting stroke and is expressed in m/min.

Feed

The movement of the work in the direction parallel to the axis of the ram per double stroke. It is expressed in mm.

Depth of Cut

It is the perpendicular distance measured between the machined surface and unmachined surface of the work piece. It is expressed in mm.

1.10 INTRODUCTION TO METAL CUTTING

Metal cuttings are the secondry manufacturing processes used in machining to obtain the desired shape size and surface finish by removing the excess material in the form of chips with the help of a cutting tool through mechanical contact.

Metal cutting processed are costlier processes and used only when it is necessary because it has been estimated that about one-half of the material is removed from the work piece during maching as a chip. There are various methods of machining like turning, shaping, milling, grinding, drilling, planing boring, honing and laping processes.

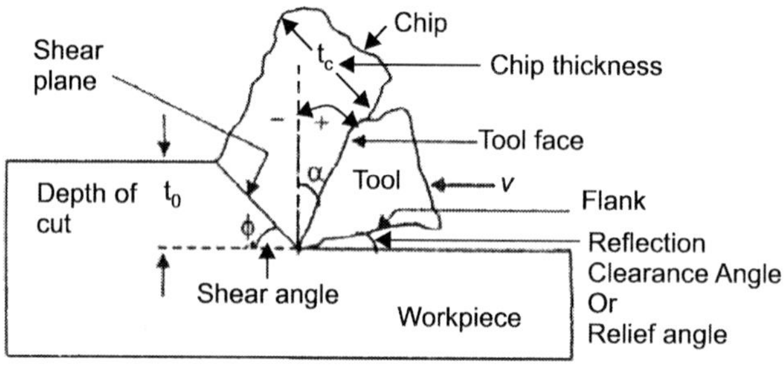

Fig.1.100 Basic machining operation and important

During the metal cutting operation, the tool moves redially inward and cut the work piece when the feed is provided to the cutting tool.

It represents the metal cutting mechanisms model which is proposed by pailspanen. It shows the chip formation that can be understood by visualising the slipping of a desk of playing cards placed over one another on a slight push one card slips over another moving a rising step pattern as shown in Fig. 1.101 Metal cutting depends on the following factors :

1. Dependent variables
2. Independent variables

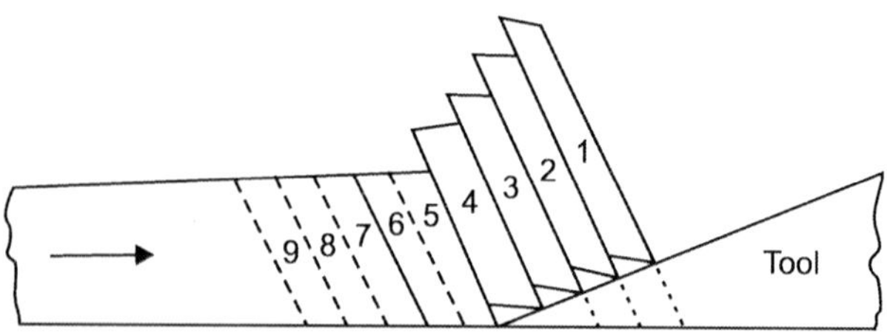

Fig. 1.101 Metal cutting

Dependent Variables

- Type of chip produced
- Surface finish of the work piece produced
- Force and energy dissipation
- Temprature rise
- Wear and failure of tool

Independent Variables

- Work piece material (hard or soft)
- The properties and geometry of the cutting tool.
- Tool material
- Cutting parameters (speed, feed and depth of cut)
- Tool shape
- The contact between the tool and work during cutting operation.

1.10.1 Introduction to Machine Tools

Machine tools are non-portable power driven machines which hold, support and guide the machine tool along with its accessories. The relative motion between work piece and cutting tool causes the cutting operation which produce the part of desired shape size and finish. In machine tool the surfaces are generated by the combination of movements.

There are two basic movements in the form of motions one is rotary motion and the other is straight line motion which are obtained by reciprocating action. The combination of both rotary and linear motions are generally used in machines like lathe, drilling, milling machine, wherever the reciprocating motion is obtained in shaper and planer in relation with cutting tool and work piece.

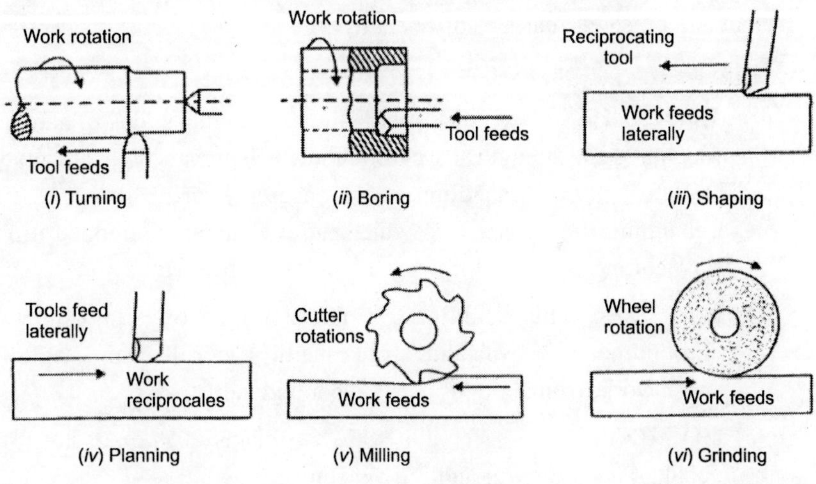

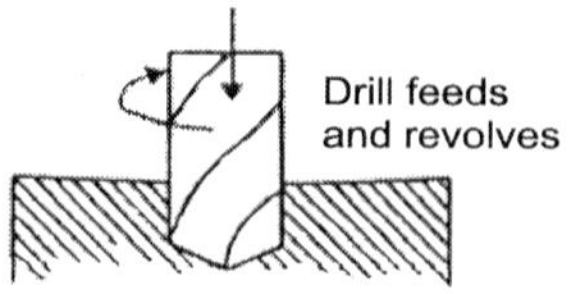

(vii) Drilling

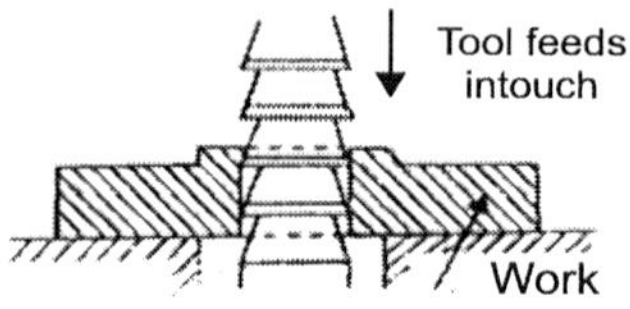

(viii) Broacting

Fig. 1.102 Machine tools

1.10.2 Tool Material

1. **Plain Carbon Steels**

   They contain only carbon in addition to the iron. The various plain carbon steels used for cutting tools are as follows:

   (*a*) **Medium Carbon Steel:** 0.45 to 0.35 % carbon

   like that use – dies, punch, cold chisel.

   (*b*) **High Carbon Steel:** 0.85% to 1 % carbon

   use – broaches and reamers

   (*c*) **High Carbon Tool Steel:** 1 to 1.5 % carbon

   use – taps, knives, threading dies, twist drill, moving cutter, files engraving tools, and cutting tools for, shaper, slotter and shaper plain carbon steels are suitable upto a temperature of 250°C beyond which they fail.

2. **High Speed Steel**

   Fast production at high speeds during World War II gave rise to development of high speed steels. HSS material can work of twice the speed that of plain carbon steels, can be softly cut as.

   There are 3 types of HSS:

   (*a*) **18 : 4 : 1 HSS :**This contains 18° tungusten which adds to hardness, toughness and strength, 4% chromium which imparts high temperature stability and 1% vanadium which ensures wear resistance to steel containing 0.8% carbon to which these elements are added use all type cutting tools.

   (*b*) **Mo - HSS :** This contains 6% molybcunum, 6% tungusten, 4% chromium and 2% vanadium. This steel is also called super HSS use super work for making of gear shaper and hub.

   (*c*) **CO - HSS :**This is a for mo-HSS. It contains 20% tungusten, 12% cobalt, 4% chromium and 2% vanadium.

Factors in Favour of HSS

1. They can be sharpened easily in both soft and hardned state.
2. They can work up to a trip of 650°C.
3. Their hardness is up to 750 BHN which is a adequate for most of materials to be cut.
4. Cutting speed up to 1.5 m/sec is sufficient.
5. They can be regrind to have fresh cutting easily before they are discarded.

3. **Cemented Carbide:** These are products of carbon of tungusten or titanium, tantalum, with some percentage of cobalt obtained by processing of their powders through powder metallurgy.

   Properties of tungusten carbide

   1. It is very good in compression, a desired characterstic for cutting tool.
   2. It is highly hard (1200 – 1600 BHN) and brittle.
   3. It can withstand a very high temperature (up to 1000° C)
   4. It provides better surface finish.

4. **Ceramic Tool Material**

   Aluminium oxide, silicon carbide, boron carbide, titanium make them suitable for cutting tools. They have zero rake angle.

   Properties

   (*a*) Poor thermal conductivity and high refractoriness

   (*b*) High compressive strength

   (*c*) Superior hardness

   (*d*) No problem of BUE (built up edge)

   (*e*) Resistant to oxidation and corrosion

   Ceramic tools can be used at funtastic cutting speed as much as 800 rpm to achieve super surface finish with longer tool life and reduced BUE. There is no need of coolant.

5. **Diamond**

   Diamond tipped tools are used for special applications such as production of surface of high finish soft materials that are normally difficult to machine.

   (*a*) Hardest known substance (BHN = 4000)

   (*b*) Lowest thermal expansion

   (*c*) High heat conductivity

   (*d*) Poor electrical conductor

   (*e*) Very low coefficient of friction against metals

Applications

1. Carbon has affinity with material, so it is suitable for non-ferrous metals alloy like plastics, hard rubber etc.
2. It is used for turning grinding wheel etc.

*Steels having hardness up to 300 BHN are easily machinable, steels having hardness more than 300 BHN are machinable with superior grade of HSS of carbide tool. High nickel, chrome alloy steel are hard to machine due to their toughness.

1.10.3 Classification of Cutting Tools

It is the simplest process of unwanted metal removed by single point or multipoint cutting tool in the form of chips by a hard metarial cutting tool. This tool is called cutting tool.

<table>
<tr><th>Single Point Cutting Tool</th><th>Multi Point Cutting Tool</th></tr>
<tr><td>Single point tool has only one cutting edge such as lathe, shaper, planer Threading tool, parting off tool
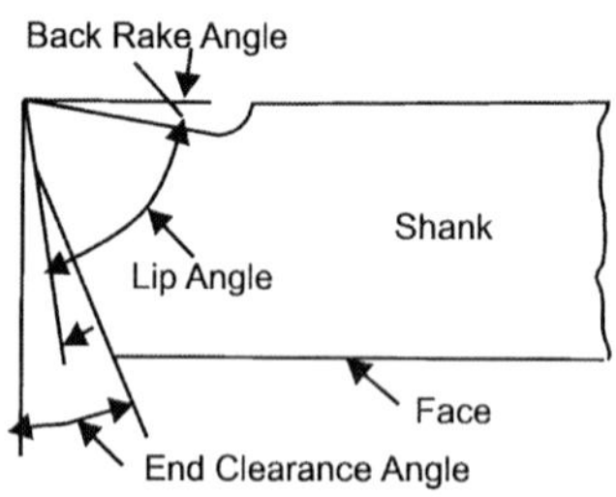
</td>
<td>Multi point tool having a number of cutting edges and used in machines like milling, broaching, drilling boring, tap, reamer knurling tool.
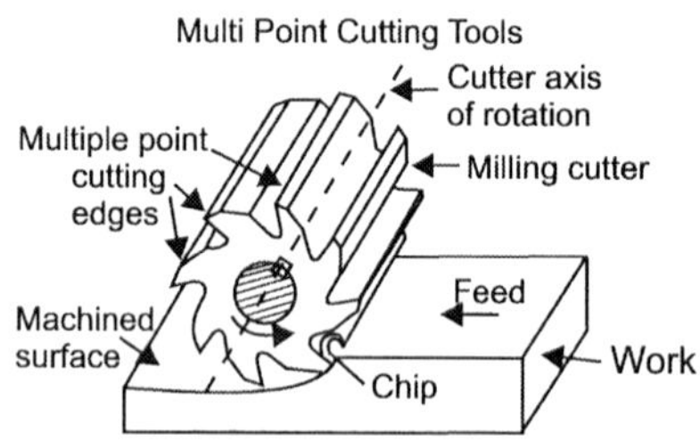
</td></tr>
</table>

1.10.4 Position of Cutting Tool

<table>
<tr><th>Left Hand Tool</th><th>Right Hand Tool</th></tr>
<tr><td>A left hand tool is that in which the feed is given from left to right hand end of lathe bed and cut the material from left to right.
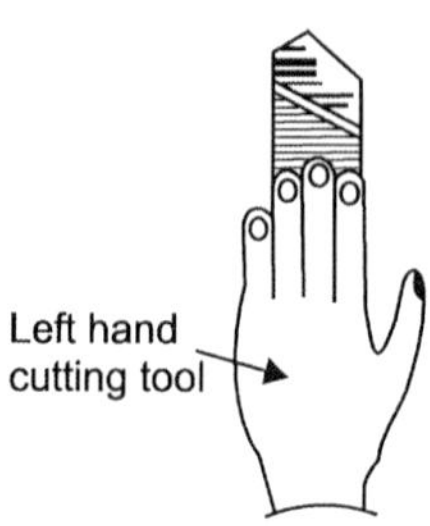
</td>
<td>A right hand tool is that in which the feed is given from right to left hand end of the lathe bed and cuts the material from right to left.
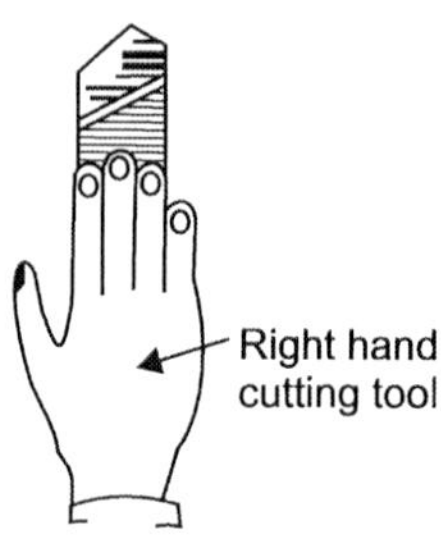
</td></tr>
</table>

1.10.5 Nomenclature of a Single Point Cutting Tool

A tool having primarily one cutting edge is known single point cutting tool. *e.g.*, lathe tool, shaper tool planer tool, boring tool etc.

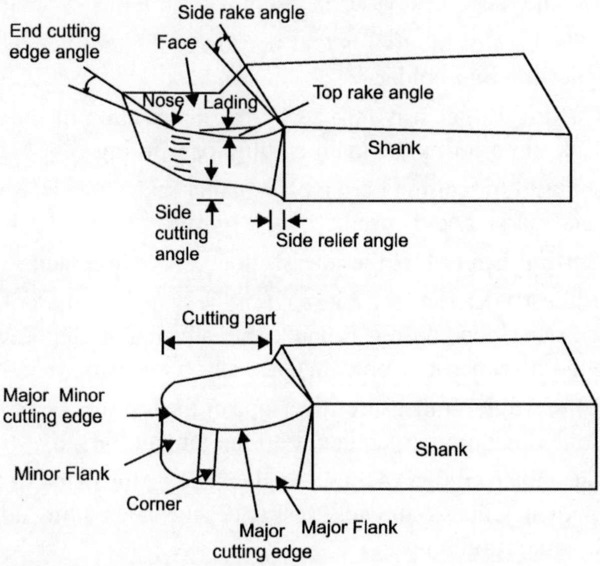

Fig. 1.103 Isometric viewes of a single point cutting tool.

Major Cutting edge usually removes the bulk of metal. Minor cutting edge usually controlls the final surface finish at the workpiece.

Three views of single point cutting tool

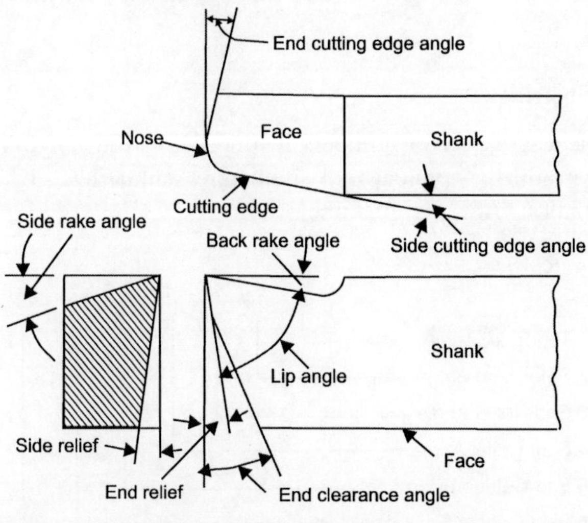

Fig. 1.104 Standard tool geometry and terminology of a single point cutting tool.

1. **Shank :** It is the body of a tool and is usually rectangular or square in cross-section used for gripping in the tool holder.

2. **Face:** It is the surface over which the chip flows.
3. **Flank:** It is the surface below the cutting edge.
4. **Nose Redius:** Nose is the junction of side cutting edge and the end cutting edge and the range lies between 0.4 mm – 1.6 mm.
5. **Base:** The base of the tool is that portion of the tool which bears against the supporting tool holder.
6. **Side Cutting Edge:** It is formed by the intersection of the flank and the side flank. It performs the main cutting for working.

   The tool used for cutting hard metal smaller rake angle is used and cutting soft metals large rake is used.
7. **End Cutting Edge:** It is the intersection of face and flank.
8. **Back Rake Angle:** The back rake angle is the slope from the tool point fowards the back side. Its purpose is to guide the direction of chip flow. Rake angle may be positive, negative or zero. Generally, it varies from 5° – 10°.
9. **Side Rake Angle:** It measures the slope of the top surface of the tool to the side in the direction perpendicular to the longitudral axis.
10. **Side Clearance (Relief) Angle:** Angle made by the blank of the tool and a plane perpendicular to the nose just under the side cutting edge. It can cut the job without rubbing.
11. **End Relief Angle:** It is the secondary relief angle between a plane perpendicular to the base and the end flank.
12. **End Cutting Edge:** It is the angle between the end or auxilary cutting edge and the work axis and it varies betwen 0° - 30°.
13. **Tip Angle:** It is the angle between the tool face and the ground end surface or flank.

1.10.6 Tool Signature

Tool signature is a sequence of numbers listing the various angle in degrees and the size of nose radius. A typical tool signuature standardized by the American Standard Association (ASA) is given below :

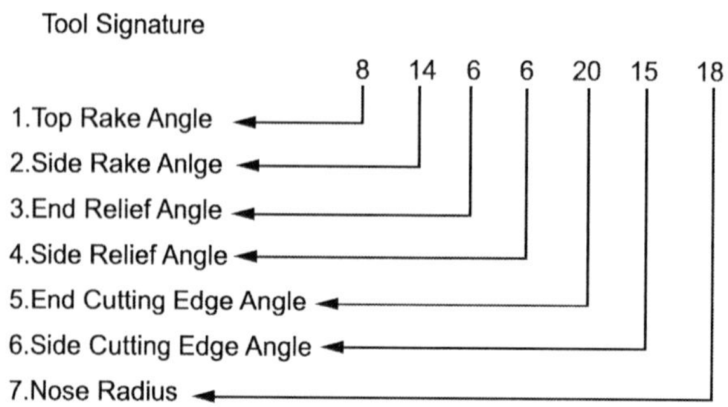

Fig. 1.105 Tool signature

Rake Angles:

1. **Positive Rake Angle**
   - It helps in chip flow from the machined surface.
   - Sharpens the cutting edge and reduces the cutting force required.
   - Produces better surfaces finish.
   - Increases tool life and decreases power consumption.
2. **Negative Rake Angle**
   - Increases the strength of the cutting point.
   - Increasees the pressure between job and cutting tool.
   - Direct the chips on to the machined surface.
   - Causes more chatter.
   - Decreases the rigidity of the machine.
   - Provides better surface finish.
   - Decrease the temperature rise.
3. **Zero Rake Angle**

   More force required to penetrate into the job.
   - More chatter and noise
   - Poor surface finish
   - Rare use.

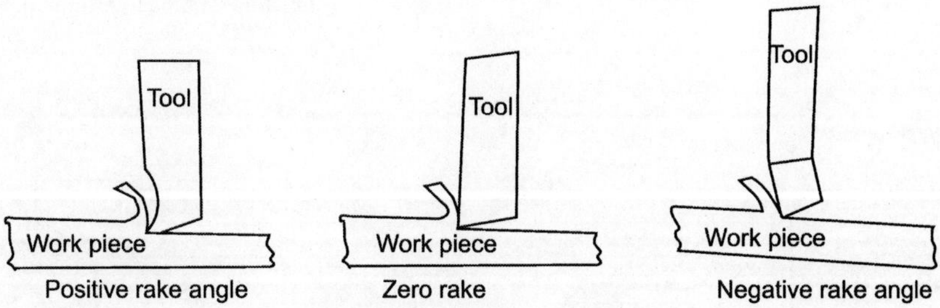

Fig. 1.106 Rake angles.

(*i*) Ceramic tools have zero rack.

(*ii*) Parting tools have zero rack.

(*iii*) Tinecous tools have zero rake.

Zero rack angle tools use more power and cutting is difficult, negative type is suitable for brittle material and heavy metal removal.

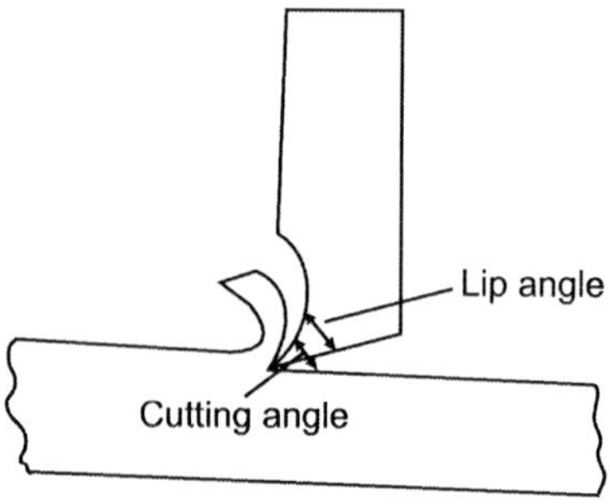

Fig. 1.107 Tinecous metal alloys like brasses are machined with zero rake angle.

1.10.7 Orthogonal and Oblique Cutting

The process of metal cutting is divided into the following main two categeories:

1. Orthogonal cutting
2. Oblique cutting

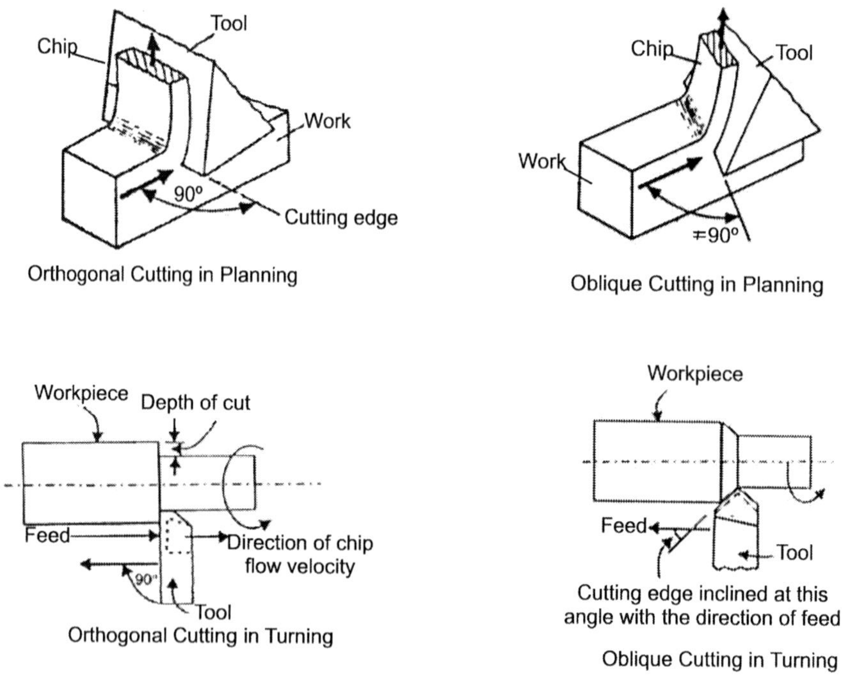

Fig. 1.108 (*a*) Orthogonal and Oblique Cutting.

Orthogonal Cutting	Oblique Cutting
(*a*) The cutting edge of the tool is set perpendicular to the cutting direction	(*a*) The cutting edge of the tool always remains inclined at an acute angle to the direction of tool feed or work feed.
(*b*) Cutting edge clears the width of the work pieces on either ends.	(*b*) The tool may or may not generate a surface parallel to work piece.
(*c*) The direction of chip flow velocity is normal to the cutting edge of the tool.	(*c*) The direction of chip flow velocity is atom. Angle with the normal to the cutting edge of the tool. The angle known as chip flow angle. Three components of the cutting forces are acting on the cutting edge. The components are mutually perpendicular to each other.
(*d*) Only two components of the cutting forces are acting on the tool, which are perpendicular to each other and can be represented in a plane.	(*d*) More than one cutting edges are in action.
(*e*) The cutting edge is larger than the width of cut.	(*e*) Maximum chip thickness may not occur at the middle.

1.10.8 Mechanics of Chip Formation

In all metal cutting operations, a wedge shaped tool is moved relative to the work piece in such a way that a larger of meterial is removed in the form of chips. During the cutting process, tool will penetrate into the material being cut.

The wedge shaped cutting tool basically consists of two intersecting surfaces to form the cutting edge. The surface along which the chip flows is known as rake face and the newly machined surface is known as the flank.

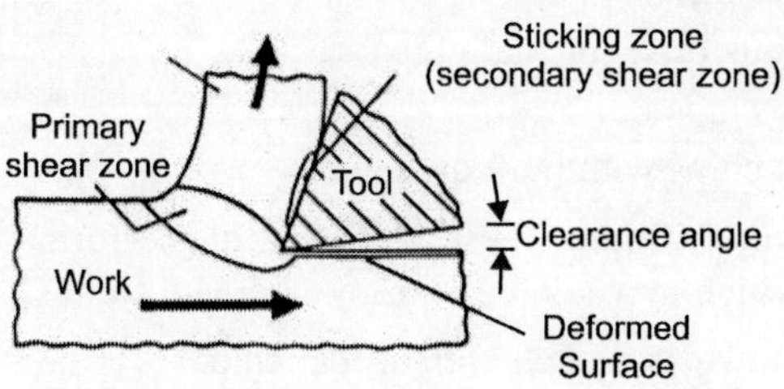

Fig. 1.108 (*b*) Mechanics of chip formation.

Chip formation can be understood by visualising the slipping of a deck of playing cards placed over one another and on a slight push one card slips over another making a rising step pattern as shown in Fig.1.109.

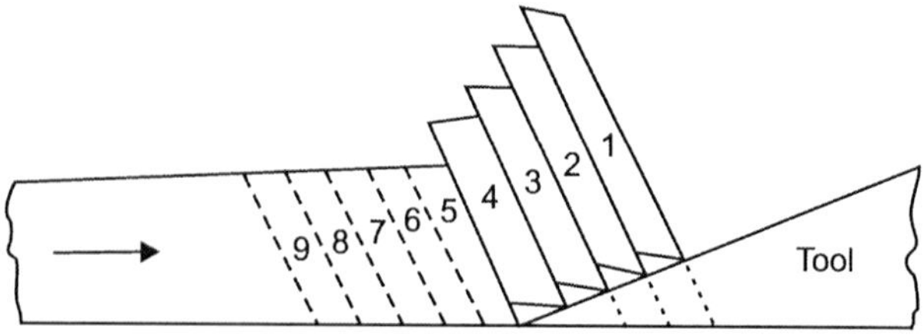

Fig. 1.109 Chip formation

Due to the action of force, the sharp edge of the tool will gradually inserted into the work piece metal. If the force is on tool is continue, the stress in the metal will also increase continuously, first it will exceed the elastic limit and then the ultimate strength of the metal which will causes sheer and consequently a piece of layer of metal gets seprated in the form of chip. Further if the force is continue the chips are formed largely by sheering action and compressive stresses are also greatest at the points which are farthest from the cutting tool and balanced by tensile stresses in the nearest zone of the tool. Hence the chip cuts outwardly away from the cut surface.

The chips formation greatly influence the surface finish produced and over all cutting conditions like tool life vibration and chatter etc.

1.10.9 Chip Formation

All machiming processes involve formation of chips by deforming the work material on the surface of the job with the help of a cutting tool. Depending upon the tool geometry, cutting conditions and work material, a large variety of chip shapes and sizes are produced during different machining operations.

1.10.10 Types of Chips

Types of chips formation is basically a function of work material and cutting conditions. The type of chip formation is an indication of deformation and surface quality produced during metal cutting.

1. Discontinuous or segmental chips
2. Continuous chips
3. Continuous chips with built up edge.

1. **Discontinuous Chips:** These chips are in the form of small individual segments which may adhere to each other and form slightly larger length.

   It is also known as broken chip or tear chip.

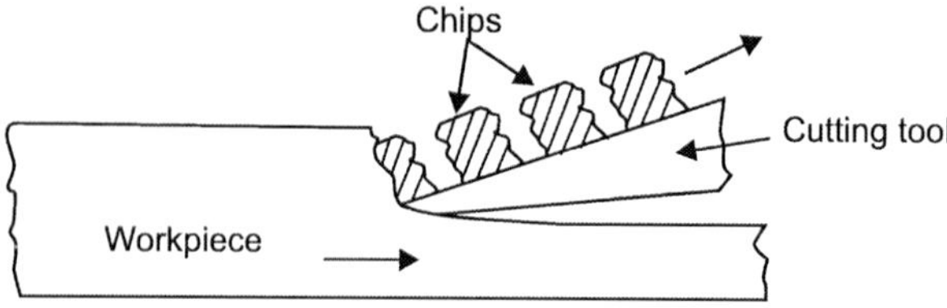

Fig. 1.110 Discountinuous chips.

Factors Responsible for Discontinuous Chips :

(*i*) Cutting or machining of low ductile, brittle material like CI, hard steel, brass, bronze.

(*ii*) Low cutting speed

(*iii*) Small rake angle

(*iv*) Large chip thickness

Advantages : Handling and disposal of these chips is easy.

2. **Continuous Chips:** This type of chips resulting good surface finish and remains in contact with the tool for a larger period, resulting functional heat. This type of problem can be solved by providing a chip breaker at the tool face. Earlier it was known as ribbon type.

These chips glide along the rake face of tool and take the shapes of coils. They are produced due to plastic deformation of metal without rupture.

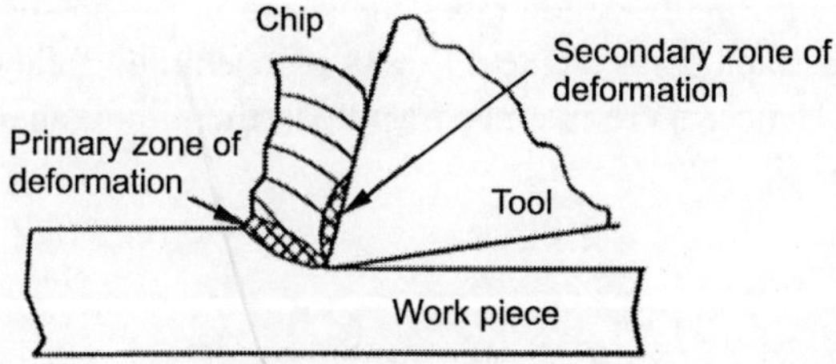

Fig. 1.111 Continuous chips

Factors responsible for continuous chips :

1. Ductile material
2. Large rake angle and sharp cutting edge of cutting tool
3. High cutting speed
4. Efficient cutting fluids
5. Low friction between tool face and chips

3. **Continuous Chips with Built up Edge:** A small climp of metal is welded on the chip tool contact area due to high temprature and pressure at cutting edge. The lump of metal is known as built up edge as shown in Fig.1.111.

Factors responsilbe for Built-up-edge (BUE)

(*i*) Ductile material

(*ii*) Coarse feed

(*iii*) Small rake angle

(*iv*) Low cutting speed

(*v*) Insufficient cutting fluid

(*vi*) A finity between tool and chip material

Consequences of built-up-edge (BUE)

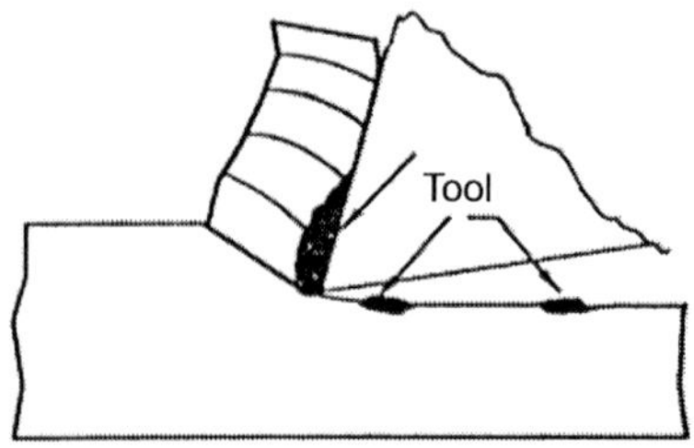

Fig. 1.112 Continuous chip with BUE

1.10.11 Chip Breakers

Need: Chip breakers are important components of tool design particularly when tool has to cut ductile materials like low carbon steel, copper, aluminium, low zinc, brass, etc.

A continuous chip normally which takes the shape of a long spriral and becomes quite troublesome. Such chips create diffculty in machining and become a problem in chip disposal. Hence, there is a need to break them into small pieces for easy disposal.

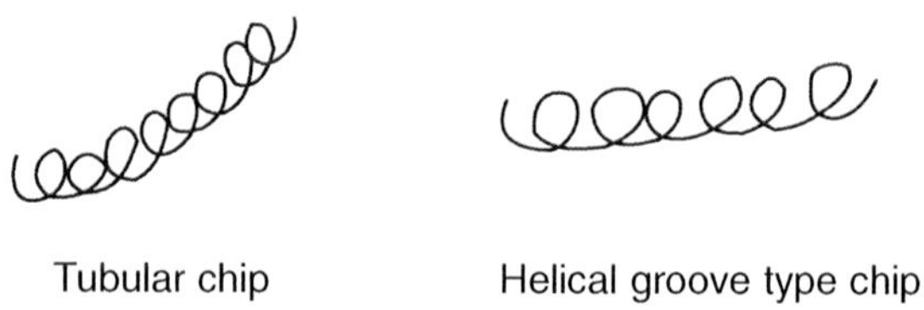

Fig. 1.113 Chip breakers.

Types of chip breakers

1. (*a*) Groove ground chip breaker

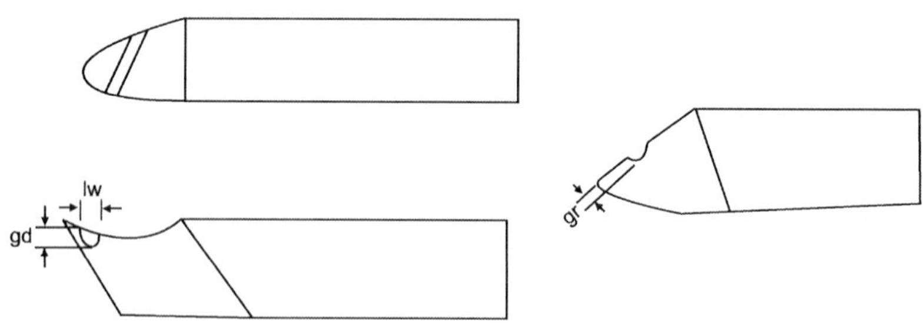

Fig. 1.114 Groove ground chip.

lw = chip breaker land width

gd = chip breaker groove depth

gr = chip breaker groove radius

(*b*) Cemented carbide have been pressed in groove

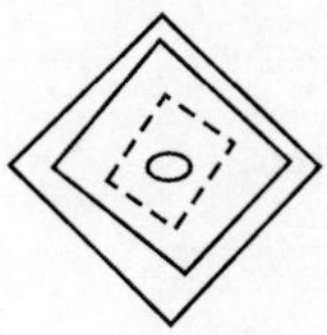

Fig. 1.115 Cemented carbide.

2. Clamp on type chip breaker

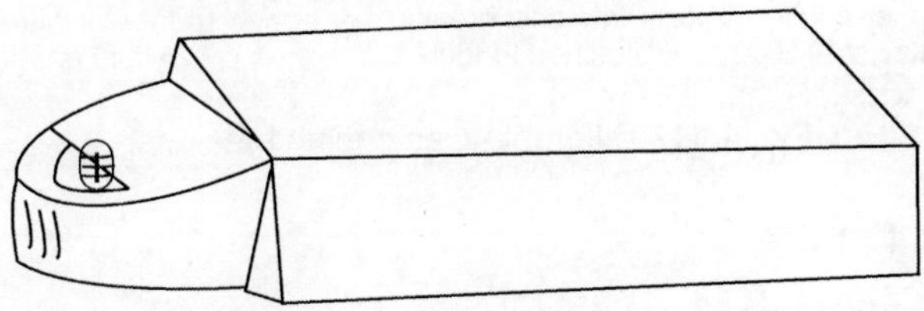

Fig. 1.116 Clamp on chip breaker.

3. Step ground type chip breaker

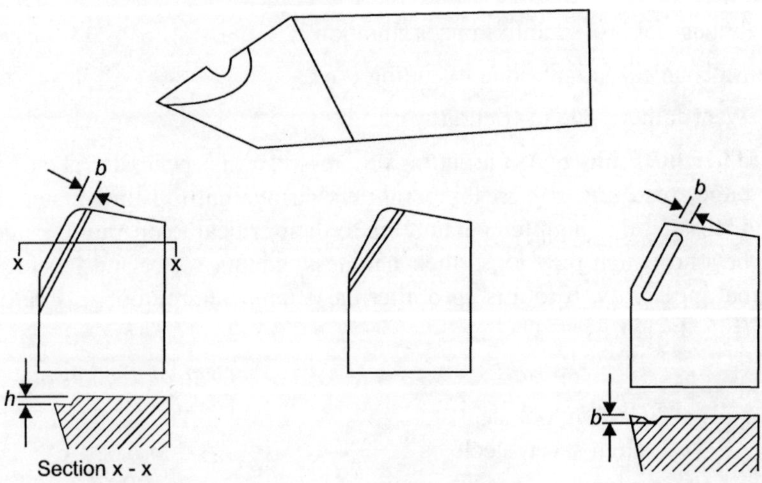

(*a*) Parallel shoulder chip breaker

(*b*) Angular shoulder chip breaker

(*c*) Groove type chip breaker

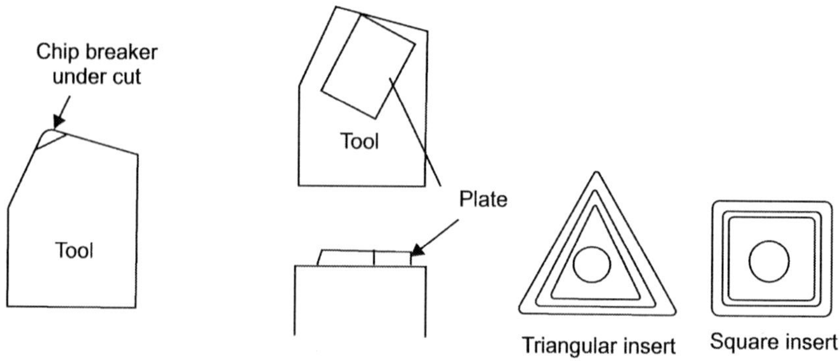

(*d*) Chip breaker for light cuts (*e*) Plate chip breaker attached to tool (*f*) Inserts with built in chip breaker

Fig. 1.117 Different types of chip breakers.

1.10.12 Tool Failure

Tool failure implies that the tool has reached a point beyond which it will not function satisfactorily until it is reground. Tool failure is due to many causes and as per a particular cause it is given a particular name as:

(*a*) Thermal failure: due to rise in temperature

(*b*) Impact failure: due to impact/vibration

(*c*) Mechanical failure: due to cutting force

(*d*) Wear failure: due to friction/wear

(*a*) **Thermal Failure:** As the cutting progresses, heat is generated at the cutting edge, resulting rise in temperature. Various cutting tool material can withstand this high temperature up to their critical temperature range and beyond which they loose their hardness, getting softer and loosing their cutting ability. If tool is used after these temperature, tool gets deformed (Fig. 1.118).

Material	*Critical Temperature*
Carbon tool steel	200 – 250°C
High speed steel	560 – 600°C
Cemented carbide	1000°C

Depression

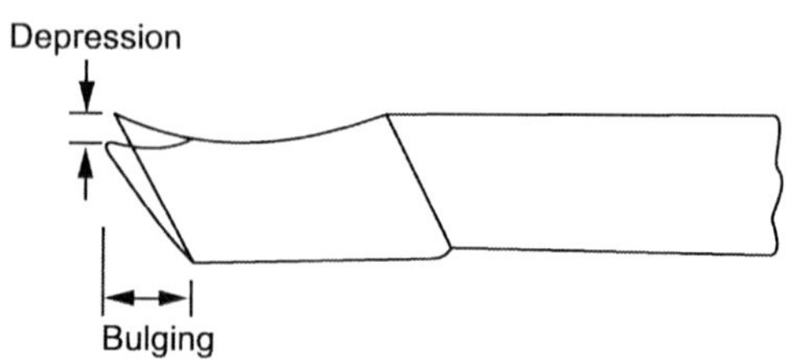

Fig. 1.118 Deformed cutting tool

(*b*) **Impact Failure:** Repeated jerks/impact during loading, unloading and machining give rise to this failure resulting in notch wear and fracture of cutting edge, development of notch (Fig. 1.119).

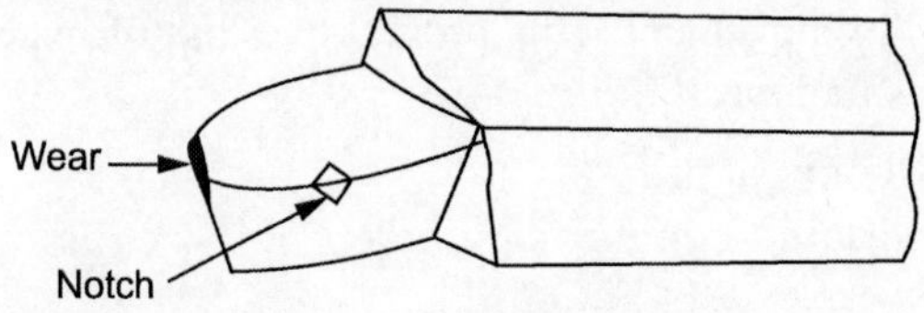

Fig. 1.119 Impact failure

(*c*) **Mechanical Failure:** Cutting force exerted on cutting tool, they may result in break of tool. Tool may suffer chatter, due to tool workpiece interruption and fail. Tool may also fail. Mechanical chipping of the nose and / or the cutting edge of the tool are commonly observed causes of tool failure (Fig. 1.120).

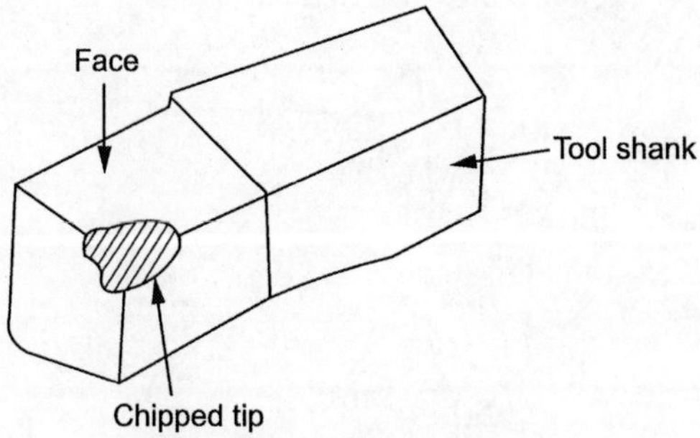

Fig. 1.120 Mechanical failure

(*d*) **Wear Failure:** Cutting tool normally reaches to their life through gradual wear.

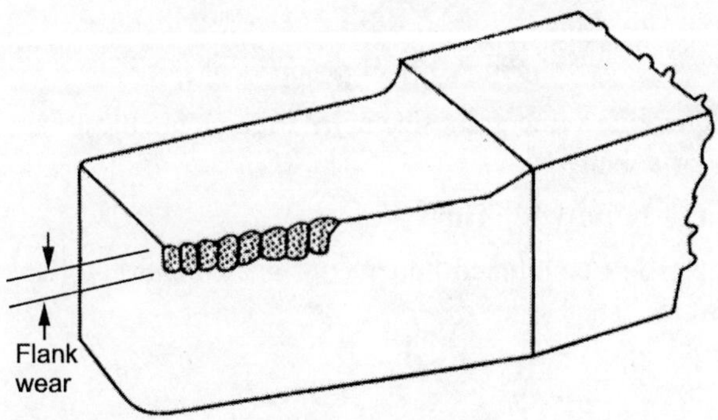

Fig. 1.121 Wear failure

1.10.13 Tool Wear

Initially the cutting tool has shape cutting edges and smooth flanks but these are lost during the maching processes because higher cutting forces causes higher temperature and chip movement. This progressive loss of material from the tool surface is known as tool wear.

Basic types of the tool wear are :

1. Abrasion wear
2. Adhesion wear
3. Diffusion wear
4. Chemical or electrolytic wear
5. Oxidation wear
6. Fatigue wear

1. **Abrasion Wear :** The wear caused by by small particles of the work piece rubbing against the tool surface. These particles plough into the safer matrix of tool for producing crater and flank wear.

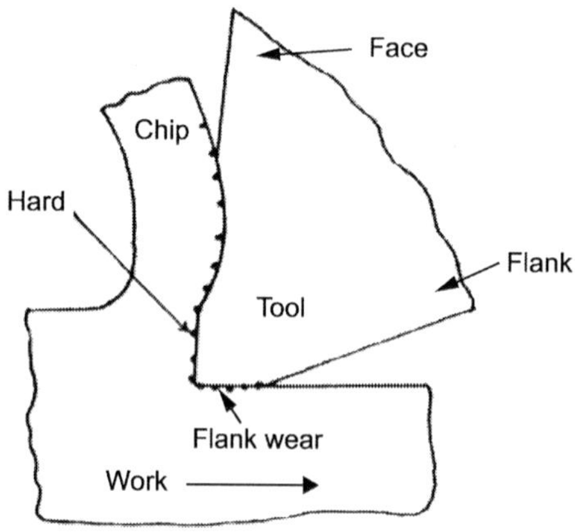

Fig. 1.122 Abrasion wear

In case of carbide tools, a fine grain size helps in reducing the abrasive wear due to low percentage of cobalt.

Progress of Flank Land with Time

When cutting tests are conducted under constant conditions, flank wear land is seem to increase in 3 stages

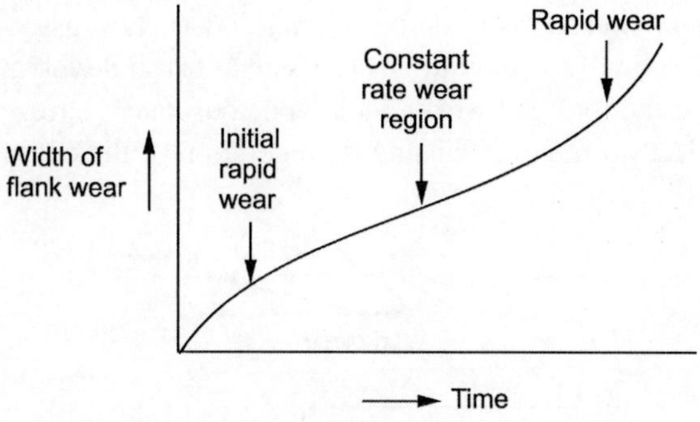

Fig. 1.123 Development of flank wear with time

(*a*) *Primary stage:* Wear is rapid as cutting tool looses its initial sharp edge leading to breaking off sharp cutting edge and establish a fine wear land.

(*b*) *Secondary stage:* Tool becomes sensitive to the increasing tool temperature and wear progressing at a slow and uniform rate.

(*c*) *Tertiary stage:* Wear occurs at gradually increasing rate. It is due to rapid rise in temperature at cutting edge.

Effect of Tool Wear

(*i*) Production is effected as time of machining will be more.

(*ii*) Where will be reduction in surface finish.

(*iii*) *Workpiece dimensional change:* A cylindrical piece on turning make it truly cylindrical may become tapered or uneven surface in place of cylindrical surface.

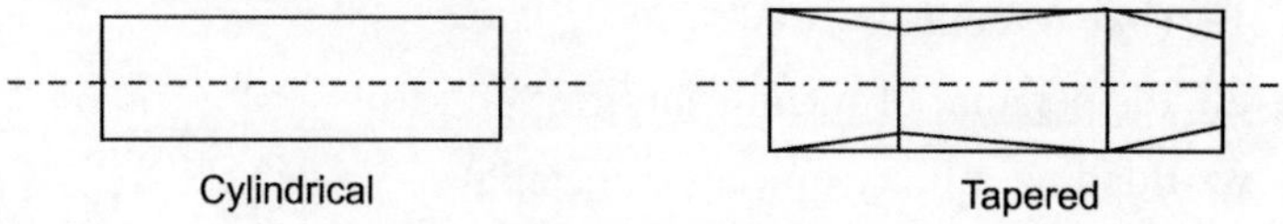

Fig. 1.124 Dimensional change

(*iv*) Higher forces are required.

(*v*) Safety and efficiency of machine operator is affected.

(*vi*) Overall power consumption increases.

Sand particles entrapped in casting will plough in the cutting tool edge.

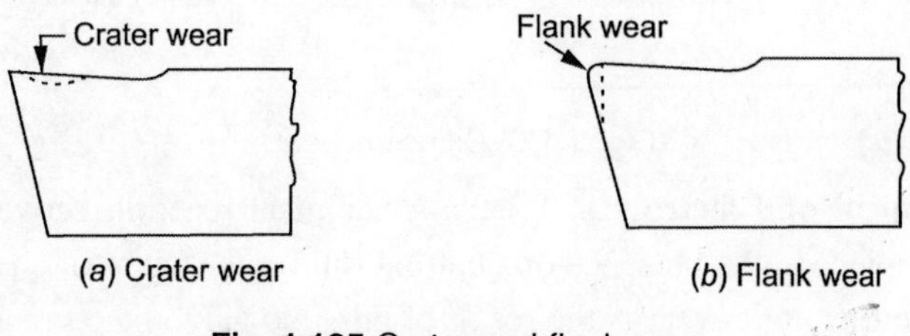

(*a*) Crater wear (*b*) Flank wear

Fig. 1.125 Crater and flank wear

2. **Adhesion Wear:** Plastic deformationa and friction associated with high temprature involved in cutting process can cause a welding action on the surface of the tool and work piece. The consequent stresses of the cutting process lead to fracture causing regreneration of the cutting tool.

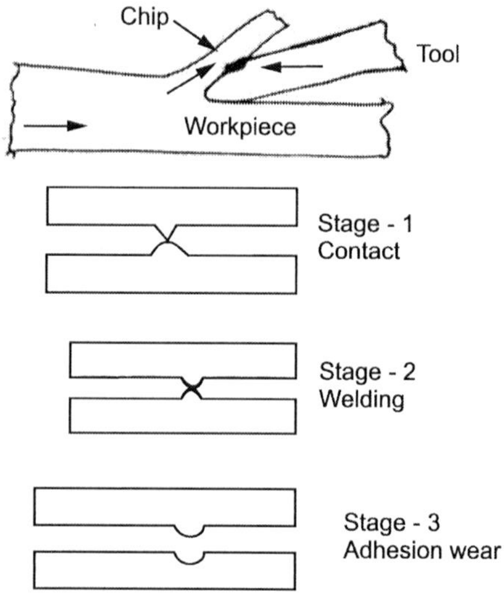

Fig. 1.126 Adhesion wear

3. **Diffusion Wear :** Diffusion wear caused by a displacement of atoms in the metallic crystal of the cutting element from one lattice point to another. This results in a gradual deformation of the tool surface. It depends upon.
    1. Time of contact
    2. Temperature of mating surface
    3. Bonding affinity of mating material
    4. Degree of atomic diffusion.

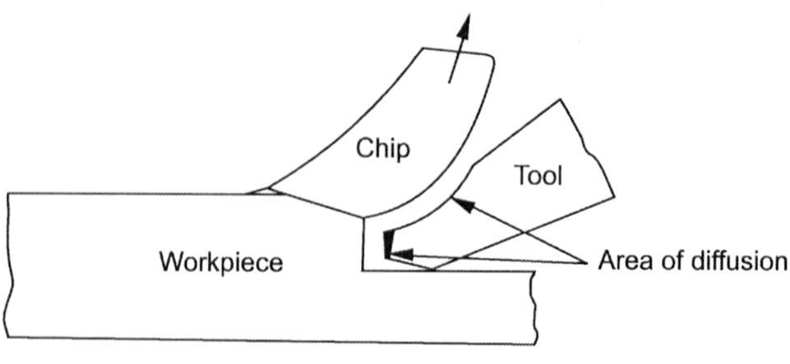

Fig. 1.127 Diffusion wear

4. **Chemical or Electrolytic Wear:** A chemical reaction between tool and workspiece in the presence of a cutting fluid is the cause of chemical wear. This electrolytic wear is the result of possible galvonic corrosion between the tool and work piece.

5. **Oxidetion Wear:** At high temprature oxidation of the carbide in cutting tool decreases its strength and causes wear of the edge.
6. **Fatigue Wear:** When two surface, of chip and cutting tool face slide in contact with each other under pressure projections of one surface interlock with another resulting in frictional stage. This frictional stage result in compressive stress on one side of interlocking and tensile stress on other side. As chip slide over further these stresses are relived, but soon another pair of interlocking is formed leading to repetition of stress cycle. The cycle gives rise to fatigue phenomena which causes surface cracks which eventually enlarge and combine to adopt a form of wear.

1.10.14 Tool Life

Tool life is defined as the time between the two suceessive grinding or resharpening the cutting tool.

1. In term of number of components machined between two successive grinding during the continuous tool working.
2. Time period in minutes between two successive grindings when tool wear is subjected to combination F.W. Taylor's developed the following relationship betwen tool life and cutting speed. Tool life (T_L) based upon the criterion of the volume of metal removed is given by the relation:

$T_L = 1000.\ t.f.T.v\ \text{mm}^3$

where T_L = Tool life in mm^3/min.

t = Depth of cut in mm

f = Feed in mm / revolution

T = Time of tool failure in min.

v = Cutting speed in m/min.

$$VT^n = C$$

where V = Cutting speed (m/min)

T = Tool life (minutes)

n = Exponent, the value depends on the cutting tool

n = 0.1 – 0.15 for high speed steel (H.S.S.)

n = 0.2 – 0.5 for cemented carbide tool

0.6 – 1.0 for ceramic tools

c = Constant – machine constant.

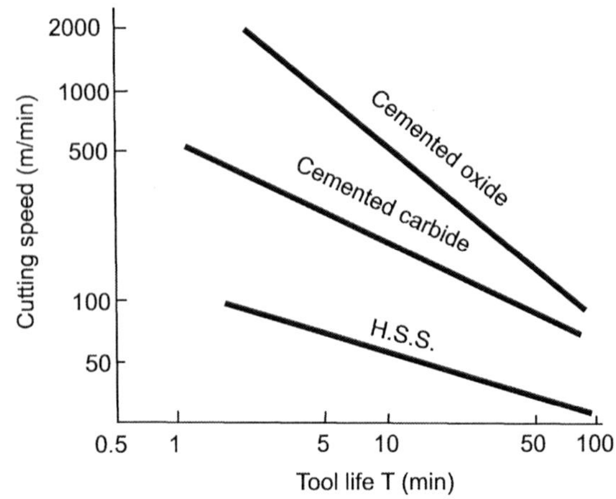

Fig. 1.128 Cutting speed v tool life

Factors Affecting Tool Life

There are the following factors:

1. Cutting speed
2. Feed and depth of cut
3. Tool geometry
4. Tool material
5. Work material
6. Nature of cutting
7. Rigidity of machine tool and work
8. Use of cutting fluids.

1.10.15 Cutting Fluids

Cutting fluids are used to denote those coolants and lubricants which are used in metal machining and their allied operations. These are applied to chip tool interface to facilitate cutting or machining operations:

Use of Cutting Fluids: Cutting fluids are used for the following reasons:

(*i*) To tool the cutting tool and the workpiece

(*ii*) To lubricate the chip, tool and workpiece.

(*iii*) To help carry away the chips.

(*iv*) To lubricate some of the moving parts of the machine tool.

Role of Cutting Fluid

1. **Primary Role:** The primary role of cutting fluid is as a coolant by which it leads to following:
   (*a*) Tool retains its hardness and hence tool life is extended.
   (*b*) Increased speed can be utilised, maintaining tool life.
   (*c*) Heat removal from cutting zone, reduces thermal distortion of workpiece and prevent expansion which might result is in accurate measurement and thus it helps in better dimensional control.
2. **Secondary Role:**Secondary role of cutting fluids is as an lubricant. As a lubricant, it reduces friction between working tool and chip by maintaining hydrodynamic film between tool and chip face and effective lubricant can modify geometry of chip formation.
   (*a*) Less deformed and cooler chip and thinner sections.
   (*b*) By discouraging formation of Built up edges.
   (*c*) By promoting improved surface finish.

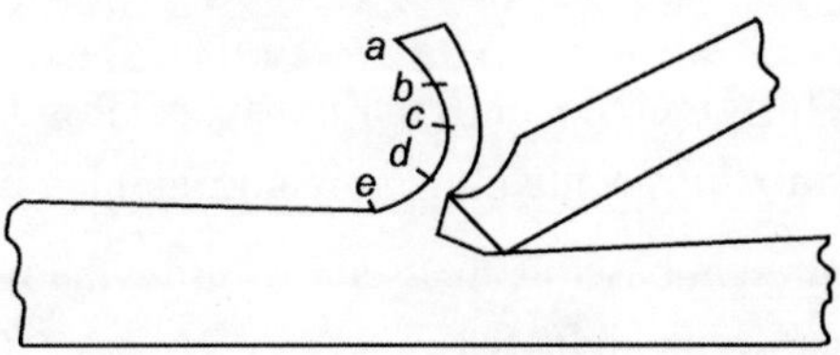

Fig. 1. 129 Top-dry cut and bottom-wet cut

3. **Tertiary Role:** It is used to remove on flush away chips and thus helps in their easy disposal.

1.10.16 Types of Cutting Fluids

1. **Air:** It is most widely available and used cutting fluid. It is commonly used.
2. **Water:** Water as a coolant is very effective, cheap and easily available with no fire hazard, however it has following demerits:
   – It is a rust hazard to both workpiece and cutting tool
   – It is ineffective as a lubricant.
3. **Oil Emulsions:**These are oil compounds which form emulsion when mixed with water. An emulsifying agent is required to obtain a mixture of oil and water. Soap has proved to be an affective emulsifying agent.

Soluble oils are easily mixed, but since emulsion is of oil in water type. It should always be prepared by pouring of oil into water instead of pouring water into the oil. The mixture may range form 1 part of soluble oil to 5 parts of water up to 1 part soluble oil to 50 parts of water.

Applications: They are extensively used for machining ferrous to non-ferrous metals where:

(*a*) cooling quality is paramount

(*b*) chip breaking pressure is not excessive.

4. **Aqueous Solution:** Water containing some alkali carbonate and soda say caustic soda.

   Applications: For cooling workplace and flush away the chips

5. **Straight Mineral oils:**

   (*a*) *Mineral oil:* This group includes oil types of oils extracted from petroleum such as mineral sea oil and kerosene. Straight base cutting oils are used for light machining operations for free machining steels and non-ferrous.

   (*b*) *Base fatty oil:* This group contains various types of high sulphurised and chlorinated mixtures containing inorganic, animal, fatty minerals.

   Applications: Straight base oil are used for cutting alloy steels where chip bearing pressure is high.

   (*c*) *Mixture of mineral and fatty oil:* Base oils are often blanded or cut back with mineral oil which are lighter.

   Application: Light to medium machining.

   (*d*) *Mineral lard oil:* May be used to obtain a cutting fluid having greater lubricarting values than straight base type oil and much lower cost than straight lard oil.

   The preperation of two would depend upon character of machining operation and need for both cooling and lubricating effects. For heavy machining, 10 to 40 lard oil is used.

Workpiece Material	Machining Operation	Types of Cutting Fluids
1. Cast Iron		Air (Normally machined dry) soluble oil emulsions (may be used, tool execessive amount of dusts around machining)
2. Low carbon steel (*a*) Free cutting steel (such as machinery steel)	Turning (fine)	Soluble oil emulsions mixed with straight mineral oils of low viscosity.
	Rough Turning Drilling	75% mineral oil, 25% lard oil, soluble oil
	Tapping	Mineral oil with 25-40%, lard oil
	Milling	Soluble oil
	Thread cutting	1 part soluble oil and 15 parts water
	Broachig	Heavy mineral oil
	Grinding	Soluble oil emulsion
	Cylindrical grinding	1 part soluble oil and 70-80 part water
(*b*) Alloy steels	Turning	75% mineral oil, 25% sulphur base water
	Drilling	Soluble oil
	Tapping	70% mineral oil, 30% lard oil
	Milling	90% mineral oil, 10%, lard oil
3. Brass	Turning	Straight mineral oil 90%, mineral oil, 10% lard oil
	Drilling water	Soluble oil with 75-90%
	Tapping	80-90% mineral oil with 20 to 10% lard oil
	Milling	Soluble oil with about 90% water

Workpiece material	Machining Operation	Types of Cutting Fluids
4. Cast brass		Usually machined dry
5. Aluminium	Turning	90% mineral oil, 10% lard oil
	drilling water	Soluble oil with 75-90%
	Tapping	Lard oil or mole grease
	milling	90% water with soluble oil
6. Magnesium	Grinding should not be used.	Normally machined dry Soluble oil-emulsive oil

HIGHLIGHTS OF THE CHAPTER

1. A machine tool which removes the material in the form of chips by rotating a workpiece against a single point cutting tool is known as lathe.

2. A power operated machine tool used for shaping, sizing and processing a product to the desired accuracy by removing the excessive metal in the form of chips known as machine tool.

3. Types of lathe are :

(*i*) Engine lathe (*ii*) Speed lathe
(*iii*) Bench lathe (*iv*) Tool room lathe
(*v*) Turet lathe (*vi*) Capston lathe
(*vii*) Gap bed lathe (*viii*) Hollow spindle lathe
(*ix*) Automative lathe (*x*) Special purpose lathe

4. Principal parts of a lathe are :

(*i*) Bed (*ii*) Head stock
(*iii*) Tail stock (*iv*) Carriage
(*v*) Feed mechanism (*vi*) Threed cutting mechanism

5. Lathe accessories :

(*i*) Chuck (*ii*) Lathe centres
(*iii*) Face plate (*iv*) Angle plate
(*v*) Mandrels (*vi*) Steady rest
(*vii*) Lathe carriers and dogs

6. Lathe operations are:

(*i*) Facing (*ii*) Turning
(*iii*) Shoulder turning (*iv*) Taper turning
(*v*) Chembering (*vi*) Forming

(*vii*) Grooving (*viii*) Parting off
(*ix*) Knurling (*x*) Spinning
(*xi*) Filling (*xii*) Polishing
(*xiii*) External thread cutting (*xiv*) Drilling
(*xv*) Boring (*xvi*) Counter boring
(*xvii*) Taper boring (*xviii*) Reaming
(*xix*) Internal threading (*xx*) Ecentric turning
(*xxi*) Under cutting (*xxii*) Spring winding
(*xxiii*) Grinding (*xxiv*) Milling

7. Methods of taper turning are :
(*i*) By form tool
(*ii*) By swivelling the compound slide
(*iii*) By taper turning attachment
(*iv*) By setting over the tail stock.

8. Shaper is a versatile machine tool which is used for producing flat, slots, grooves, convex, concave, and irregular surface with a reciprocating tool by using single point cutting tool.

9. Principal parts of a shaper are:
(*i*) Base (*ii*) Column
(*iii*) Cross rail (*iv*) Table
(*v*) Ram (*vi*) Tool Head
(*vii*) Vice

10. Shaper operations are :
(*i*) Horizontal shaping (*ii*) Vertical shaping
(*iii*) Grooving shaping (*iv*) Angular shaping
(*v*) Cutting of gear teeth (*vi*) Angular cutting

11. Planer is a machine tool used to remove metal by moving the work against a single point cutting tool.

12. Types of planers are :
(*i*) Double housing (*ii*) Openside planer
(*iii*) Pit planer (*iv*) Plate planer
(*v*) Divided table planer

13. Principal parts of a planer are :
(*i*) Bed (*ii*) Table
(*ii*) Housing (*iv*) Cross rail
(*v*) Tool heads

14. Planer operations are :

(*i*) Planing horizontal plate surface
(*ii*) Planing vertical plate surface
(*iii*) Planing angular surface
(*iv*) Planing contours and curved surface

15. Work holding devices in a planer are :

(*i*) Vices (*ii*) Clamps
(*iii*) T-bold, nut, washer (*iv*) Stops
(*v*) Stop blocks (*vi*) Planer Jacks
(*vii*) V blocks (*viii*) Angle plate
(*ix*) Holding down dogs (*x*) Parallel strips etc.

16. A machine tool which is used to produce flat, curved, grooves and stock with the help of a revolving multiple tool cutter is known as milling machine.

17. Types of milling machine

(*i*) Column and knee types (*ii*) Bed type
(*iii*) Planer type (*iv*) Special purpose type

18. Principal parts of a milling machine

(*i*) Base (*ii*) Column
(*iii*) Knee (*iv*) Saddle
(*v*) Table (*vi*) Over arm
(*vii*) Spindle (*viii*) Arbor

19. Milling methods

(*i*) Conventional or up milling (*ii*) Climb or down milling

20. Milling operations are :

(*i*) Plain milling (*ii*) Face milling
(*ii*) Side milling (*iv*) Straddle milling
(*v*) Angular milling (*vi*) Gang milling
(*vii*) Form milling (*viii*) Profile milling
(*ix*) End milling (*x*) Saw milling
(*xi*) Key way, groves and slot, milling (*xii*) Gear cutting
(*xiii*) Helical milling (*xiv*) Cam milling
(*xv*) Thread milling

21. Work holding and clamping devices

(*i*) Plain clamp (*ii*) Plain clamp with pin
(*iii*) Offset clamp (*iv*) Stop clamp
(*v*) Offset conical clamp (*vi*) Finger clamp
(*vii*) Offset clamp (*viii*) Offset set clamp
(*ix*) Wedge block (*x*) U clamp
(*xi*) Nut and bolt (*xii*) C clamp

22. Drilling is a metal cutting operation which produces round holes in metal to take up rivets, screws, bolts and shafts etc.

23. Types of drilling machines.

(*i*) Portable drilling machine
(*ii*) Sensitive drilling machine
(*iii*) Upright drilling machine
(*iv*) Radial drilling machine
(*v*) Gang drilling machine
(*vi*) Multi spindle drilling machine
(*vii*) Automatic production drilling machine
(*viii*) Deep hole drilling machine

24. Principal parts of a drilling machine are :

(*i*) Base
(*ii*) Column
(*iii*) Table
(*iv*) Power head
(*v*) Spindle and spindle sleeve assembly
(*vi*) Drill chuck

25. Drill operations are :

(*i*) Drilling
(*ii*) Boring
(*iii*) Step drilling
(*iv*) Boring
(*v*) Counter boring
(*vi*) Spot facing
(*vii*) Counter sinking
(*viii*) Reaming
(*ix*) Trapenning
(*x*) Tapping
(*xi*) Surface grinding
(*xii*) Buffing
(*xiii*) Hole milling

26. Work holding device for drilling machine are :

(*i*) Vice
(*ii*) Parallel
(*iii*) T-bolt
(*iv*) Drill chuck
(*v*) Stop blocks
(*vi*) Jack screw
(*vii*) Angle plate
(*viii*) V-block
(*ix*) Drill chuck

27. A machine tool used for producing, slots, grooves, key ways and work in a vertically reciprocating single point cutting tool known as slotter.

28. Types of slotter machines are :

(*i*) Puncher slotter
(*ii*) Production slotter
(*iii*) Tool room slotters
(*iv*) Key slotter

29. Principal parts of a slotter are :

(*i*) Base
(*ii*) Column
(*iii*) Table
(*iv*) Ram
(*v*) Cross rail

30. Slotter operations are :

(*i*) Machining cylindrical surfaces

(*ii*) Machining irregular surfaces

(*iii*) Machining flat surfaces

(*iv*) Machining slots, grooves, key ways and recesses.

31. Metal cutting or machining is a process of producing a desired shape, size and surface finish by removing the excess material in the form of chips.

32. Principal elements of a metal cutting :

(*i*) Cutting speed
(*ii*) Feed
(*iii*) Depth of cut

33. Tool materials

(*i*) Plain carbon steel
- (*a*) Medium carbon steel
- (*b*) Low carbon steel
- (*c*) High carbon steel

(*ii*) High speed steel
- (*a*) 18 : 4 : 1 H.S.S.
- (*b*) Mo - H.S.S.
- (*c*) Co - H.S.S.

(*iii*) Cemented carbide
(*iv*) Ceramic tool steel
(*v*) Diamond

34. Sequence of tools in tool signature

(*i*) Top rack
(*ii*) Side rake angle
(*iii*) End relief angle
(*iv*) Side relief angle
(*v*) End cutting edge angle
(*vi*) Side cutting edge angle
(*vii*) Nose radius

35. Metal cutting processes are :

(*i*) Orthogonal cutting
(*ii*) Oblique cutting

36. Types of chips

(*i*) Discontinuous chip
(*ii*) Continuous chip
(*iii*) Continuous with builtup edge.

37. Tool failure implies that a tool reacts a point beyond which it will not work satisfactorily until it is regrind.

38. Causes of tool failure :

(*i*) Thermal failure
(*ii*) Impact failure
(*iii*) Mechanical failure
(*iv*) Wear failure

39. Types of tool wear are :

(*i*) Abrasion wear
(*ii*) Adhesion wear
(*iii*) Diffusion wear
(*iv*) Chemical wear

40. Tool life may be defined as the effective cutting time between two successive grindings (resharpenings)

41. Cutting fluids are used to denote those coolants and lubricants which are used in metal machining and their allied operations.

42. Types of cutting fluids :

(*i*) Air
(*ii*) Water
(*iii*) Oil emulsions
(*iv*) Aqueous solution
(*v*) Straight mineral oil

REVIEW QUESTIONS

1. What do you mean by machine tool?

2. What are the functions of a machine tool?

3. What are the specifications of a lathe machine?

4. What is a lathe machine? Sate the principle of working of a lathe?

5. Explain the various parts of a lathe by neat sketch.

6. Explain various lathe accessories with neat sketches.

7. State the function of following accessories in a lathe:

(*i*) lathe centres
(*ii*) chucks
(*iii*) steady rest
(*iv*) lathe dogs

8. What machining operations can be performed on a lathe machine?

9. Write short notes on the following operations performed on a lathe:

(*a*) Drilling
(*b*) Reaming
(*c*) Knurling
(*d*) Facing
(*e*) Parting off

What do you mean by chip breaker and its type?

10. Explain the difference between three jaw chuck and four jaw chuck.

11. Name the various taper turning attachments.

12. Explain the method of taper turning with compound slide.

13. Define cutting speed, feed and depth of cut for a machining operation in a lathe.

14. Explain the mechanism of chip formation with neat sketches.

15. What are the commonly cutting tool materials used in machining?

16. Explain various safety precautions associated with lathe.

17. What is a working principle of shaper?

18. Give classification of shapers.

19. Classify the shapers.

20. Explain principal parts of a shaper with neat sketch.

21. How can you specify a shaper?
22. Explain with neat sketch, the working principle of a shaper.
23. Explain the principle of quick return motion mechanism of a shaper.
24. What are the common operators performed on a shaper?
25. What are the safety precautions associated with shaper?
26. What is a planer? State the working principle of a planer.
27. How do you classify planer?
28. What are the main operations performed on a planer?
29. Show by neat sketch various types of planer tools.
30. Describe with neat sketch the principle parts of a double housing planer.
31. How does a planer differ from a shaper?
32. Explain cutting speed, feed and depth of cut in a planer.
33. What are the safety precautions associated with planer?
34. Describe the working principle of a milling machine.
35. Explain principle parts of a column and knee type a milling machine by neat sketches.
36. How you will classify the milling machines?
37. Explain various types of milling operations with neat sketches.
38. How is a milling machine specified?
39. What do you mean by up milling and down milling.
40. What are the indexing or dividing heads? What are their functions?
41. Sketch and describe the use of following milling cutters:
    (*a*) face milling cutter
    (*b*) shell end mill
    (*c*) slitting sew
    (*d*) staggered teeth side mill
42. What are the elements of a plain milling cutter?
43. What do you mean by indexing in a milling machine?
44. Differentiate between up milling and down milling.
45. What are the safety precautions associated with milling machine.
46. What type of work holding and clamping devices used in a milling machine?
47. How a slotting machine is specified?
48. What are the principal parts of a slotting machine? Explain with neat sketch.
49. How do you classify the slotting machines?
50. Define cutting speed, feed and depth of cut for a slotting machine.
51. What is a drill machine? State the working principle of a drill machine.
52. How do you classify the different types of drilling machines?
53. Explain the principal parts of a drilling machine.
54. How do you classify the different types of drills?

55. What is a twist drill? Make a neat sketch of a twist drill and show its different parts on it.

56. What are special drills? Briefly describe them with sketches.

57. What are the operations performed on a drilling machine, explain them with neat sketch?

58. How do you specify drilling machine?

59. What do you understand by the terms drilling boring and reaming? How do they differ from each other?

60. How slotting machines are classified ?

61. Explain cutting speed, feed and depth of cut in a drilling machine.

62. What do you mean by work holding devices naming the some work holding devices with neat sketch?

63. What safety precautions associated with drilling machine?

64. What is the working principal of a slotter?

65. What are the principle parts of a slotter?

66. What type of operations are performed on a slotter?

67. What are the sizes and specifications of a slotter?

68. What is the difference between shaper and slotter?

69. (*a*) Introduction to metal cutting

(*b*) What do you mean by orthogonal cutting and oblique cutting?

(*c*) Define tool life and name the variables which effect tool life.

70. Write short note on metal cutting?

71. What is the principal element of metal cutting?

72. What is tool failure?

73. What type of tool material used for maching purpose?

74. Differentiate between single point and multi point cutting tool.

75. Differentiate between orthogonal and oblique cutting with neat sketch.

76. Show the nomenclature of a single point cutting tool used in a lathe machine.

77. What is a crater and flank wear?

78. Explain the role of cutting fluid in machining.

■■■

2

Basic Concept of Thermodynamics

2.1 INTRODUCTION

The word 'Thermodynamics' is derived from (Thermo + dynamics) where thermo means heat and dynamics means movement. Hence, it is a science which deals with the energy possessed by gases and vapours or in other words the science which deals with heat and work (*i.e.*, mechanical energy) means transformation of energy which is governed by the various laws of thermodynamics based on experimentation which deals with the physical behaviour of the system by logical reasoning like a machine which is used for converting heat into mechanical work known as heat engine.

The thermodynamics is based on the four laws of thermodynamics known as zeroth, first, second, and third laws.

The application of thermodynamics is extremely wide and its principles are used in the designing of energy converting devices like steam engines. *i.e.*,engines, steam and gas turbines, as well in refrigerators and air conditioning etc.

2.2 THERMODYNAMIC SYSTEM

Thermodynamic system is defined as any portion of matter under consideration, which is separated from the surroundings by real or imaginary boundaries. The thermodynamic system may be homogeneous or heterogeneous, but has to be macroscopic.

2.2.1 Boundary

It is an envelope around the system known as boundary of the system.

The boundary may be real physical surface like *i.e.* engine cylinder or walls of a vessel. The boundary may be fixed or movable having some matter such as steam, gas and vapours on compression or expansion it can take any shape.

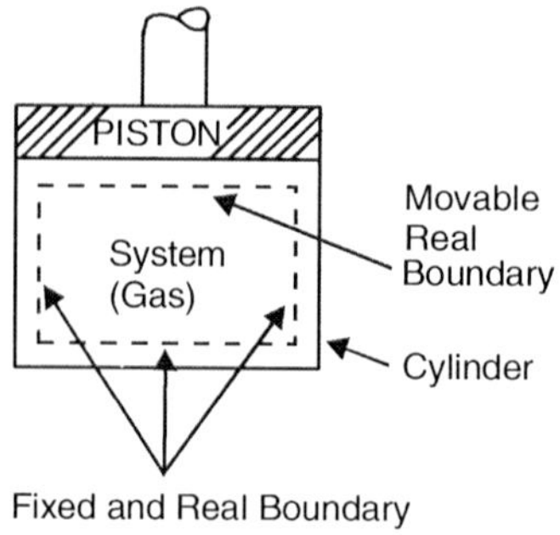

Fig.2.1 Representation of System Boundary and Surrounding

2.2.2 Surrounding

Everything outside the system boundary is called the surrounding.

2.2.3 Universe

When system and surrounding put together is called universe. Example piston cylinder arrangement.

2.3 TYPES OF SYSTEMS

There are three types of systems:

1. **Open System:** It is the system which is capable of exchanging both mass and energy (as heat) with the surrounding or across the boundary called open system.
2. **Closed System:** A system which is capable of exchanging only energy (as a heat or work) and mass remains constant with the surrounding called closed system. Example, a sealed gas container or piston cylinder.
3. **Isolated System:** It is the system which can exchange neither matter nor energy with the surroundings called isolated. Example, sealed thermoflask.

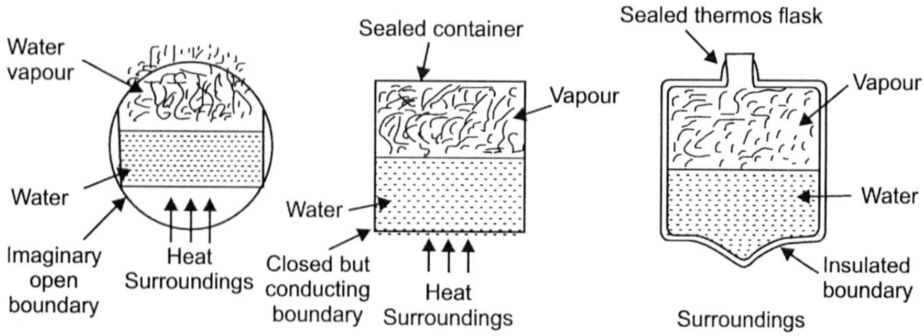

Fig.2.2 (*a*) An open system, (*b*) A closed system, (*c*) An isolated system

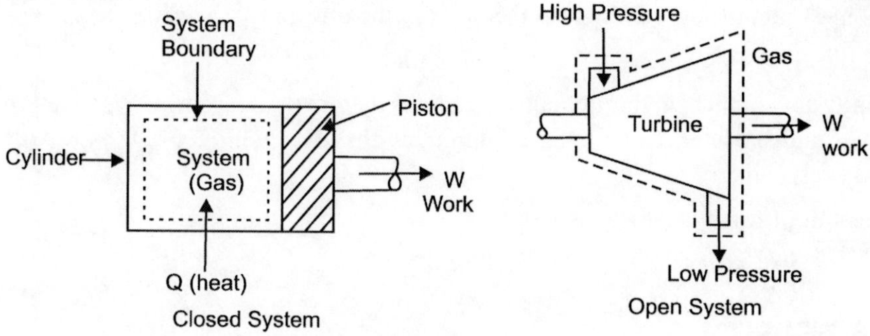

Fig. 2.3 Closed and open systems

2.4 PROPERTY

The quantitities which characterise the given state of the system are called properties or parameters. A property can be measured directly or indirectly while the system is in equilibrium. The value of the property depends only on the state and not on the process or path by which the state is achieved.

Properties are classified into two categories:

1. **Extensive Property:** It is the property which depends upon the amount of the substance present in the system. For example, total mass, volume, energy etc. of a system, since they cannot be determined without considering the system as a whole.
2. **Intensive Property:** It is the property which is independent of amount of substance present in the system. For example, viscosity, surface tension, refractive index, temperature, pressure, dielectric constant, density, specific gravity, specific heat, vapour pressure of a liquid, etc., are some of the important intensive properties.

Table 2.1. Some Intensive and Extensive Properties

Intensive properties		Extensive properties	
Temperature	Pressure	Mass	Enthalphy
Density	Refractive index	Number of moles	Entropy
Concentration	Surface tension	Energy	Internal energy
Specific heat	Boiling point	Volume	
Viscosity	Freezing point	Free energy	

2.5 FORCE

Thermodynamic properties can be measured either directly or indirectly. Properties, such as force, mass, length, and time have the basic units and are related to Newton's Second law of motion which states that the force acting on a body is the product of the mass and the acceleration of the body in the direction of the force.

$$F = ma$$

The system of units in throughout the world is the International system and referred as S.I. units. Meter, second and kilogram are the basic units for length, time and mass respectively.

The unit of force is Newton (N) and

$$1 \text{ N} = 1 \text{ kg m/s}^2$$

2.6 DENSITY

Density is defined as the ratio of mass per unit volume. The density varies significantly for a gas with pressure and temperature but the variation is negligible for liquids.

$$\text{Density, } \rho = \frac{\text{mass}}{\text{volume}} \text{ in kg/m}^3$$

2.7 SPECIFIC WEIGHT

Specific weight (w) is defined as the ratio of weight per unit volume.

$$\text{Specific weight, w} = \frac{\text{weight}}{\text{volume}} \text{ in N/m}^3$$

2.8 SPECIFIC GRAVITY

Specific gravity of a liquid is defined as the ratio of specific weight of that liquid to the specific weight of water.

2.9 SPECIFIC VOLUME

Specific volume of a substance is defined as the volume per unit mass and the density is defined as the mass per unit volume, *i.e.,* the reciprocal of the specific volume. Specific volume and density are intensive properties. The SI unit for volume is Cubic metre. The other unit of volume is the litre and is equal to 0.001 cubic metres.

2.10 VISCOSITY

The property which controls the rate of flow of a liquid is known as viscosity. Viscosity is due to the cohesive force between the liquid particles and it is exhibited when the liquid is in motion.

Newton's law of viscosity states that the shear stress on a layer of a fluid is directly proportional to the rate of shear strain.

Shear stress, $\tau \; \alpha \; \dfrac{dv}{dy}$

$$\Rightarrow \qquad \tau = \mu \, \frac{dv}{dy}$$

where $\dfrac{dv}{dy}$ is the rate of shear strain or velocity gradient and μ is the proportionality constant known as coefficient of viscosity. μ is also known as coefficient of absolute viscosity or dynamic viscosity. The unit of μ is Ns/m^2.

also $\quad$ 1 poise = 0.1 Ns/m^2 and

1 centipoise = (1/100) poise

$$\tau = \mu \frac{dv}{dy} = \frac{NS}{m^2} \frac{m/s}{m} = \frac{N}{m^2}$$

Kinematic viscosity of a liquid is the ratio of its absolute viscosity to its density.

Kinematic viscosity $\gamma = \mu/\rho$, m^2/s

1 stoke = 1/10000 $m^2/s = 10^{-4}$ m^2/s

1 centistoke = 1/100 stoke.

2.11 PRESSURE

Pressure is defined as the normal component of force per unit area. Pressure at a point in a fluid in equibrium is the same in all directions.

A fluid filled in a container exerts a force at all points on the sides and bottom of the container. This force per unit area is called pressure. The direction of this pressure is always normal to the surface with which the fluid comes in contact.

Atmospheric Pressure: The pressure exerted by the weight of air column on the surface of the earth is called as atmospheric pressure.

$$\text{Atmospheric Pressure} = \frac{\text{Weight of air column}}{\text{Area}}$$

$$= \frac{\text{Mass of air column} \times \text{Acceleration due to gravity}}{\text{Area}}$$

$$= \frac{\text{Volume of air column} \times \text{Density of air column} \times \text{Acceleration due to gravity}}{\text{Area}}$$

= Height of air column × Density of air × Acceleration due to gravity

$$= h\rho \, g \left[m \frac{kg}{m^3} . \frac{m}{s^2} = \frac{N}{m^2} \right]$$

Since the acceleration due to gravity g is constant ($g = 9.81$ m/s ^{2}), the atmospheric pressure at any location depends on the height of air column and density of air at that place. The value of atmospheric pressure is considered as standard at the surface of the sea as it is horizontal everywhere. The maximum air column height exists at the sea level and value of standard atmospheric pressure is 1.03×10^5 N/rn^2 or 1.03 kg /cm^2. The value of atmospheric pressure will be less than this at any other place on earth due to its higher altitude from the sea level. The density of air is inversely proportional to temperature.

Gauge Pressure: Pressure measuring instruments are usually calibrated with respect to atmospheric pressure. They will read zero at atmospheric pressure. Pressure measured with these gauges is known as gauge pressure. Pressure gauges are used to measure pressure above atmospheric level and vacuum gauges are used to measure pressure below atmospheric level.

Absolute Pressure: Absolute pressure is the sum of gauge pressure and atmospheric pressure. Absolute pressure is zero at absolute vacuum.

Absolute Pressure = Gauge Pressure + Atmospheric Pressure

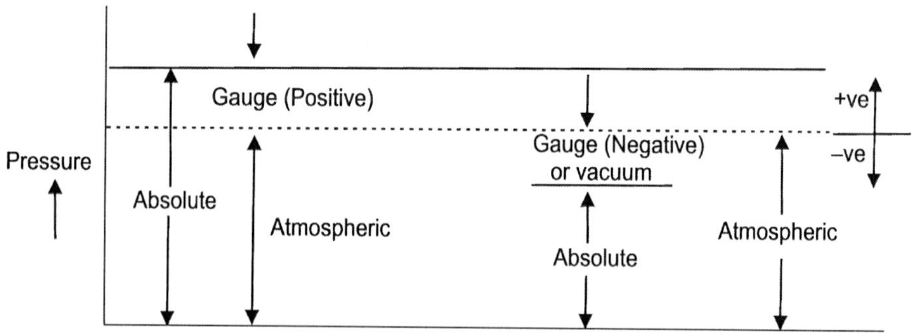

Fig. 2.4 Pressure Measurement

$$P_{abs} = P_{atm} + P_{vacuum} \quad P_{abs} = P_{atm} - P_{vacuum}$$

The unit for pressure in the International System is (Force in Newton acting on a square metre area) called a Pascal (Pa).

$$\text{I Pa} = 1\ \text{N/m}^2$$

The other units are:

$$1\ \text{bar} = 10^5\ \text{Pa} = 0.1\text{MPa}$$

$$1\ \text{atm} = 101325\text{Pa} = 14.696\ \text{lbf/in}^2\ \text{or Psi}$$

$$1\ \text{kgf/cm}^2 = 0.980665\ \text{bar.}$$

$$1\ \text{torr} = 1\ \text{mm Hg.}$$

$$1\ \text{micron} = 1\ \mu = 10^{-6}\ \text{m Hg} = 10^{-3}\ \text{mm Hg}$$

$$1\ \text{mm of Hg} = 133.3\ \text{N/m}^2$$

$$1\ \text{N / mm}^2 = 10^6\ \text{N/m}^2$$

2.12 THERMODYNAMIC EQUILIBRIUM AND NON-EQUILIBRIUM

A system is said to be in a state of thermodynamics equilibrium when the state variables remain constant throughout the system *i.e.*, state variables do not change with time and space. Thermodynamic equilibrium may be of following types:

1. **Thermal Equilibrium:** When the temperature in all parts of the system is uniform and is the same as that of the surroundings, the system said to be in thermal equilibrium.
2. **Mechanical Equilibrium:** A system is said to be in mechanical equilibrium when no mechanical work is done by one part of the system on any other part of the system *i.e.*, mechanical properties are uniform throughout the system.
3. **Chemical Equilibrium:** When the chemical composition of the system remains uniform and does not change with time, the system, is said to be in chemical equilibrium.

A system is said to be in a non-equilibrium state when the state variables have different values in different parts of system.

2.13 STATE

The state of the system is described by its properties such as pressure, temperature, density etc. This is shown on pressure-volume diagram as state 1 and 2.

Each property has a single value of each state *i.e.*, all properties are state or point functions.

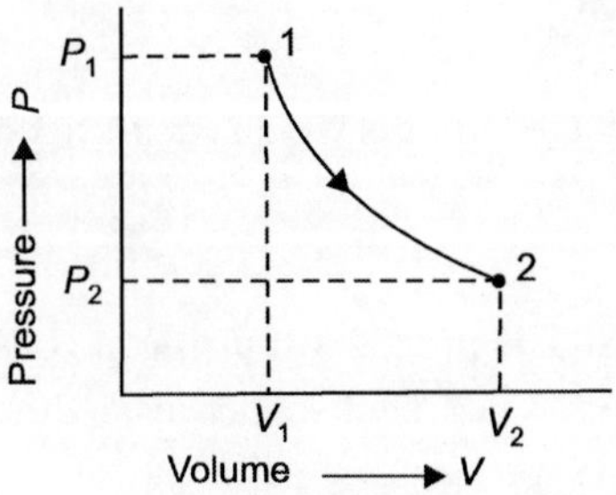

Fig. 2.5 State of a System

2.14 PATH

A thermodynamic system passing through a series of states is called a path.

2.15 PROCESS

A process is defined when a system changes its state from one state to another state and the path is completely specified, such change of state is called a process.

In Fig. 2.6 state 1 is the initial state of the system. Then due to expansion of the system, the final state is 2. The line 1,2 is the process that has taken place.

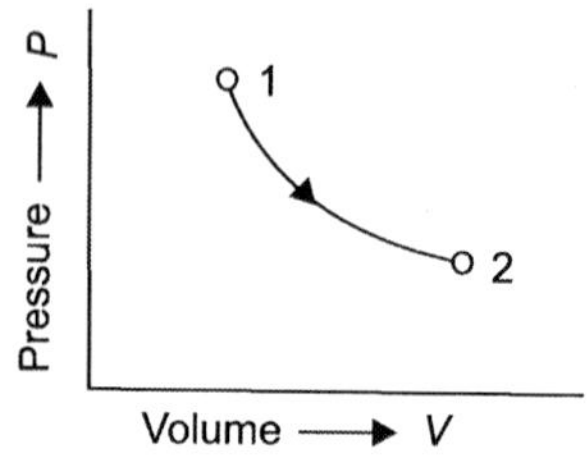

Fig.2.6 Representation of Process

2.16 TYPES OF THERMODYNAMIC PROCESS

An operation which brings about changes in the state of a system is called a thermodynamic process or simply a process.

A thermodynamic system may undergo following types of processes.

1. **Isothermal Process:** The process occurring at constant temperature is termed as an isothermal process. For such a process, δT (change in temperature) = 0.

2. **Adiabatic Process:** The process in which no heat is exchanged between the system and the surroundings is called adiabatic process. In such a process, the heat energy of the system remains constant and change in heat *i.e.*, $\delta q = 0$.

3. **Isobaric Process:** The process which occurs at constant pressure is termed as isobaric process. For such a process, change in pressure *i.e.*, $\delta P = 0$.

4. **Isochoric Process:** The process occurring at constant volume is known as a isochoric process. For such a process, the change in volume *i.e.*, $\delta V = 0$.

5. **Cyclic Process:** When initial and final states of a system in a process are the same, the process is called a cyclic process. In such a process, a system undergoes various processes and returns back to its initial state. Since internal energy, E and enthalhy, H are functions of state, we have $\delta E = 0$ and $\delta H = 0$ in a cyclic process.

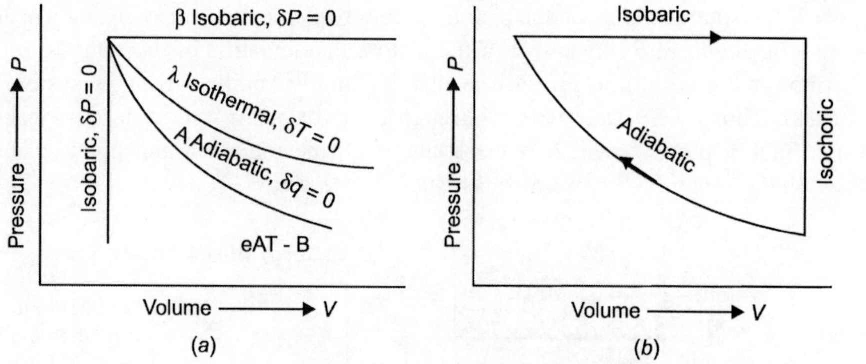

Fig. 2.7 (*a*) Isothermal, adiabatic and isobaric processes (*b*) a cyclic process

A graphical representation of these processes is shown in Fig. 2.7

2.17 CYCLIC PROCESS

When a process or processes are performed on a system in such a way that the final state is identical with the initial state. It is then known as cyclic process.

A cyclic process may be represented on P-V diagram as shown in Fig. 2.8 if the various states and paths are known.

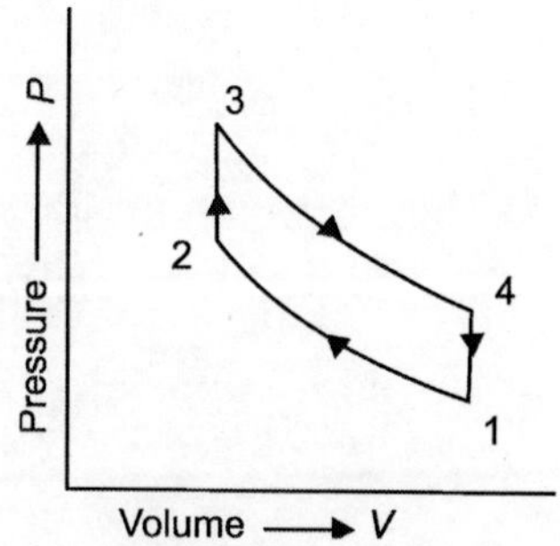

Fig.2.8 Cyclic process

2.18 QUASI-STATIC PROCESS (QUASI EQUILIBRIUM)

A quasi-static process is one in which the system deviates from one equilibrium state by only infinitesimal amounts throughout the entire process.

Consider a system (gas) enclosed in a cylinder and piston arrangement as shown in Fig.2.9. Let the system is initially in equilibrium state when the piston is at *A*, where the pressure is P_A, volume V_A and temperature T_A as shown in Fig.

The weight (W) on the piston is composed of a number of small weights which balances the upward force exerted by the system. If the whole weight is removed from the piston, then there will be unbalanced force between the system and the surroundings and the piston will be more upwards till it hits the stops at *B*. At this point *B*, the system again comes to an equilibrium state where the pressure is P_B, volume V_s and temperature T_s. But the intermediate state through which the system gets passed, are non-equilibrium states whose properties (pressure, volume and temperature) are not uniform throughout the system and thus the state of the system cannot be well-defined.

Now, if the small weights on the piston are removed one by one slowly, then at any instant of the upward movement of the piston, the deviation of the state from the thermodynamic equilibrium will be infinitesimally small, if the gas system is isolated. Thus, every state passed through by the system will be in equilibrium state. Such a process, which is the locus of all these equilibrium points passed through the system is known as quasi-static process.

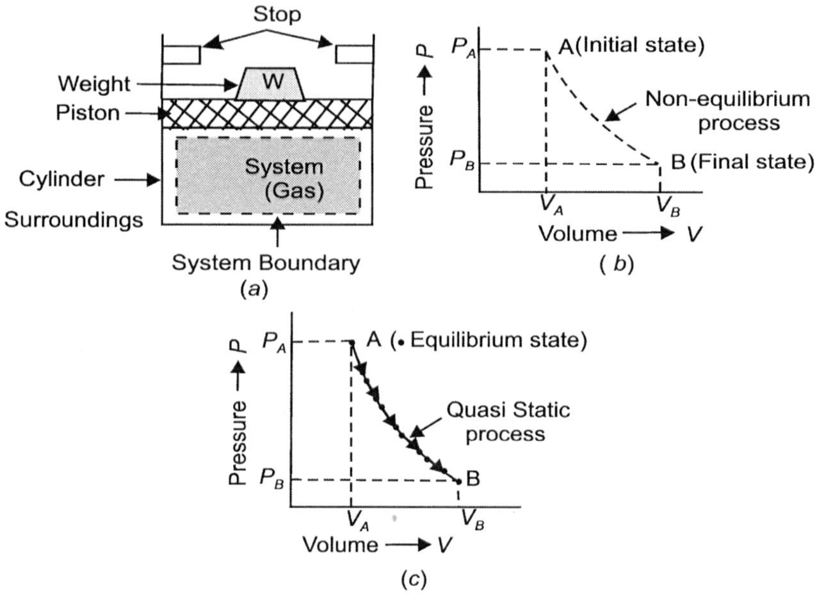

Fig.2.9 Quasi Static Process

2.19 REVERSIBLE AND IRREVERSIBLE PROCESSES

When a process is carried out infinitesimally slowly in such a way that each step remains in equilibrium with the previous step and the process could be reversed at any stage, it is called a reversible process.

A reversible process is supposed to take place in a series of infinitesimally small steps. If the process is carried out this way, the system remains in equibrium at each step and an infinitesimally change in the system can bring it back to the previous step.

Irreversible process is one in which the system or surroundings are not restored to their initial state at the conclusion of the reverse process. All naturally occurring processes are irreversible. They always tend to proceed in a definite direction and if left to themselves, they will never proceed in the opposite direction. For example, a clock always runs down and cannot rewind itself. Other common examples of irreversible process are : (1) flow of heat from a hotter to a colder body; (2) expansion of gases from high pressure regions to low pressure regions through orifices; (3) muting of two gases.

Irreversible process are also called "spontaneous processes"; which means that the direction of the process is spontaneous.

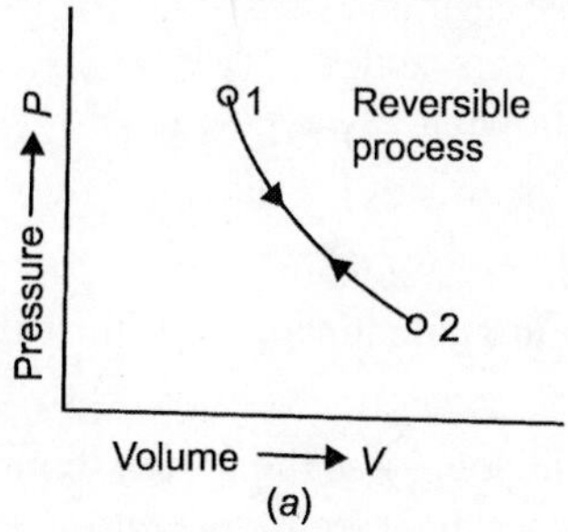

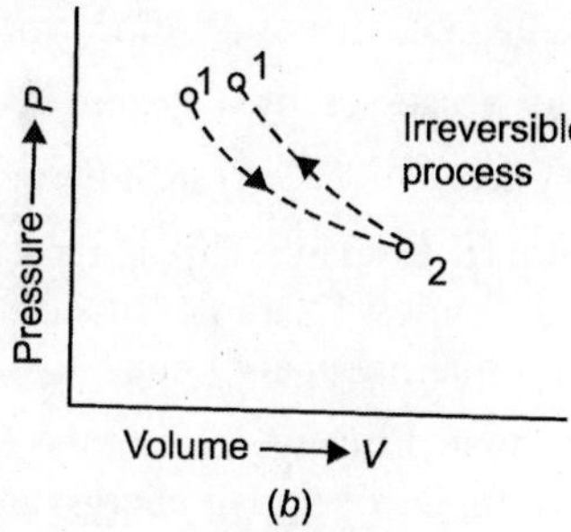

Fig. 2.10 Reversible and irreversible processes

2.20 MICROSCOPIC AND MACROSCOPIC POINT OF VIEW

There are two points of views, microscopic and macroscopic, from which the behaviour of matter or working of a system can be studied.

(*a*) **Microscopic Point of View:** It is considered that the matter is not continuous, but it is made up of a large number of identical particles called molecules.

For Example, consider a gas in a cylinder as system. Then this gas is made up of a number of molecules. Each molecule of a gas, at a given instant has certain position, energy, etc. and for each molecules. Each molecules of a gas, at a given instant has certain position, energy, etc. and for each molecule these changes rapidly because of the collisions (striking). The behaviour of the gas is described by summing up the behaviour of the molecules, statistical methods are employed and hence it is also called as statistical thermodynamics.

(*b*) **Macroscopic Point of View:** Certain quantity of matter is considered and the events that are taking place at the molecular level are not taken into account.

For example, let us consider a system of an I.C. engine, consisting of a charge in the engine cylinder. At any instant, the system has a certain volume depending upon position of the piston, this volume is easily measurable. Another quantity to describe the system is the pressure of the gas inside the cylinder. A pressure gauge can be used to measure the same. Similarly, temperature, chemical composition etc. may be described. Thus, in macroscopic point of view the system will be described large scale properties. It is also called "Classical Thermodynamics".

2.21 ENERGY

It is the capacity to do work. In other words, a system is said to possess energy when it is capable of doing some work.

The energy possessed by a system is of the following two types:

1. Stored energy 2. Transit energy

1. **Stored Energy:** This is the energy which is contained within the system boundaries. Examples of stored energy are potential energy, kinetic energy and internal energy etc.
2. **Transit Energy:** This is the energy which crosses the system boundaries. Examples of transit energy are heat, work and electric energy.

2.21.1 Potential Energy

It is the energy, possessed by a body by virtue of its height above the earth surface (ground level) *e.g.*, a body raised to some height above the ground level possesses potential energy because it can do some work by falling on earth's surface.

Let W = Weight of the body

m = Mass of the body

z = Distance through which the body falls, and

g = Acceleration due to gravity = 9.81 m/s^2

$\therefore$ Potential energy, P.E. = Wz Nm

or P.E. = mgz Nm $\left[\begin{array}{l} \text{kg} \times \frac{\text{m}}{\text{s}^2} \times \text{m} = \text{N} - \text{m} \\ \because 1\text{N} = \frac{1\text{kg} - \text{m}}{\text{s}^2} \end{array}\right]$

2.21.2 Kinetic Energy

It is the energy possessed by a body by virtue of its mass and velocity of motion.

Let m = Mass of the body, and

v = Velocity of the body

Kinetic Energy,

$$\text{K.E.} = \frac{1}{2} mv^2 Nm \qquad \left[\begin{array}{l} \text{kg} \times \frac{\text{m}}{\text{s}^2} \times \text{Nm} \\ \because 1\text{N} = \frac{1\text{kg} - \text{m}}{\text{s}^2} \end{array}\right]$$

2.21.3 Internal Energy

It is the energy possessed by a body due to its molecular arrangement and motion of the molecules. It is usually represented by U.

2.21.4 Total Energy

It is sum of potential, kinetic and internal energy.

$$E = P.E. + K.E. + U$$

$$= mgz + \frac{1}{2}mv^2 + U$$

2.21.5 Law of Conservation of Energy

It states that the energy can neither be created nor destroyed. This means that the total energy possessed by a body remains constant. Energy can be changed from one form to another form as explained in the following examples.

(*i*) Water flows from a tank through a vertical pipe -Potential energy is converted into Kinetic energy.

(*ii*) Brake is applied in an automobile-Kinetic energy of the wheel is converted into heat energy due to friction.

2.21.6 Law of Conservation of Mass

This law states that the mass can neither be created nor destroyed but can only be changed from one form to another form.

A substance can be changed from solid to liquid and liquid to gas or *vice-versa*.

2.22 HEAT

Heat is defined as this energy transformed without transfer of mass across the boundary of a system due to a temperature difference between the system and surrounding. It is expressed in Joule (J) or kilo joule (kJ) or in calorie for English system.

The amount of heat transferred during a process of work done by the system during the some process is dependent on the path followed during the process means the system from its initial state 1 to final state 2 by the two different paths 1-A-2 and 1-B-2. Thus, the heat and work are said to be path functions and not a point function.

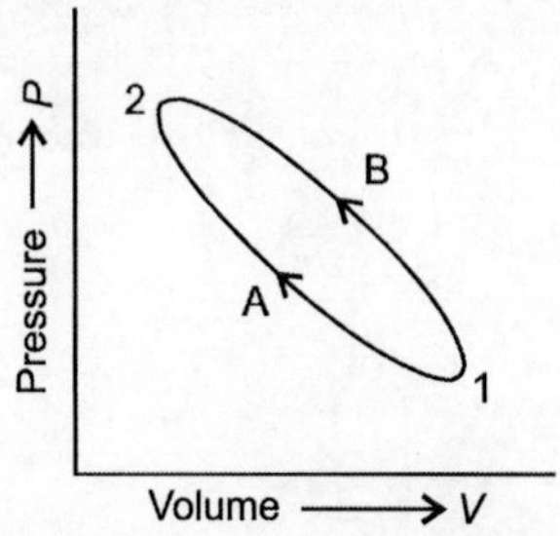

Fig. 2.11 Presure *v* volume

The heat energy is generally transferred to three ways namely conduction, conservation and radiation.

1. **Conduction:** The transfer of energy from hot end to the cold end by molecules activity. Thus the mode of heat transfer without change of position of the molecules is known as conduction.
2. **Convection:** The transfer of heat only occurs in fluids. This mode of heat transfer takes place when hotter molecules colliding with the cooler molecules is known as convection.
3. **Radiation:** The transfer of heat by electromagnetic wave propagation from a hot body to a cold body is known as radiation.

2.23 SPECIFIC HEAT

The specific heat of a substance is defined as "the amount of heat required to raise the temperature of a unit mass of any substance through one degree."

It is generally denoted by c. In S.I., system of units, the unit of specific heat is taken as kJ /kg K. If m kg of a substance of specific heat, c is required to raise the temperature from an initial temperature T_1, to a final temperature of T_2, then

$$\text{Heat required} = mc\ (T_2 - T_i)\ \text{kJ}$$

where T_1 and T_2 may be either in °C or in K.

Since solid and liquid do not change the volume on heating, therefore they have only one specific heat. But the gases have the following two specific heats depending upon the process adopted for heating the gas.

1. Specific heat at constant pressure (c_p)
2. Specific heat at constant volume (c_v)

2.23.1 Specific Heat at Constant Volume

It is the amount of heat required to raise the temperature of a unit mass of gas through one degree when it is heated at a constant volume. It is generally denoted by c_v.

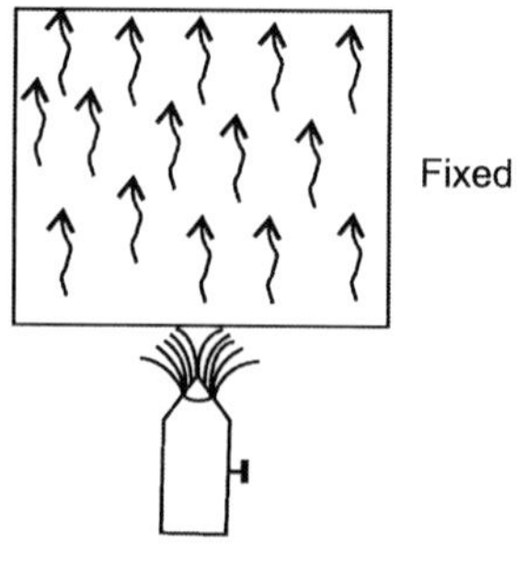

Fig.2.12 Heat being supplied at constant volume

Consider a gas contained in a container with a fixed lid as shown in Fig. 2.12. Now, if this gas is heated, it will increase the temperature and pressure of the gas in the container. Since the lid of the container is fixed, therefore the volume of gas remains unchanged.

Let m = Mass of the gas

T_1 = Initial temperature of the gas and

T_2 = Final temperature of the gas

∴ Total heat supplied to the gas at constant volume

Q = Mass × Specific heat at constant volume × Rise in temperature

$= mc_v(T_2 - T_1)$

It maybe noted that whenever a gas is heated at constant volume, no work is done by the gas.

We know that,

work done, $W = PdV = P(V_2 - V_1)$

where P = Pressure of the gas

dV = Change in volume = $V_2 - V_1$

When there is no change in volume, then

$dV = 0$

Therefore, $W = 0$

2.23.2 Specific Heat at Constant Pressure

It is the amount of heat required to raise the temperature of a unit mass of a gas through one degree, when it is heated at constant pressure. It is generally denoted by c_p.

Consider a gas contained in a container with a movable lid as shown in Fig. 2.13. Now, if this gas is heated, it will increase the temperature and pressure of the gas in the container. Since the lid of the container is movable, therefore it will move upwards, in order to counterbalance the tendency for pressure to rise.

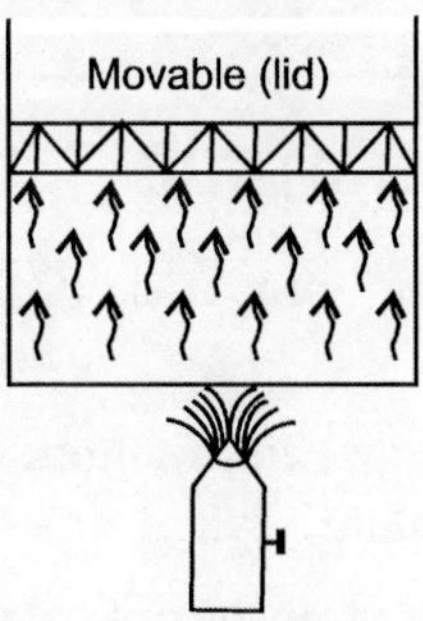

Fig. 2.13 Heat being supplied at constant pressure

Let m = Mass of the gas

T_1 = Initial temperature of the gas

T_2 = Final temperature of the gas

v_1 = Initial volume of the gas

$v_2 =$ Final volume of the gas

∴ Total heat supplied to the gas, at constant pressure

Q = Mass × Specific heat at constant pressure × Rise in temperature

$= m\, c_p(T_2 - T_1)$

2.24 TEMPERATURE

It is an intensive thermodynamic property, which determines the degree of hotness or coolness due to change in level of heat intensity. If the body is at high temperature. It shows high level heat intensity similarly if the body is at low temperature, it shows a low level of heat intensity.

An instrument used to measure the temperature is known as thermometer. Basically three scales are used to measure the temperature of a body.

1. Celsius or centigrade, 2. Fahrenheit scale, 3. Kelvin scale.

1. **Celsius Scale:** The freezing point of water is given zero value and boiling point is 100. It is represented by °C.
2. **Fahrenheit Scale**: The freezing point of water is 32 and boiling point is given 212. It is represented by °F.
3. **Kelvin Scale:** The freezing point of water is 273 and boiling point is given 273. It is represented by °K.

The relationship between Celsius, Fahrenheit and Kelvin scale is given as :

$$\frac{C-0}{100-0} = \frac{F-32}{212-32} = \frac{K-273}{373-273}$$

$$\boxed{\frac{C}{5} = \frac{F-32}{9} = \frac{K-273}{5}}$$

2.25 ABSOLUTE TEMPERATURE

The zero readings on the ordinary centigrade (Celsius) and Fahrenheit scale are chosen arbitarily for simplicity.

The temperature, below which the temperature of any substance cannot fall, is known as absolute zero temperature.

The absolute zero temperature, for all sorts of calculations, is taken as –273°C in case of Celsius scale and –460°C in case of Fahrenheit scale. The temperature measured from this zero are called absolute temperatue. The asbolute temperature in Celsius scale is called degree Kelvin (°K), such that K = °C + 273. Similarly, absolute temperature in Fahrenheit scale is called degree Rankine (°R) such that °R = °F + 460.

2.26 WORK

Work is defined as a force acts upon a body causing a body to move along the direction of force or it is product of force (F) and distance moved (x) in the direction of force

$$W = F \times x$$
$$= \text{Force} \times \text{Distance Moved}$$
$$= \text{Pressure} \times \text{Area of Piston} \times \text{Distance}$$

The unit of work is Nm in S.I. system.

The work of (1Nm) is known as Joule because 1Nm = 1J

Thermodynamics Definition of the Work.

1. *"The energy transformed (without transfer of mass) across the boundary of a system due to the intensive propoerty difference other than temperature. That exits between the system and sourrounding"* Here the pressure difference means mechanical work and intensive property difference mean electrical work.
2. *"Work is energy in transition which flows from a system to the surroundings during a given process if the sole effect external to the system can be reduced to raising of a weight the work done, by the system is considered as positive work. While the work done on the system is considered as negative work."*

   Consider a sysem of battery and motor as shown Fig 2.14. The motor drives the fan in arrangement (*a*) The fan is replaced with the pulley and weight in arangement (*b*) When the motor runs, the weight is raised. The work is crossing the boundary of the system as the sole effect external to the system is for raising of a weight.

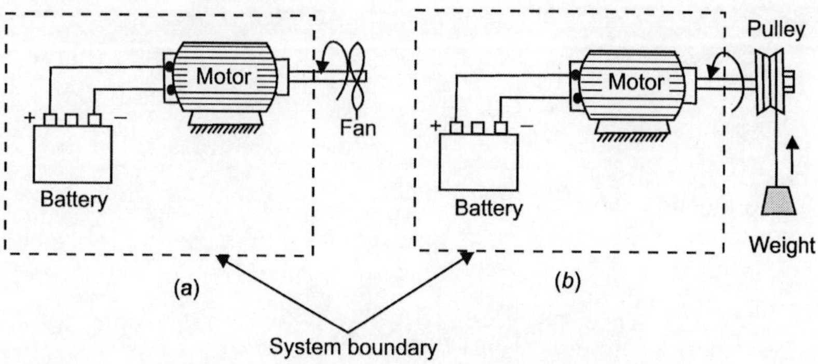

Fig.2.14 Work crossing the boundary of a system

2.27 MOVING BOUNDARY WORK

Work can be done by the movement of the system boundary as the case of moving piston in a cylinder. Consider a system of the gas contained in a cylinder and piston as shown in Fig.2.15.

The total force on the piston is PA. Where P is the pressure of the fluid and A is the area of the piston.

The work done is

$$\delta W = PA\,dl$$

The change in volume

$$dV = A.dl$$

$$\therefore \quad \delta W = P.dV$$

The work done at the moving boundary can be found by integrating the above equation if the relationship between P and V is known during this process.

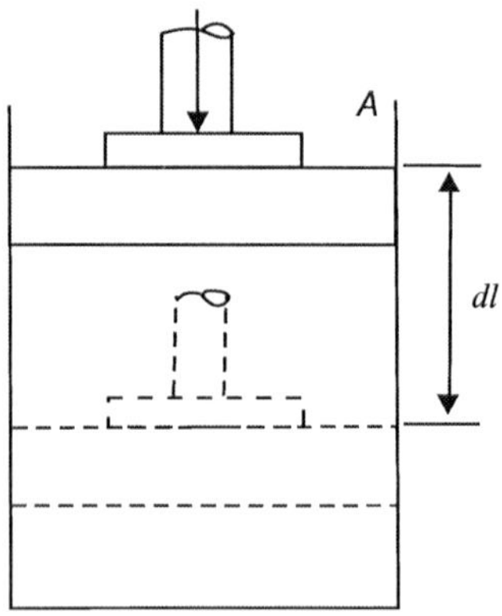

Fig. 2.15

The relationship between P and V in a polytropic process is

$$PV^n = \text{Constant}$$

$$P = \frac{\text{Constant}}{V^n}$$

$$\text{work done} = \int_1^2 PdV = \text{Constant}\int_1^2 \frac{dV}{V^n}$$

$$= \text{Constant}\left|\frac{V^{-n+1}}{-n+1}\right|_1^2$$

$$= \frac{\text{Constant}}{1-n}(V_2^{1-n} - V_1^{1-n})$$

$$= \frac{P_2 V_2^n V_2^{1-n} - P_1 V_1^n V_1^{1-n}}{1-n}$$

$$= \frac{P_2 V_2 - P_1 V_1}{1-n}$$

The equation is valid for any exponent of n, except $n = 1$ when $n = 1$

$$PV = \text{Constant} = P_1V_1 = P_2V_2$$

and $$\int_1^2 PdV = P_1V_1 \int_1^2 \frac{dV}{V}$$

$$= P_1V_1 ln \frac{V_2}{V_1}$$

2.28 PATH AND POINT FUNCTIONS

Path functions have inexact differentials represented by the symbol δ, the magnitude of a path function depends on the path followed during a process as well as on the end states. Therefore, a differential amount of heat or work is represented by δQ or δW, instead of δQ or δW.

The total workdone during the process 1-2 is given by

$$\int_1^2 \delta W = W_{12} \text{ or } W_{1-2}$$

$$\neq W_2 - W_1 \text{ or } \delta W$$

Similarly,

$$\int_1^2 \delta Q = Q_{12} \text{ or } Q_{1-2}$$

$$\neq Q_2 - Q_1 \text{ or } \delta Q$$

Point functions have exact differentials designated by the symbol d.

Properties are point functions, they depend on the state only, and not on how a system reaches that state.

Property changes are also designated by the symbol d, *e.g.*, a differential amount of pressure or temperature is represented by dp or dT.

A small change in volume is represented by dV and the total volume change during a process between state 1 and 2 is given by

$$\int_1^2 dV = V_2 - V_1 = \delta V$$

2.29 SIGN CONVENTIONS FOR HEAT AND WORK

The commonly used sign convention for work

- Work done by a system is +ve (positive)
- Work done on a system is –ve (negagive)

The commonly used sign convention for heat

- Heat transfer to a system is +ve (positive)
- Heat transfer from a system is –ve (negative)

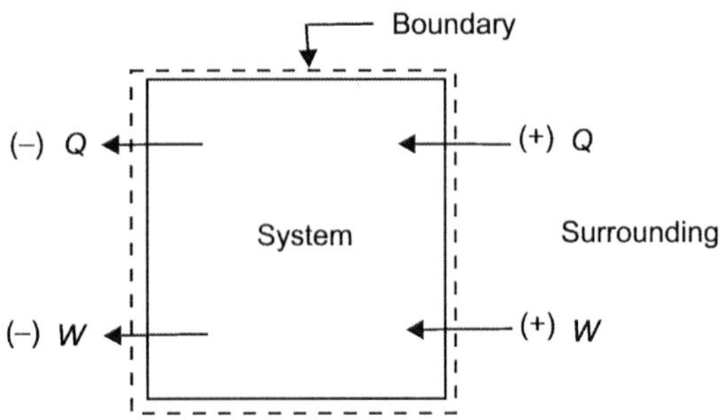

Fig. 2.16 Sign conventions

2.30 LAWS OF PERFECT GASES

There are some laws of perfect gases. These are as follows:

2.30.1 Boyle's Law

"*The absolute pressure of a given mass of a perfect gas varies inversely as its volume, when the temperature remains constant.*"

This law was formulated by Robert Boyle in 1662.

Mathematically,

$$P \propto \frac{1}{V}$$

or $$PV = \text{Constant}.$$

2.30.2 Charles's Law

This law was formulated by Jacques A.C. Charles in 1787.

This law is stated in the two different forms.

(*i*) "*The volume of a given mass of a perfect gas varies directly as its absolute temperature, when the absolute pressure remains constant.*"

Mathematically,

$$V \propto T$$

or $$\frac{V}{T} = \text{Constant}$$

(*ii*) "*All perfect gases change in volume by* $\frac{1}{273}$ *th of its original volume at* 0°C *for every* 1°C *change in temperature, when the pressure remains constant.*"

V_0 = Volume of a given mass of gas at 0°C

V_1 = Volume of the same mass of gas at t°C

Now

$$V_t = V_o + \frac{1}{273} V_0 t$$

$$= V_0 \left[\frac{273+t}{273}\right]$$

$$= V_0 \frac{T}{T_0}$$

or $$\frac{V_1}{T} = \frac{V_0}{T_0}$$

where T = Absolute temperature corresponding to t°C

T_0 = Absolute temperature corresponding to 0°C.

The temperature at which the volume of a gas becomes zero is called absolute zero temperature.

2.30.3 Gay-Lussac Law

"*This law states that the absolute pressure of a given mass of a perfect gas varies directly as its absolute temperature, when the volume remains constant.*"

Mathematically,

$$P \propto T$$

or $$\frac{P}{T} = \text{Constant}$$

2.30.4 General Gas Equation

In actual practice all the three variables *i.e.*, pressure, volume and temperature change simultaneously. So, Boyle's law and Charles's law are combined together which give us a general gas equation.

According to Boyle's Law

$$P \propto \frac{1}{V} \text{ or } V \propto \frac{1}{P} \qquad (T \text{ constant}) \ldots(i)$$

and according to Charles's Law

$$V \propto T \qquad (P \text{ constant}) \ldots (ii)$$

From Equations (*i*) and (*ii*)

$$V \propto \frac{1}{P} \text{ and T both}$$

or $$V \propto \frac{T}{P}$$

$\therefore$ $$PV \times T$$

or $$\frac{PV}{T} = \text{Constant}$$

Useful form of the general gas equation is

$$\frac{P_1V_1}{T_1} = \frac{P_2V_2}{T_2} = \ldots = \text{Constant}$$

2.30.5 Joule's Law

"*This law state that the change of internal energy of a perfect gas is directly proportional to the change of temperature.*"

Mathematically,

$$dU \propto dT$$

or $$dU = mc\ dT = mc\ (T_2 - T_1)$$

where m = Mass of gas

c = constant, known as specific heat.

2.30.6 Characteristic Equation of a Gas

Let a given mass '*m*' of a perfect gas expands from its initial condition (1) to the final condition (2).

Let the first part of expansion *i.e.*, from 1 to 1'occurs at constant temperature and the second part *i.e.*, 1' to 2 occurs at constant pressure, as shown in Fig. 2.17.

Now, for the first part of expansion, at constant temperature, applying Boyle's Law

$$P_1V_1 = P_2V_1' \qquad (\because P_1' = P_2) \quad \cdots (i)$$

for the second part of expansion, at constant pressure, *i.e.* P_1' or P_2 applying Charle's Law

$$\frac{V_1'}{T_1} = \frac{V_2}{T_2} \qquad (\because T_1' = T_1) \quad \ldots (ii)$$

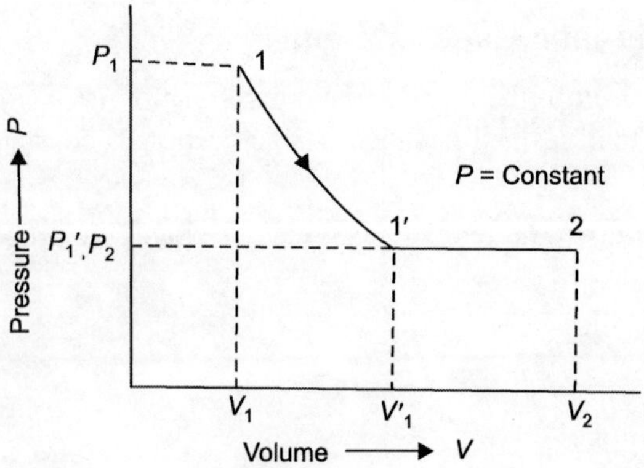

Fig. 2.17

$$\Rightarrow \qquad V_1' = \frac{V_2 \times T_1}{T_2}$$

Substituting this value of V_1' in Equation (*i*) We get

$$P_1V_1 = P_2 \times \frac{V_2}{T_2} \times T_1$$

or $$\frac{P_1V_1}{T_1} = \frac{P_2V_2}{T_2}$$

or $$\frac{PV}{T} = \text{Constant}$$

or $$\frac{PV}{T} = R \qquad (\text{'}R\text{' is a constant})$$

or $$PV = RT \qquad \ldots (iii)$$

Equation (*iii*) is called the equation of a perfect gas or the characteristic equation of a gas.

If m = mass of the given gas

Then, the equation (*iii*) becomes

$$PV = mRT \quad ...(iv)$$

where

P = Absolute pressure of the given gas in N/m^2

V = Volume of this gas in m^3

m = Mass of gas in kg

R = Characteristic gas constant in Nm/kg °K or J/kg °K

T = Absolute temperature of gas in °K

2.30.7 Characteristic Gas Constant

The constant 'R' used in the characteristic gas equation $PV = RT$, has a constant value for a particular gas and is called characteristic gas constant or specific gas constant.

Its value depends upon the temperature scale used and the properties of the gas, under consideration.

From the equation

$$PV = mRT$$

the units of 'R' will be

$$R = \frac{PV}{mT} = \frac{N}{m^2} \times \frac{m^2}{kg \times K} = Nm/\text{kg}\,^\circ\text{K}$$

$$= \text{J/kg K} \qquad (\because 1 \text{ Nm} = 1 \text{ Joule})$$

If the unit of pressure 'P' are bar and the volume V, m^3 then equation becomes

$$10^5 PV = mRT$$

for atmospheric air,

$$R = 287\text{J/kg °K}$$

2.30.8 Universal Gas Constant or Molar Constant

The universal gas constant or molar constant of a gas is the product of the gas constant and the molecular mass of the gas. It is denoted by R_u

Mathematically,

$$R_u = MR$$

where M = Molecular mass of the gas expressed in Kg-Mole

R = Gas constant

The value of R_u is same for all gases

In S.I. units, value of R_u is 8314 J/kg mol K

or 8.314 KJ/kg mol K

2.30.9 Regnault's Law

This law states that "the two specific heats of a gas (*i.e.*, specific heat at constant pressure, C_p and specific heat at constant volume, C_v) do not change with the change in pressure and temperature of the gas."

2.30.10 Avogadro's Law

This law states that "equal volumes of all gases, at the same temperature and pressure, contain equal number of molecules."

Consider two different ideal gases having the same volume, same pressure, and same temperature and their molecular masses be denoted by m_1 and m_2 respectively.

Now according to the Avagadro law, each gas will contain the same no. of molecules, say n and so masses of the gasess would be:

$$m_1 \propto nm_1 = Knm_1$$

$$m_2 \propto nm_2 = Kn\, m_2 \qquad \ldots(i)$$

where K is constant of proportionality.

P_1 and P_2 are the densities of the gases then

$$m_{12} P_1 V, m_2 = P_2 V$$

where $V_1 = V_2 = V$, volume of each gas

By putting the values of masses in equation (i)

$$\frac{P_1 V}{P_2 V} = \frac{Knm_1}{Knm_2} : \frac{P_1}{P_2} = \frac{m_1}{m_2}$$

i.e., the densities of the gases are directly proportional to their molecular masses at the same temperature and pressure of

$$P = \frac{1}{V}$$

then $$\frac{V_2}{V_1} = \frac{m_1}{m_2},\ V_1 m_1 = V_2 m_2 \qquad Vm_2 \text{ constant.}$$

Hence, the specific volume of the gases are inversely proportional to their molecular masses at the same temperature and pressure.

2.31 ZEROTH LAW OF THERMODYNAMICS

Zeroth law states that when the two bodies are in thermal equilibrium with the third body individually, then these two will be in thermal equilibrium with each other.

Consider three bodies A, B, C as shown out when *A* and *B* are in thermal equilibrium with *C*. Then the two bodies *A* and *B* also be in thermal equilibrium with each other.

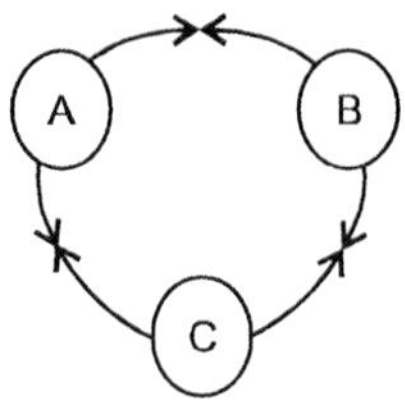

Fig.2.18 *A, B, C,* in thermal equilibrium

2.32 FIRST LAW OF THERMODYNAMICS

(*a*) **"Heat and mechanical work are mutually convertible:** It states that when closed system undergoes a cyclic process then net heat transfar is equal to net work transfer *i.e.*, the cyclic integral of heat transfer is equal to cyclic integral of work transfer."

Mathmatically.

$$\oint \delta Q = \oint \delta \omega$$

where $\oint$ = Cyclic integral

δQ = Amount of heat transfer

$\delta\omega$ = Amount of work transfer

(*b*) **Energy Conservation Law:** It state that "energy can neither be created nor destroyed but it can be converted from one form to another". When a system undergoes a change of state or energy exchange takes place (*i.e.*, heat and work transfer) during any process (not cyclic) then whole of the heat supplied is not converted in to work but some part of it is stored by the system itself called total energy of the system. It is denoted by dE.

Mathematically,

$$\delta Q - \delta W = dE$$

The symbol δ is used for a quantity which is inexact differential and d is used for a quantity which is an exact differential.

The quantity E is an extensive property and represents the total energy of the system at a particular state.

On integrating the above expression for a change of state from 1 to 2.

We have

$$\int_1^2 \delta Q - \int_1^2 \delta W = \int_1^2 \delta E$$

$$Q_{1-2} - W_{1-2} = E_2 - E_1 \quad \text{... (i) (Q, W and E are in same units)}$$

For a unit mass

$$Q_{1-2} - W_{1-2} = E_2 - E_1$$

where

Q_{1-2} = Heat transfer to the system during the process from state 1 to state 2.

W_{1-2} = Work done by the system on the surroundings during the process, and

E_1 = Total energy of the system at state 1

$$= PE_1 + KE_1 \times U_1$$

$$= mgz_1 + \frac{mV_1^2}{2} + U_1$$

where

PE is Potential Energy, KE is Kinetic energy and U is internal energy

E_2 = Total energy of the system at state 2

$$= PE_2 + KE_2 + U_2$$

$$= mgz_2 + \frac{mV_2^2}{2} + U_2$$

Thus, the above expression *i.e.*, Equation (*i*) may be written as

$$Q_{1-1} - W_{1-2} = (PE_2 + KE_2 + U_2) - (PE_1 + KE_1 + U_1)$$

$$= (PE_2 - PE_1) - (KE_2 - KE_1) + (U_2 - U_1)$$

$$= (mgz_2 - mgz_1) + \frac{mv_2^2}{2} - \frac{mv_1^2}{2} + (U_2 - U_1)$$

$$= mg(z_2 - z_1)\, m\left(\frac{v_2^2}{2} - \frac{v_1^2}{2}\right) + (U_2 - U_1)$$

For unit mass, the expression is written as

$$q_{1-2} - w_{1-2} = (gz_2 - gz_1) + \left(\frac{v_2^{\,2}}{2} - \frac{v_1^{\,2}}{2}\right) + (U_2 - U_1)$$

When there is no change in potential energy of the system (*i.e.*, when the height of the system from the datum level is same), then $PE_1 = PE_2$. Thus, equation (*ii*) becomes

$$Q_{1-2} - W_{1-2} = (KE_2 - KE_1) = (U_2 - U_1) \qquad (iii)$$

when there is no change of PE and also there is no flow of the mass into or out of the system, then $PE_1 = PE_2$ and $KE_1 = KE_2$. Thus, equation (*ii*) becomes

$$Q_{1-2} - W_{1-2} = U_2 - U_1 = dU \qquad (iv)$$

For a closed or non-flow thermodynamic system,

$$PE = 0 \text{ and } KE = 0$$

Thus, equation (*iv*) is known as non-flow energy equation.

For an islolated system for which $Q_{1-2} = W_{1-2} = 0$ the equation (*i*) becomes $E_2 = E_1$.

This shows that the first law of thermodynamic is the law of conservation of energy.

Limitations of the First Law of Thermodynamics:

(*i*) It is unable to tell direction of flow of heat.

(*ii*) It does not explain the physical and chemical processes occurring in a nutural direction.

(*iii*) It does not tell about the feasibility and direction of a process.

(*iv*) It is unable to tell about the extent of change of one form of energy into another.

2.33 APPLICATION OF FIRST LAW OF THERMODYNAMICS TO A CLOSED SYSTEM (NON-FLOW PROCESSES) INVOLVING PERFECT GASES

From first law of thermodynamics, equation for a closed system is given by

$$Q_{1-2} - W_{1-2} = E_2 - E_1 = dE$$

where $$dE = dU + dKE + dPE$$

If dKE and dPE are negligible, then

$$Q_{1-2} - W_{1-2} = dU = U_2 - U_1$$

The analysis for first law of thermodynamics, for closed system (*i.e.*, non-flow system) which changes from one state to another and to develop some relations for thermodynamic are as given ahead:

(*i*) Constant volume (Isochoric) process

(*ii*) Constant pressure (Isobaric) process

(*iii*) Constant temperature (Isothermal) Process

(*iv*) Adiabatic (Isoentropic) process

(*v*) Polytropic process.

2.33.1 Constant Volume (Isochoric) Process

The process in which the volume of the given mass of the gas is not allowed to change, is known as "constant volume process".

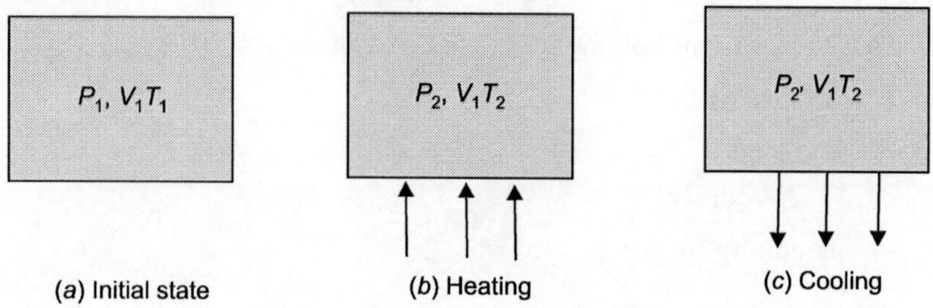

Fig. 2.19 Constant volume process

If gas is continued in a closed cylinder, its volume cannot change when heated or cooled. The temperature of the gas increases with the addition of heat and the absolute pressure increases in direct proportion to the increase in absolute temperature. If cooled, the absolute pressure of the gas will decrease in direct proportion to the decrease in absolute temperature as shown in Fig. 2.19.

When any process is taking place between initial (P_1, V_1, T_1) and final (P_2, V_2, T_2) conditions represented by 1 and 2, then general expressions for some quantities are as given for different processes for m kg of a gas.

Constant Volume Process: The process is given by $\frac{P}{T}$ = constant. It is governed by Gay Lussac law drawn on $P - V$, $P - T$ and $T - S$ as shown

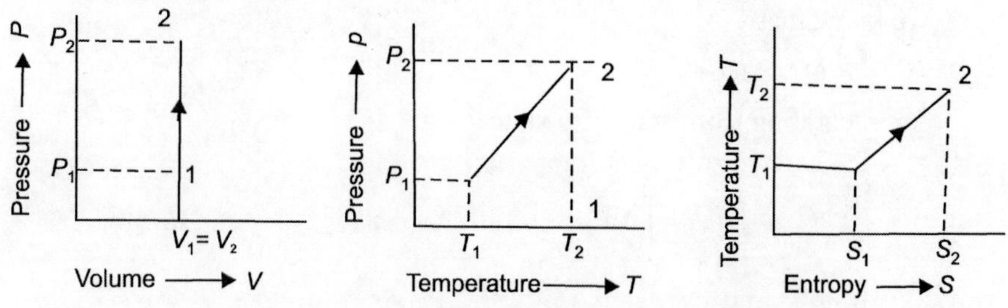

Fig. 2.20 Pressure, volume, temperature, relationship

(*i*) Pressure – Volume – Temperature ($P - V - T$) Relationship

We known that the general gas equation is

$$\frac{P_1V_1}{T_1} = \frac{P_2V_2}{T_2}$$

Since the gas is heated at constant volume, therefore $V_1 = V_2$

$$\therefore \quad \frac{P_1}{T_1} = \frac{P_2}{T_2}$$

$$\text{or} \quad \frac{P}{T} = \text{constant}$$

Thus the constant volume process is governed by Gay Lussac Law.

(*ii*) **Work done by the Gas**

We know that

$$\delta W = PdV$$

On integrating from state 1 to state 2

$$\int_1^2 \delta W = \int_1^2 PdV = P\int_1^2 dV$$

$$\text{or} \quad W_{1-2} = P(V_2 - V_2) = 0 \qquad (\because V_1 = V_2)$$

(*iii*) **Change in Internal Energy**

We know that change in internal energy

$$dU = mc_v\, dT \qquad \text{(from Joule's Law)}$$

On integrating from state 1 to state 2

$$\int_1^2 dU = mc_v \int_1^2 dT$$

$$\text{or} \quad U_2 - U_1 = m\,c_v\,(T_2 - T_1)$$

(*iv*) **Heat Supplied or Heat Transfer**

We know that

$$\delta Q = dU + \delta W$$

On integrating from state 1 to state 2

$$\int_1^2 \delta Q = \int_1^2 dU + \int_1^2 \delta W$$

$$\text{or} \quad Q_{1-2} = U_2 - U_1 + W_{1-2}$$

since $W_{1-2} = 0$

$\because$ Heat supplied

$$Q_{1-2} = U_2 - U_1 = mc_v (T_2 - T_1)$$

This shows that all the heat supplid to the gas is utilised in increasing the internal energy of the gas.

(*v*) **Change in Enthalpy**

We know that the change in enthalpy

$$dH = dU + d(PV)$$

On integrating from state 1 to state 2

$$\int_1^2 dH = \int_1^2 dU + \int_1^2 d(PV)$$

or $$H_2 - H_1 = U_2 - U_1 + (P_2V_2 - P_1V_1)$$ ($\because P_1V_1 = mRT_1$ and $P_2V_2 = mRT_2$) ($U_2 - U_1 = mc_v (T_2 - T_1)$)

$$= mc_v (T_2 - T_1) + mR (T_2 - T_1)$$

$$= m(T_2 - T_1)(c_v + R)$$

$$= mc_v (T_2 - T_1)$$

($\therefore c_p - c_v = R$)

Results:

Workdone $W_{1-2} = \int_1^2 PdV = P(V_2 - V_1) = 0$

Change in internal energy: $\delta U = mC_v (T_2 - T_1)$

(C_v specific heat at constant volume)

Heat supplied: $Q_{1-2} = W_{1-2} + (U_2 - U_1)$

Change in enthalpy: $\Delta H = C_p (T_2 - T_1)$

(C_p-specific heat at constant pressure)

Change in Entropy: $\delta S = 2.3 \text{ In } C_v \log \left(\frac{P_2}{P_1}\right)$

2.33.2 Constant Pressure (Isobaric) Process

The process in which the pressure of the given mass of a gas is not allowed to change is known as constant pressure process.

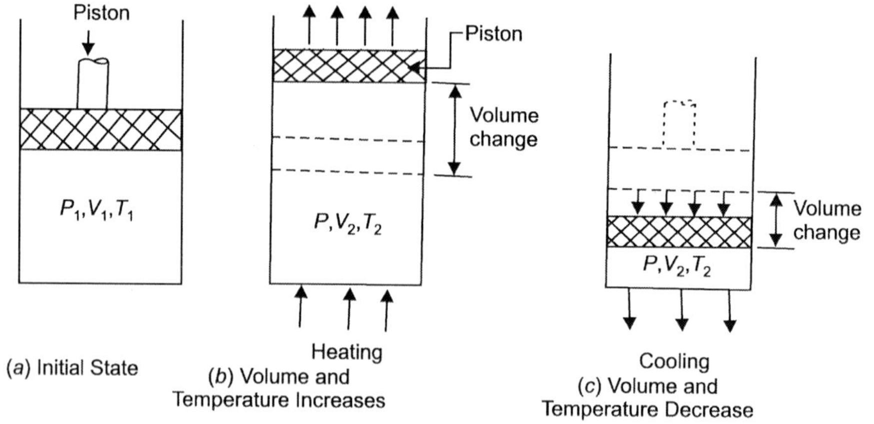

Fig.2.21 Constant Pressure Process

If heat is supplied to a gas under constant process, the volume of the gas will increase in direct proportion to the change in the absolute temperature of the gas and *vice versa* as shown in Fig. 2.21.

Since there is a change in its volume, therefore the heat supplied to the gas is utilised to increase the internal energy of the gas and for doing some external work. This process is governed by Charles law according to law.

The constant pressure process $\frac{P}{V}$ = constant can be represented on $P - V$, $P - T$, $T - S$ diagram as shown in Fig. 2.22.

It is governed by Charles Law.

Let us consider mass of the gas as m kg.

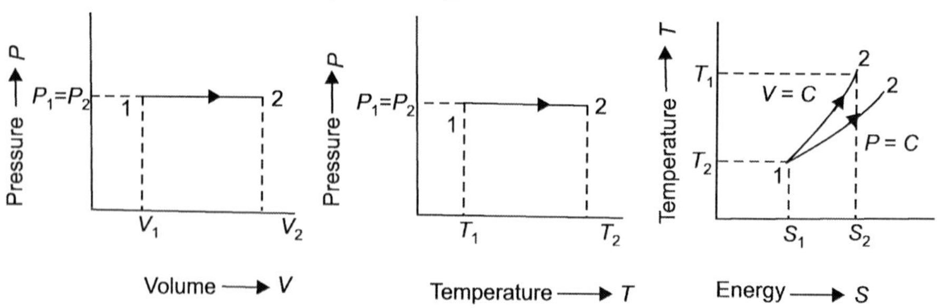

Fig.2.22 *PV, PT, TS* diagrams

(*i*) **Pressure – Volume – Temperature ($p - V - T$) Relationship**

We known that the general gas equation is

$$\frac{P_1V_1}{T_1} = \frac{P_2V_2}{T_2}$$

Since the gas is heated at constant pressure, therefore $P_1 = P_2$

$$\therefore \quad \frac{V_1}{T_1} = \frac{V_2}{T_2}$$

or $$\frac{V}{T} = \text{constant}$$

Thus, the constant pressure process is governed by Charle's Law.

(*ii*) **Work done by the Gas**

We know that

$$\delta W = PdV$$

On integrating from state 1 to state 2

$$\int_1^2 \delta W = \int_1^2 PdV = P\int_1^2 dV$$

or $$W_{1-2} = P(V_2 - V_1) = (PV_2 - PV_1)$$

or $$W_{1-2} = mR\,(T_2 - T_1) \qquad \left[\begin{matrix}(\because PV_1 = mRT_1, \\ PV_2 = mRT_2)\end{matrix}\right]$$

(*iii*) **Change in Internal Energy**

Change ininternal energy (dU) is same for all the processes. So, change in internal energy,

$$dU = U_2 - U_1 = mc_v\,(T_2 - T_1)$$

(*iv*) **Heat Supplied or Heat Transfer**

We know that

$$\delta Q = dU + \delta W$$

On integrating from state 1 to state 2

$$\int_1^2 \delta Q = \int_1^2 dU + \int_1^2 \delta W$$

or $$Q_{1-2} = U_2 - U_1 + W_{1-2}$$

since $$= mc_v\,(T_2 - T_1) + mR\,(T_2 - T_1) \qquad ...(i)$$

$$= m(T_2 - T_1)\,(c_v + R)$$

$$= mc_p\,(T_2 - T_1) \qquad (\because c_p - c_v = R)$$

Equation (*i*) shows that the heat supplied to the gas is utilised in increasing the internal energy of the gas and for doing some external work.

(*v*) **Change in Enthalpy**

Change in enthalpy is same for all the processes. So,

Change in enthalpy, $dH = H_2 - H_1 = m\, C_p\, (T_2 - T_1)$

Change in enthalpy is equal to the heat supplied or heat transferred.

Results:

Workdone $W_{1-2} = mR\,(T_2 - T_1)$

Change in internal energy

$$\delta U = mC_v\,(T_2 - T_1)$$

Heat supplied $Q_{1-2} = mC_p\,(T_2 - T_1)$

Change in enthalpy: $\delta H = mC_p\,(T_2 - T_1)$

Change in entropy $\Delta S = 2.3\; C_p\, \log\left(\frac{V_2}{V_1}\right)$

2.33.3 Constant Temperature (Isothermal) Process

The process in which, the temperature of the given mass of a gas is not allowed to change, is known as constant temperature process or isothermal process. If the volume of a gas is increased or decreased with constant temperature, the absolute pressure will vary inversely with the volume as shown in Fig.2.23.

In isothermal process, there is no changes in temperature, internal energy and in enthalpy. This process is govenred by Boyle's law. Let us consider a mass of a gas as *m* kg.

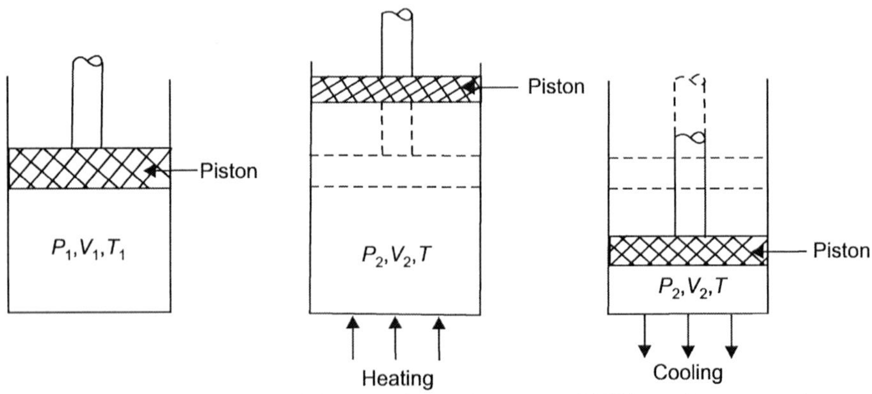

(*a*) Initial Condition (*b*) Volume increasing and pressure decrease. (*c*) Volume decrease and pressure increase

Fig. 2.23 Constant temperature process

2. The constant temperature process pV = constant. It is represented on $P - V$, $P - T$ and $T - S$ diagrams and shown in Fig. 2.24

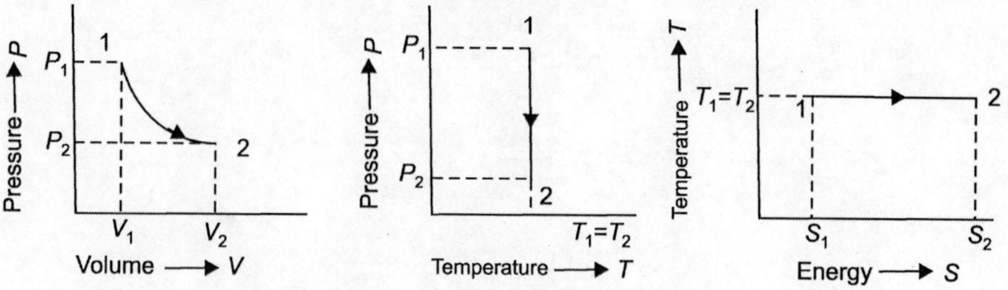

Fig.2.24 *PV*, *PT* and *TS* diagrams

(i) Pressure – Volume – Temperature (P – V – T) Relationship

We know that the general gas equation is

$$\frac{P_1V_1}{T_1} = \frac{P_2V_2}{T_2}$$

since the gas is heated at constant volume, therefore $T_1 = T_2$

$$\therefore \quad P_1V_1 = P_2V_2$$

or $\quad PV =$ constant

Thus, the constant temperature process or isothermal process is governed by Boyle's law.

(ii) Work done by the Gas

We know that

$$\delta W = PdV$$

On integrating from state 1 to state 2

$$\int_1^2 \delta W = \int_1^2 PdV$$

or $$W_{1-2} = \int_1^2 PdV$$

Since the expansion of the gas is isothermal

i.e. $\quad PV = C$

therefore, $\quad PV = P_1V_1$

or $$P = \frac{P_1V_1}{V}$$

Substituting this value of P in equation (*i*), we have

$$W_{1-2} = \int_1^2 \frac{P_1V_1}{V} dV$$

$$= P_1V_1 \int_1^2 \frac{dV}{V}$$

$$= P_1V_1 \left[\log_e V_1\right]_1^2$$

$$= P_1V_1 \log_e \frac{V_2}{V_1}$$

The above equation may be expressed in terms of corresponding logarithm to the base 10 *i.e.*

$$W_{1-2} = 2.3\ P_1V_1 \log \left(\frac{V_2}{V_1}\right)$$

$$= 2.3\ P_1V_2 \log r$$

(where $r = \frac{V_2}{V_1}$, Expansion ratio)

or $$W_{1-2} = 2.3\ mRT \log \left(\frac{V_2}{V_1}\right)$$

$$= 2.3\ mRT \log r \qquad (\because P_1V_1 = P_2V_2 = mRT)$$

(*iii*) **Change in Internal Energy**

We know that change in internal energy

$$dU = U_2 - U_1 = mc_v (T_1 - T_1)$$

Since it is a constant temperature process *i.e.* $T_1 = T_2$

Therefore, $$dU = U_2 - U_1 = 0$$

or $$U_1 = U_2$$

(*iv*) **Heat Supplied or Heat Transfer**

We know that heat supplied from state 1 to state 2

$$Q_{1-2} = dU + W_{1-2}$$

$$= W_{1-2} \qquad (\because dU = 0)$$

This shows that all the heat supplid to the gas is equal to the work done by the gas.

(*v*) **Change in Enthalpy**

We know that the change in enthalpy

$$dH = H_2 - H_1$$
$$= mC_p (T_2 - T_1)$$

Since it is a constant temperature process *i.e.* $T_1 = T_2$

Therefore, $dH = H_2 - H_1 = 0$

or $H_1 = H_2$

Results:

Workdone $W_{1-2} = 2.3\, P_1V_1 \log\left(\frac{V_2}{V_1}\right)$ $(\because P_1V_1 = P_2V_2 = PV)$

Change in internal energy: $\delta U = mC_v (T_2 - T_1) = mC_v \times 0 = 0$

Heat supplied $Q_{1-2} = \delta U + W_{1-2} = 0 + W_{1-2} = W_{1-2}$

Change in enthalpy : $\delta H = mC_p (T_2 - T_1) = 0$

Change in Entropy: $\delta S = 2.3 \text{ in } m(C_p - C_v) \log\left(\frac{P_1}{P_2}\right)$

2.33.4 Adiabatic (Isentropic) Process

The process in which no heat is added or withdrawn from the system under consideration is known as an adiabatic process.

In an adiabatic process, the gas changes its condition without the transfer of heat to or from the surroundings.

Pressure, volume and temperature of the gas vary during an adiabatic process.

In adiabatic or isentropic process.

1. No heat leaves or enters the gas.
2. The temperature of the gas changes as the work is done at the cost of internal energy.
3. The change in internal energy is equal to the work done.

When there is no heat exchange with the system, the process is called adiabatic. Thus, by 1st law

$$\delta Q = \delta W + \delta U$$
$$\delta W = -\delta U$$
$$PdV = -m\, C_v\, dT$$

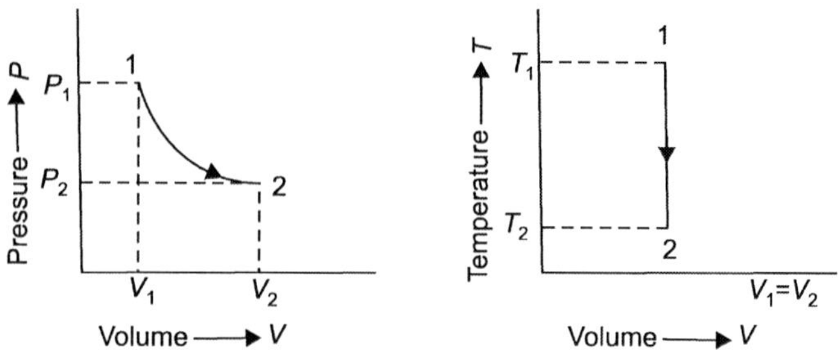

Fig. 2.25 *P, V, T* diagrams

Let m = mass of the gas

$P_1 V_1$ and T_1 = Pressure, Volume and Temperature at the initial state 1, and

$P_2\ V_2$ and T_2 = Pressure, Volume and Temperature at the final state 2.

Following relation derived from this process:

1. Pressure – Volume – Temperature (*P* – *V* – *T*) Relationship

We know that the general gas equation is

$$\delta Q = \delta dW + dU \qquad \ldots(i) \text{ (First Law of Thermodynamics)}$$

Since in an adiabatic process, no heat transfer takes place, therefore

$$\therefore \quad \delta Q = 0$$

$$\text{or} \quad \delta W + dU = 0$$

$$\text{or} \quad pdV + mc_v dT = 0$$

$$\therefore \quad dT = \frac{-PdV}{mC_v}$$

We know that

$$PV = mRT$$

Differentiating this expression, we get

$$PdV + VdP = mR\,dT$$

$$\therefore \quad dT = \frac{PdV + Vdp}{mR}$$

$$= \frac{PdV + VdP}{m(c_p - c_v)} \qquad (\because R = c_p - c_v) \ldots(iii)$$

Equating equation (*ii*), and (*iii*)

$$\frac{-PdV}{mc_v} = \frac{PdV + VdP}{m(c_p - c_v)}$$

$$\frac{c_p - c_v}{c_v} = \frac{PdV + VdP}{-PdV} = -1 - \frac{VdP}{PdV}$$

$$\left(\frac{c_p}{c_v} - 1\right) = -1 - \left(\frac{V}{dV} \times \frac{dP}{p}\right)$$

$$\gamma = -\left(\frac{V}{dV} \times \frac{dP}{P}\right) \qquad \left(\because \frac{c_p}{c_v} = \gamma\right)$$

$$\therefore \qquad \gamma \times \frac{dV}{V} = -\frac{dP}{P}$$

$$\gamma \times \frac{dV}{V} + \frac{dP}{P} = 0$$

Integrating both sides

$$\gamma \log_e V + \log_e P = \text{constant}$$

or $\log_e PV^\gamma = \log_e C$

$\therefore$ $PV^\gamma = C$

or $P_1V_1^{\gamma} = P_2V_2^{\gamma} \ldots = C$ ...(*iv*)

The equation (*iv*) may also be expressed in the following term,

$$\frac{P_2}{P_1} = \frac{P_2V_2}{T_2}\left(\frac{V_1}{V_2}\right)^\gamma \qquad \ldots(v)$$

From the general gas equation, we know that

$$\frac{P_1V_1}{T_1} = \frac{P_2V_2}{T_2}$$

or
$$\frac{P_2}{P_1} = \frac{T_2 \times V_1}{V_2 \times T_1} \qquad \ldots(vi)$$

Equating equation (*v*) and (*vi*)

$$\left(\frac{V_1}{V_2}\right)^\gamma = \frac{T_2 \times V_1}{V_2 \times T_1}$$

or
$$\frac{T_2}{T_1} = \left(\frac{V_1}{V_2}\right)^\gamma \times \left(\frac{V_1}{V_2}\right)$$

or $$\frac{T_2}{T_1} = \left(\frac{V_1}{V_2}\right)^{\gamma} \times \left(\frac{V_1}{V_2}\right)^{-1}$$

or $$\frac{T_2}{T_1} = \left(\frac{V_1}{V_2}\right)^{\gamma-1} \qquad \ldots(vii)$$

From equation (*iv*), we also know that

$$\left(\frac{V_1}{V_2}\right) = \left(\frac{P_2}{P_1}\right)^{1/\gamma} \qquad \ldots(viii)$$

From the general gas equation, we know that

$$\frac{P_1V_1}{T_1} = \frac{P_2V_2}{T_2}$$

or $$\frac{P_1V_1}{T_1} = \frac{P_2}{T_2} \times \frac{T_1}{P_1} \qquad \ldots(ix)$$

Equating equations (*viii*) and (*ix*)

$$\left(\frac{P_2}{P_1}\right)^{1/\gamma} = \frac{P_2 \times T_1}{T_2 \times P_1}$$

or $$\frac{T_2}{T_1} = \frac{P_2}{P_1} \times \left(\frac{P_1}{P_2}\right)^{1/\gamma}$$

or $$\frac{T_2}{T_1} = \left(\frac{P_2}{P_1}\right)\left(\frac{P_2}{P_1}\right)^{-1/\gamma}$$

or $$\frac{T_2}{T_1} = \left(\frac{P_2}{P_1}\right)^{1-1/\gamma} = \left(\frac{P_2}{P_1}\right)^{\frac{\gamma-1}{\gamma}}$$

2. Workdone during Adiabatic Expansion

We know that

$$\delta W = PdV$$

On integrating from state 1 to state 2

$$\int_1^2 \delta W = \int_1^2 PdV$$

or $$W_{1-2} = \int_1^2 PdV \qquad ...(xi)$$

Since adiabatic expansion of the gas follows the law

i.e. $$PV^\gamma = P_1V_1^\gamma$$

therefore, $$P = \frac{P_1V_1^\gamma}{V^\gamma}$$

Substituting this value of p in equation (*xi*), we have

$$W_{1-2} = \int_1^2 \frac{P_1V_1^\gamma}{V^\gamma} dV$$

or $$W_{1-2} = P_1V_1^\gamma \int_1^2 V^{-\gamma} dV$$

or $$W_{1-2} = P_1V_1^\gamma \left[\frac{V^{-\gamma+1}}{-\gamma+1}\right]_1^2$$

or $$W_{1-2} = \frac{P_1V_1^\gamma}{1-\gamma}\left[V_2^{1-\gamma} - V_1^{1-\gamma}\right]$$

or $$W_{1-2} = \frac{P_1V_1^\gamma V_2^{1-\gamma} - P_1V_1^\gamma V_1^{1-\gamma}}{1-\gamma} \qquad (\because P_1V_1^\gamma = P_2V_2^\gamma)$$

or $$W_{1-2} = \frac{P_2(V_2^\gamma V_2^{1-\gamma} - P_1(V_1^\gamma V_1^{1-\gamma})}{1-\gamma}$$

or $$W_{1-2} = \frac{P_2V_2 - P_1V_1}{1-\gamma}$$

or $$W_{1-2} = \frac{P_1V_1 - P_2V_2}{1-\gamma}$$ For expansion

or $$W_{1-2} = \frac{P_2V_2 - P_1V_1}{1-\gamma}$$ For compression

or $$W_{1-2} = \frac{mR(T_1 - T_2)}{\gamma - 1}$$ For expression $$\left(\begin{array}{l}\because P_1V_1 = mRT_1 \\ P_2V_2 = mRT_2\end{array}\right)$$

or $$W_{1-2} = \frac{mR(T_2 - T_1)}{\gamma - 1}$$ For compression

3. Change in Internal Energy

We know that change in internal energy

$$dU = U_2 - U_1 = mC_v (T_2 - T_1)$$

4. Heat Supplied or Heat Transfer

We know that heat supplied or heat transferred in case of adiabatic process is zero, therefore

$$Q_{1-2} = 0$$

5. Change in Enthalpy

We know that the change in enthalpy

$$dH = H_2 - H_1 = mC_p (T_2 - T_1)$$

Results:

Workdone $$W_{1-2} = \frac{P_2V_2 - P_1V_1}{\gamma - 1}$$

Change in internal energy: $\delta U = mC_v (T_2 - T_1)$

Heat supplied $Q_{1-2} = 0$

Change in enthalpy: $\delta H = mC_p (T_2 - T_1)$

Change in Entropy: $\delta S = 0$.

2.33.5 Polytropic Process

The process in which the relation pV^n = Constant is known as polytropic process. where p is the pressure, V volume of the gas and 'n' is the polytropic exponent, the value of which varies from zero to infinity for various gases as shown in Fig. 2.26.

An expansion process in which the energy to do the, work is supplied partly from an external source and partly from the gas itself is known as polytropic process that will follow a path which will fall in between those of the isothermal and adiabatic process as shown in Fig. 2.26. The various equations for polytropic process may be expressed by changing the index n for Y in the adiabatic process.

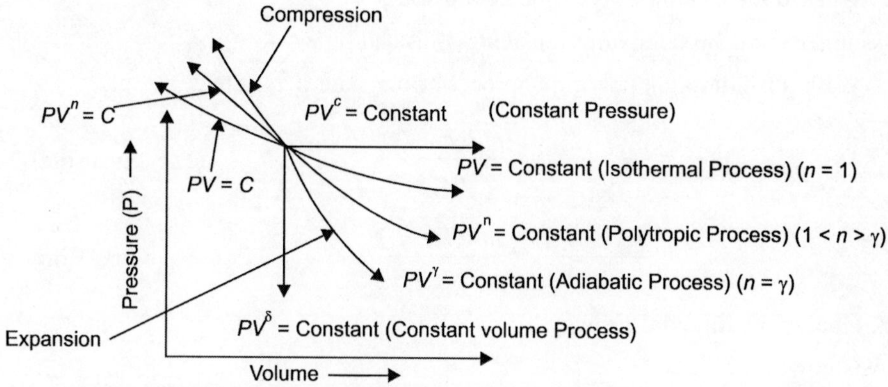

Fig. 2.26 Polytropic Process

Let us consider a mass 'm' of a gas be heated polytropically from an initial state 1 to a final state 2 as shown in Fig. 2.27.

Let P_1, V_1, T_1 = Pressure, Volume and Temperature at the initial state 1,

P_2, V_2, T_2 = Pressure, Volume and Temperature at the final state 2.

Following relation derived from this process.

1. Pressure – Volume – Temperature ($P – V – T$) Relationship

The following relations for polytropic process are derived in the similar way as in adiabatic process.

(*i*) $P_1V_1^n = P_2V_2^n ... = C$ (*ii*) $\frac{T_2}{T_1} = \left(\frac{V_1}{V_2}\right)^{n-1}$

(*iii*) $\frac{V_1}{V_2} = \left(\frac{P_2}{P_1}\right)^{1/n}$ (*iv*) $\frac{T_2}{T_1} = \left(\frac{P_2}{P_1}\right)^{\frac{n-1}{n}}$

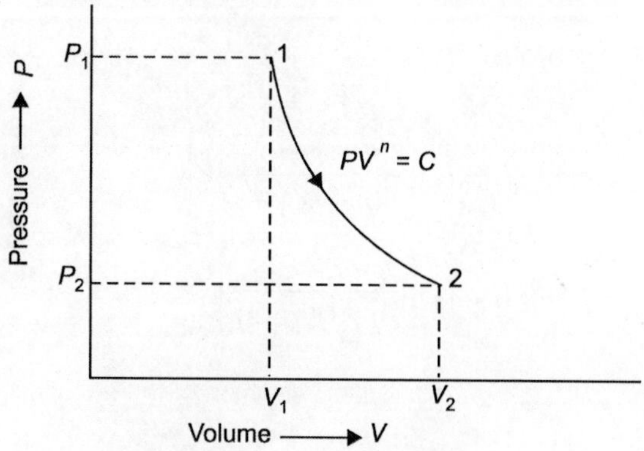

Fig. 2.27 Pressure and volume

2. Work done during Polytropic expansion

As in adiabatic process, after replacing n instead of γ

$\therefore$ Work done during a polytropic process from state 1 to state 2

$$W_{1-2} = \frac{P_1V_1 - P_2V_2}{n-1} = \frac{mR(T_1 - T_2)}{n-1} \quad \text{(For expansion)}$$

$$= \frac{P_2V_2 - P_2V_1}{n-1} = \frac{mR(T_2 - T_1)}{n-1} \quad \text{(For compression)}$$

3. Change in Internal Energy,

We know

$$dU = U_2 - U_1 = mC_v (T_2 - T_1)$$

4. Heat supplied or Heat transferred

We know that

Heat supplied,

$$Q_{1-2} = W_{1-2} + dU$$

$$= \frac{P_1V_1 - P_2V_2}{n-1} + mc_v(T_2 - T_1)$$

$\because c_p - c_v = R$

Dividing on both side by c_v

$$\frac{c_p}{c_v} - 1 = \frac{R}{c_v}$$

$$= \frac{mR(T_1 - T_2)}{n-1} + m \times \frac{R}{\gamma - 1}(T_2 - T_1)$$

$$\gamma - 1 = \frac{R}{c_v}$$

$$\therefore c_v = \frac{R}{\gamma - 1}$$

$$= mR(T_1 - T_2)\left[\frac{1}{n-1} - \frac{1}{\gamma - 1}\right]$$

$$= mR(T_1 - T_2)\left[\frac{(\gamma - 1) - (n-1)}{(n-1)(\gamma - 1)}\right]$$

$$= mR(T_1 - T_2)\left[\frac{\gamma - n}{(n-1)(\gamma - 1)}\right]$$

$$Q_{1-2} = \frac{\gamma - n}{\gamma - 1} \times \frac{mR(T_1 - T_2)}{n-1}$$

5. Change in Enthalpy

We know that change in enthalpy

$$dH = H_2 - H_1 = mc_p (T_2 - T_1)$$

Result:

Work done $(W_{1-2}) = \dfrac{P_2V_2 - P_1V_1}{n-1}$

Change in internal energy: $\delta U = mc_v(T_2 - T_1)$

Heat transfer $Q_{1-2} = \dfrac{\gamma - n}{1-\gamma} \times \left[\dfrac{(P_2V_2 - P_1V_1)}{n-1}\right]$

Change in enthalpy: $\delta H = mc_v(T_2 - T_1)$

Change in entropy: $\delta S = 2.3\ mc_v \times \left(\dfrac{\gamma - n}{n}\right) \log\left(\dfrac{P_1}{P_2}\right)$

2.33.6 Isentropic Process

The process, such as reversible adiabatic process, in which the entropy of the system remains constant is known as an isentropic process.

2.33.7 Hyperbolic Process

The process in which the product of the pressure of the gas and its volume remains constant V known as hyperbolic process. In the above case in isothermal process, PV = Constant.

So, the work done and heat transferred during hyperbolic process can be calculated by using the formula for the above isothermal process.

2.34 FLOW PROCESS

The process occurring in open system (or control volume) which permit the transfer of mass to and from the system, are known as flow process.

In a flow process, the mass *i.e.*, (working substance) enters the system and leaves after doing the work.

The flow process may be classified as:

1. Steady flow process
2. Unsteady flow process

2.34.1 Steady Flow Process

Steady flow process is defined as a process in which the conditions within the control volume do not vary with time.

The following conditions must be satisfied in a steady flow process:

1. The mass rate within the system remains constant, *i.e.*, the mass rate at entrance and exit is same and do not vary with time.
2. The state of working substance at any point within the system is same at all times.
3. The rate at which heat and work transfer, across the system, does not vary with time, *i.e.* the rate of heat and work transfer is constant.
4. There is no change in the chemical composition of the system so no change in chemical energy is involved.
5. The state of energy of the working substance at the inlet or outlet of system does not vary with time.

2.34.2 Unsteady Flow Process

A process which does not satisfy the condition of steady flow *i.e.*, conditions vary with time is known as unsteady flow process.

2.35 FLOW ENERGY

The flow or displacement energy is the energy required to flow or move the working substance against its pressure. It is also known as flow work.

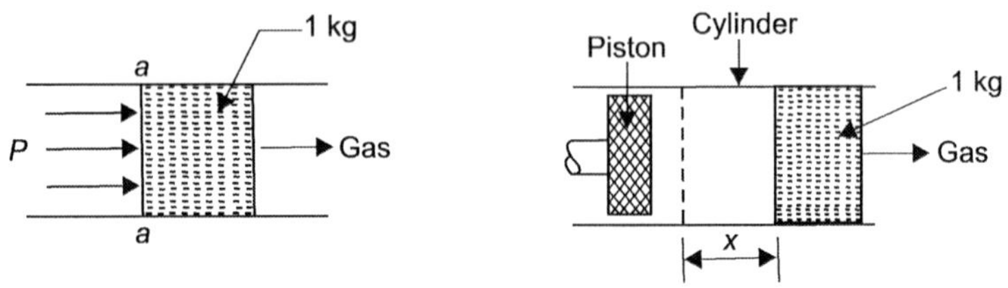

Fig. 2.28 Flow work

For Example, let the working substance with pressure in P (in N/ m^2) flow through area A (in m^2) and moves through a distance x (in metre).

∴ Energy or work required to flow the working substance,

$$\text{Flow energy} = \text{Force} \times \text{Distance}$$

$$= (P \times A) \times x \ \text{kJ/kg}$$

$$= PV \ \text{J/kg} \qquad (\because V = A \times x)$$

where V = Specific volume of the working substance in m^3/kg.

P = Pressure at section a, kN/m^2.

2.36 APPLICATION OF FIRST LAW OF THERMODYNAMICS TO OPEN SYSTEM (STEADY FLOW PROCESS)

Consider an open system through which the working substance flows at a steady rate, as shown in Fig. 2.29.

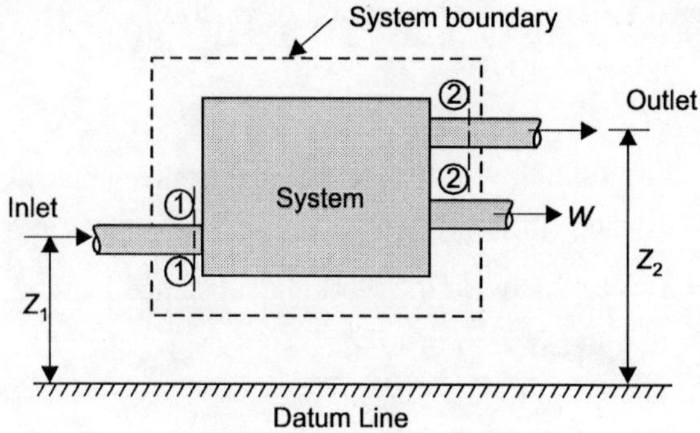

Fig. 2.29 Steady flow process

Let P_1 = Pressure of the working substance entering the system in N/m^2.

V_1 = Specific volume of the working substance entering the system in m^3/kg.

v_1 = Velocity of the working substance entering the system in m/s.

u_1 = Specific internal energy of the working substance entering the system in J/kg.

Z_1 = Height above datum line for inlet in metres.

P_2, V_2, V_2, u_2, and Z_2 = Corresponding values for the working substance leaving the system.

q_{1-2} = Heat supplied to the system in J/kg

W_{1-2} = Work delivered by the system in J/kg.

Consider 1 kg of mass of the working substance,

We know that

Total energy entering the system per kg of the working substance

e_1 = Internal energy + Flow energy + Kinetic energy + Potential energy + Heat supplied.

$$= u_1 + P_1V_1 + \frac{v_1^2}{2} + gZ_1 + q_{1-2} \text{ J/kg}$$

Similarly, Total energy leaving the system per kg of the working substance

$$e_2 = u_2 + P_2V_2 + \frac{v_2^2}{2} + gZ_2 + W_{1-2}\text{J/kg}$$

Assuming no loss of energy during flow, then according to first Law of Thermodynamics (*i.e.* Law of Conservation of Energy)

$$e_1 = e_2$$

We know that

$u_1 + P_1V_1 = h_1$ = Enthalpy of the working substance entering the system in J/kg and

$u_2 + P_2V_2 = h_2$ = Enthalpy of the working substance leaving the system in J /kg.

Thus, the above expression may be written as

$$h_1 + \frac{v_1^2}{2} + gZ_1 + q_{1-2} = h_2 + \frac{v_2^2}{2} + gZ_2 + W_{1-2} \quad \ldots(i)$$

or $$h_1 + ke_1 + Pe_1 + q_{1-2} = h_2 + ke_2 + Pe_2 + W_{1-2}$$

In equation (*i*) all terms represent the energy flow per unit mass of the working substance (*i.e.* in J /kg) and multiplied through *m* of the working substance to equation (*i*) which will represent the energy flow per unit time *i.e.* in J/sec.

Thus, the equation (*i*) may also be written as

$$m\left(h_2 + \frac{v_1^2}{2} + gZ_1 + q_{1-2}\right) = m\left(h_2 + \frac{v_2^2}{2} + gZ_2 + W_{1-2}\right) \quad \ldots(ii)$$

Both the equations (*i*) and (*ii*) are known as steady flow energy equations.

The steady flow energy equation (*i*) for unit mass flow may be written as:

$$q_{1-2} - W_{1-2} = (h_2 - h_1)\left(\frac{v_2^2}{2} - \frac{v_1^2}{2}\right) + (gZ_2 - gZ_1) \quad \ldots(iii)$$

$$= (h_2 - h_1) + (ke_2 - ke_1) + (Pe_2 - Pe_1)$$

In differential form, this expression is written as

$$\delta q - \delta W = dh + d\,(ke) + d\,(Pe) \quad \ldots(iv)$$

2.37 WORK DONE IN A STEADY FLOW PROCESS

We know that the steady flow equation for unit mass flow, in the differential form is

$$\delta q - \delta W = dh + d\,(ke) + d\,(Pe) \quad \ldots(i)$$

We also know that

$$h = u + PV$$

Differentiating this expression

$$dh = du + d\,(PV)$$

$$= du + PdV + VdP \qquad \ldots(ii)$$

According to first law of thermodynamics for a closed system.

We know that

$$\delta q = du + PdV$$

So, $$du = \delta q - PdV$$

Substituting this value of du in equation (ii)

$$dh = \delta q - PdV + pdV + VdP$$

$$dh = \delta q + VdP$$

Substituting this value of dh in equation (i), we have

$$\delta q - \delta W = (\delta q + VdP) + d\,(ke) + d\,(Pe)$$

$$-\delta W = VdP + d(ke) + d\,(Pe) \qquad \ldots(iii)$$

If the changes in kinetic and potential energies are negligible [*i.e.* $d\,(ke) = 0$ and $d\,(Pe) = 0$] then. Equation (iii) may be written as

$$-\delta W = VdP$$

$$\delta W = -VdP$$

On integrating,

$$\int_1^2 \delta W = \int_1^2 VdP$$

or $$W_{1-2} = \int_1^2 VdP$$

Thus, in steady flow process, the work done W_{1-2} is $-\int_1^2 VdP$ instead of $\int_1^2 PdV$ in a non flow process.

Fig. 2.30 shows the difference between work done in a non-flow and steady flow process.

$$= P_1V_1 \log\left(\frac{V_2}{V_1}\right) = 2.3P_1V_1 \log\left(\frac{V_2}{V_1}\right) \qquad \left(\begin{array}{c}\because P_1V_1 = P_2V_2 \\ or\ \dfrac{P_1}{P_2} = \dfrac{V_2}{V_1}\end{array}\right)$$

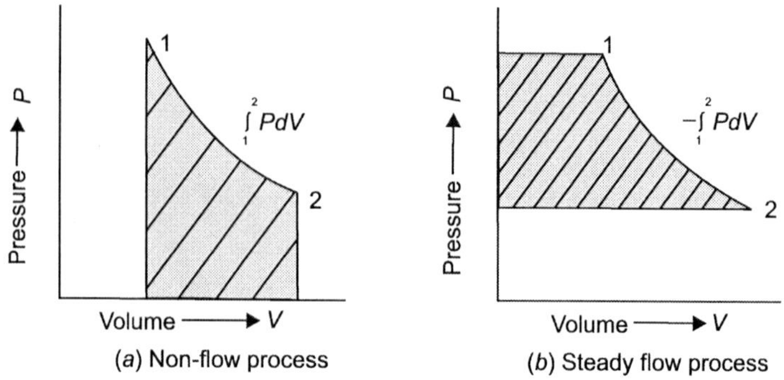

Fig. 2.30 Work done in a steady flow process

(*iv*) Polytropic Process

$$PV^{\gamma} = P_1V_1 = P_2V_2 = \text{Constant, } C$$

or
$$V = V_1\left(\frac{P_1}{P}\right)^{1/\gamma}$$

We know that, work done

$$W_{1-2} = -\int_1^2 VdP = -\int_1^2 V_1\left(\frac{P_1}{P}\right)^{1/\gamma} dP$$

$$= -V_1P_1^{1/\gamma}\int_1^2 P^{-1/\gamma}dP$$

$$= -V_1P_1^{1/\gamma}\left|\frac{P^{-1/\gamma}+1}{-1/\gamma+1}\right|_1^2$$

$$= -V_1P_1^{1/\gamma}\left|\frac{P^{\frac{\gamma-1}{\gamma}}}{\frac{\gamma-1}{\gamma}}\right|_1^2$$

$$= \frac{\gamma}{\gamma-1}\times -V_1P_1^{1/\gamma}\left(P_2^{\frac{\gamma-1}{\gamma}} - P_1^{\frac{\gamma-1}{\gamma}}\right)$$

$$= \frac{\gamma}{\gamma-1}\left(-V_1P_1^{1/\gamma}P_2^{\frac{\gamma-1}{\gamma}} + P_1^{\gamma-1/\gamma}V_1P_1^{1/\gamma}\right) \qquad (\because P_1V_1^{\gamma} = P_2V_2^{\gamma})$$

$$= \frac{\gamma}{\gamma - 1}(P_1V_1 - P_2V_2)$$

(v) Polytropic Process

$$PV^n = P_1V_1^{\,n} = P_2V_2^{\,n} = \text{constant}, C$$

∴ Work done

$$W_{1-2} = \frac{n}{n-1}(P_1V_1 - P_2V_2)$$

(Substituting $\gamma = n$, for polytropic process)

2.38 WORK DONE FOR VARIOUS STEADY FLOW PROCESS

(i) Constant Volume Process

$$W_{1-2} = -\int_1^2 VdP$$

$$= -\int_1^2 dP = -V(P_2 - P_1)$$

$$= V(P_2 - P_1)$$

(ii) Constant Pressure Process

$$W_{-2} = -\int_1^2 VdP$$

$$= -V\int_1^2 dP$$

$$= -V(P_2 - P_1) = 0 \qquad (\because P_1 = P_2 = \text{constant})$$

(iii) Constant Temperature Process

$$PV = P_1V_1 = P_2V_2 = \text{Constant, C}$$

$$W_{1-2} = -\int_1^2 VdP = -\int_1^2 \frac{P_1V_1}{P}dP$$

$$= -P_1V_1\int_1^2 \frac{dP}{P} \qquad \left(\because V = \frac{P_1V_1}{P}\right)$$

$$= -P_1V_1\,[\log_e P_2 - \log_e P_1]$$

$$= P_1V_1 [\log_e P_1 - \log_e P_2]$$

$$= P_1V_1 \log_e\left(\frac{P_1}{P_2}\right)$$

$$= 2.3P_1V_1 \log\left(\frac{P_1}{P_2}\right)$$

2.39 LIMITATIONS OF FIRST LAW OF THERMODYNAMICS

According to first law of thermodynamics:

1. When a closed system undergoes a thermodynamic cycle, the net heat transfer is equal to the network transfer.

   This statement does not specify the direction of flow of heat and work (*i.e.*, whether the heat flows from a hot body to a cold body or from a cold body to a hot body). It also does not give any condition under which these transfers take place.

2. The heat energy and mechanical work are mutually convertible.

   Though the mechanical work can be fully converted into heat energy, but only a part of heat energy can be converted into mechanical work. This means that the heat energy and mechanical work are not fully mutually convertible. In other words, there is a limitation on the conversion of one from of energy into another form.

   A machine which violates the first law of thermodynamics (*i.e.*, energy can neither be created nor destroyed, but can be transformed from one form to another) is known as perpetual motion machine of first kine (PMM-1).

   It is defined as a machine which produces work energy without consuming as equivalent of energy from other source. Such a machine as shown in Fig. 2.31 is impossible to obtain in actual practice, because no machine can produce energy of its own without consumingany other form of energy.

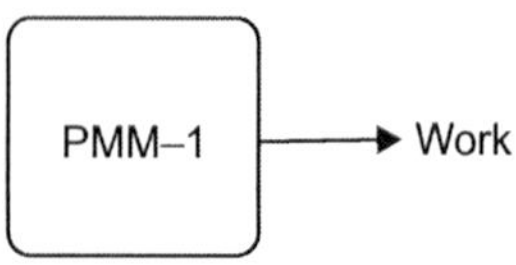

Fig. 2.31 Perpetual motion machine of the first kind

2.40 SECOND LAW OF THERMODYNAMICS

The second law of thermodynamics depends on the statements.

(*a*) Kelvin Planck statement

(*b*) Clausius statement

(*a*) **Kelvin Planck Statement:** It is impossible to construct an engine working on a cyclic process, whose sole purpose is to convert heat energy from a single thermal reservoir into an equivalent amount of work. According to the statement, no actual heat engine working on a cyclic process can convert whole of the heat supplied in to mechanical work. It means degradation of the energy in producing mechanical work from heat supplied. Hence, it is also known as the law of degredsion of energy.

A heat engine which violates this statement of the 2nd law is known as perpetual motion machine of the second kind (ppm–11) which is impossible for the actual amount of work.

Thus, a heat engine is device used for converting heat energy into mechanical work. Practically two reservoirs are needed one at higher temperature and other at lower temperature.

The second law can be stead in several ways.

It is impossible to convert the heat completely into an equivalent amount of work, without producing some other changes in the some parts of the system.

It is impossible for a self-acting machine without any external source to supply heat form a body at a lower temperature to a body at higher temperature.

Heat cannot be supplied itself from a colder body to a hotter body.

Only a fraction of a giving quantity of heat may be converted into work due to temperature difference all natural or spontaneous pressure tends to take place in the direction and they cannot be reversed *i.e.*, irreversible.

The energy of the universe is constant *i.e.*, entropy is constant in a reversible pressure and increases when the process is irreversible maximum randomness.

For example, heat energy Q_1 is supplied at temperature T_1 forms higher reservoir to the engine as shown in Fig (*b*). A part of this energy is rejected to a lower temp reservoir (sink) at a temperature T_2 and heat Q_2 then the difference of both the heats $(Q_1 - Q_2)$. Converted in to mechanical work $(WE = Q_1 - Q_2)$. Hence, the thermal efficiency is the ratio of maximum. Mechanical work to the total heat supplied to engine.

$$\eta_{\text{maximum}} = \frac{\text{Maximum mechanical work}}{\text{Total heat supplied}}$$

$$\eta = \frac{Q_1 - Q_2}{Q_1} = 1 - \frac{Q_2}{Q_1} = 1 - \frac{T_2}{T_1}$$

$$\because \text{ for a reversible heat engine} = \frac{Q_1}{T_1} = \frac{Q_2}{T_2}$$

$$h_{max} = \frac{\text{Maximum mechanical work}}{\text{Total heat supplied}} = \frac{Q_1 - Q_2}{Q_1}$$

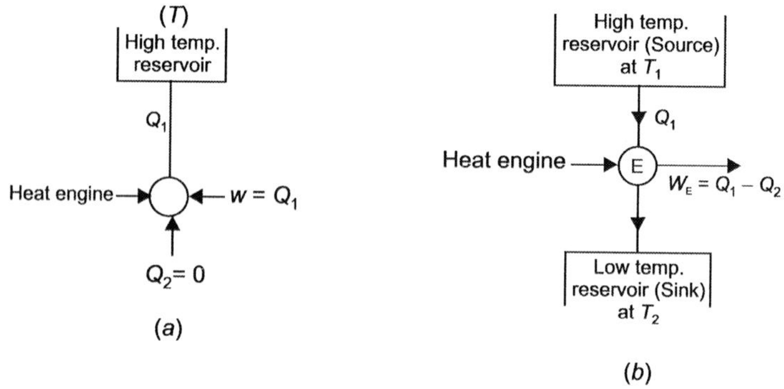

Fig. 2.32 (*a*) Perpetual motion machine of second kind (impossible) (*b*) Heat engine.

(b) Clausius Statement: It is impossible for a self-acting machine without any external source to supply heat from a body at a lower temperature to a body at higher temperature or in other words, heat cannot flow itself (without the expenditure of mechanical work.)

The device (such as refrigerator or a heat pump) violates the Clausius statement because no input work is supplied to the device to transfer heat from a colder body to a hotter body. Such a work is called perpetual motion machine of the second kind.

The refrigerator and heat pump while operating in a cyclic process, require an input work as shown in figs. (*b*) and (*c*). Hence, there is no difference between the cyclic of operation of the refrigerator and a heat pump to achieve the same objective, but the purpose is quite different. A refrigerator is a device which is used to maintain the temperature of cold body (refrigerated space) at a temperature lower than the temperature of the sorrounding while the, pump is a device operate in a cyclic process to maintain the temperature of a hot body (heated space) at a temperature higher than the temperature of sorrounding. The performance of refrigerator and heat pump is measured in terms of (COP), coefficient of performance.

COP : It is a ratio of maximum heat transfer, *i.e.*, (heat taken from cold body) to the amount of heat required to produce the desired effect.

$$(\text{C.O.P.}) = \frac{Q_2}{W_R} = \frac{Q_2}{Q_1 - Q_2} = \frac{T_2}{T_1 - T_2} \qquad \text{where Ref. - Refrigerator}$$

$$\text{C.O.P.} = \frac{Q_1}{W_p} = \frac{Q_1}{Q_1 - Q_2} = \frac{T_1}{T_1 - T_2} = \frac{T_2}{T_1 - T_2} + 1 = (\text{C.O.P.})_R + 1$$

where HP. - Heat Pump

Hence, the C.O.P. of heat pump is greater then C.O.P. of refrigerator by unity.

Carnot Refrigerator

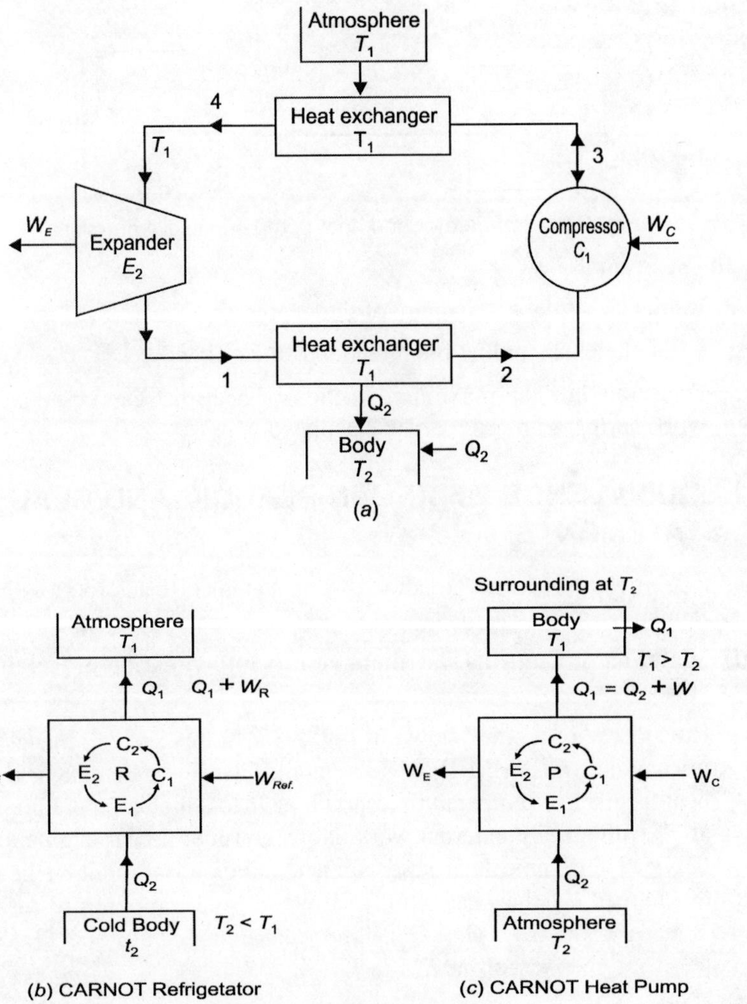

Fig. 2.33 Carnot refrigerator

OR

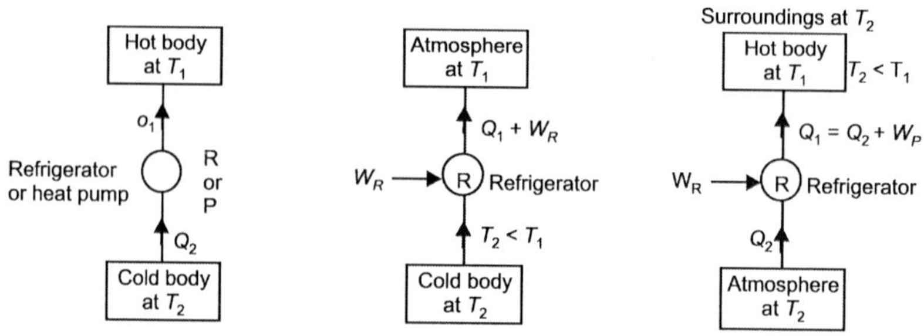

(*a*) Perpetual motion machine of the second kind (*b*) Refrigerator (*c*) Heat Pump

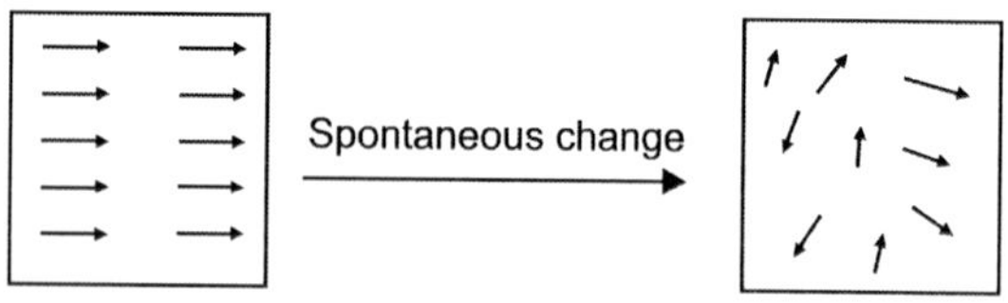

Fig. 2.34 Refrigerator and heat pump in a cyclic process.

Benefits of Second Law

1. It tells about the spontaneity of the processes.
2. It also helps in finding the direction of heat flow.
3. From this law the maximum fraction of heat that can be converted into work during a process can be calculated.

2.42 EQUIVALENCE OF KELVIN-PLANCK AND CLAUSIUS STATEMENTS

In this section we prove that violation of one statement leads to violation of the other statement and *vice versa*.

(*i*) **Violation of Clausius statement results into violation of Kelvin Planck statement.**

Consider a system as shown in Fig. 2.35. In this system, a heat pump or refrigerator (*i.e*. PMM-II) is violating the Clausius statement as it transfers heat from a low temperature reservoir at T_2 to a high temperature reservoir at T_1 without any external work. Now, consider a heat engine, operating between the same heat reservoirs, receives an amount of heat Q_t (as discharged by the heat pump) from the high temperature reservoir at T_1 does work ($W_E = Q_1 - Q_2$) and rejects an amount of heat Q_2 to the low temperature reservoir at T_2.

If the combination of the heat pump (or refrigerator) and the heat engine is considered as a single system as shown in Fig. 2.35, then the result is a device that operates in a cycle whose sole effect is to remove heat at the rate of $(Q_1 - Q_2)$ and convert it completely into an equivalent amount of work, thus violating the Kelvin-Planck statement. Hence a violation of Clausius statement leads to a violation of Kelvin-Planck statement.

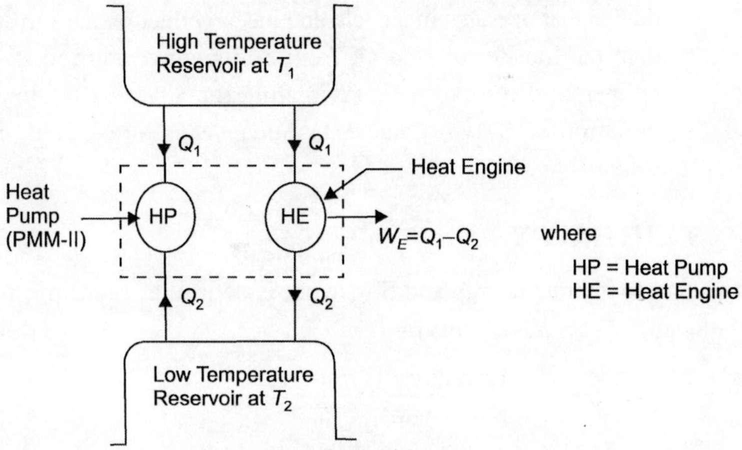

Fig. 2.35 A heat pump

(ii) **Violation of Kelvin-Planck statement results into violation of Clausius statement.**

Consider a system as shown in Fig. 2.36. In this system, a heat engine having 100 per cent thermal efficiency (*i.e.* PMM-II) is violating the Kelvin-Planck statement as it converts the heat energy (Q_1) from a single high temperature reservoir at T_v into an equivalent amount of work (*i.e.* $W = Q_5$). This work output of the heat engine can be used to drive a heat pump (or refrigerator) which receives an amount of heat Q_2 from a low temperature reservoir at T_2 and

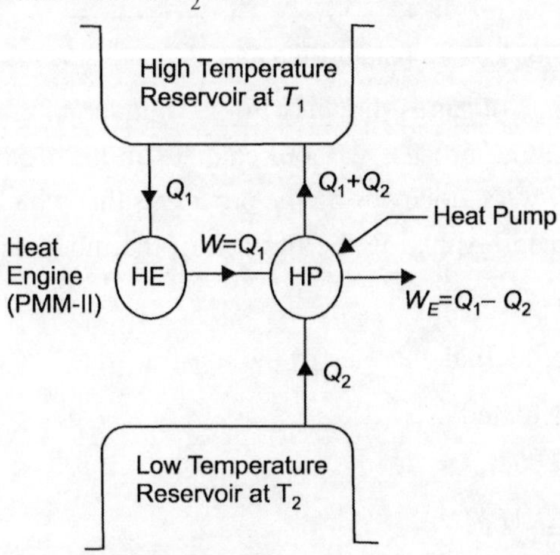

Fig. 2.36 Equivalence of Kelvin-Planck and Clausius Statements

rejects an amount of heat ($Q_1 + Q_2$) to a high tempera-ture reservoir at T_t. If the combination of a heat engine and a heat pump (or refrigerator) is considered as a single system, as shown in Fig. 2.36, then the result is a device that operates in a cycle and has no effect on the surroundings other than the transfer of heat Q_2 from a low temperature reservoir to a high temperature reservoir; thus violating the Clausius statement. Hence, a violation of Kelvin-Planck statement leads to a violation of Clausius statement.

2.43 ENTHALPY

The sum of internal energy and flow work is defined as Enthalpy,

Enthalpy, $h = u + PV$ J/kg

where h = Enthalpy

k = Internal energy, and

PV = Flow work on work done

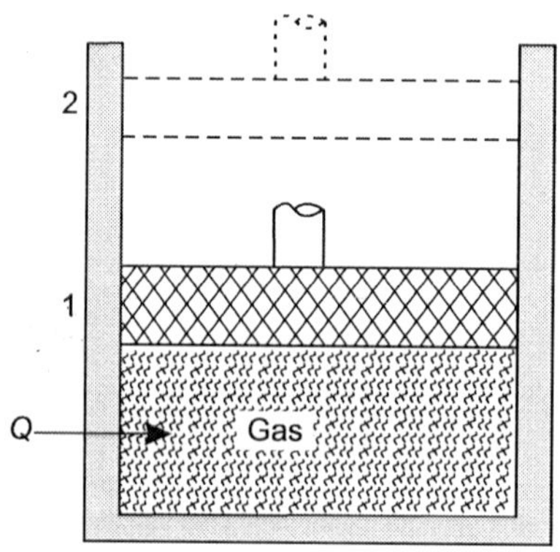

Fig. 2.37 Constant Pressure Heating Process

Consider a control mass of gas as shown in Fig.2.37 undergoing a quasi-equilibrium constant-pressure heating process without changes in kinetic or potential energy. Assume that the only work done during the process is that which is associated with the boundary movement. Applying the firs law of thermodynamics.

$$Q_{1-2} = U_2 - U_1 + W_{1-2}$$

where U_2 and U_1 are internal energies of the mass at state 2 and 1 respectively.

The work can be calculated as:

$$W_{1-2} = \int_1^2 PdV$$

As the pressure is constant

$$W_{1-2} = \int_1^2 PdV$$

$$= P(V_2 - V_1)$$

Hence

$$Q_{1-2} = (U_2 - U_1) + (P_2V_2 - P_1V_1)$$
$$= (U_2 + P_2V_2) - (U_1 + P_1V_1)$$

The heat transfer during the process is represented in terms of the change in the quantity $(U + PV)$ between the initial and final states. A new extensive property, enthalpy is defined as:

$$H = U + PV$$

and for unit mass

$$h = u + PV$$

The heat transfer in the constant-pressure quasi-equilibrium process is equal to the change in enthalpy, which includes both the change in internal energy and the work for this particular process.

Many tables and charts of thermodynamic properties give values for enthalpy and not for internal energy. Hence, to calculate the internal energy at a state from the table values.

$$u = h - PV$$

The enthalpy of saturated liquid is h_f saturated vapour is h_g, and the latent heat of vaporization to h_{fg}. The saturation state enthalpy can be calculated by using the following equation:

$$h = h_f + xh_{fg}$$

where x = dryness fraction of steam.

2.44 ENTROPY

Entropy of a system is a measure of randomness or disorder of the system and is denoted by the symbol S. It is a state on point function (*i.e.*, the change in entropy is independent of path). It can also be regarded as the measure of “unavailable energy”. The change in entropy of a system (or a reaction) is given by:

$$\Delta S = D_{\text{final}} - S_{\text{initial}} = S_2 - S_1$$

Entropy is expressed in Joule deg^{-1} mole^{-1} (JK^{-1}mol^{-1}).

$$ds = \frac{\delta Q}{T} \int \delta Q = \int T.ds. \int^2 ds$$

$$= \int_1^2 \frac{\delta Q}{T} \text{ – so entropy of a substance is zero at zero temperature.}$$

It is defined as the disorderness or randomness of any substance and is given by the amount of heat supplied per degree change in absolute temperature. Entropy is an extensive property of the system and hence a point function depending on the initial and final states only. Mathematically,

$$ds = \frac{\delta Q}{T} \text{ or } \int \delta Q = \int T.ds$$

If a reversible process is represented on temperature, entropy diagram then the area below the curve line will give the total heat change during that process. Now if the initial position is marked by 1 and the final by 2. Then the entropy change will be given by

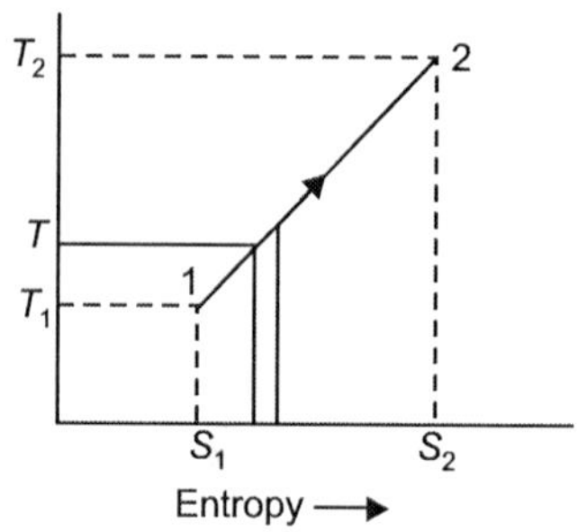

Fig. 2.38 Entropy change

$$\int_1^2 ds = \int_1^2 \frac{\delta Q}{T}$$

The entropy of a substance is zero at absolute zero temperature (0°V)

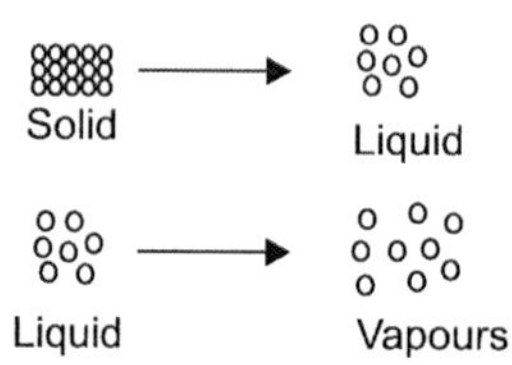

Fig. 2.39 Process showing the increase in entropy of the system

The Unit of Entropy: If the heat supplied or rejected is taken in kJ and the temperature in °K, then the unit of entropy is kJ/°K.

Clausius Inequality: When a closed system undergoes a cyclic process, the cyclic integral of $\delta Q/T$ is equal to zero for reversible process and less than zero for irreversible process.

$$\oint \frac{\delta Q}{T} = 0, \text{ Reversible process}$$

$$\oint \frac{\delta Q}{T} < 0, \text{ Irreversible process}$$

Combining the above two equations we get

$$\oint \frac{\delta Q}{T} \leq 0, \text{ is known as Clausius inequality.}$$

Thus for only system,

$$S_{\text{Gas}} > S_{\text{Liquid}} > S_{\text{Solid}}$$

2.45 PRINCIPLE OF INCREASE OF ENTROPY

By Clausius inequality we have

$$\oint \frac{\delta Q}{T} \leq 0$$

We also know that

$$ds = \frac{\delta Q}{T}$$

Since the entropy is a thermodynamic property and we know the cyclic integral of it is zero.

$$\oint \frac{\delta Q}{T} \leq \oint ds$$

or $$\frac{\delta Q}{T} \leq ds$$

or $$ds \geq \frac{\delta Q}{T}$$

$$ds = \frac{\delta Q}{T} \text{ for reversible process}$$

$$ds > \frac{\delta Q}{T} \text{ for irreversible process}$$

$$ds \geq \frac{\delta Q}{T} \text{ for isolated system } (\delta Q = 0)$$

For reversible cyclic process

$$\delta ds = 0 \text{ or } s = \text{constant}$$

For irreversible cyclic process.

Since almost all the processes are irreversible, therefore entropy will go on increasing like universe.

2.46 RELATION BETWEEN PRESSURE-VOLUME AND WORK

Consider a certain amount of a gas enclosed in a cylinder of cross-sectional area 'A' and fitted with a hypothetical, weightless and frictionless piston. Allow the gas to expand against the opposing pressure, P (*see* Fig 2.40) If the pressure is reduced by a very small amount dP, the piston will be forced to move upward, due to the expansion of gas by a small distance dt. The corresponding work associated with the change, dw, is given by:

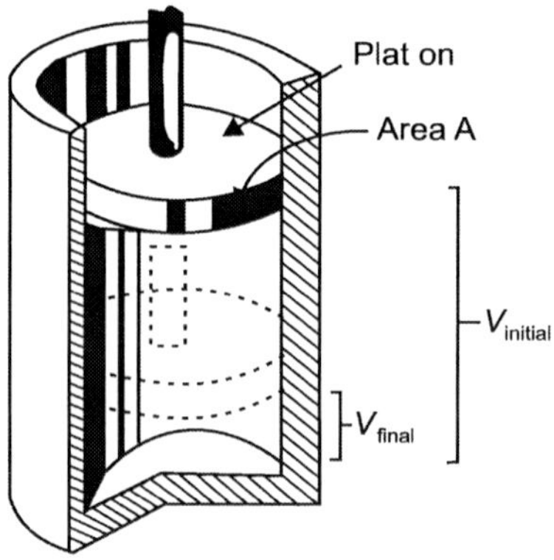

Fig. 2.40 Work in *P–V* form.

$$dw = \text{Force} \times \text{Distance}$$

But $\quad \text{Pressure} = \text{Force/Area}$

$\therefore \quad dw = \text{Pressure} \times \text{Area} \times \text{Distance}$

$$= P \times A \times dl$$

But the product $A \times dl$ is the volume swept by the piston, dV.

$\therefore \quad dw = P.dV$

This is nothing but the work done when the volume changes by a very small amount dV against pressure P.

If the volume changes from V_{initial} to V_{final} then total work done by the pass is obtained by integration.

$$\therefore \quad \text{Total work done} \int_{Initial}^{Final} dw = w = \int_{V_{initial}}^{V_{final}} P.dV$$

If the external pressure (P) is constant. then :

$$w = P\int_{V_1}^{V_2} dV = P(V_{final} - V_{initial}) = P\Delta V$$

SOLVED EXAMPLES

Example 2.1: A force of 100N is uniformly distributed on a piston crown having a diameter of 3 cm. Determine the pressure on the piston.

Solution:

Given: Force, F = 100N, diameter, d = 3 cm

Area of the piston crown.

$$A = \frac{\pi}{4} d^2$$

$$= \frac{\pi}{4} (3)^2 = 7.07 \text{ cm}^2 = 7.07 \times 10^{-4} \text{ m}^2$$

We know that

$$\text{Pressure} = \frac{\text{Force}}{\text{Area}}$$

So,
$$P = \frac{f}{A}$$

$$= \frac{100}{7.07 \times 10^{-4}} = 14 \times 10^4 \text{ N/m}^2 \text{ or } 14 \times 10^4 \text{ Pa.}$$

Example 2.2: The pressure measured by a pressure gauge is 5000 kN/m^2. Barometric pressure of the atmosphere is 760 mm of mercury. Find the absolute pressure.

Solution: Given: Gauge pressure = 5000 kN/m^2, Barometric pressure = 760 mm of Hg

Barometric pressure = Atmospheric pressure

= 760 mm of Hg

= 13600 × 9.81 × 0.76 N/m^2 $(\because p = \rho g h)$

= 101.4 kN/m

∴ Absolute pressure = Atmospheric pressure + Gauge pressure

= 101.4 + 5000 = 5101.4 kN/m^2

= 5101.4 kPa = 51.014 Bar.

Example 2.3: Determine the specific volume and density of 5 kg of gas contained in a tank of volume 25 m^3.

Solution:

Given: mass = 5 kg, Volume = 25 m^3,

We know that

$$\text{specific volume, } V = \frac{\text{Volume}}{\text{mass}}$$

so, $$V = \frac{V}{m} = \frac{25}{5} = 5 \text{ m}^3/\text{kg}$$

and $$\text{Density, } \rho = \frac{1}{v} = \frac{1}{5} = 0.2 \text{ kg/m}^3.$$

Example 2.4: In insulated container, water is heated by an electric heated of 5KW capacity to maintain the temperature uniform inside the container, a mechanical stirrer of 0.3 KW capacity is used. Find the change in internal energy of the water after 5 minutes assuming that the work and heat are entrirely transferred to water.

Solution:

Work done by mechanical stirrer = 0.3 kJ/s × 5 min

$$= 0.3 \times 300 = 90 \text{ kJ}$$

From first law of thermodynamics state that

Change in internal energy of water

$$U = Q - W$$

$$U = 5 \times 300 - 90 \text{ kJ}$$

$$= 1410 \text{ kJ}$$

Example 2.5: 2.5 m^3 of a gas at 8 bar absolute pressure expand at constant temperature to a volume of 10m^3. Find the final pressure of the gas.

Solution:

Given: P_1 = 8 bar absolute, V_1 = 2.5 m^3, V_2 = 10m^3, P_2 = ?

Since the temperature remains constant, so applying Boyle's Law

$$\therefore \quad P_1V_1 = P_2V_2$$

or $$P_2 = \frac{P_1V_1}{V_2} = \frac{8\times 2.5}{10} = 2 \text{ bar absolute.}$$

Example 2.6: 0.14 m^3 of a gas at 40°C receive heat at constant pressure so that the final temperature is 200°C. Find the final volume and write down the increase in volume during the supply of heat.

Solution:

Given: T_1 = 273 + 40 = 313 K, T_2 = 273 + 200 = 473 K,

V_1 = 0.14m^3, V_2 = ?

Now, since heat is supplied at constant pressure.

So applying Charles' Law

$$\therefore \qquad \frac{V_1}{T_1} = \frac{V_2}{T_2}$$

$$\text{or} \qquad V_2 = \frac{V_1 \times T_2}{T_1}$$

$$\text{or} \qquad V_2 = \frac{0.14 \times 473}{313} = 0.212 \text{ m}^3$$

$\therefore$ Increase in volume = $V_2 - V_1 = 0.212 - 0.14 = 0.072 \text{ m}^3$.

Example 2.7: One kg of an ideal gas is heated from 18.3°C to 93.4°C. Assuming R = 0.287 kJ/kg K and Y = 1.4 for the gas. Find specific heats at constant pressure and at constant volume both.

Solution:

We know that

$$c_p = c_v = R \qquad \text{...}(i)$$

$$\text{also} \qquad \frac{c_p}{c_v} = \gamma \qquad \text{...}(ii)$$

From (*ii*) $\qquad c_p = \gamma\, c_v = 1.4\, c_v \qquad$ ...(*iii*)

Put this volume of c_p in equation (*i*)

$$1.4c_v - c_v = R$$

$$0.4\, c_v = R$$

$$c_v = \frac{R}{0.4}$$

$$= \frac{0.287}{0.4} = 0.7175 \text{ kJ/kg K}$$

From equation (*iii*)

$$c_p = 1.4\, c_v = 1.4 \times 0.7125 = 1.0 \text{ kJ/kg K}$$

Hence, $\qquad c_p = 1.0$ kJ/kg K, and

$\qquad c_v = 0.7175$ kJ/kg K.

Example 2.8: The volume of c_p and c_v for an ideal gas are 1 kJ/kg K and 0.7128 kJ/kg K respectively. Find the volumes of R and ratio of specific heats, γ for the gas. If one Kg of this gas is heated at constant pressure from 25°C to 200°C, estimate the heat added, ideal work done and change in internal energy. Also, calculate the pressure and final volume if initial volume is 2 m^3.

Solution:

Given: $c_p = 1$ kJ/kg K, $c_v = 0.7128$ kJ/kg K, $V_1 = 2\text{m}^3$,
$T_1 = 25 + 273 = 298\text{K}$, $T_2 = 200 + 273 = 473$ K, $m = 1$ kg.

We know that

$$c_p - c_v = R$$

$$\therefore \quad 1 - 0.7128 = R$$

So,
$$R = 0.2872 \text{ kJ/kg K}$$
$$= 0.287 \text{ kJ/kg K}$$

and,
$$\gamma = \frac{c_p}{c_v} = \frac{1}{0.718} = 1.4$$

During the constant pressure process:

Heat transferred,

$$Q_{1-2} = mc_p (T_2 - T_1)$$
$$= 1 \times 1 \times (473 - 298)$$
$$= 175 \text{ kJ}$$

Change in internal energy,

$$dU = mc_v (T_2 - T_1)$$
$$= 1 \times 0.7128 \times (473 - 298)$$
$$= 124.74 \text{ kJ}$$

We know that

$$Q_{1-2} - W_{1-2} = dU$$

$\therefore$ Workdone, $W_{1-2} = Q_{1-2} - dU$

$$= 175 - 124.74 = 50.28 \text{ kJ}$$

From the characteristic gas equation

$$10^5 P_1V_1 = mRT_1$$

$$P_1 = \frac{mRT_1}{V_1 \times 10^5} = \frac{1 \times 287 \times 298}{10^5 \times 2} = 0.4276 \text{ bar}$$

Final pressure, $P_2 = P_1$ as it is constant pressure process,

Further during constant pressure process

$$V_2 = \frac{T_2}{T_1} \times V_1 = \frac{473}{298} \times 2 = 3.1744 \text{ m}^3/\text{kg}$$

Example 2.9: Gas at 1.5 bar absolute and 20°C in a closed vessel is compressed to 10 bar absolute. Its temperature then becomes 180°C. If the compression follows the law $PV^n = C$. Find the volume of n.

Solution:

Given: $P_1 = 1.5$ bar absolute, $T_1 = 20 + 273 = 293$K, $P_2 = 10$ bar absolute.

$$T_2 = 180 + 273 = 453 \text{ K}$$

We know that, Polytropic process

$$\frac{T_2}{T_1} = \left(\frac{P_2}{P_1}\right)^{\frac{n-1}{n}}$$

or
$$\frac{453}{293} = \left(\frac{10}{1.5}\right)^{\frac{n-1}{n}}$$

or
$$1.546 = (6.67)^{\frac{n-1}{n}}$$

Taking log of both sides

$$\log 1.546 = \frac{n-1}{n} \log 6.67$$

or
$$\frac{n-1}{n} = \frac{\log 1.546}{\log 6.67} = \frac{0.1892}{0.8241}$$

or
$$\frac{n-1}{n} = 0.2295$$

$$n - 1 = 0.2295\, n$$

$$n - 0.2295\, n = 1$$

$$0.7705\, n = 1$$

$$n = \frac{1}{0.7705} = 1.297$$

$$n = 1.3$$

Example 2.10: A gas occupies a volume of 0.1 m^3 at a temperature of 20°C and a pressure of 1.5 bar. Find the final temperature of the gas, if it is compressed to a pressure of 7.5 bar and occupies a volume of 0.04 m^3.

Solution:

Given: $v_1 = 0.1$ m^3 ; $T_1 = 20°C = 20 + 273 = 293$ K;

$P_1 = 1.5$ bar $= 0.15 \times 10^6$ N/m^2; $P_2 = 7.5$ bar $= 0.75 \times 10^6$ N/m^2; $v_2 = 0.04$ m^3.

let T_2 = Final temperature of the gas

We know that

$$\frac{P_1V_1}{T_1} = \frac{P_2V_2}{T_2}$$

$$T_2 = \frac{P_2V_2T_1}{P_1V_1} = \frac{0.75\times10^6\times0.04\times293}{0.15\times10^6\times0.1} = 586\,\text{K}$$

$$= 586 - 273 = 313°\text{C}.$$

Example 2.11: A certain gas occupies a space of 0.3 m^3 at a pressure of 2 bar and a temperature of 77°C. It is heated at a constant volume, until the pressure is 7 bar. Determine: 1. temperature at the end of the process; 2. mass of the gas; 3. change in internal energy; and 4. change in enthalpy during the process.

Assume c_p = 1.005 kJ/kg K; c_v = 0.712 kJ/kg K; and R = 287 J/kg K.

Solution:

Given: v_1 = 0.3 m^3; P_1 = 2 bar = 0.2 × 10^6 N/m^2;

T_1 = 77°C = 77 + 273 = 350K; P_2 = 7 bar = 0.7 × 10^6 N/m^2; c_p = 1.005 kJ/kg K;

c_v = 0.712 kJ/kg K; R = 287 J/kg K.

1. Temperature at the end of the process

Let T_2 = Temperature at the end of the process.

We know that $\dfrac{P_1}{T_1} = \dfrac{P_2}{T_2}$

$$\therefore \quad T_2 = \frac{P_2T_1}{P_1} = \frac{0.7\times10^6\times350}{0.2\times10^6}$$

$$= 1225\text{ K} = 1225 - 273 = 952°\text{C}.$$

2. Mass of the gas

Let m = Mass of the gas.

We know that $P_1V_1 = mRT_1$

$$\therefore \quad m = \frac{P_1V_1}{RT_1} = \frac{0.2\times10^6\times0.3}{287\times350} = 0.597\text{ kg}$$

3. Change in internal energy

We know that change in internal energy.

$$dU = U_2 - U_1 = mc_v\,(T_2 - T_1)$$

$$= 0.597 \times 0.712\,(1225 - 350) = 372\text{ kJ}$$

4. Change in enthalpy

We know that change in enthalpy.

$$dH = H_2 - H_1 = mc_p (T_2 - T_1)$$
$$= 0.597 \times 1.005 (1225 - 350) = 525 \text{ kJ}.$$

Example 2.12: A quantity of gas has a volume of 0.14 m^3, pressure 1.5 bar and a temperature of 100°C. If the gas is compressed at a constant pressure, until its volume becomes 0.112 m^2, determine: 1. the temperature at the end of compression; 2. work done in compressing the gas; 3. decrease in internal energy; and 4. heat given out by the gas.

Assume c_p = 1.005 kJ/kg; c_v = 0.712 kJ/kg K; and R = 285 K.

Solution:

Given: V_1 = 0.14 m^3; P = 1.5 bar = 0.15 × 10^6 N/m^2;

T = 100°C = 100 + 273 = 373K; V_2 = 0.112 m^3; c_p = 1.005 kJ/kg K;

c_v = 0.712 kJ/kg K; R = 285 J/kg K.

The compression of gas at constant pressure is shown in Fig.2.41

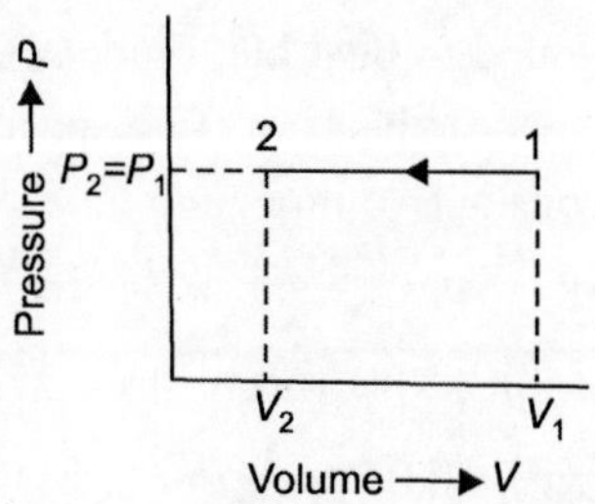

Fig. 2.41

1. Temperature at the end of compression

Let T_2 = Temperature at the end of compression.

We know that $\dfrac{V_1}{T_1} = \dfrac{V_2}{T_2}$

$$\therefore \quad T_2 = \frac{V_2 T_1}{V_1} = \frac{0.112 \times 373}{0.14} = 298.4 \text{ K}$$

$$= 298.4 - 273 = 25.4°\text{C}$$

2. Workdone is compressing the gas

We know that workdone in compressing the gas,

$$W_{1-2} = P (V_1 - V_2) = 0.15 \times 10^6 (0.14 - 0.112)$$
$$= 4200\text{J} = 4.2 \text{ kJ}$$

3. Decrease in internal energy

First of all, let us find the mass of gas (m) admitted for compression. We know that

$$P_1 V_1 = m\,R\,T_1$$

$$\therefore \qquad m = \frac{P_1 V_1}{RT_1} = \frac{0.15\times10^6\times0.14}{285\times373} = 0.197 \text{ kg} \qquad ...(\because P_1 = P)$$

We know that decrease in internal energy,

$$dU = U_1 - U_2 = mc_v\,(T_1 - T_2)$$

$$= 0.197 \times 0.712\,(373 - 298.4)\;10.46 \text{ kJ.}$$

4. Heat given out by the gas

We know that heat given out by the gas,

$$Q_{1-2} = mc_p\,(T_1 - T_2)$$

$$= 1.97 \times 1.005\,(373 - 298.4) = 14.77 \text{ kJ.}$$

Example 2.13: The initial volume of 0.18 kg of a certain gas was 0.15 m^3 at a temperature of 15°C and a pressure of 1 bar. After adiabatic compression to 0.056 m^2, the pressure was found to be 4 bar. Find: 1. Gas constant; 2. Molecular mass of the gas; 3. Ratio of specific heats; 4. Two specific heats, one at a constant pressure and the other at a constant volume; and 5. Change of internal energy.

Solution:

Given: $m = 0.18$ kg; $V_1 = 0.15\ m^3$; $T_1 = 15°\text{C} = 15 + 273 = 288$ K;

$P_1 = 1$ bar $= 0.1 \times 10^6$ N/m^2; $V_2 = 0.056$ m^3; $P_2 = 4$ bar $= 0.4 \times 10^6$ N/m^2.

The P-V diagram is shown in Fig. 2.42

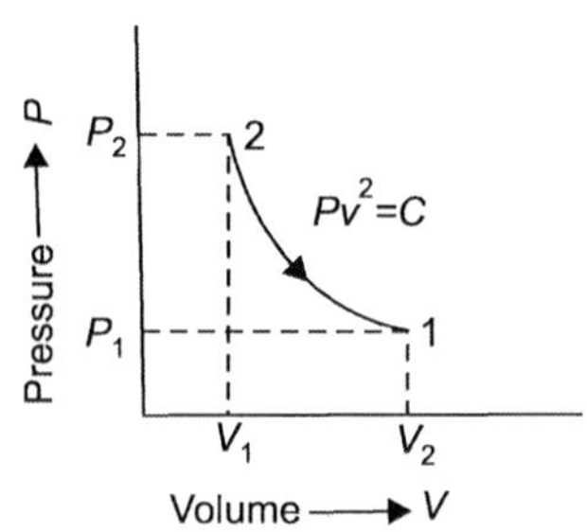

Fig. 2.42

1. **Gas constant**

   Let $\qquad R =$ Gas constant,

   We know that

$$P_1 V_1 = mRT_1$$

$$\therefore \qquad R = \frac{P_1V_1}{mT_1} = \frac{0.1\times10^6\times0.15}{0.18\times288}$$

$$= 289.4 \text{ J/kg K} = 0.2894 \text{ kJ/kg K}$$

2. Molecular mass of the gas

We know that molecular mass of the gas,

$$M = \frac{\text{Universal gas constant } (R_u)}{\text{Characteristic gas constant } (R)} = \frac{8314}{289.4} = 28.73 \text{ kg}$$

...($\because R_u$ = 8314 J/kg K, for all gases)

3. Ratio of specific heats

We know that ratio of specific heats,

$$\gamma = \frac{\log\left(\frac{P_2}{P_1}\right)}{\log\left(\frac{V_1}{V_2}\right)} = \frac{\log\left(\frac{0.4\times10^6}{0.1\times10^6}\right)}{\log\left(\frac{0.15}{0.056}\right)} = \frac{\log 4}{\log 2.678} = \frac{0.6020}{0.4278}$$

$$= 1.407.$$

4. Two specific heats, one at a constant pressure and the other at a constant volume

5. Change of internal energy.

Example 2.14: A gas undergoes a reversible non-flow process according to the relation $P = (-3V + 15)$ where V is the volume in m^3 and P is the pressure in bar. Determine the work done when the volume changes from 3 to 6 m^3.

Solution:

$$P = (-3V + 15) \text{ bar} = (-3V + 15) \times 10^5 \text{ N/m}^2$$

Work done $\quad W_{1-2} = \int p\,dV = \int_{V_1}^{V_2} (-3V + 15)\times10^5\,dV$

$$= \left|-3\frac{V^2}{2} + 15V\right|_3^6 \times 10^5 = \left[\frac{-3}{2}(6^2 - 3^2) + 15(6-3)\right]\times10^5$$

$$= (-40.5 + 45) \times 10^5 = 40.5 \times 10^5 \text{ Nm (or J)}$$

Example 2.15: Consider a gas enclosed in a piston-cylinder arrangement. The gas is initially at 150 kPa and occupies a volume of 0.03 m^3. The gas is now heated until the volume of the gas increases to 0.1 m^3. Calculate the work done by the gas if volume of the gas is inversely proportional to the pressure.

Solution: The path followed by the system is $V \propto \frac{1}{p}\cdot$ That is

$$PV = \text{constant } C = P_1V_1 = P_2V_2$$

or $$P = \frac{C}{V}$$

Work done, $$W_{1-2} = \int P\,dV = \int_{V1}^{V2} \frac{C}{V}\,dV$$

$$= C \log_e \frac{V_2}{V_1} = P_1V_1 \log_e \frac{V_2}{V_1}$$

$$= 150 \times 10^3 \times 0.03 \log_e \frac{0.1}{0.03}$$

$$= 5.41 \times 10^3 \text{ Nm (or J)} = 5.41 \text{ kJ.}$$

Example 2.16: Consider a steam turbine power plant system where the work and heat interactions across the system boundary are:

(*i*) Power available at the turbine shaft is 600 kW and the feed pump requires 3 kW of work to pump the condensate back into the boiler.

(*ii*) Steam generation in the boiler requires 2500 kJ/kg of heat, and in the condenser 1800 kJ/kg of heat is rejected to cooling water.

Make calculations for the steam flow rate in kg/hr round the cycle.

Solution:

$$\oint \delta W = 600 - 3 = 597 \text{ kW} = 597 \times 3600 \text{ kJ/hr}$$

If m is the steam flow rate in kg/hr, then

$$\oint \delta Q = m\ (2500 - 1800 = 700 \text{ m kJ/hr}$$

From the first law of thermodynamics,

$$\oint \delta Q = \oint \delta W\ ;\ 700\ m = 597 \times 3600$$

$$m = \frac{597 \times 3600}{700} = 3070 \text{ kg/hr.}$$

Example 2.17: A domestic refrigerator is loaded with fresh food and the door closed. During a certain period, the machine consumers 1.25 kWh of electrical energy in cooling the food and the internal energy of the system decreases by 4500 kJ. Find the magnitude and sign of net heat transfer for the system. The refrigerator and its contents may be considered as the system.

Solution:

Work supplied to refrigerator,

$$\delta W = 1.25 \text{ kWh} = 1.25 \times 3600 = 4500 \text{ kJ}$$

This work interaction is negative as work has been supplied to the system.

Change in internal energy,

$$dU = -4500 \text{ kJ (since decreases)}$$

From non-flow energy equation

$$\delta Q - \delta W + dU; \ \delta Q = -4500 + (-4500) = -9000 \text{ kJ}$$

Since this is negative, heat flows out of the system.

Example 2.18: A reversed heat engine absorbs 250 kJ of heat from a low temperature region and has a mechanical input of 100 kJ. What would be the heat transferred to the high temperature region? Also evaluate the coefficient of performance of the reversed engine when working as a refrigerator and as a heat pump.

Solution:

Refer Fig.

From the principle of energy conservation (1st law)

$$Q_1 = W + Q_2$$

∴ Heat transferred to the high temperature region,

$$Q_1 = W + Q_2 = 100 + 250 = 350 \text{ kJ}$$

$$(COP)_{ref} = \frac{Q_2}{W} = \frac{250}{100} = 2.5$$

$$(COP)_{ref\ pump} = \frac{Q_1}{W} = \frac{350}{100} = 3.5$$

Example 2.19: An inventor claims to have developed an engine that operates between a source at 450K and a sink at 280 K, and is capable of delivering 0.15 kWh of work for every 1200 kJ of heat received. As a patent officer, would you issue a patent for such an engine?

Solution:

For the engine proposed by the inventor,

Work output, $W = 0.15 \text{ kWh} = 0.15 \times 3600 = 540 \text{ kJ}$

Thermal efficiency,

$$\eta_{th} = \frac{\text{work output}}{\text{heat supplied}} = \frac{540}{1200} = 0.45 = 45\%$$

Thermal efficiency of a reversible engine working between the same temperture limits,

$$\eta_{rev} = \frac{T_1 - T_2}{T_1} = \frac{450 - 280}{480} = 0.378 \text{ or } 37.8\%$$

Since no engine can be more efficient than a reversible engine, the claim is false and the patent is not to be issued.

Example 2.20: A domestic food refrigerator maintains a temperature of –10°C whilst the ambient air temperature is –30°C. The heat leakage into the freezer is estimated to be at the continuous rate of 2kJ/s. Determine the least power needed to pump out this heat continuously.

Solution:

The refrigerator is required to remote heat from the freezer at the same rate at which heat leaks into it. That is

$$Q_2 = 2\text{kJ/s}$$

For minimum power requirement, the refrigerator operators on reversed Carnot cycle. Accordingly,

$$\text{COP} = \frac{T_1 - T_2}{T_1} \qquad ...(i)$$

In terms of heat and work interactions

$$\text{COP} = \frac{Q_2}{W} = \frac{Q_2}{Q_1 - Q_2} \qquad ...(ii)$$

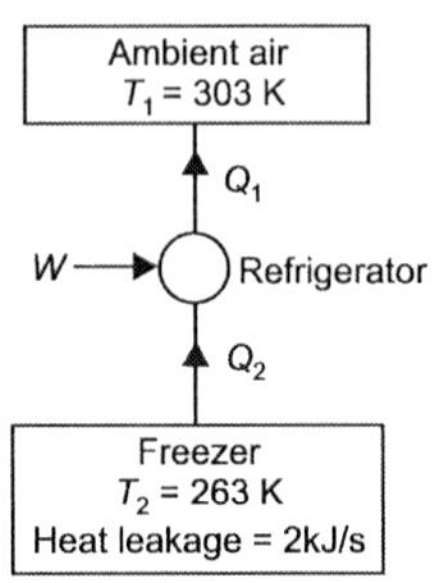

Fig. 2.43

From expressions (*i*) and (*ii*)

$$\frac{T_1 - T_2}{T_1} = \frac{Q_2}{Q_1 - Q_2} \text{ or } \frac{T_1}{T_2} = \frac{Q_1}{Q_2}$$

$\therefore$
$$Q_1 = \frac{T_1}{T_2} \times Q_2 = \frac{303}{263} \times 2 = 2.304 \text{ kJ/s}$$

and
$$W = Q_1 - Q_2 = 2.304 - 2 = 0.30 \text{ kJ/s}$$

Hence, the least power required to pump the heat continuously is 0.304 kW.

Example 2.21: It is proposed to design a refrigeration plant for a food store which is to be maintained at –5°C. The ambient temperature is 25°C and the estimated heat transfer from the store is at the rate of 5 kW. If the system operates on reversed Carnot heat engine cycle, determine thye performance index and the minimum power required to operate the refrigeration plant.

Solution:

$$T_1 = 25°C = 273 + 25 = 298K$$

$$T_2 = -5°C = 273 + (-5) = 268 \text{ K}$$

The performance of a refrigeration system is expressed in terms of coefficient of performance, and the COP of the refrigerator working between two thermal reservoirs T_1 (298K) and T_2(268 K) is given by

$$(COP)_{ref} = \frac{T_2}{T_1 - T_2} = \frac{268}{298 - 268} = 8.93$$

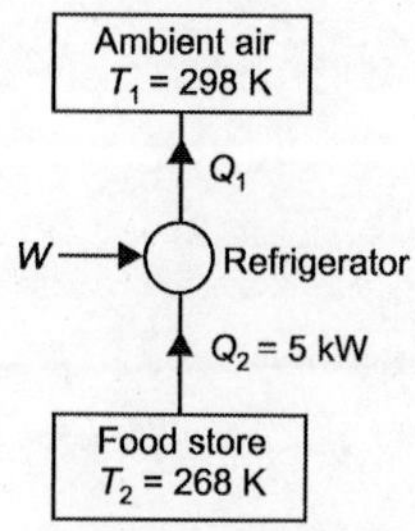

Fig. 2.44

In terms of heat interactions,

$$(COP)_{ref} = \frac{Q_2}{Q_1 - Q_2}; \; 8.93 = \frac{5}{Q_1 - 5}$$

or $8.93\, Q_1 - 44.65 = 5$

$$Q_1 = \frac{49.65}{8.93} = 5.56 \text{ kW}$$

From principle of energy conservation (1st law):

$$Q_1 = W + Q_2$$

∴ Power required to operate the refrigerator,

$$W = Q_1 - Q_2 = 5.56 - 5 = 0.56 \text{ kW}.$$

A reversible refrigerator requires the minimum power to operate it. Therefore, the minimum power required is 0.56 kW.

Example 2.22: A perfect reversed heat engine is used for making ice at –5°C from water available at 25°C. The temperature of freezing mixture is –10°C. Calculate the quantity of ice formed per kWh.

For ice: specific heat = 2.1 kJ/kgK and latent heat = 335 kJ/kg.

Solution:

The temperature limits for the reversed heat engine are:

$$T_1 = 25°C = 298 \text{ K and } T_2 = -10°C = 263 \text{ K}$$

A perfect reversed heat engine works on Carnot cycle for which COP in terms of temperatures is given by:

$$\text{COP} = \frac{T_2}{T_1 - T_2} = \frac{263}{298 - 263} = 7.514$$

$$\text{Work input} = 1 \text{ kWh} = 3600 \text{ kJ}$$

In terms of heat and work interactions,

$$\text{COP} = \frac{Q_2}{Q_1 - Q_2} = \frac{Q_2}{W}$$

∴ Total heat extracted from ice,

$$Q_2 = \text{W} \times \text{COPO} = 3600 \times 7.514 = 27050 \text{ kJ/hr}$$

Removal of heat from 1 kg of water at 25°C so that it becomes ice at –5°C is given by

$$= mc_{pw}\,\Delta T + \text{mL} = mc_{pi}\,\Delta T$$

$$= 1 \times 4.186\,(25 - 0) + 1 \times 335 + 1 \times 2.1\,[0 - (-5)]$$

$$= 104.65 + 335 + 10.5 = 450.15 \text{ kJ}$$

$$\therefore \text{ Ice formed per kWH} = \frac{27050}{450.15} = 60.09 \text{ kg.}$$

Example 2.23: A refrigerator system operating on a reversed Carnot cycle produces 400 kg/hr of ice at –5°C from water at 30°C. Make calculations for

(*a*) the power required to drive the machine,

(*b*) the heat rejected from the system.

Take latent heat of freezing = 335 kJ/kg and specific heat of ice = 2.1 kJ/kgK.

Solution:

$$T_1 = 25°C = 298K \text{ and } T_2 = 5°C = 268 \text{ K}$$

In terms of operating temperatures, the COP of the machine is given as

$$\text{COP} = \frac{T_2}{T_1 - T_2} = \frac{268}{298 - 268} = 8.93$$

Heat absorbed or net refrigerating effect

$$= (5 \times 2.1 + 335 + 4.186 \times 30) \times 400$$

$$= 188432 \text{ kJ/hr} = 52.34 \text{ kJ/s}$$

Now, $$\text{COP} = \frac{\text{refrigerating effect}}{\text{work input}}; \; 8.93 = \frac{52.34}{W}$$

∴ Work required to run the machine,

$$W = \frac{52.34}{8.93} = 5.86 \text{ kJ/s} = 5.86 \text{ kW}$$

(*b*) From the principle of energy conservation (1st law)

heat rejected = heat absorbed + work input

$$= 52.34 + 5.86 = 58.2 \text{ kJ/s}$$

Example 2.24: The air supplied to an air-conditioned room is noted to be at temperature 20°C and specific humidity 0.0085. Corresponding to these conditions, determine the partial pressure of vapour, relative humidity and dew point temperature.

Take barometric or total pressure - 1.0132 bar.

Solution:

Specific humidity

$$\omega = 0.622 \frac{P_v}{P_a} = 0.622 \frac{P_v}{P_t - P_v}$$

Thus; $$0.0085 = 0.622 \frac{P_v}{1.0132 - P_v}$$

∴ Partial pressure of vapour

$$P_v = \frac{1.0132 \times 0.0085}{0.622 + 0.0085} = 0.01366 \text{ bar}$$

(*b*) The relative humidity is defined as $\phi = \frac{P_v}{P_{vs}}$. From steam tables, the saturation vapour pressure at 20°C = 0.0234 bar.

Then: $$\phi = \frac{0.0136}{0.0234} = 0.5837 \text{ or } 58.38\%$$

(*c*) The dew point temperature is the saturation temperature of water at a pressure of 0.01366 bar.

DPT (from steam tables by interpolation)

$$= 11 + (12 - 11) \times \frac{(0.01366 - 0.01312)}{0.01401 - 0.01312)}$$

$$= 11 + 0.607 = 11.607°\text{C}.$$

HIGHLIGHTS OF THE CHAPTER

1. Thermodynamics is the brand of science that deals with heat energy and work possessed by vapours.

2. There are three systems of Thermodynamics

(*i*) Open (*ii*) Closed (*iii*) Isolated

Open systems: The system in which mass and energy transfer across its boundry.

Closed: The system in which mass remains constant and only energy may transfer across its boundry.

Isolated: The system in which both mass and energy remains constant within the its boundry

3. **Property:** The quantities which characterise the given state of the system are called its properties and parameters.

4. **Intensive:** The property which is independent from the substance (mass) present in the system.

Extensive: The property which is dependent upon the substance (mass) present in the system.

5. Force is a product of mass and the acceleration acted upon a body in the direction of force. $F = ma$.

5. Density is defined as re-ratio of mass per unit volume – unit-kg/m^3

7. Specific weight defined as the ratio of weight per unit volume unit-n/m^3

8. Specific gravity of the liquid is defined as the ratio of specific weight of the liquid to the specific weight of water.

9. Specific volume defined as volume per unit mass.

10. Viscosity is the cohesive force between the liquid particles and exhibited when liquid is in motion.

11. Pressure is defined as the normal component of force per unit area.

12. Gauge pressure is defined as a pressure measured with messing instrument with respect to atmospheric pressure 1.0132 bar.

13. Thermodynamic equilibrium is that in which state variables remains constant throughout the system and do not change with time and space.

14. The instantauous condition of a thermodynamic system is called its state.

15. A thermodynamic system which is passing through a series of states called a path.

16. A system in which final state is identical with the initial state when the path completely specified is called a process.

17. Isothermol process is that process in which temperature remains constant $\delta T = 0$.

18. Adiabatic process is that process in which heat remains constant $\delta q = 0$.

19. Isobaric process is that process in which pressure remains constant $\delta P = 0$.

20. Isochoric process is that process in which volume remains constant $\delta V=0$

21. Cyclic process is that process in which final state is identical with the initial state called cyclic process.

22. Quasi-static process is that process in which the system deviates from one equilibrium state by only infinitesimal amount throughout the entire process.

23. Reversible process is that in which each step remainus in equilibrium with the previous step.

24. Irreversible process is that which is not identical with the previous step.

25. Capacity of doing work is known as energy.

26. The energy possessed by a body by virtue of its mass and velocity of motion is called kinetic energy.

$$\text{KE} = \frac{1}{2}\, mv^2\text{-Nm}$$

27. The energy possessed by a body due to its molecular arrangement and motion of the molecules is called internal energy.

28. The energy possessed by a body by virtue of its height above the earth surface is called potential energy.

29. Law of conservation of mass states that the mass can neither be created nor destroyed but it can be changed from one form to another form.

30. Heat is the energy transformed without transformer of mass across the boundary of a system due to temperature difference between the system and surrounding.

31. Specific heat is defined as the amount of heat required to raise the temperature of a unit mass of any substance through one degree- unit kJ/kg °K.

32. Temperature is a thermodynamic property which determines the degree of hotness or coolness due to change in level of heat intensity.

33. Work is a force acted upon a body causing a body to move along the direction of force called work.

$$W = F \times x$$

34. Law of Perfect Gases

(*a*) *Boyle's low*: The absolute pressure of a given mass of a perfect gas is inventory proportional to its volume.

$$P \propto \frac{1}{V}$$

$$PV = \text{Constant}$$

(*b*) *Charles's low*: The volume of a given mass of perfect gas is directly proportional with Temperature

$$V \propto T$$

$$\frac{V}{T} = \text{Constant}$$

(*c*) *Gay-lussac law*: The absolute pressure of a given mass of a perfect gas varies directly as its absolute temperature.

$$P \propto T$$

$$\frac{P}{T} = \text{Constant}$$

(*d*) *Joule's law*: The change of internal energy of a perfect gas is directly proportional to the change of temperature.

$$\partial u \propto \partial T$$

$$\partial u \propto mc\,\partial T$$

$$\partial u = mc\,(T_2 - T_1)$$

35. First law of thermodynamics defined as

(*a*) The heat and mechanical work are mutually convertible.

(*b*) The energy can neither be created nor destroyed but it can be transformed from one form to another form.

36. Second law of Thermodynamics defined as

(*a*) According to Kelvin planck statement it is impossible to construct an engine working on a cyclic process whose sole purpose is to convert heat energy supplied to it into an equivalent amount of work.

(*b*) According to Clausius statement it is impossible for a self-acting machine without any external source to supply heat from a body at a lower temperature to a body at higher temperature.

37. Third law of thermodynamics defined as

According to the Third law when a system is at zero absolute temperature, the entropy of the system is $2\rho W^{n}$

REVIEW QUESTIONS

1. Define system, boundary and surrounding.

2. Differentiate between

(*i*) open and closed system

(*ii*) extensive and intensive properties

(*iii*) reversible and irreversible processes

(*iv*) closed and isolated system

(*v*) adabatic and isothermal process

(*vi*) efficiency and COP

(*vii*) heat engine and heat pump

(*viii*) heat pump and refrigerator

(*ix*) PPM_1 and PPM_2

3. Give the thermodynamic definition of work

4. State the zeroth law of thermodynamics. What is its importance?

5. Define the first law of thermodynamics. Give mathematical relationship for this law.

6. State the laws of perfect gas.

7. What is ment by thermodynamic equilibrium?

9. What are the limitations of first law of thermodynamics?

10. State the 2nd law of thermodynamics as per Kelvin-Planck and Clausius.

11. Show that whenever a system executes a complete cyclic process, if $\oint \frac{\delta Q}{T}$ is less than zero or in the limit is equal to zero. Hence, prove that entropy is a property of the system.

12. Define the term coefficient of performance as applied to a refrigerator and a heat pump. Show that $COP_{H.P.} = COP_{ref} + 1$.

13. What is its importance in thermodynamics?

16. A system execurtes a cyclic process. The heat transfer in each process are: 224 kJ, –20 kJ, -200 kJ and 37 kJ. **Ans.** 41 kJ.

17. The working fluid in an engine executes a cyclic process, and two work intractions are involved: 10 kJ to the working fluid and 30 kJ from the working fluid. The working cycle also involve their heat transfer two of which are 75 kJ to the working fluid and 40 kJ from the working first. Determine the magnitude and direction of the third heat interaction.

Ans. = –15 kJ (work flow out of the system).

■■■

3

Properties of Steam and Steam Generator

3.1 INTRODUCTION

Steam is a vapour of water and it is invisible when pure and dry. Like any other pure substance it can be converted in to three stages *i.e.* solid, liquid and gas. It is a most common working substance as working medium like steam engines, steam turbines and in atomic power plants for power generation steam does not obey laws of perfect gases, until it is perfectly dry. It uses in the process industries as well as heat carrying agent in heat exchange equipments.

3.2 THERMODYNAMIC PROPERTIES OF STEAM

3.2.1 Wet Steam

When the steam contains moisture particles of water in suspension means there is no complete evaporation of water and whole of latent heat is not absorbed by steam.

3.2.2 Dry Saturated Steam

When the wet steam is further heated and does not contain any moisture content, in the steam known as dry saturated steam. The dry saturated steam has absorbed its full latent heat and behaves like a perfect gas is known as dry saturated steam.

3.2.3 Superheated Steam

When the dry steam is further heated at constant pressure by raising its temperature and volume called superheated steam.

3.2.4 Wetness Fraction

It is the ratio of mass of water vapour or moisture in suspension to the total mass of steam. It is generally denoted by y.

$$y = \frac{mf}{mf + mg} = \frac{1 - mg}{mf + mg}$$

where

mf = mass of fluid *i.e.* mass of water in suspension.

mg = mass of actual dry steam.

The wetness fraction is expressed in per cent is called priming.

3.2.5 Dryness Fraction of Steam or Quality of Wet Steam

It is the ratio of mass of actual dry steam to total mass of steam. It is generally denoted by x

$$x = \frac{\text{mass of actual dry steam}}{\text{mass of mixture}} = \frac{mg}{mg + mf} = \frac{mg}{m}$$

where

mg = mass of actual dry steam

mf = mass of fluid *i.e.* mass of water in suspension

m = mass of wet steam = $(mg + mf)$

The dryness fraction when expressed in per cent is called quality of steam.

If value of

$x < 1$ wet steam

$x = 1$ dry steam

$x > 1$ superheated steam.

3.2.6 Enthalpy of Saturated Water or Sensible Heat (h_f)

The amount of heat energy, absorbed by 1 kg of water from 0°C to boiling point or saturation temperature at a given pressure. It is also called enthalpy of saturated water. It is denoted by (h_f). Sensible heat increases as the pressure increases. The value of enthalpy can directly be obtained from steam table.

The specific heat of water at constant pressure is usually taken as 4.2 kJ/kg

= Mass × Sp. heat × Rise in temperature

$= 1 \times 4.2\,[(t + 273) - (0 + 273)]$

$= 4.2\,t$ kJ/kg

3.2.7 Enthalpy of Vaporisation or Latent Heat (h_{fg})

The amount of heat required to convert 1 kg of water at saturation temperature (boiling point) for a given pressure into dry and saturated steam without change of temperature. The value is not constant, the valve of h_{fg} decreases as pressure of water increases and becomes zero at critical point. The valve of h_{fg} and directly can be obtained from the steam table. The Lateat Heat (L.H.) of steam is 2257 kJ/kg at atm. pressure.

3.2.8 Enthalpy of Steam or Total Heat (h_g)

The amount of heat required to convert 1 kg of water at °C into superheated steam at constant pressure.

∴ Total heat of steam = sensible heat + latent heat

$$h_{fg} = (h_f + h_{fg})$$

It is denoted by h_g and its valve can be obtained directly from the steam table.

The expressions for the enthalpy of steam.

(*i*) Wet steam

$$h = h_f + xh_{fg}$$

where h_f = sensible heat

h_{fg} = latent heat

x = dryness fraction

(*ii*) Dry steam

$$h = h_f + h_{fg} - h_g \qquad x = 1 \text{ (for dry steam)}$$

(*iii*) Superheated steam:

h_{sup} = Total heat for dry steam + Heat of superheated steam.

$$= h_f + h_{fg} + C_{ps}\,(t_{sup} - t_{sat})$$

$$= h_g + c_p\,(t_{sup} - t_{sat})$$

where

C_p = Means specific heat at const pressure for superheated steam.

t_{sup} = Temperature of superheated steam

t_{sat} = Saturation temperature at given const pressure.

$(t_{sup} - t)$ = Known as degree of superheat

The valve of C_p for steam lies between 1.67 kJ/kg°K of 2.5 kJ/kg/°K.

3.2.9 Specific Volume

It is defined as the volume occupied by the steam per unit mass at a given temperature and pressure and it is expressed in m^3/kg and represented by V.

Note: The value of specific volume decreases with the increase in pressure.

Expression for the volume occupied by steam.

(*i*) **Wet Steam:** Consider 1 kg of wet steam of dryness fraction x and having x kg of dry steam and $(1 - x)$ kg of water.

Let v_f is the volume of 1 kg of water, then volume of 1 kg of wet steam

$$= xV_g + (1 - x)\, V_g$$

$$= xV_g \text{ m}^3 \qquad [\because V_f \text{ is very small as compared to } v_g.]$$

Hence $(1 - x)\, V_f$ may be is neglected

Volume of wet steam V_g = specific volume of dry saturated steam.

V_f = specific volume of water

$V = xV_g$ m^3/kg

∴ Specific volume of 1 kg of wet steam.

(*ii*) **Dry Steam:** In case of dry steam, the mass of water in suspension is zero and (dryness fraction) is unity

∴ specific volume of dry steam $V = V_g$ m^3/kg

(*iii*) **Superheated Steam:** The superheated steam behaves more or less like a perfect gas by keeping the pressure as a constant.

Hence, when the steam is further heated in to dry steam under a constant pressure. There is a increase in volume with the rise in temp. The superheated steam behaves more or less like a perfect gas.

According to Charles' Law

$$\frac{V_{sup}}{T_{sup}} = \frac{V_g}{T_{sup}} \qquad V \propto T \text{ or } \frac{V}{T} = \text{constant}$$

$$V_{sup} = \frac{V_g \times T_{sup}}{T_{sup}} \qquad (\because P = \text{constant gas})$$

where

V_{sup} = Specific volume of superheated steam

V_g = Specific volume of dry steam

T_{sup} = Absolute temperature of superheated steam

T_{sat} = Absolute saturation temperature at the pressure of steam formation

3.2.10 Entropy of Steam

The heating and evaporation of water during the steam formation process takes place at a constant pressure so the superheating of steam is also done at constant pressure. Hence, entropy of vaporization is denoted by S_{fg}.

3.2.11 Entropy of Evaporation

Evaporation of water completely takes place at constant pressure and constant temprature (T_s) by absorbring its full latent heat. (h_{fg}).

$$\text{Entropy} = \frac{\text{Heat absorbed, (latent heat of vaporization)}}{\text{Absolute temperature or saturation temperature}}$$

$$S_{fg} = \frac{h_{fg}}{T_s}$$

If the steam is wet, partial evaporation will take place with dryness fraction x.

$$= xh_{fg}$$

$$S_{fg} = \frac{xh_{fg}}{T_{sat}} \text{ (for wet saturated steam)}$$

3.2.12 Expression for Entropy of Steam

(*i*) **Wet Steam:** Entropy of water + Entropy change during evaporation

$$s = s_f + xs_{fg} \quad \text{or} \quad s = s_f + \frac{xh_{fg}}{T_{sat}}$$

(*ii*) **Dry Steam**

$$s_g = s_f + \frac{h_{fg}}{T_{sat}} = s_f + s_{fg} = s_g \qquad [\because x = 1]$$

The value of s_g can be directly read from the steam table.

(*iii*) **Superheated Steam**

Entropy change during superheating at constant pressure

$$ds = \frac{\text{Heat of superheat}}{T_{sat}} = \frac{\delta q}{T_{sat}}$$

$$\delta q = C_{ps} dT \text{ per kg of steam}$$

$$\frac{\delta q}{T_{sat}} = \frac{C_{ps} dT}{T_{sat}}$$

(dividing both sides by T_s)

$$\therefore \quad ds = \frac{C_{ps} \times dT}{T_{sat}} \qquad C_{ps} = \text{specific heat}$$

$$\left[\because \frac{\delta f}{T} = ds\right]$$

(The value of specific heat varies from = 2 kJ/kg °K – 2.5 kJ/kg °K)

The total increase in entropy of superheated steam above freezing point of water for T°K to T_{sup}. Hence entropy from T to t_{sup} may be obtained by integrating the above expression between the proper limit.

$$\int_{Sg}^{S_{sup}} d_S = C_{ps} \int_{Ts}^{T_{sup}} \frac{dT}{T}$$

or $$S_{sup} - S_g = C_{ps} \log_e \left(\frac{T_{sup}}{T_{sat}}\right) = 2.3\, C_{ps} \left(\frac{T_{sup}}{T_{sat}}\right)$$

$(S_{Sup} - S_g)$ is increase in entropy

$$S_{Sup} = S_g + 2.3\, C_{ps} \log_e \left(\frac{T_{sup}}{T_{sat}}\right)$$

3.2.13 Internal Energy

It is the actual heat energy stored in steam above the freezing point of water is known as internal energy of the steam. It can be calculated by subtracting the external work done during evoporation from the total heat. It is denoted by u.

Internal energy of steam = Enthalpy – External work done during evaporation

Expression for the Internal Energy of Steam

1. Wet Steam

$$u = h - 100\, PV_g$$

$$u = (h_f + xh_{fg}) + 100\, PxV_g$$

$$u = h_g - 100\, PrV_g \text{ kJ/kg}$$

2. Dry Saturated Steam

$$u = h_g - 100\, PrV_g$$

$$u = h_f + h_{fg} - 100\, Prg \text{ kJ/kg}$$

3. Superheated Steam

$$u = h_{sup} - 100\, PV_{sup}$$

$$= [h_g + C_p\,(t_{sup} - t_{sat}] - 100\, PV_{sup} \text{ kJ/kg}$$

3.2.14 External Work Done during Evaporation

When the water at boiling temperature is heated at a constant pressure, it gets converted into steam after absorbing the latent heat the work done in overcoming the external resistance to the movement of the piston due to increase in its volume. This work is called external work of evaporation and represents the energy which has been taken out of the steam.

External work = $P(V_g - V_f)$

W = Pressure × change in volume

$W = P \times \delta V$

where

W - external work done

P - pressure on the piston in bar = $P \times 10^5$ N/m^2

V_f - volume of water in m^3 at pressure P.

V_g - volume of steam in m^3 at pressure P.

so $W = P \times 10^5 (V_g - V_f)$J

$= 100\, P\, (V_g - V_f)$ kJ

At low pressures, the volume of water (V_f) is very small, as compared to the volume of steam (V_g).

Hence neglecting the value of V_f in above equation.

$W = 100\, PV_g$

Expression for the external work done during evaporation.

wet steam $W = 100\, PxV_g$ kJ where x-dryness fraction

Dry steam $W = 100\, PV_g$ kJ (work done during evoporation)

superheated steam

$W = 100\, PV_{sup}$ kJ.

or

Total work done = W.D. during evoporation + W.D. during superheating

$= 100\, PV_g + 100\, P(V_{sup} - V_g)$

$= 100\, PV_{sup}$ kJ.

3.3 FORMATION OF STEAM AT CONSTANT PRESSURE FROM WATER

Consider 1 kg of ice under a pressure say p bar and temperature of –10°C is kept in a vessel and heated by keeping the pressure constant. The volume of ice will increase slightly with increase in temprature and reaches up to freezing temprature at °C at atmospheric pressure. Now water at 0°C contained in the piston cylinder arrangement as shown in Fig. 3.1(*a*). If heating of water at 0°C is continued in the cylinder, its temperature rises and continues to rise until the boiling point is reached.

1. As the temperature increases there is slightly increase in the volume of water as shown in Fig.3.1(*a*) and will cause the piston to move slightly upward and hence work is obtained at this stage increase in volume is considerably small. It is neglected for all type of calculations.
2. With the addition of further heat, the temperature of water rises up to boiling point corresponding to pressure P exerted by the piston and the boiling point of water at (NTP), pressure is 1.013 bar (76 mm of Hg) and temperature is 100°C. But it increases as increase in pressure as shown in Fig. 3.1(*b*).
3. During evaporation process, the temperature remains constant and piston pushes upward against the constant pressure. Hence, the specific volume of steam increases as shown in Fig. 3.1(*c*) at this stage steam have same particles of water and is termed as wet steam or sensible heat (S.H.) of water.
4. On further heating the temperature remains constant and the water particles in suspension will be converted into steam and it is known as dry steam or latent heat of steam (L.H.) denoted by (h_{fg}) as shown in Fig.3.1 (*d*)

5 On further heating, the temperature of the steam starts rising (heating above its saturation temperature). Hence, the sensible heat of water increases with increasing pressure while the latent heat decreases with increase in pressure the steam in such a state is termed as superheated steam as shown in Fig. 3.1(*e*)

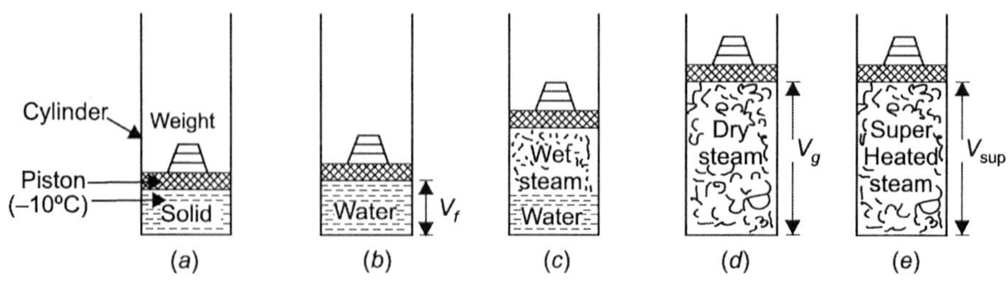

Fig. 3.1 Formation of steam at a Constant Pressure from Water.

3.4 ADVANTAGES OF SUPERHEATED STEAM

The advantages of superheated steam are as follows:

1. The heat content of superheated steam is more than saturated steam, hence the capacity to do work is more without increase in pressure.
2. The overall thermal efficiency of the plant increases by using superheated steam.
3. Increase in wetness during expansion of superheated steam is less than saturated or wet steam, hence the condensation gets reduced at the last stage of engine or turbine.
4. Superheating is done in superheaters located inpath of waste hot flue gases without any additional cost.
5. Superheated steam is free from moisture and it is used to avoid corrosion and erosion of the surfaces which are coming in direct contact of the steam.

3.5 TERMINOLOGY OF PURE SUBSTANCE

1. *Saturation temperature:* The temperature at which a pure substance starts to evaporate at a given pressure.
2. *Saturation pressure:* The pressure at which a pure substance starts to evaporate at a given temperature.
3. *Saturated steam:* Steam is about to condense.
4. *Super heated steam:* The steam at higher temperature than its saturation temperature.
5. *Wet steam:* Mixture of dry steam and moisture content of water.
6. *Sublimation:* Transformation of solid phase directly into vapour phase called sublimation.
7. *Degree of superheat:* It is the difference between ($T_{sup} - T_{sat}$) is called degree of superheat.

3.6 TEMPERATURE *vs* TOTAL HEAT GRAPH DURING STEAM FORMATION

The process of steam formation, as discussed above, may also be represented on a graph taking heat (enthaply) on *X* axis and temperature on *Y* axis. The point A represents the initial condition of water at 0°C and represents pressure P in (bar) as shown in the figure.

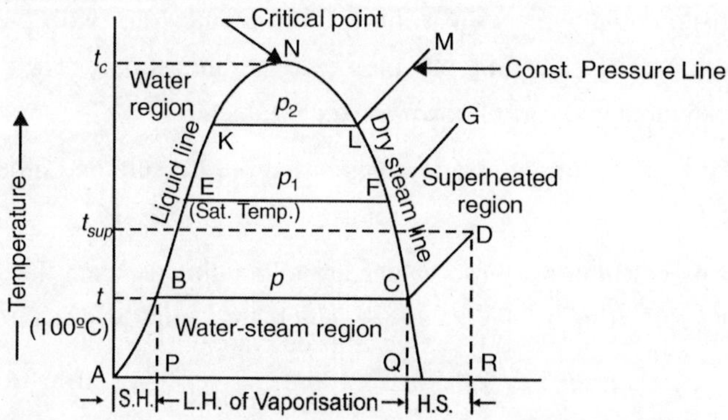

Fig. 3.2 Temperature-total heat graph during steam formation.

A – Initial condition of water at 0°C and pressure P (bar)

ABCD – Relation between temperature and heat at specific pressure.

AB – Heating of water up to boiling temp. or saturation temperature.

AP – Heat absorbed by water *i.e.* sensible heat or liquid heat.

BC – Change of state from liquid to steam.

PQ – Heat absorbed during this state is latent heat of vaporisation.

CD – Superheating process

QR – Heat absorbed during this state.

AR – Total heat of the superheated steam and during this state as temperature increases pressure also increases.

E – Boiling temperature or saturation temperature at pressure p_1

F – Dry saturated steam point

FG – Constant pressure process (superheated steam)

N – Critical point (at this point liquid heat becomes zero and liquid directly flashes to vapour).

at critical point the values for water are:

Critical Temperature $t_c = 374.15°C$

Critical pressure $P_c = 221.2$ bar.

Critical volume $V_c = .003155$ m^3/kg

3.7 STEAM TABLES

The properties of dry saturated steam like saturation temperature, pressure, sensible heat, latent heat, enthalpy specific volume, entropy etc. vary with pressure. The values of these properties can be obtained by experiments only. These properties are available in tabulated form known as steam tables.

There are two type important steam tables based on pressure and other saturated temperature.

For engineering calculations these values are taken directly from the steam table and calculated each time by the equations which are complex in nature.

Saturated Steam Table

Temp.	Pressure	Specific Volume, m³/kg			Internal Energy, kJ/kg			Enthalpy, kJ/kg			Entropy, kJ/kg K		
T °C	P kPa	Sat. Liquid V_f	Evap. V_{fg}	Sat. Vapor V_g	Sat. Liquid U_f	Evap. U_{fg}	Sat. Vapour U_g	Sat. Liquid h_f	Evap. vfg h_{fg}	Sat. Vapour h_g	Sat. Liquid s_f	Evap. s_{fg}	Sat. Vapour s_g
0.01	0.6113	0.001000	206.131	206.132	0	2375.33	2375.33	0.00	2501.35	2501.35	0	9.1562	9.1562
5	0.8721	0.001000	147.117	147.118	20.97	2361.27	2382.24	20.98	2489.57	2510.54	0.0761	8.9496	9.0257
10	1.2276	0.001000	106.376	106.377	41.99	2347.16	2389.15	41.99	2477.75	2519.74	0.1510	8.7498	8.9007
15	1.705	0.001001	77.924	77.925	62.98	2333.96	2396.04	62.98	2465.393	2528.91	0.2245	8.5569	8.7813
20	2.339	0.001002	57.7887	57.7897	83.94	2319.98	2402.91	83.94	2454.12	2538.06	0.2966	8.3706	8.6671
25	3.169	0.001003	43.3583	43.3593	104.86	2304.90	2409.76	104.87	2442.30	2547.17	0.3673	8.1905	8.5579
30	4.246	0.001004	32.8922	32.8932	125.77	2290.81	2416.58	125.77	2430.48	2556.25	0.4369	8.0164	8.4533

Saturated Steam Table

Pressure	Temp.	Specific Volume, m³/kg			Internal Energy, kJ/kg			Enthalpy, kJ/kg			Entropy, kJ/kg K		
kPa P	°C T	Sat. Liquid V_f	Evap. V_{fg}	Sat. Vapor V_g	Sat. Liquid U_f	Evap. U_{fg}	Sat. Vapour U_g	Sat. Liquid h_f	Evap. h_{fg}	Sat. Vapour h_g	Sat. Liquid s_f	Evap. s_{fg}	Sat. Vapour s_g
0.625	0.01	0.001000	206175	206.132	0	2375.70	0000.00	0.06	2501.3	2501.30	0	9.1562	9.1563
1	6.98	0.001000	29.20702	129.20802	29.29	2355.69	2394.98	29.29	2484.89	2514.18	0.1059	8.8697	8.9756
1.5	13.03	0.001001	87.97913	87.98013	54.70	2338.64	2393.32	54.80	2320.59	2525.30	0.1956	8.6322	8.8278
2	17.50	0.001001	67.00285	67.00385	73.47	2326.02	2399.48	73.47	2460.02	2533.49	0.2607	8.4629	8.7236
2.5	21.08	0.001002	54.25285	54.25385	88.47	2315.93	2404.40	88.47	2451.56	2540.03	0.3120	8.3310	8.6431
3	24.08	0.001003	45.66402	45.66502	101.03	2307.48	2408.51	101.03	2444.47	2545.50	0.3545	8.2231	8.5775
3.5	28.96	0.001004	34.79915	34.80015	121.44	2293.73	2415.17	121.44	2432.93	2554.37	0.4226	8.0520	8.4746

Table 1.1: Conversion Factors for Pressure

	bar	dyne/cm^2	h_{gf}/cm^2 ata	N/m^2 or Pa	mm Hg (21°C) or torr	mm H_2O (21°C)	atm
bar=	1	10^6	1.01972	10^5	750.062	10197.2	0.986923
1 dyne/cm^2 =	10^{-0}	1	1.01972 $\times 10^{-6}$	0.1	750.062 $\times10^{-6}$	10197.2 $\times10^{-6}$	0.986923 $\times10^{-6}$
1 kgf/cm^2 =	0.980665	0.9800665 $\times10^6$	1	0.980006 $\times10^{-5}$	735.559 $\times10^{-5}$	1000 $\times10^{-5}$	0.967838
1 mm Hg= (21°C)	1.333223 $\times10^{-3}$	1.323223 $\times10^3$	1.3595 $\times10^{-3}$	1.333223 $\times10^2$	1	13.5951	1.31578 $\times10^{-3}$
1 mm H_2O = (21°C)	98.0665 $\times10^{-6}$	98.0665	10^{-4}	9.80665	0.073556	1	96.7838 $\times10^{-6}$
1 atm =	1.01325	1.01325 $\times10^6$	1.03323	1.01325 $\times10^5$	760	10332.3	1

Superheated Steam Table

Temp. °C	*V* (m^3/kg)	*U* (kJ/kg)	*h* (kJ/kg)	*s* (kJ/kg K)	*V* (m^3/kg)	*U* (kJ/kg)	*h* (kJ/kg)	*s* (kJ/kg K)
	P = 10 kPa (45.81)				P = 50 kPa (81.33)			
Sat.	14.67355	2437.89	2584.63	8.1501	3.24034	2483.85	2645.87	7.5939
50	14.86920	2443.87	2592.56	8.1749	—	—	—	—
100	17.19561	2515.50	2687.46	8.4479	3.41833	2511.61	2682.52	7.6947
150	19.51251	2587.86	2782.99	8.6881	3.88917	2585.61	2780.08	7.9400
200	21.82507	2661.27	2879.52	8.9037	4.35595	2659.85	2877.64	8.1579
250	24.13559	2735.95	2977.31	9.1002	4.82045	2734.97	2975.99	8.3555
300	26.44509	2812.06	3076.51	9.6076	6.20929	2968.43	3278.89	8.86141
400	31.06252	2968.89	3279.51	9.6076	6.20929	2968.43	3278.89	8.8611
500	35.67896	3132.26	3489.05	9.8977	7.13364	3431.94	3488.62	9.1545
600	40.29488	3302.45	3705.49	40.4608	8.05748	3302.22	3705.10	9.4177
700	44.91952	3479.63	3928.73	10.4028	8.98104	3479.45	3928.51	9.6599
800	49.52599	3663.84	4159.10	10.6281	9.90444	3663.70	4158.92	9.8852
900	54.14127	38.3303	4.84244	10.8395	10.34773	4854.93	4396.30	10.0967
1000	58.75669	4053.01	4640.58	11.0392	11.75097	4052.91	4640.46	10.2964

Compressed Liquid Water

Temp. °C	V (m³/kg)	U (kJ/kg)	h (kJ/kg)	s (kJ/kg °K)	V (m³/kg)	U (kJ/kg)	h (kJ/kg)	s (kJ/kg °K)
	50 bar (263.99)				100 bar (311.06)			
Sat.	0.001286	1147.78	1154.21	2.9201	0.001452	1393.00	1407.53	3.3593
0	0.000998	0.03	5.02	0.001	0.000995	0.10	10.05	0.0003
20	0.00100	83.64	88.64	0.2955	0.000997	83.35	93.32	0.2945
40	0.001006	160.93	171.95	0.5705	0.001003	166.33	176.36	0.5685
60	0.001015	250.31	255.28	0.8284	0.001013	249.34	259.47	0.8258
80	0.001027	333.69	338.83	1.0719	0.001025	332.56	342.81	1.0687
100	0.001011	417.50	422.71	1.3034	0.001039	416.09	426.48	1.2992
120	0.001058	501.79	507.07	1.5232	0.001055	500.07	510.61	1.5188
140	0.001027	586.74	592.13	1.7342	0.001074	584.69	595.40	1.7291
160	0.001099	672.61	678.10	1.9374	0.000195	670.11	681.07	1.9316
180	0.001124	759.62	765.24	2.3341	0.001129	756.63	767.83	2.1274
200	0.001153	848.08	853.85	2.3254	0.001148	844.49	855.97	2.3178
220	0.001187	938.43	944.36	2.3328	0.001184	934.07	945388	2.5038
240	0.001226	1031.34	1037.47	2.6978	0.001219	1025.94	1038.13	2.6872
260	0.001275	1127.92	1134.30	2.8829	0.00265	1121.03	1133.68	2.8698

3.8 MOLLIER DIAGRAM (ENTHALPY *v.* ENTROPY DIAGRAM)

Entropy (*h-s*) diagram for water and steam. It is a graphical representation of steam tables which was first introduce by Richard Mollier in 1904. In this diagram enthalpy (*h*) is plotted along *y* axis and entropy along *x* axis first of all enthalpy and entropy of water and dry saturated steam for any particular pressure are obtained from the steam table. The given values of enthalpies and entropies are plotted and then liquid line and dry saturated line is obtained. The diagram is divided into two portions by a thick curve line termed as saturation line. In the lower *i.e.* (wet region) and upper region (*i.e.* super heat) region. The temperature of steam increases at the given pressure.

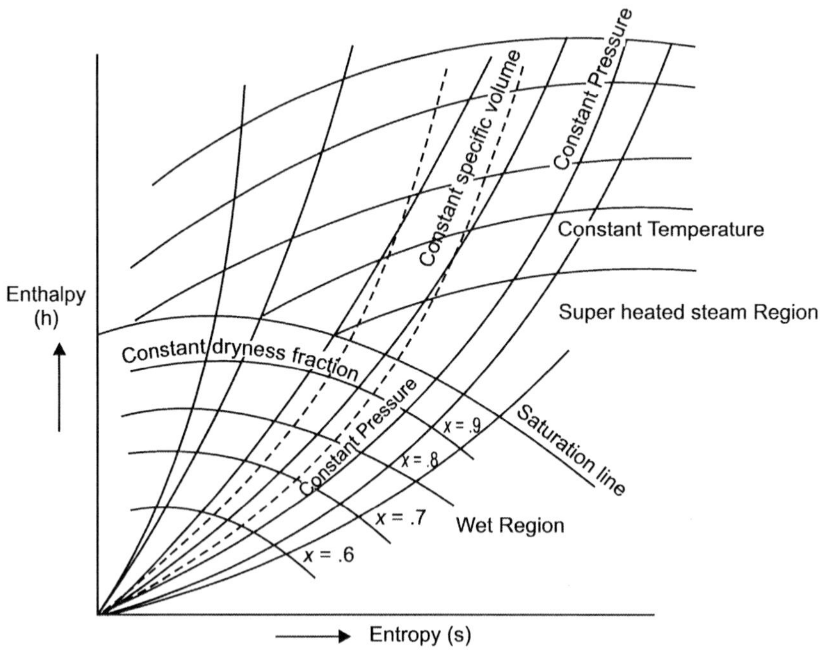

Fig. 3.3 Enthalpy *v.* entropy (mollier chart).

The enthalpy-entropy chart, like temperature entropy, is also very useful in solving the problems on is entropic expansion and compression of steam and it is also a useful tool in the analysis of steady flow devices such as nozzles and turbines. The mollior diagram has the following lines:

1. Dryness fraction line
2. Constant volume lines
3. Constant pressure lines
4. Isothermal lines.
5. Isentropic lines
6. Throttling lines

1. **Dryness Fraction Lines on *h-s* Diagram:** The dryness fraction lines are drawn only below the saturation line and represent the condition of wet steam between various values of enthalpy and entropy.
2. **Constart Volume Lines on *h-s* Diagram:** The constant volume lines are drawn in both wet steam region and superheated steam region. These lines are straight in wet steam region and curved upward in superheated region.
3. **Constant Pressure Lines on *h-s* Diagram:** The constant pressure lines are drawn in both the wet steam region and superheated region. These lines are straight in wet steam region, because during vaporisation the increase of enthalpy is directly proportional to the increase in quality and

hence increase in entropy. The constant pressure lines are curved slightly upward above the saturation curve.

4. **Isothermal Lines on *h-s* Diagrams:** Isothermal means constant temperature line above the saturation line and represent the condition of superheated steam values of enthalpy and entropy.
5. **Isentropic Lines of *h–s* Diagram:** Process means there is no change of entropy *i.e.* entropy remains constant.
6. **Throttling Process:** Throttling process means, there is no change of enthalpy *i.e.* enthalpy remains constant.

3.9 TEMPERATURE ENTROPY (T-S) DIAGRAM

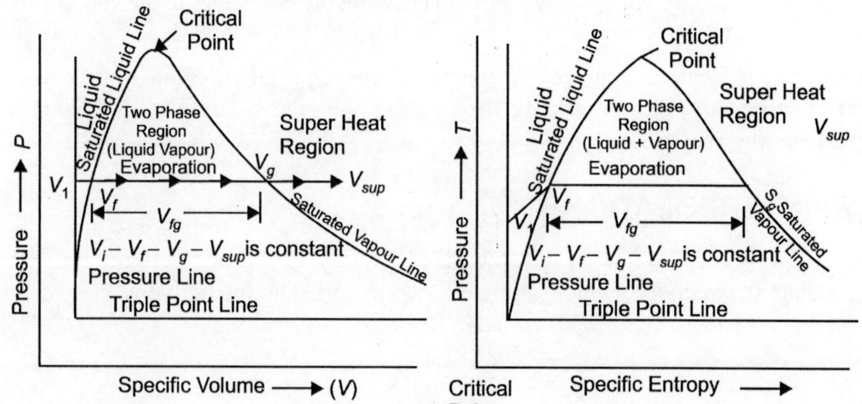

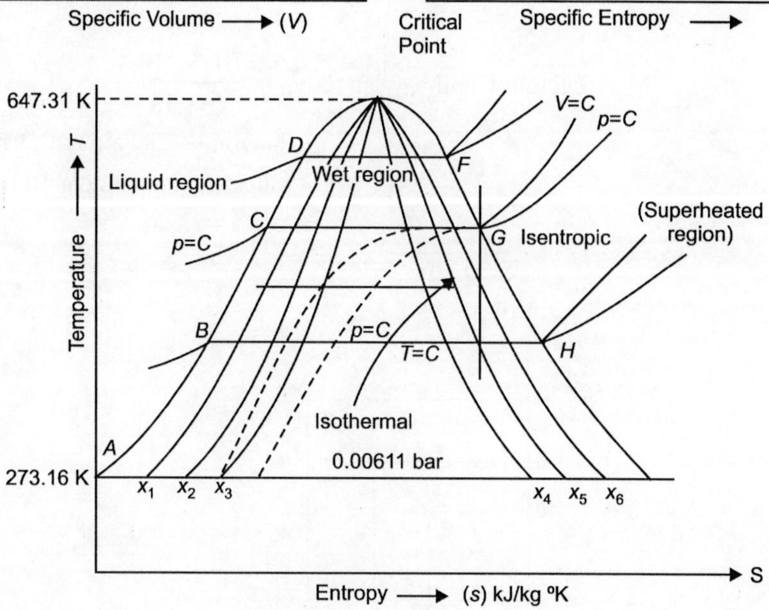

Fig.3.4 Temperature *v.* Entropy.

Temprature-entropy diagram is very useful in solving the problems on adiabatic expansion and compressions of steam. Entropy of 1 kg of water and steam above freezing point of water is represented on x axis and temprature of 273.16°K on y axis and entropy of liquid at the critical point becomes equal to zero. Liquid line *ABCDE* and Voapour line *EFGHI* divide the diagram into three main parts.

Left of *AE i.e.*, liquid region, between *AE* and *EI* is wet vapour region (liquid + vapour) and right of *EI* is superheated region. *E* is the critical point with (P_c) critical pressure 221.2 bar, (T_c) critical temprature 374.15°C and critical volume (V_c) is .003155 m^3/kg.

In the diagram constant pressure lines are parallel in the wet region and curved in superheated region and wet region represent dryness fraction and constant volume line.

The constant volume lines are steeper than constant pressure lines in the superheated zone.

Advantage of T-S diagram over P-V diagram is that area under T-S diagram represent heat absorbed or rejected during the process whereas area under PV diagram represents the workdone.

3.10 INTERPOLATION

For the properties which are not tabulated exactly in the tables is necessary to interpolate between the values tabulated, *e.g.* to calculate the enthalpy of 15.5 bar

$$y = y_1 + \frac{P - P_1}{P_2 - P_1}(y_2 - y_1)$$

$$y = \text{Saturation enthalpy at 10 bar} + \frac{(10.5 - 10)}{11 - 10}$$
$$\times \text{(Saturation enthalpy at 11 bar} - \text{Saturation enthalpy at 10 bar)}$$

$$y \text{ or } h = 2778 + \frac{(10.5 - 10)}{11 - 10} \times (2780 - 2778)$$
$$= 2778 \text{ kJ/kg}$$

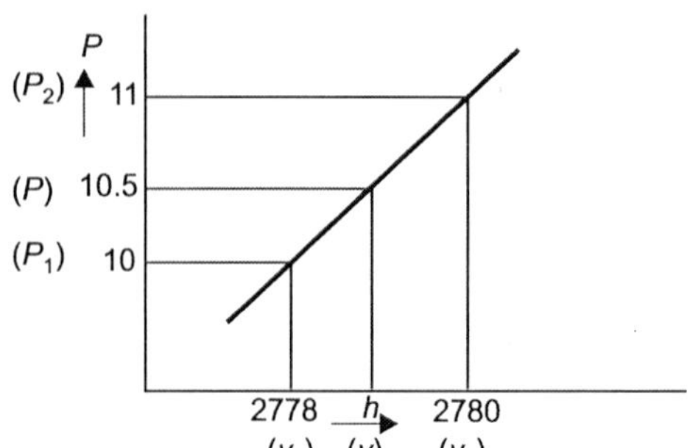

Fig. 3.5 *P. v. Y.*

3.11 MEASUREMENT OF DRYNESS FRACTION OF STEAM BY THROTTLING CALORIMETER

Throttling: Throttling is a flow process during which pressure drops but enthalpy remains constant.

Measurement of dryness fraction is done by a device called calorimeter. Here the throttling is done by an appratus is known as throttling calorimeter. During throttling there is a drop in pressure but the enthalpy remains constant. Throttling reduces the temperature of steam but increase the dryness fraction of wet steam.

A sample of wet steam at pressure P_1 is taken from the steam main through a perporated tube through a control value. The pressure P_1 and tempratures, are measured by pressure gauge and thermometer. The wet steam is then throttled through a narrow restricted valve opening, (total heat remains constant) to a pressure P_2 and temprature T_2 in such a way that steam reaches in superheated state.

Since throttling is an isoentropic (constant enthalpy) process.

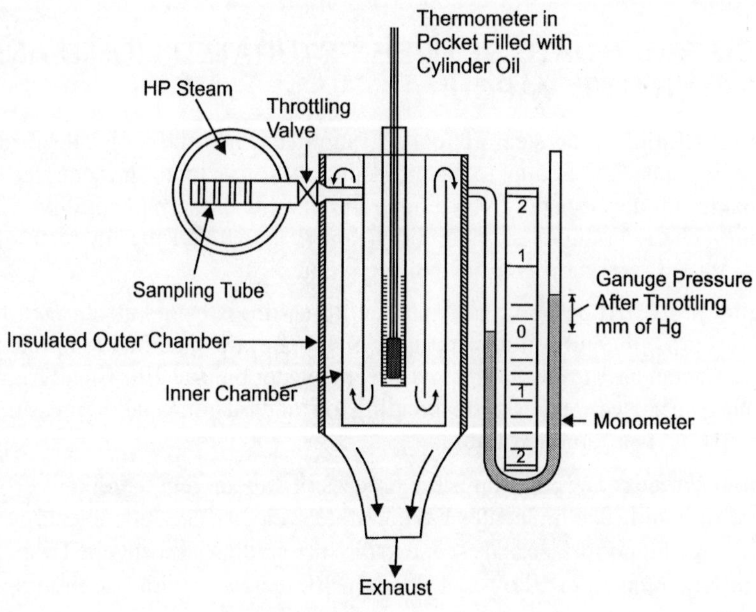

Fig. 3.6 Throttling calorimeter.

Hence,

P_1 = pressure of steam before throttling

P_2 = pressure of steam after throttling

h_{f_1} = sensible heat of water at pressure P_1

h_{fg_1} = latent heat of steam at pressure P_1

h_{g_2} = total heat of dry steam at pressure P_2

t_{sup} = temprature of superheated steam

t_2 = saturation temprature at pressure P_2

cp = specific heat of superheated steam

x = dryness fraction of steam before throttling

During Throttling Process

Enthalpy (total heat) before throttling = Enthalpy (Total heat) after throttling

$$(h_{f_1} + xh_{fg_1})P_1 = h_f + (h_{fg})P_2 \qquad P_1 = P_2$$

$$(h_{f_1} + xh_{fg_1}) = h_{g_2} + Cp_s\,(t_{sup} - \mathrm{t}_{sat})$$

∵ After throttling steam becomes super heated.

Hence
$$x = \frac{h_{g2} + C_{ps}(T_{sup} - T_{sat}) - h_{f1}}{h_{fg_1}}$$

3.12 DIFFERENTIATE BETWEEN STURATED LIQUID AND SATURATED VAPOUR

Saturated Liquid : The steam is called saturated when the molecules escaping from the liquid becomes equal to the molecules returning to it. The saturated steam has a maximum density at a given temperature and is in thermal equilibrium with the liquid. The temprature of saturated steam (vapour) and the liquid in contact with it is called saturation point or boiling point.

The liquid at its boiling point at a specified pressure is called a saturated liquid. Such a liquid is in equilibrium with its own vapour at a specified temprature and pressure. At standard atmospheric pressure, the water boils at 100°C and formation of vapour takes place. This is corresponding to standard atmospheric pressure. The water at 100°C is a saturated liquid.

Saturated Vapour: Evaporation is the process of vapour generation only from the surface of a liquid. The molecules having grater velocity *i.e.,* kinetic energy break away from the liquid surface and escape to the surrounding atmosphere. Evaporation occurs at any temperature, however its intensity increases with rise in temprature of the liquid.

When heat is added to a liquid, it starts boling and vapour gets generated both from the face surface of liquid and in side it. Boiling refers to be process of vapour formation that takes place in the whole mass of the liquid. All the liquid gets converted into vapour, when vaporisation takes place in an infinite space *i.e.,* in an open vessel. However, when the vessel is closed, vapours escaping from the liquid fill the free space above it and a fraction of these vapours return back into the liquid.

3.13 SATURATION TEMPERATURE AND SATURATION PRESSURE

Boiling takes place at a definite temperature which depends upon physical properties of the liquid and pressure. This temperature, denoted by t_{sat} is called saturated temperature and it remains constant until all the liquid has transformed in to vapour. At the saturation temperature, the boiling liquid is in equilibrium with its own vapour. The pressure at a specific temperature at which the liquid boils or in equilibrium with its own vapour, is called saturation pressure. At standard atmosphere pressure (P = 1.013 bar), water boils at 100°C. A such, at P = 1.01325 bar, the saturation temperature of water is 100°C. Likewise, the saturation pressure at 100°C boiling is 1.0325 bar and saturation pressure is referred to as vapour pressure.

3.14 WHAT DO YOU UNDERSTAND BY THE DEGREE OF SUPER HEAT AND DEGREES OF SUB-COOLING?

Saturated steam may be dry or wet, when the saturated vapour contains particles of liquid evenly distributed over the entire mass of vapour, it is called net saturated steam. Wet steam is characterised by its dryness fraction which is defined as ratio of mass of dry saturated vapour to the total mass of mixture.

$$\text{Dryness fraction } X = \frac{\text{Mass of dry saturated vapour}}{\text{Mass of mixture}}$$

$$= \frac{mg}{m} = \frac{mg}{mf + mg}$$

where m is mass of mixture

mg - mass of dry saturated steam.

mf - mass of water vapour in suspension.

The dryness fraction when expressed in percentage is called quality of steam.

Wetness Fraction: It is the ratio of mass of water vapour or moisture in suspension with the mass of steam.

$$\text{Wetness fraction} \qquad y = \frac{mf}{mf + mg} = 1 - \frac{mf}{mf + mg} = 1 - x$$

When wetness fraction is expressed in percentage is called primary quality + primary quantity = 100%.

Saturated liquid which contains no liquid particle is called dry saturated steam.

for dry saturated steam $x = 1$.

If steam at a given pressure, its temprature is higher then that of solution temprature. The steam becomes superheated. The difference ($t_{sup} - t_{sat}$) is known as degree of dipeheat higher the degree of superheat, greater distance between molecules and force of attraction (*i.e.*, molecular force).

High superheated steam behaves like a ideal gas.

SOLVED EXAMPLES

Example 3.1: Determine the quantity and heat required to produce 1kg of steam at a pressure of 6 bar and at a temperature of 25°C. Determine:

1. When the steam is wet having a dryness fraction 0.9
2. When the steam is dry saturated.
3. When the steam is superheated at a constant pressure at 250°C, assuming specific heat of superheated steam to be 2.3 kJ/kg°K.

Given $P = 6$ bar $\quad T_{sup} = 250°C$

$t_w = 25°C \quad CP = 2.3$ kJ/kg °K

$x = 0.9$

Solution:

From the steam table, corresponding to a pressure of 6 bar.

$h_f = 670.4$ kJ/kg $\quad h_{fg} = 2.85$ kJ/kg

$t = 158.8°C$

1. **When the steam is wet**

   Total heat (enthalpy) of 1 kg of wet steam

   $$h = h_f + xh_{fg} = 670.4 + 0.9 \times 2085 = 2546.9 \text{ kJ}$$

   The initial temperature of water is 25°C

   Hence

   Heat already in water $= 4.2 \times 25 = 105$ kJ

   ∴ Heat actually required $= 2546.9 - 105 = 2441.9$ kJ

2. **When the steam is dry saturated**

   Total heat (enthalpy) of 1 kg of dry saturated steam

   $$h_g = h_f + h_{fg} = 670.4 + 2085 = 2755.4 \text{ kJ}$$

   Net heat required $= 2755.4 - 105 = 2650.4$ kJ

3. **When the steam is superheated**

   Total heat (enthalpy) of 1 kg of superheated steam

   $$\begin{aligned} h_{sup} &= h_f + h_{fg} + CP\,(t_{sup} - t_{sat}) \\ &= hg + CP\,(t_{sup} - t_{sat}) \\ &= 2755.4 + 2.3\,(250 - 158.8) \\ &= 2965.16 \text{ kJ} \end{aligned}$$

   Net heat required $= 2965.16 - 105 =$ **2860.16 kJ**

Example 3.2: Determine the volume of 1 kg of superheated steam at a pressure of 20 bar and a temprature of 300°C.

Given:

$$P = 20 \text{ bar}$$

$$T_{sup} = 300°C = 300 + 273 = 573 \text{ K}$$

Solution:

From steam table corresponding to a pressure of 20 bar

$$t = 212.4°C$$

$$V_g = 0.1 \text{ m}^3/\text{kg}$$

$$T_s = 212.4 + 273 = 485.4 \text{ K}$$

If V_{sup} = volume of 1 kg of superheated steam

so $$\frac{V_{sup}}{T_{sup}} = \frac{V_g}{T_{sat}}$$

$$V_{sup} = \frac{V_g}{T_s} \times T_{sup} = \frac{0.1}{485.4} \times 573 = \mathbf{0.118 \text{ m}^3}$$

Example 3.3: Find the external workdone during evaporation per kg of steam at a pressure of 15 bar when the steam is

1. 90% dry
2. Dry saturated.

Given: $P = 15$ bar

Solution:

From the steam table corresponding to pressure of 15 bar

$$V_g = 0.1317 \text{ m}^3/\text{kg}$$

1. **Work done**

   when steam is wet (*i.e.* $x = 0.9$)

$$W = \text{Pressure} \times \text{Vol.}$$

$$= 100\, P_{vg} = 100 \times 15 \times 0.9 \times 0.1317$$

$$= \mathbf{177.8 \text{ kJ/kg}}$$

2. **When steam is dry saturated**

$$W = \text{Pressure} \times \text{Vol.}$$

$$= 100\, P_{vg} = 100 \times 15 \times 0.1317$$

$$= \mathbf{197.5 \text{ kJ/kg}}$$

Example 3.4: Calculate the internal energy of 1kg of steam at a pressure of 10 bar when the condition of steam is

1. 0.9 dry
2. Dry saturated.

Given: $P = 10$ bar

Solution:

From the steam table corresponding to a pressure of 10 bar

$$h_f = 762.6 \text{ kJ/kg}$$
$$h_{fg} = 2013.6 \text{ kJ/kg}$$
$$v_g = 0.1943 \text{ m}^3/\text{kg}$$

Internal energy

(*i*) When steam is wet (*i.e.* $x - 0.9$)

$$\text{Internal energy } (\mu) = \text{Total heat} - \text{work done}$$
$$= h_f + xh_{fg} - 100\, Px_{vg}$$
$$= 762.6 + 0.9 \times 2013.6 - 100 \times 10 \times 0.9 \times 0.1943 \text{ kJ}$$
$$= 2574.8 - 174.8$$
$$= \mathbf{2400 \text{ kJ}}$$

(*ii*) When the steam is dry saturated

$$\text{Internal energy } (\mu) = h_f + h_{fg} - 100\, P_{vg}$$
$$= 762.6 + 2013.6 - 100 \times 10 \times 0.1943$$
$$= 2776.2 - 194.3$$
$$= \mathbf{2581.9 \text{ kJ}}$$

Example 3.5: Determine the entropy per kg of superheated steam at a pressure of 20 bar and a temprature of 250°C. Assume C_f for superheated steam as 2.2 kJ/kg °K.

Given:

$$P = 20 \text{ bar}$$
$$T_{sup} = 250°\text{C} = 250 + 273 = 523 \text{ K}$$
$$C_p = 2.2 \text{ kJ/kg °K}$$

Solution:

From steam table corresponding to a pressure of 20 bar.

$$T = 212.4°\text{C} = 212.4 + 273 = 485.4 \text{ K}$$
$$S_g = 6.337 \text{ kJ/kg°K}$$

Entropy of superheated steam:

$$Sup = S_g + 2.3\ C_p \log\left(\frac{T_{sup}}{T}\right)$$

$$= 6.337 + 2.3 \times 2.2 \log\left(\frac{523}{485.4}\right)$$

$$= 6.5 \text{ kJ/kg °K}$$

Example 3.6: Steam at 18 bar and dryness 0.9 heated at constant pressure until dry and saturated. Find increase in volume, heat supplied and workdone per kg of steam.

If the volume is nowkept constant, find how much heat must be extracted to reduce the pressure to 14 bar.

Given:

$$P = 18 \text{ bar}$$

$$P_1 = 14 \text{ bar}$$

$$n = 0.9$$

Solution:

From the steam table corresponding to a pressure of 18 bar

$$h_f = 884.5 \text{ kJ/kg}$$

$$h_{fg} = 1910.3 \text{ kJ/kg}$$

$$h_g = 2794.8 \text{ kJ/kg}$$

$$V_g = 0.1103 \text{ m}^3\text{/kg}$$

we know that

Volume of wet steam,

$$V_f = xV_g = 0.9 \times .1103 = 0.0983 \text{ m}^3\text{/kg}$$

$$V_g = 0.1103 \text{ (from steam table)}$$

$$\therefore \quad \text{Increase in volume} = V_g - V_f = 0.1103 - 0.0993 = .011 \text{ m}^3\text{/kg}$$

Total heat of wet steam

$$h = h_f + xh_{fg} = 884.5 + 0.9 \times 19103$$

$$= 2603.8 \text{ kJ/kg}$$

$$\text{Heat supplied} = h_g - h = 2794.8 - 2603.8 = 191 \text{ kJ/kg}$$

$$\text{Work done} = \text{pressure} \times \text{change in volume}$$

$$W = 100\text{p}\ (V_g - V_f) = 100 \times 18 \times 0.011$$

$$= \mathbf{19.8\ kJ/kg}$$

At pressure 14 bar from the steam table corresponding to 14 bar

$$h_{f_1} = 830.1 \text{ kJ/kg}$$

$$h_{fg_1} = 1957.7 \text{ kJ/kg}$$
$$V_{g_1} = 0.1407 \text{ m}^3\text{/kg}$$
$$x_1 = \text{dryness fraction}$$

If the volume is kept constant

$$V_g = x_1 V_{g_1}$$
$$x_1 = V_g / V_{g_1} = 0.1103/0.1407 = 0.78$$

Let, Internal energy of steam at a pressure of 18 bar

$$u = h_g - 100\, PV_g = 2794.8 - 100 \times 18 \times 0.1103$$
$$= 2596.3 \text{ kJ/kg}$$

Now internal energy of steam at a pressure of 14 bar

$$u_1 = h_{f_1} + h_{fg_1} - 100\, P_1 x_1 V_{g_1}$$
$$= 830.1 + 0.78 \times 1957.7 - 100 \times 14 \times 0.78 \times 0.1407$$
$$= 2357 - 154 = 2203 \text{ kJ/kg.}$$

Hence, heat extracted to reduce the pressure

$$= \mu - \mu_1 = 2596.3 - 2203$$
$$= \mathbf{393.3 \text{ kJ/kg}}$$

Example 3.7: Determine the enthalpy, volume, internal energy and entropy of superheated steam at 15 bar pressure and 220°C. The volume of water may be neglected and take specific heat of superheat equal to 2.2 kJ/kg°K.

Given:

$$P = 15 \text{ bar}$$
$$T_{sup} = 220°C$$
$$cp = 2.2 \text{ kJ/kg °K}$$

Solution :

From the steam table the values corresponding to pressure 15 bar

$$V_g = 0.132 \text{ m}^3\text{/kg}$$
$$h_f = 844.7 \text{ kJ/kg}$$
$$h_{fg} = 1945.2 \text{ kJ/kg}$$
$$s_f = 2.314 \text{ kJ/kg °K}$$
$$s_{fg} = 4.1261 \text{ kJ/kg°K}$$
$$T_{sat} = 198.3°C$$

(*i*) Total heat of 1 kg of superheated steam

$$h_{sup} = h_f + h_{fg} + C_{ps}(T_{sup} - T_{sat})$$
$$= 844.7 + 1945.2 + 2.2\,(220 - 198.3)$$
$$= 2837.64 \text{ kJ/kg}$$

(*ii*) Volume:

$$\frac{V_{\text{sup}}}{T_{\text{sup}}} = \frac{V_g}{T_{\text{sat}}}, \ V_{\text{sup}} = \frac{T_{\text{sup}} \times V_g}{T_{\text{sat}}}$$

$$= 0.132 \times \frac{220+273}{198.3+273} = 0.1381 \ \text{m}^3\text{/kg}$$

(*iii*) External work $= 100\ PV_{\text{sup}}$

$= 100 \times 15 \times .1381 = 207.15$ kJ

(*iv*) Internal energy $u = h_{\text{sup}}$ – ext. workdone

$= 2837.64 - 207.15 = 2630.49$ kJ/kg

(*v*) Entropy of superheated steam

$$S_{\text{sup}} = S_f + S_{fg} + C_{ps} \log_e \frac{T_{\text{sup}}}{T_{\text{sat}}}$$

$$= 2.314 + 4.1261 + 2.2 \log_e \frac{220+273}{198.3+273}$$

$= 6.539$ kJ/kg°K

Example 3.8: The steam is stored in a container of 0.45 Jm^3 capacity. The pressure of steam is 8 bar and temprature is 200 °C. The container is cooled and pressure drops to 2.5 bar.

Find:

1. Quality of steam after cooling
2. Heat lost to atmosphere due to cooling.

Neglect the volume of water

Given: V = constant

$V_1 = v_2 = 0.45$ m^2

$P_1 = 8$ bar

$T_1 = 200$°C

$P_2 = 2.5$ bar

Solution:

From the steam table corresponding to pressure 8 bar

$V_1 = 2608$ m^3/kg

$h_g = 2839.25$ kJ/kg

from steam table values corresponding to pressure 2.5 bar

$$T_{at} = 12743°C$$
$$V_{g_2} = 0.71871$$
$$h_{f_2} = 535.34 \text{ kJ/kg}$$
$$h_{fg} = 2181.55 \text{ kJ/kg}$$

(*i*) Quality of steam:

$$V = \text{constant}$$
$$V_1 = V_2$$

when the steam is wet

$$V_1 = xVg_2 \qquad V_{g_2} > V_1 \text{ at pressure 2.5 bar}$$
$$.2608 = x\ .7187$$
$$x = .3638 \text{ Ans.}$$

(*ii*) Heat transfer during cooling

Mass of steam in container

$$m = \frac{V}{V_1} = \frac{0.45}{.2608}$$

Heat transfer at constant volume

$$H = \text{change in internal energy}$$
$$u_1 = h_1 - P_1V_1 = 2839.25 - (8 \times 100) \times .2608$$
$$u_2 = h_2 - P_2V_2 = (hf_2 + h_{fg2}) - p_2\,(x.V_{g2})$$
$$= (535.34 + .3628 \times 2181.55)$$
$$- (2.5 \times 100) \times (.3628 \times .7187)$$
$$= 1261.60 \text{ kJ/kg}$$
$$H = 1.725 \times (1261.6) - 2630.61$$
$$= -2361.53 \text{ kJ (Rejected)}$$

Example 3.9: 11 kg of steam undergoes a reversible isothermal process from 20 bar and 250°C to a pressure of 30 bar. Calculate the heat flow, stating whether it is supplied or rejected and sketch the process on a T-S diagram.

Given: $P_1 = 20$ bar $\quad T_1 = 250°C$

$P_2 = 30$ bar $\quad T_2 = 250°C$

Process is isothermal so temperature remains constant

Solution:

From the steam table values of enthalpy corresponding to 20 bar.

$$h_1 = 2902.5 \text{ kJ/kg}$$

Similarly from the steam table values of enthalpy corresponding to 30 bar

$$h_2 = 2855.8 \text{ kJ/kg}$$

$$\text{So net heat flow } = h_2 - h_1 = 2855.8 - 2902.5 \text{ kJ/kg} = -46.7 \text{ kJ}$$

$$q = mxh \quad = -467 \text{ kJ/kg} \times \text{J}$$

Enthalpy is decreasing so the heat is rejected.

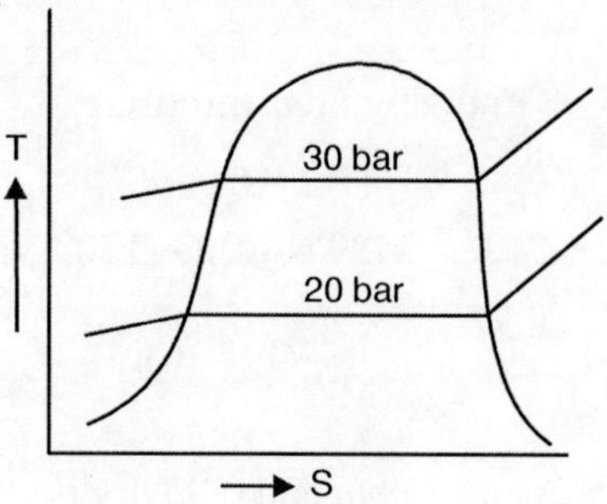

Fig. 3.7

Example 3.10: Steam flows in a pipe line at 1.5 MPa. After expending to 0.1 MPa in a throttling calorimeter, the temperature is found to be 120°C. Find the quality of steam in a pipe line. What is the maximum moisture at 1.5 MPa that can be determined with this set up if at least 5°C of superheat is required after throtting for accurate reading?

Given:

$$P_1 = 1.5 \text{ MPa} = 15 \text{ bar}$$

$$P_2 = 0.1 \text{ MPa} = 1 \text{ bar}$$

$$t_{sup} = 120°\text{C}$$

Solution:

From the steam table values corresponding to pressure 15 bar

$$h_{f_1} = 844.7$$

$$h_{fg_1} = 1945.2$$

Similarly from the steam table values corresponding to pressure 1 bar

$$h_{f_2} = 417.5$$

$$h_{fg_2} = 2257.9$$

$$Cp = 2.09$$

$$t_{saturation} = 99.6°\text{C}$$

Enthalpy (Total heat) before throttling = Enthalpy (total heat) after throttling

$$h_{f_1} + x_1 h_{fg_1} = h_{f_2} + h_{fg_2} + Cp\,(t_{sup} - t_{sat})$$

$$844.7 + x_1\ 1945.2 = 417.5 + 2257.9 + 2.09\,(120° - 99.6°)$$

$$x_1 \times 1945.2 = 1873.33$$

$$x_1 = \frac{1873.33}{1945.2} = 0.96$$

Hence, quality of steam $x_1 = 96\%$

Moistures (when t_{sat} 5°C)

Hence,

Enthalpy before throttling = Enthalpy after throttling

$$h_{f_1} + xh_{fg_1} = h_{f_2} + h_{fg_2} + Cps(t_{sup} - t_{sat})$$

$$844.7 + x_1 \times 1945.2 = 21.0 + 2489.6 + 209\ (120° - 5°)$$

$$x_1 = \mathbf{0.97}$$

Hence the steam is wet.

Example 3.11: Steam at pressure 15 bar and 0.95 dry is generated in a boner and is made to pass through its superheater where additional quantity of heat is supplied to steam at constant pressure. Consequently the temperature of steam increases to 350°C. Determine the following:

(*a*) heat supplied in the superheater

(*b*) change in internal energy.

Solution :

Given: Pressure, p_1 = 15 bar; Dryness fraction, x = 0.95 bar; Pressure, $p_2 = p$ = 15 bar; Temperature, t_{sup}= 350°C = 350 + 273 = 623 K.

From steam table

For 15 bar pressure, values are

$$h_f = 844.9 \text{ kJ /kg}$$

$$h_{fx} = 1947.3 \text{ kJ/kg}$$

$$V_g = 0.132 \text{ m}^3\text{/kg}$$

For 15 bar pressure and 350°C temperature values are

$$t_{sat} = 198.3°\text{C}$$

$$h_{sup} = 3147.5 \text{ kJ/kg}$$

$$V_{sup} = 0.1866 \text{ m}^3\text{/kg}$$

For 15 bar and 0.95 dryness fraction

Total heat of 1 kg of steam at 15 bar and 0.95 dryness fraction

$$h_1 = h_{f_1} + xh_{fg_1}$$

$$= 844.9 + 0.95 \times 1947.3$$

$$= 2694.83 \text{ kJ/kg}$$

Specific volume

$$V_1 = xV_g$$
$$= 0.95 \times 0.132$$
$$= 0.1254 \text{m}^3/\text{kg}$$

Internal Energy,

$$u_1 = h_1 - 100P_1V_1$$
$$u_1 = 2694.83 - 100 \times 15 \times 0.1254$$
$$= 2506.73$$

For 15 bar pressure and temperature 350°C

Total heat of 1 kg of steam at 15 bar and 350ºC temperature

$$h_2 = 3147.5 \text{ kJ/kg}$$
$$V_2 = 0.1866 \text{m}^3/\text{kg}$$
$$u_2 = h_2 - 100\ P_2V_2$$
$$= 3147.5 - 100 \times 15 \times 0.1866$$
$$= 2867.6 \text{ kJ/kg}$$

(*a*) Quantity of heat supplied in the superheater $= h_2 - h_1$

$$= 3147.5 - 2694.83 = 452.67 \text{ kJ/kg}$$

(*b*) Change in internal energy $= u_2 - u_1$

$$du = 2867.6 - 2506.73$$
$$= 360.8 \text{ kJ/kg}$$

Example 3.12: Calculate the internal energy per kg of superheated steam at a pressure of 10 bar and a temperature of 300ºC. Also find the change of internal energy if this steam is expanded to 1.4 bar and dryness fraction 0.8.

(*a*) heat supplied in the superheater

(*b*) change in internal energy.

Solution:

At 10 bar, 300ºC. From steam tables for superheated steam.

$$h_{sup} = 3051.2 \text{ kJ/kg } (T_{sup} = 300 + 273 = 573 \text{ K})$$

and corresponding 10 bar (from tables of dry saturated steam)

$$T_s = 179.9 + 273 = 452.9 \text{ K}; V_g = 0.194 \text{ m}^3/\text{kg}$$

To find V_{sup}, using the relation

$$\frac{V_g}{T_s} = \frac{V_{sup}}{T_{sup}}$$

$$\therefore \qquad V_{sup} = \frac{V_g \times T_{sup}}{T_s} = \frac{0.194 \times 573}{452.9} = 0.245 \text{ m}^3/\text{kg}$$

Internal energy of superheated steam at 10 bar,

$$u_1 = h_{sup} - pV_{sup}$$
$$= 3051.2 - 10 \times 10^5 \times 0.245 \times 10^{-3}$$
$$= 2806.2 \text{ kJ/kg.}$$

At 1.4 bar. From steam tables;

$$h_f = 458.4 \text{ kJ/kg}, h_{fg} = 2231.9 \text{ kJ/kg}; V_g = 1.236 \text{ m}^3/\text{kg}$$

Enthaply of wet steam (after expansion)

$$h = h_f + xh_{fg}$$
$$= 458.4 + 0.8 \times 2231.9 = 2243.92 \text{ kJ.}$$

Internal energy of this steam,

$$u_2 = h - PxV_g$$
$$= 2243.92 - 1.4 \times 10^5 \times 0.8 \times 1.236 \times 10^{-3}$$
$$= 2105.49 \text{ kJ}$$

Hence, change of internal energy per kg

$$u_2 - u_1 = 2105.49 - 2806.2$$
$$= \mathbf{-700.7 \text{ kJ.}}$$

Negative sign indicates decrease in internal energy.

Example 3.13: The followin data were obtained in a test on a combined separating and throttling calorimeter:

Pressure of steam sample = 15 bar, pressure of steam at exit = bar, temperature of steam at the exit = 150°C, discharge from separating calorimeter = 0.5 kg/min, discharge from throtting calorimeter = 0 kg/min.

Determine the dryness fraction of the sample steam.

Solution :

Pressure of seam sample, $P_1 = P_2 = 15$ bar

Pressure of steam at the exist, $P_3 = 1$ bar

Temperature of steam at the exit, $t_{sup3} = 150°\text{C}$

Discharge from separating calorimeter, $m_w = 0.5$ kg/min

Discharge from throtting calorimeter, $m_s = 10$ kg/min

From steam tables:

At $P_1 = P_2 = 15$ bar: $h_{f2} = 844.7$ kJ/kg, $h_{fg2} = 1945.2$ kJ/kg

At $P_3 = 1$ bar and 150°C: $h_{sup3} = 2776.4$ kJ/kg

$$h_2 = h_3$$

$$h_{f_2} + x_2 h_{fg_2} = h_{sup3}$$

$$844.7 + x_2 \times 1945.2 = 2776.4$$

$$x_2 = \frac{27764 - 8447}{19452} = 0.993$$

Now, quality of steam supplied,

$$x_1 = \frac{x_2 m_s}{m_s + m_w} = \frac{0.993 \times 10}{10 + 0.5}$$

$$= \mathbf{0.946.}$$

Example 3.14: In a throttling calorimeter, the steam is admitted at a pressure of 10 bar. If it is discharged at atmospheric pressure (1.013 bar) and 110°C after throttling, determine the dryness fraction of steam. Assume specific heat of steam as 2.2 kJ/kg °K.

Solution: Given

$$P_1 = 10 \text{ bar}$$

$$P_2 = 1.013 \text{ bar}$$

$$t_{sup} = 110°\text{C}$$

$$C_p = 2.2 \text{ kJ/kg °K}$$

from the steam table values corresponding to a pressure of 10 bar

$$h_{f_1} = 762.6 \text{ kJ/kg}$$

$$h_{fg_1} = 2013.6 \text{ kJ/kg}$$

Similarly from the steam table values corresponding to a pressure of (1.013 bar)

$$h_{g_2} = 2676 \text{ kJ/kg}$$

$$t_2 \text{ or } t_{sat} = 100 \text{ °C}$$

We know

Enthalpy (total heat) before throttling = Enthalpy (total heat) after throttling

$$h_{f_1} + nx_{fg_1} = h_{g_2} + C_p(t_{sup} - t_{sat})$$

$$762.6 + x \times 2013.6 = 2676 + 2.2\,(110 - 100)$$

$$x = \frac{2698 - 762.6}{2013.6}$$

$$= \mathbf{.961.}$$

Example 3.15: A sample of steam from a boiler drum at 3 MPa. is put through a throttling calorimeter in which the pressure and temprature are found be 0.1 MPa, 120°C. Find the quality of sample taken from the boiler.

Solution:

Given

$$P_1 = 3 \text{ MPa} = 30 \text{ bar}$$
$$P_2 = 0. \text{ MPa} = 1 \text{ bar}$$
$$t_{sup} = 120°\text{C}$$

from the steam table values, corresponding to 30 bar

$$h_{f_1} = 1008.4$$
$$h_{fg_1} = 1799.9$$

Similarly from the steam-table values crorresponding to 1 bar

$$h_{f_2} = 417.5$$
$$h_{fg_2} = 2257.9$$
$$t_{sup} = 120°\text{C}$$
$$t_{sat} = 99.63 \qquad [C_p = 2.09 \text{ kJ/kg}]$$

Enthalpy before throttling = Enthalpy after throttling

$$1008.4 + x_1\, 1799.9 = 417.5 + 2257.9 + 2.09\,(120 - 99.63)$$

$$x_1 = \frac{2717.97 - 1008.4}{1793.4} = 0.953$$

Hence,

quality of steam x_1 = **93%.**

HIGHLIGHTS OF THE CHAPTER

1. Steam is a vapour of water which is invisible when pure and dry. It can be converted into three stages solid, liquid and gas.
2. *Wetsteam*: The steam contains moisture content and when latent heat is not absorbed by steam.
3. *Dry steam*: If the well steam is farther heat and there is no moisture content present in side.
4. *Wetness fraction* (Y): It is the ratio of mass of water vapour to the total mass of the steam and it is denoted by Y

$$Y = \frac{1 - mg}{mf + mg} = \frac{mf}{mf + mg}$$

where mf = mass of fluid

mg = mass of actual dry steam

5. *Dryness fraction* (x): It is the ratio of actual dry steam to the total mass of the steam and is denoted by x.

$$x = \frac{mg}{mf + mg} = \frac{mg}{m}$$

for Dryness fraction of

($x < 1$ Wet steam)

($x = 1$ Dry steam)

($x > 1$ Superheated steam)

6. Sensible heat of water is the amount of heat energy absorbed by 1 kg of water from 0ºC to being t or or saturation temperature at a gross pressure It do moreever or the pressure increases.

7. Letent heat is the amount required to convert 1 kg of water at saturation temperature for a given pressure into dry and saturated steam without change of temperature the value of latent heat is not constant:

8. Total heat is the amount of heat required to convert 1 kg of water into super heated steam at constant pressure.

Total heat of steam = Sensible heat + Latent heat.

Expression for the Enthalpy of steam.

(*i*) For wet steam

$$h = h_f + nh_{fg}.$$

(*ii*) For dry steam

$$h = h_f + h_{fg}.$$

(*iii*) For superheated steam

9. Specific volume of steam is the volume occupied by the steam per unit mass at a given temperature and pressure.

Expression for specific volume.

(*i*) For wet steam

$$V = xV_g$$

(*ii*) For dry steam

$$V = V_g$$

(*iii*) For superheated steam

$$V_{sup} = \frac{V_g T_{\text{sup}}}{T_{sat}}$$

Entropy is the degree of randomness.

(*i*) for wet steam.

$$s = s_f + xs_{fg}$$

$$s = s_f + \frac{xh_{fg}}{T}$$

(*ii*) for dry steam

$$s_g = s_f + xs_{fg}$$

or $s_g = s_f + \dfrac{h_{fg}}{T}$

(*iii*) for superheated steam

$$\delta_{\text{sup}} = s_g + 2.3\, C_{ps} \left(\frac{T_{\text{sup}}}{T}\right)$$

9. Internal energy is the actual heat energy stored in the steam above the freezing point of water is known as internal energy of steam.

Expression for internal energy

(*i*) For wet steam

$$u = h_f - xh_{fg} - 100\, PV_g$$

(*ii*) For dry steam

$$u = hg - 100\, PV_g$$

(*iii*) For super leated steam

$$u = h_{\text{sup}} - 100\, PV_{\text{sup}}$$

10. Saturation temperature is the temperature at which a pure substance starts to evaporate at a given pressure.

11. Saturated steam is the steam about to condense.

12. Superheated steam is a steam which is higher temperature than its saturation temperature.

13. Sublimation is the process where a transformation takes place of solid phase directly into vapour phase.
14. *Degree of superheat:* It is a difference between ($t_{sup} - t_{sat}$).
15. Mollier diagram is a graphical representation of steam table in which the enthalpy is plotted along the ordinate and entropy along abscisses.
16. Throttling is a flow process in which pressure drops but entropy remains constant.
17. Throttling calorimeter is a decline which is used to measure the dryness fraction of steam.

REVIEW QUESTIONS

1. What do you mean by steam generation?
2. Explain the process of steam generation (change of phase of water) at constant pressure and show the various stages on *P-V*, *T-V* & *T-S* diagrams.
4. What are the advantages of using superheated steam?
5. What is a pure substance? Can air be treated as a pure substance.
6. Explain the following properties of steam:
   (*i*) Wet steam
   (*ii*) Dry steam
   (*iii*) Superheated steam
   (*iv*) Enthalpy of steam
   (*v*) Specific volume
   (*vi*) Internal energy of steam
   (*vii*) Entropy of steam
7. What is critical point? What process is possible below critical point?
8. Define quality and dryness fraction of steam.
9. What do you mean by external work done during evaporation?
10. What do you mean by Mollier chart? Explain with neat sketch.
11. Sketch and explain the construction and working of a throttling calorimeter used for determining the dryness fraction of steam in a boiler.

12. How throttling helps in the dryness fraction of steam?

13. Differentiate between saturated liquid and saturated vapour.

14. Define the terms and saturation temperature and saturation pressure.

15. What do you understand by the degree of superheat and degree of subcooling?

16. Draw temperature *vs* total heat graph during steam formation.

■■■

4

Refrigeration and Air Conditioning

4.1 INTRODUCTION

"Refrigeration is the method of producing and maintaining the low temperature in an enclosed space below the surrounding temperature is known as refrigeration". It is also used for cooling and dehumidifying

A device used for this purpose is known as refrigerators and the process is known as refrigeration. It is extensively used in chemical industry, food industry and air conditioning plants etc..

"Air conditioning means conditioning of air for maintaining low temperature, humidity, purity air motion and human being comfort within the enclosed space is known as air conditioning". Generally it is used in theaters, residences, offices, buildings, auditorium, class rooms, hospitals, railway coaches and automobiles, etc.

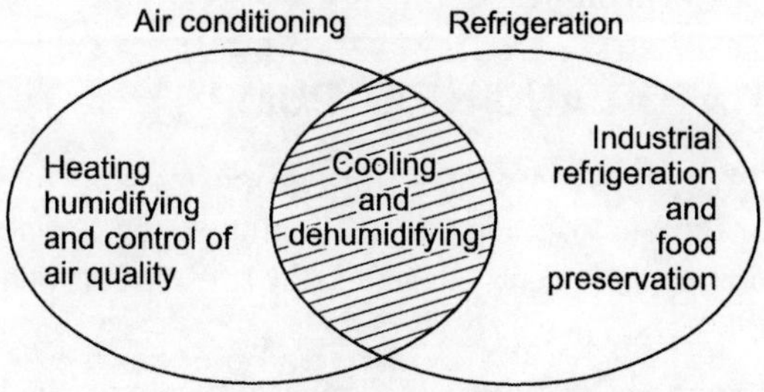

Fig. 4.1 Relationship of refrigeraton and air-conditioning

4.2 RATING OF REFRIGERATION MACHINES

Rating of refrigeration means tonne of refrigeration and it is the amount of heat to be removed in order to form one tonne of ice at 0°C in 24 hours. In S.I. units the latent heat of ice is 335 kJ/kg therefor one tonne of refrigeration

$$\text{ITR} = \frac{1000\times 335}{24\times 60} = 210 \text{ kJ/mm}$$

In metric system 72,000 kcal/24 hours = 300 kcal/hr.

In S.I. units 1 tone refrigeration = 210 kJ/min = 3.5 k. watt.

4.3 COEFFICIENT OF PERFORMANCE (COP)

To measure the performance of a refrigerating machine in term of performance known as coefficient of performance (C.O.P.).

$$C.O.P. = \frac{\text{Refrigerating Effect}}{\text{Work done}} = \frac{\text{Heat extracted at low temperature}}{\text{Work done}}$$

It is also defined as the ratio of heat entreated at low temperature from the refrigerator to the work done is known as coefficient of performance.

$$C.O.P. = \frac{Q}{W}$$

where Q = Quantity of heat extracted from the refrigeration at low temperature

W = Work done to the system.

Relative C.O.P.: It is the ratio of actual C.O.P. to the theoretical C.O.P. is known as relative coefficient of performance.

$$\text{Relative C.O.P.} = \frac{\text{Actual C.O.P.}}{\text{Theoretical C.O.P.}}$$

4.4 REFRIGERATOR AND HEAT PUMP

Refrigerator: Refrigerator is a machine which is used to remove the heat continuously from the enclosed space and maintain the temperature below the surrounding temperature and rejects the heat to the surrounding at higher temperature (*i.e.*, atmosphere).

Heat Pump: Heat pump is a device which is used to supply the heat continuously to a space for maintaining the temperature above surrounding atmospheric temperature by absorbing the heat from the surrounding at low temperature. The working of refrigerator and heat pump is shown in Fig. 4.2.

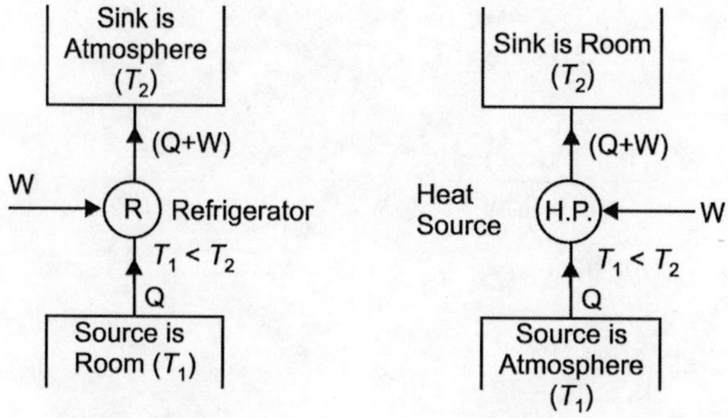

(*a*) Refrigerator (*b*) Heat-pump

Fig. 4.2 Working of refrigerator and heat pump

The C.O.P. of refrigeration system

$$\text{C.O.P.} = \frac{Q}{W}$$

The C.O.P. of heat pump

$$\text{C.O.P.} = \frac{Q+W}{W} = 1 + \frac{Q}{W}$$

The C.O.P. of the cycle is given by

$$\text{C.O.P.} = \frac{\text{Heat absorbed}}{\text{Work done}}$$

$$= \frac{\text{Heat absorbed}}{\text{Heat rejected} - \text{Heat absorbed}}$$

Heat absorbed

$$dQ_2 = T_2.dS$$

Heat rejected

$$dQ_1 = T_1 dS$$

$$= \frac{T_2(S_2 - S_1)}{T_1(S_3 - S_4) - T_2(S_2 - S_1)}$$

$$= \frac{T_2(S_2 - S_1)}{T_1(S_2 - S_1) - T_2(S_2 - S_1)} \qquad (\because S_3 - S_4 = S_2 - S_1)$$

$$\text{C.O.P.} = \frac{Q_2}{Q_1 - Q_2}$$

$$= \frac{T_2}{T_1 - T_2} \quad \text{where } T_2 < T_1$$

we know that $= \dfrac{\text{Heat extracted}}{\text{Work done}}$

The refrigerator absorb heat-

$$\text{C.O.P.} = \frac{Q}{W} = \frac{T_2}{T_1 - T_2}$$

where T_1 = Heat rejection (refrigerator)

T_2 = Heat absorbed

4.5 PRESSURE-ENTHALPY (*P-h*) CHART

The *P-h* (pressure-enthalpy) chart is used for studying the behaviour of a refrigerant. In this chart the ordinates (*y*-axis) represent pressure and the abscissa (*x*-axis) represent enthalpy.

A *P-h* chart is shown in which important lines and region are drawn.

The saturated liquid and the saturated vapour line meets into one another at the critical point.

A saturated liquid is in whih tempreture is equal to saturation temperature corresponding to its pressure. A liquid line exists in sub-cooled liquid region and a space between liquids vapour lines is called wet vapour region in right side of the saturated vapour line in a superheated vapour region.

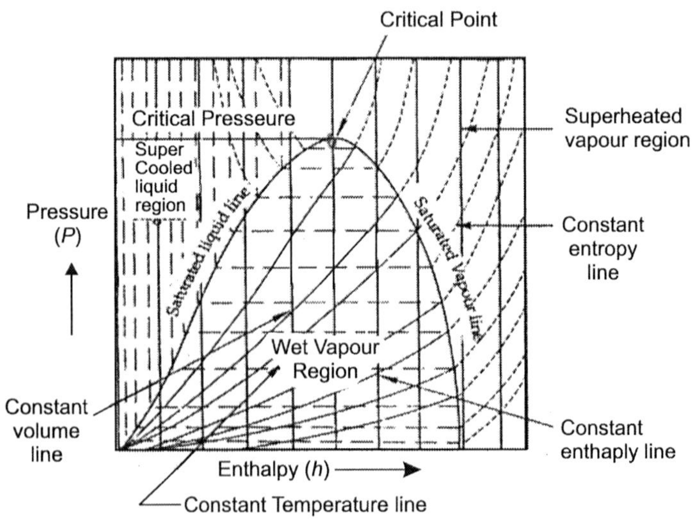

Fig. 4.3 Pressure-Enthalpy chart.

4.6 REVERSE CARNOT CYCLE

A refrigerating system working on the reversed Carnot cycle will have the maximum possible coefficient of performance like a heat engine working on Carnot cycle has the highest possible efficiency. But it is not possible to make a refrigerating machine working on the reversed Carnot cycle same with heat engine working on the Carnot cycle.

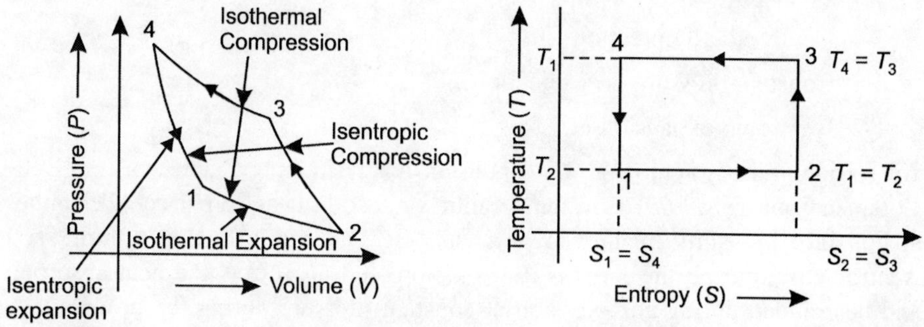

Fig. 4.4 Carnot-cycle on *P-V* diagram

Processes:

Isothermal expension (1-2)	Isentropic compression (2-3)
Isothermal compression (3-4)	Isentropic expension (4-1)

The refrigerant absorbs heat at constant temperature during the process (1 – 2) and is compressed isentropically during the process (2 – 3). Heat rejection from the refrigerant to the cooling medium takes place isothermally during the process (3 – 4) and then finally the refrigerant is expanded back to its original position through an isentropic expansion process (4 –1).

The performance of the cycle is expressed as the ratio of output to input. In refrigeration cycle, output is to extract the maximum amount of heat in the evaporator for a net expenditure of work in compressor.

The C.O.P. of cycle is given by

$$\text{C.O.P.} = \frac{\text{Heat absorbed}}{\text{Work done}}$$

$$= \frac{\text{Heat absorbed}}{\text{Heat rejected} - \text{Heat absorbed}}$$

$$= \frac{T_2(S_2 - S_1)}{T_1(S_3 - S_4) - T_2(S_3 - S_4)} = \frac{T_2(S_2 - S_1)}{T_1 - T_2(S_3 - S_4)} = \frac{T_2}{T_1 - T_2} \quad \text{where } T_2 < T_1$$

$$\text{C.O.P.} = \frac{Q}{w} = \frac{T_2}{T_1 - T_2}$$

where T_1 = Temperature of refrigeration system (*i.e.*, heat absorbed)

T_2 = Temperature of the sorrounding (*i.e.*, heat rejected)

4.7 BELL-COLEMAN CYCLE (REVERSED BRAYON CYCLE OR JOULE CYCLE)

In the Bell Coleman Cycle, heat absorption and rejection follow constant pressure processes at shown in Fig. 4.5 (*a*, *b* and *c*). The other two processes are isentropic.

Following four processes are as follows:

2-3 Constant pressure heat absorption (Refrigeration)

3-4 Isentropic compression

4-1 Constant pressure heat rejection

1-2 Isentropic expansion

Air refrigeration system (Fig. 4.5*a*) is used mostly in aircraft and work on Bell-Coleman Cycle (Fig 4.6*b*). Low temperature air absorbs heat from the cold chamber at constant pressure during the process 2-3. This air is further compressed isentropically during the process 3-4. Compressed air rejects the heat absorbed and heat added during compression at constant pressure during the process 4-1. Isentropic expansion during process 1-2 brings back the air to the initial position.

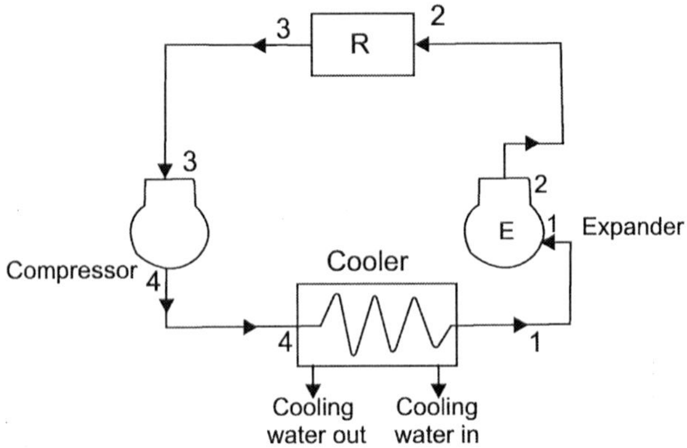

(*a*) Air Refrigerator system on a Bell Coleman cycle

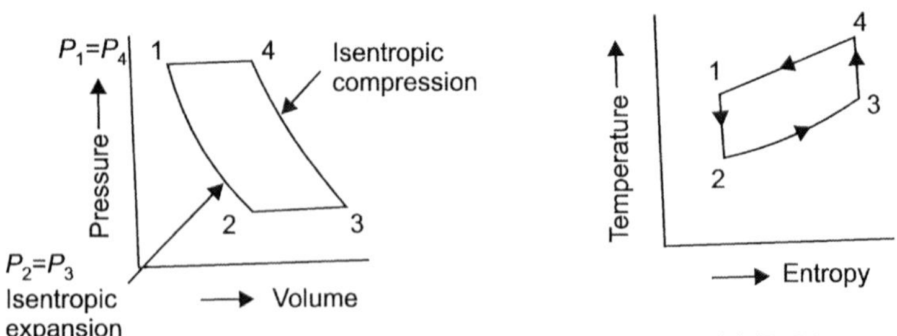

(*b*) *P-V* Diagram (*c*) *T-s* Diagram

Fig. 4.5 Bell-Coleman cycle

The C.O.P. of the Bell Coleman cycle is

$$\text{C.O.P.} = \frac{\text{Heat absorbed or Refrigerating effect}}{\text{Work done}}$$

$$= \frac{\text{Heat absorbed}}{\text{Heat rejected} - \text{Heat absorbed}}$$

$$= \frac{mc_p(T_3 - T_2)}{mc_p(T_4 - T_1) - mc_p(T_3 - T_2)}$$

$$= \frac{T_3 - T_2}{(T_4 - T_1) - (T_3 - T_2)}$$

$$= \frac{T_2\left[\frac{T_3}{T_2} - 1\right]}{T_1\left[\frac{T_4}{T_1} - 1\right] - T_2\left[\frac{T_3}{T_2} - 1\right]} \quad \ldots(i)$$

We know that for isentropic expansion process 1-2,

$$\frac{T_2}{T_1} = \left(\frac{P_2}{P_1}\right)^{\frac{\gamma-1}{\gamma}} \quad \ldots(ii)$$

Similarly, for isentropic compression process 3-4,

$$\frac{T_3}{T_4} = \left(\frac{P_3}{P_4}\right)^{\frac{\gamma-1}{\gamma}} \quad \ldots(iii)$$

Since $P_1 = P_4$ and $P_2 = P_3$, therefore, from equations (*ii*) and (*iii*)

But $$\frac{T_2}{T_1} = \frac{T_3}{T_4} \text{ or } \frac{T_3}{T_2} = \frac{T_4}{T_1} \quad \ldots(iv)$$

Substituting these values in equation (*i*), we get,

$$\text{C.O.P.} = \frac{T_2}{T_1 - T_2}$$

4.8 SIMPLE VAPOUR COMPRESSION REFRIGERATION SYSTEM

Vapour compression refrigeration systems are the most commonly used refrigeration systems. In this refrigeration system working fluid is refrigerant which undergoes phase change at least during one process, *i.e.* it evaporates and condenses or changes alternately between vapour and liquid phases without leaving the refrigeration system. In evaporation, refrigerant absorbs latent heat from the cold body. This latent heat is used for converting the liquid to vapour, while condensing, it rejects latent heat to external body to create a cooling effect in the working fluid.

It is generally used for all industrial purposes from a small domestic refrigeration to a big air conditioning plant.

4.9 SIMPLE VAPOUR COMPRESSION REFRIGERATION CYCLE

In a simple vapour-compression cycle, there are four fundamental processes required to complete one cycle.

These are as follows:

1. Compression
2. Condensation
3. Expansion
4. Vaporization

These are explained in relation to flow diagram shown in Fig. 4.6.

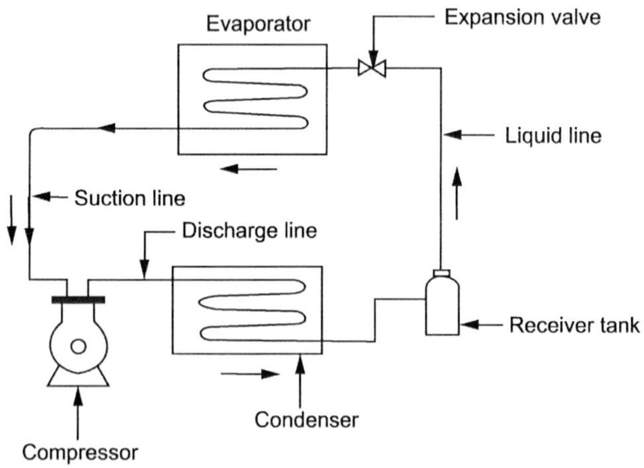

Fig. 4.6 Flow diagram of simple vapour compression system

1. **Compression:** The low temperature and pressure vapour in dry state enters the compressor during the suction stroke of the compressor. During compression stroke, it compressed isentropically and subsequently its

temperature, pressure increases until the vapour temperature is greater than the temperature of condenser cooling medium (air or water).

2. **Condensation:** When the high pressure refrigerant vapour enters the condenser, heat flows from condenser to cooling medium thus allowing the vaporized refrigerant to return to liquid state.
3. **Expansion:** After condensation, the liquid refrigerant is stored in the receiver tank. From the receiver, it passes through an expansion valve where, it is throttled down to a lower pressure to allow the vaporization of liquid at a low temperature of about –10°C.
4. **Vaporization:** The low pressure refrigerant vapour after expansion in the expansion valve enters the evaporator or refrigerated space, where heat is absorbed by it and refrigeration is furnished.

4.10 PRINCIPAL PARTS OF A SIMPLE VAPOUR COMPRESSION REFRIGERATION SYSTEM

The principal parts of a simple vapour compression refrigeration system shown in the flow diagram of figure. These parts are as follows:

1. **Evaporator:** Its function is to provide a heat transfer surface through which heat can pass from the refrigerated space into the vaporizing refrigerant.
2. **Suction Line:** It carries the low pressure vapour from the evaporator to the suction inlet of the compressor.
3. **Compressor:** The function of the compressor is to draw refrigerant vapour from the evaporator and then raise its temperature and pressure to such a point so that it may be easily condensed with normally available condensing media. It also maintains continuous flow of the refrigerant through the system.
4. **Discharge Line:** It carries the high pressure and high temperature refrigerant from the compressor to the condenser.
5. **Condenser:** The function of the condenser is to provide a heat transfer surface through which heat passes from the refrigerant to the condensing medium which is either air or water.
6. **Receiver Tank:** It acts as a reservoir which stores the liquid refrigerant coming from the condenser and supplies it to the evaporator according to the requirement.
7. **Liquid Line:** It carries the liquid refrigerant from the receiver tank and conveys it to the expansion valve.
8. **Expansion Value:** Its function is to supply a proper amount of refrigerant

to the evaporator after reducing its pressure considerably so that the refrigerant may take sufficient amount of heat from the refrigerating space during evaporation.

4.11 TYPES OF SIMPLE VAPOUR COMPRESSION CYCLES

A simple vapour compression cycle consists of compression, condensation, expansion and vaporization process.

To increase the coefficient of performance of the cycle there are different arrangements, such as:

1. Cycle with dry saturated vapour after compression.
2. Cycle with wet vapour after compression.
3. Cycle with superheated vapour after compression.
4. Cycle with superheated vapour before compression, and
5. Cycle with under cooling or sub-cooling of refrigerant.

1. **Simple Vapour Compression Cycle with Dry Saturated Vapour after Compression:** A simple vapour compression cycle with dry saturated vapour after compression is shown on T - S (Temperature - Entropy) and P - h (Pressure - Enthalpy) diagrams in Fig. 4.7 (*a*) and (*b*) respectively. Following are the processes of simple vapour compression cycle with dry saturated vapour after compression:

   1-2: Isentropic compression of refrigerant vapour (Compression process)

   2-3: Constant pressure heat rejection (Condensation process).

   3-4: Constant Enthalpy expansion or Throttling Process (Expansion process)

   4-1: Constant pressure heat absorption or refrigeration process. (Vaporization process)

   Process 1-2 (*Compression* Process): The vapour refrigerant at low pressure p_1 and temperature T_1, is compressed isentropically to dry saturated vapour as shown in Fig. 4.7 by the vertical line 1-2 on T - S diagram and by the curve 1 - 2 on P - h diagram. The pressure and temperature rises from P_1 to P_2 and T_1 to T_2 respectively.

   The work done during isentropic compression per kg of refrigerant is given by:

   $$W = h_1 - h_2$$

   where h_1 = Enthalpy of vapour refrigerant at temperawre T_1, *i.e.* at suction of the compressor, and

   h_2 = Enthalpy of the vapour refrigerant at temperature T_2 *i.e.* at discharge of the compressor.

Process 2-3 (*Condensation Process*): The high pressure and temperature vapour refrigerant from the compressor is passed through the condenser where it is condensed at constant pressure p_2 and temperature T_2 as shown in Fig. 4.7 by the horizontal line 2 - 3 on T - S and P-h diagrams. The vapour refrigerant is charged into liquid refrigerant.

The refrigerant while passing through the condenser gives its latent heat to the surrounding condensing medium.

Process 3-4 (*Expansion Process*): Condensation process is constant pressure and temperature process so temperature $T_2 = T_3$ and pressure $P_2 = P_3$.

The liquid refrigerant at pressure P_3 and temperature T_3 is expanded by throttling process. Through a expansion valve to a low pressure P_1 and temperature T_5 as shown in Fig. 4.7 by the curve (3 - 4) on T - S diagram and by vertical line (3 - 4) on P-h diagram.

Process 4-1 (*Vaporizing Process*): The liquid and vapour mixture of the refrigerant from expansion valve at pressure P_1 and temperature T, is evaporated and changed into vapour refrigerant at constant pressure and temperature as shown in Fig.4.7 by the horizontal line 4 -1 on T - S diagram and P-h diagram.

During evaporation liquid-vapour refrigerant absorb its to latent heat of vaporization from the medium which is to be cooled. This process continues upto point 1 which is the starting point and thus the cycle is completed.

We know that

$$\text{Refrigerating effect } (R_E) = h_1 - h_4$$

$$= h_1 - h_{f_3} \qquad (\because h_4 = h_{f_3})$$

Where h_{f_3} = Sensible heat at temperature T_3 *i.e.* enthalpy of liquid refrigerant leaving the condenser.

$$\therefore \text{ Coefficient of Performance, C.O.P.} = \frac{\text{Refrigerating effect}}{\text{Work done}}$$

$$= \frac{h_1 - h_4}{h_2 - h_1}$$

$$= \frac{h_1 - h_{f_3}}{h_2 - h_1}$$

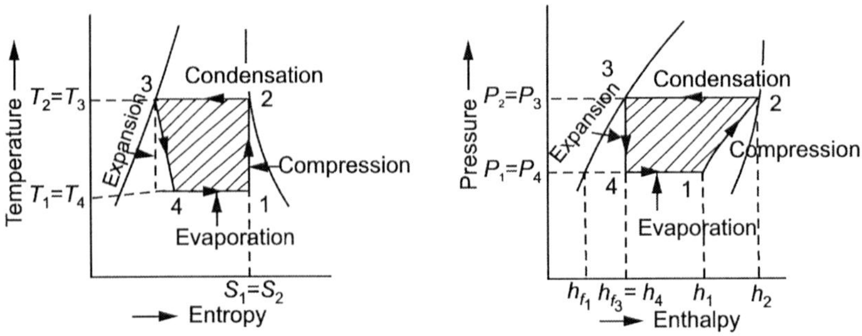

(*a*) *T-S* diagram (*b*) *P-h* diagram

Fig. 4.7 Simple vapour compression cycle with dry saturated vapour after compression

2. **Simple Vapour Compression Cycle with Wet Vapour after Compression:** A simple vapour compression cycle with wet vapour after compression is shown in Fig. 4.8 (*a*) and (*b*) as *T - S* diagram and *P - h* diagram respectively.

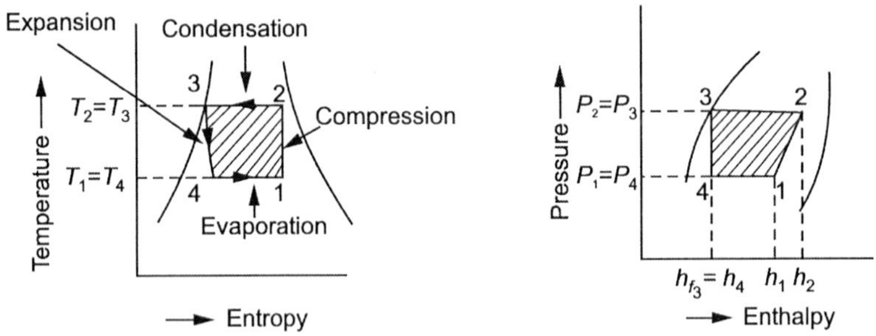

(*a*) *T-S* diagram (*b*) *P-h* diagram

Fig. 4.8 Simple vapour compression cycle with wet saturated vapour after compression.

In this cycle, the enthalpy at point 2 is found out with the help of dryness fraction at this point. The dryness fraction at points 1 and 2 may be obtained by equating entropies at points 1 and 2.

$$\therefore \text{ Coefficient of Performance, C.O.P.} = \frac{\text{Refrigerating effect}}{\text{Work done}}$$

$$= \frac{h_1 - h_4}{h_2 - h_1}$$

$$= \frac{h_1 - h_{f_3}}{h_2 - h_1}$$

3. **Simple Vapour Compression Cycle with Wet Vapour after Compression:** A simple vapour compression cycle with wet vapour after compression is shown in Fig. 4.9 (*a*) and (*b*) as *T* - *S* diagram and *P* - *h* diagram respectively. In this cycle, the enthalpy at point 2 is found out with the help of dryness fraction at this point. The dryness fraction at point 1 and 2 may be obtained by equating entropies at point 1 and 2.

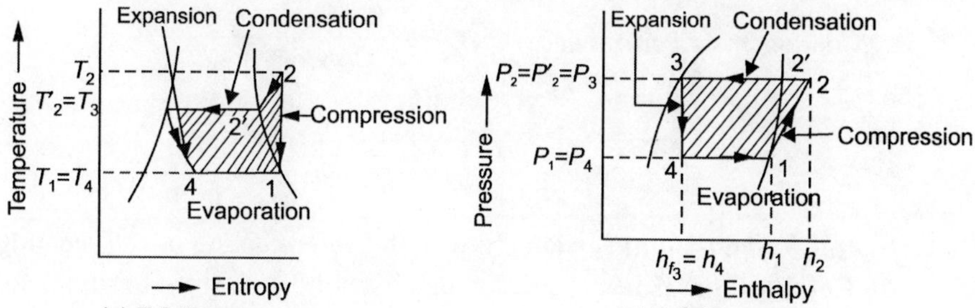

(*a*) *T-S* diagram (*b*) *P-h* diagram

Fig. 4.9 Simple compression cycle with superheated vapour after compression.

$$\therefore \text{Coefficient of Performance, C.O.P.} = \frac{\text{Refrigerating effect}}{\text{Work done}}$$

$$= \frac{h_1 - h_4}{h_2 - h_1}$$

$$= \frac{h_1 - h_{f3}}{h_2 - h_1}$$

4. **Simple Vapour Compression Cycle with Superheated Vapour before Compression:** A simple vapour compression cycle with superheated vapour before compression is shown in Fig. 4.10 (*a*) and (*b*) as *T* - *S* diagram and *P* - *h* diagram respectively.

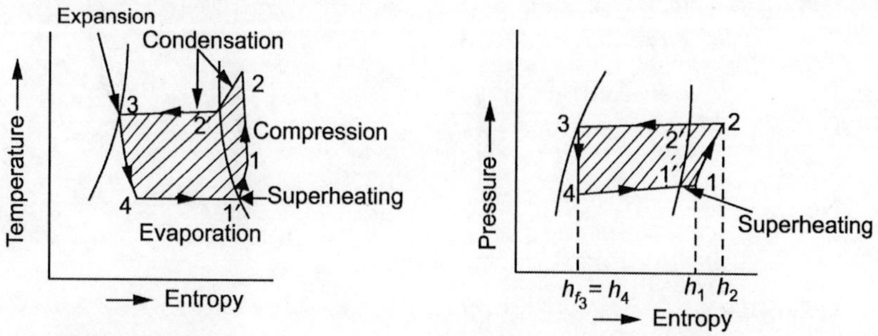

(*a*) *T-S* diagram (*b*) *P-h* diagram

Fig. 4.10 Simple vapour compression cycle with superheated vapour before compression.

In this cycle, the enthalpy at point 4 and continues upto point 1′ when it is dry saturated. The vapour is now superheated before entering the compresor upto point 1′ when it is dry saturated The vapour is now superheated before entering the compresor upto the point 1, Basically in this cyle heat is extracted or absorbed in two stages, *i.e.*, 4 to 1′ and 1′ to point.

$$\therefore \text{ Coefficient of Performance, C.O.P.} = \frac{\text{Refrigerating effect}}{\text{Work done}}$$

$$= \frac{h_1 - h_4}{h_2 - h_1} = \frac{h_1 - h_{f3}}{h_2 - h_1}$$

5. **Simple Vapour Compression Cycle with Under-cooling or Sub-cooling of Refrigerant:** The process of under-cooling is brought about by circulating more quantity of cooling water through the condenser or by using water colder than the main circulating water.

   Refrigerant after condensation process 2' - 3' is cooled below the saturation temperature (T_3') before expansion by throttling such a process is called under-cooling or sub-cooling is shown in Fig.4.11.

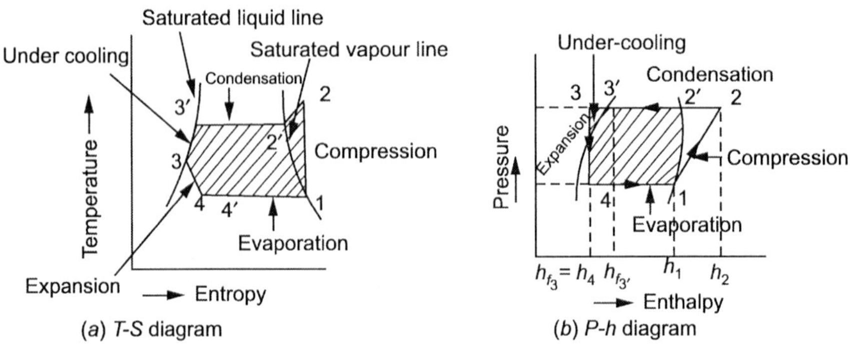

(*a*) *T-S* diagram (*b*) *P-h* diagram

$$\text{C.O.P.} = \frac{\text{Refrigerating effect}}{\text{Work done}}$$

$$= \frac{h_1 - h_{f3}}{h_2 - h_1}$$

Where $h_{f_3} - c_p \times$ Degree of under cooling

Fig. 4.11 Simple compression cycle with under-cooling or sub-cooling of refrigerant.

Merits and Demerits of Vapour Compression System over Air-refrigeration System:

Merits:

1. C.O.P. is quite high as the working of the cycle is very near to that of reversed Carnot cycle.
2. When used on ground level the running cost of vapour-compression refrigeration system is only l/5th of air refrigeration system.
3. For the same refrigerating effect the size of the evaporator is smaller.
4. The required temperature of the evaporator can be achieved simply by adjusting the throttle valve of the same unit.

Demerits:

1. Initial cost is high.
2. The major disadvantages are *inflammability, leakage of vapours* and *toxity.* These have been overcome to a great extent by improvement in design.

Functions of Parts of a Simple Vapour Compression System

Here follows the brief description of various parts of a simple vapour compression system shown in Fig. 4.11.

1. **Compressor:** The function of a compressor is to remove the vapour from the evaporator, and to raise its temperature and pressure to a point such that it (vapour) can be condensed with available condensing media.
2. **Discharge Line** (or hot gas line): A hot gas or discharge line delivers the high-pressure, high-temperature vapour from the discharge of the compressor to the condenser.
3. **Condenser:** The function of a condenser is to provide a heat transfer surface through which heat passes from the hot refrigerant vapour to the condensing medium.
4. **Receiver Tank:** A receiver tank is used to provide storage for a condensed liquid so that a constant supply of liquid is available to the evaporator as required.
5. **Liquid Line:** A liquid line carries the liquid refrigerant from the receiver tank to the refrigerant flow control.
6. **Expansion Value** (Refrigerant Flow Control): Its function is to meter the proper amount of refrigerant to the evaporator and to reduce the pressure of liquid entering the evaporator so that liquid will vapourize in the evaporator at the desired low temperature and take out sufficient amount of heat.
7. **Evaporation:** An evaporator provides a heat transfer surface through which heat can pass from the refrigerated sapce into the vapourizing refrigerant.
8. **Suction Line:** The suction line conveys the low pressure vapour from the evaporator to the suction inlet of the compressor.

4.12 COMPARISON BETWEEN AIR REFRIGERATION SYSTEM AND VAPOUR COMPRESSION

Air Refrigeration System	Vapour Compression System
1. High C.O.P.	1. Low COP.
2. Small size for the given refrigerating effect	2. Bigger size.
3. Less running cost (only 1/5th that of air-refrigeration system) when operated on ground level.	3. High running cast.
4. Simple adjustment of the throttle valve helps to achieve required temperature at the evaporator.	4. Adjustment is not so such easy.
5. Can be employed over large range of temperatures.	5. It is limited.
6. Initial cost is high.	6. Initial cost is low.
7. Suffers from problems associated with toxicity, inflamability and leakage of the refrigerant.	7. It does not suffer such type of problems.

4.13 COMPARISON BETWEEN REVERSE CARNOT CYCLE AND VAPOUR COMPRESSION REFRIGERATION SYSTEM

Reverse Carnot Cycle	Vapour Compression Refrigeration
1. Ideal cycle with highest COP	1. Practical cycle with lower COP
2. Heat absorption and heat rejection are at constant temperature.	2. Heat absorption and heat rejection are at constant pressure as well as at constant temperature.
3. There is no superheating and no under cooling of the refrigerant.	3. The refrigerant may be superheated in the compressor and undercolled in the condenser so as to improve the efficiency of the system.
4. Involves an expansion process which is isentropic (reversible adiabatic)	4. Expansion is by throttling which is irreversible and at constant enthalpy.
5. Independent of working substance.	5. Performance depends on the type of vapour used.
6. It canot be practically implemented.	6. It is most popular and widely used refrigeration system for commercial, industrial and domestic services.

Applications: Vapour compression refrigeration has been and is the most widely used method for air-conditioning large public buildings, private residences, hotels, hospitals, theatres, restaurants and automobiles. It is also used in domestic and commercial refrigerators, large-scale warehouses for storage of foods and meats, refrigerated trucks and railroad cars, and a host of other commercial and industrial services.

4.14 REFRIGERANTS

A refrigerant is the working substance or heat carrying medium used in a refrigeration cycle to absorb heat from a low temperature system and discard the heat so absorbed to another system at higher temperature. During heat absorption, the refrigerant evaporates (changes from liquid to vapour state) and during heat rejection it condenses (changes from vapour to liquid state). Obviously, the refrigerant undergoes a phase change from gas to liquid and back. These processes occur in the vapour compression system as well as in the vapour absorption system. A wide variety of refrigerants are available, each having typical properties which limit its efficient use in particular refrigerating machine.

The selection of refrigerant for an application is essentially based on certain parameters such as pressure-temperature characteristics, density, viscosity, flammability, toxicity, miscibility with oil, and heat absorbing capacity etc.

Early mechanical systems employed sulphur dioxide gas or anhydrous ammonia with small home refrigerators. Being toxic, sulphur dioxide disappeared from the market with the introduction of Freon. Ammonia is still used in large commercial plants, well away from residential areas, where a leak will not cause widespread injuries. Until concerns about depletion of ozone layer arose in 1980s, the most widely used refrigerants were the halomethanes R-12 and R-22 with R-12 being more common in automotive air conditioning and small refrigerators and R-22 being early systems used R-11 because its low boiling point allows low pressure system for which low mechanical strength components can be used.

This chapter focuses attention on classification, nomenclature and properties of different refrigerants available in the market and being put to use. Brief mention is also made of the necessity of driers and lubricants in a refrigeration system.

4.15. PROPERTIES OF AN IDEAL REFRIGERANT

1. It should be a stable gas.
2. It should be non-explosive.
3. It should be non-corrosive.
4. It should have a low boiling point.
5. It should be non-poisonous.
6. It should be non-toxic.

7. It should be non-inflammable.
8. Parts moving to the fluid should be easy to lubricate.
9. It should have high critical pressure and temperature and low freezing point.
10. It should easily available at low cost.
11. It should have high C.O.P.
12. It should have a well-balanced enthalpy of evaporation per unit mass.
13. There should be minimum difference between vaporising pressure and condensing pressures.

4.16 CLASSIFICATION OF REFRIGERANTS

The refrigerants are usually classified as follows on the basis of :

Working Principle:

Primary Refrigerant: These are the heat carrier substances that take part directly in the refrigeration system and cool the product/space by absorption of latent heat The refrigerant undergoes a cyclic process involving compression, condensation, expansion and evaporation with change in phase.

Secondary Refrigerant: These are the heat carrier substances that are first cooled by primary refrigerants, and are subsequently used as a medium for producing cooling effect in the product or space being refrigerated. The secondary refrigerants do not undergo any cyclic process: no change in phase and produce the cooling affect by absorption of sensible heat

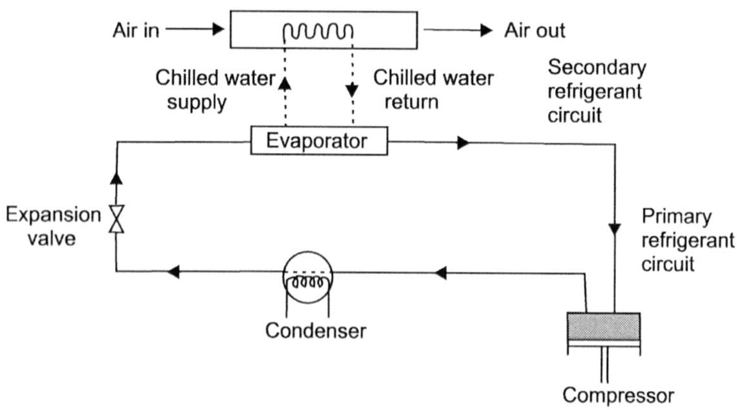

Fig. 4.12 Primary and secondary refrigerant circuits.

4.17 COMPARISON BETWEEN PRIMARY AND SECONDARY REFRIGERANTS

Parameters	Primary Refrigerant	Secondary Refrigerant
1. Cyclic/Non-cyclic operation	Operates in a cycle; alternatively evaporating and condensing (phase change occurs)	Non-cyclic operation and no phase change
2. Heat absorption	Latent heat	Sensible heat
3. Refrigerating effect	High	Low
4. Cost	Expensive	Comparatively much cheaper
5. Quantity of refrigerant required	Less	More
6. Corrosive effect	May or may not corrode	Usually corrosive
7. Requirement of anti-freeze	No need of anti-freeze	Anti-freeze is required
8. Examples	R-11, NH_3 and SO_2	Water and Brines

4.18 METHODS OF REFRIGERATION

Refrigeration is the technique of producing cooling effect by abstraction/withdrawal of heat so that temperature below that of surroundings is produced in a substance or within a space. The desired cooling effect can be produced by:

1. ***Ice refrigeration (change of phase):*** The use of ice to refrigerate and thus preserve food goes back to the prehistoric times and the ancient cultures of Chinese, Greeks, Romans and Persians. Ice and snow were stored in caves or dugouts lined with straw or other insulating materials. This practice worked well down through the centuries, with ice houses remaining in use. Greater work was done on developing better insulation products for long distance shipment of ice and the ice harvesting became a big business.

   The natural ice or artificially produced ice is brought into contact with the substance to be cooled. The ice melts and the heat required for melting of ice is supplied by the substance being cooled. The cooling effect produced by ice is

   $$Q = m \times h_{sf}$$

   where m and h_{sf} are the rate of fusion of ice and enthalpy of fusion respectively. At normal atmospheric pressure of 1.01325 bar (1 atm), h_{sf} equals 335 kJ/kg.

   Ice refrigeration prevents dehydration and preserves the fresh appearance of eatable products like fruit and vegetables. However, ice has to be fed to the refrigerator of and on and that is neither convenient nor economical. Further, water coming out of the refrigerator poses a problem in its disposal.

2. ***Dry ice refrigeration (sublimation):*** Solid carbon dioxide (called dry ice) has a peculiar characteristic that it changes from solid state to vapour state without passing through liquid state. During change of phase, it absorbs heat equivalent to its latent heat of vaporisation and produces cooling effect.

   Dry ice is used to preserve food stuff during transportation. The slabs of ice are usually packed on either side or on top of food packages in cartons. When dry ice evaporates, it absorbs heat from the food stuff and preserves it in frozen state.

   The refrigeration methods (1) to (4) as mentioned above are called the natural methods. These methods are non-cyclic and the temperatures attainable are limited. Further, there is continuous consumption of the refrigerating substance and that necessitates replenishment. However, these methods are sometimes convenient forms of cooling where small refrigeration is required such as in the laboratory and workshop.

3. ***Evaporation:*** When a liquid evaporates, it absorbs heat from the surroundings equivalent to its latent heat of vaporisation, and that results in lowering the temperature of surroundings. For example, we feel cooling effect when there is evaporation of a drop of spirit placed on the palm of hand. Likewise, the evaporation of moisture from the skin of a human body helps to keep it cool. There is a common practice to cool the water for drinking purposes by keeping it in the porous earthen pots. The water evaporates through the pores and that produces cooling effect. The army people keep small water containers made of metal and covered with water soaked namada; the walls of the metallic container get cooled and that cools the water kept inside it.

   The principle of evaporative refrigeration in employed in desert (room) coolers. The dry ionospheric air is made to pass through water soaked packings. When this water evaporates, it takes heat from the air causing it to cool.

4. ***Air or gas refrigeration:*** Expansion of gas lowers its pressure and temperature. This cooling effect results without change in the phase of gas. The different air refrigeration systems use this thermodynamic principle for producing low temperatures.

5. ***Chemical methods:*** Here the heat required for the completion of chemical reaction is taken from the substance being cooled. The chemical method for producing cooling effect cannot be followed on a commercial scale.

6. ***Throttling process:*** Throttling is the expansion of fluid from high pressure to low pressure. This process occurs when fluid passes through an obstruction (partially opened valve or a small orifice) placed in the fluid flow passage. Under certain conditions, a gas gets cooled during throttling and that produces cooling effect.

7. ***Dissolution of salts in water:*** When certain salts are dissolved in water, they absorb heat and lower the temperature of water and create a sort of refrigeration bath for cooling substance. Sodium chloride lowers the water temperature upto - 20°C while calcium chloride upto - 50°C. The salt can be recovered by evaporation of water from the solution.

8. ***Mechanical refrigeration:*** The natural and chemical methods have been successfully replaced by mechanical or heat energy refrigeration techniques. In these methods, the heat is abstracted from the substance or space (which is to be cooled) is pumped to a system (which is at high temperature level) by taking energy from an external source as input to refrigerating machine. The refrigeration system consists of a cycle of processes with the same quantity of working fluid (refrigerant) in continuous circulation.

   The first commercial hand operated refrigeration system was developed in UK by Perkins. The system consisted of a hand operated compressor, a water cooled condenser, a throttle valve and an evaporator. The working substance (ether) was compressed in the hand operated compressor and then condensed in the water cooled condensing unit. Thereafter, the liquid ether was throttled to low pressure and taken to the evaporator where heat was absorbed and cooling effect was produced. The refrigerant ether was used again and again in the cyclic process with negligible wastage. Subsequent developments took place in United States where steam engine was used as a prime mover to drive the compressor. The scope got further widened with the development ot electric motors and consequent high speeds of the compressor. Ether too was replaced by new working substances.

   Mechanical refrigeration systems are broadly classified into:

   - air or gas refrigeration
   - vapour compression system
   - vapour absorption system, and
   - steam jet refrigeration.

   Energy needed for the mechanical systems is essentially in the form of mechanical, electrical or thermal. Due to world energy crisis, concerted efforts are being made by various agencies to develop refrigeration systems which utilize waste heat, solar energy, wind energy and bio-energy etc for their functioning. A lot of research is being done to devise ways and means which make the refrigeration systems more energy efficient.

9. ***Non-conventional refrigeration systems:***

   (*a*) Thermo-electric cooling uses the Peltier effect to create a heat flux between the junction of two different types of materials.

   When a direct current is made to pass through electrical junctions between unlike metals, heating or cooling effect is produced depending on the direction of flow of current. One junction gets heated and the other gets cooled. The cold junction is located inside the space desired to be cooled and the warmed junction lies outside in the surroundings. The hot and cold junctions get reversed when a change occurs in the direction of flow of current. Apparently the same setup can be used for heating and cooling of an insulated space.

This effect is commonly used in camping and portable coolers, and for cooling electronic components and small instruments.

(*b*) Magnetic refrigeration (adiabatic demagnetisation) is a cooling technology based on magnet caloric effect, an intrinsic property of magnetic solids. A strong magnetic field is applied to the refrigerant, forcing its various magnetic dipoles to align and putting these degrees of freedom of the refrigerant into a state of lowered entropy. A heat sink then absorbs the heat released by the refrigerant due to its loss of entropy. Thermal contact with the heat sink is then broken so that the system is insulated and the magnetic field is switched off. This increases the heat capacity of the refrigerant, thus decreasing its temperature below the temperature of heat sink.

The refrigerant is often a paramagnetic salt such as cesium magnesium nitrate, and the active magnetic dipoles are those of electron shells of the paramagnetic atoms.

(*c*) Thermo acoustic refrigeration which uses sound waves in a pressurised gas to drive heat exchange.

(*d*) Vortex tube that operates with compressed air and is used for spot cooling. A high pressure gas is allowed to expand through a nozzle fitted tangentially to a pipe. This causes simultaneous discharge of cool air at the core and hot air at the periphery.

The mechanical refrigeration and the non-conventional refrigeration systems have been dealt with in details at appropriate places.

4.19 PROPERTIES AND USES OF COMMONLY USED REFRIGERANTS

1. Air

Properties:

(*i*) No cost involved ; easily available

(*ii*) Completely non-toxic

(*iii*) Completely safe

(*iv*) Free of cost

(*v*) The C.O.P. of air cycle operating between temperatures of 80°C and - 15°C is 1.67.

Uses:

(*i*) Air was one of the earliest refrigerants and was widely used even as late as World War I wherever a completely non-toxic medium was needed.

(*ii*) Director low C.O.P., it is used only where operating efficiency is secondary as in aircraft refrigeration.

2. Ammonia (NH_3)

Properties:

(*i*) Low cost.

(*ii*) It has the excellent thermal properties.

(*iii*) It has the highest refrigerating effect per kg of refrigerant.

(*iv*) Low volumetric displacement.

(*v*) It is highly toxic and flammable.

(*vi*) Low weight of liquid circulated per tonne of refrigeration.

(*vii*) High efficiency.

Uses:

(*i*) It is widely used in large industrial and commercial reciprocating compression systems where high toxicity is secondary. It is extensively used in ice plants, packing plants, large cold storages and skating rings etc.

(*ii*) It is widely used as the refrigerant in absorption systems.

Note:

(*i*) Ammonia should never be used with copper, brass and other copper alloys; iron and steel should be used in ammonia systems instead,

(*ii*) In ammonia systems, to detect the leakage a sulphur candle is used which gives off a dense white smoke when ammonia vapour is present.

3. Sulphur Dioxide (SO_2)

Properties:

(*i*) It is non-explosive and non-flammable.

(*ii*) It is extremely toxic and has a pungent irritating odour.

(*iii*) It is a colourless gas or liquid.

(*iv*) It has a liquid specific gravity of 1.36.

(*v*) Works at low pressures.

(*vi*) Possesses small latent heat of vapourisation.

Uses:

It finds little use these days. However, its use was made in small machines in early days.

Note: The leakage of sulphur dioxide may be detected by bringing aqueous ammonia near the leak, this gives off a white smoke.

4. Carbon dioxide (CO_2)

Properties:

(*i*) It is non-toxic and non-flammable.

(*ii*) It is non-explosive and non-corrosive.

(*iii*) It is a colourless and odourless gas, and is heavier than air.

(*iv*) It has liquid specific gravity of 1.56.

(*v*) It has extremely high operating pressures.

(*vi*) It gives very low refrigerating effect.

Uses:

This refrigerant has received only limited use because of the high power requirements per tonne of refrigeration and the high operating pressures. In *former years* it was selected for *marine refrigeration,* for *theater air-conditioning systems,* and for *hotel and institutional refrigeration* instead of ammonia because it is non-toxic.

At the present-time its use is limited primarily to the manufacture of dry ice (solid carbon dioxide).

Note: The leak detection of CO_2 is done by soap solution.

5. Methyl chloride (CH_3Cl)

Properties :

(*i*) It is neither flammable nor toxic.

(*ii*) It is a colourless liquid with a faint, sweet, non-irritating odour.

(*iii*) It has liquid specific gravity of 1.002 at atmospheric pressure.

Uses :

It has been used in the past in both domestic and commercial applications. It should never be used with aluminium.

6. R11 (Trichloro monofluoro methane)

Properties:

(*i*) It is composed of one carbon, three chlorine and one fluorine atoms (or parts by weight) and is non-corrosive, non-toxic and non-flammable.

(*ii*) It dissolves natural rubber.

(*iii*) It has a boiling point of –24°C.

(*iv*) It mixes completely with mineral lubricating oil under all conditions.

Uses:

1. It is employed for 50 tonnes capacity and over in small office buildings and factories.
2. A cen-trifugal compressor is used in the plants employing this refrigerant.
3. Its leakage is detected by a halide torch.

7. R12 (Dichloro-difluoro methane) or Freon-12

Properties:

(*i*) It is non-toxic, non-flammable, and non-explosive, therefore it is most suitable refrigerant.

(*ii*) It is fully oil miscible therefore it simplifies the problem of oil return.

(*iii*) It condenses at moderate pressure and under atmospheric conditions.

(*iv*) Its latent heat at –15°C is 161.6 kJ/kg.

(*v*) C.O.P. = 4.61.

(*vi*) It does not break even under the extreme operating conditions.

(*vii*) The operating pressures of R12 in evaporator and condenser under standard tonne of refrigeration are 1,9 bar abs. and 7.6 bar abs. (app.).

Uses:

1. It is suitable for high, medium and low temperature applications.
2. It is used for domestic applications.
3. It is excellent electric insulator therefore it is universally used in sealed type compressors.

8. R22 (Monochloro-difluoro methane) or Freon-22

R22 refrigerant is superior to R12 in many respects. It has the following properties and uses:

Properties :

(*i*) The compressor displacement per tonne of refrigeration with R22 is 60% less than the compressor displacement with R12 as refrigerant.

(*ii*) R22 is miscible with oil at condenser temperature but tries to separate at evaporator temperature when the system is used for very low temperature applications (–90°C). Oil separators must be incorporated to return the oil from the evaporator when the system is used for such low temperature applications.

(*iii*) The pressures in the evaporator and condenser at standard tonne of refrigeration are 2.9 bar abs. and 11.9 bar abs. (app.).

(*iv*) The latent heat at - 15°C is low and is 89 kJ/kg.

The major disadvantage ofR22 compared with R12 is the high discharge temperature which requires water cooling of the compressor head and cylinder.

Uses:

R22 is universally used in commercial and industrial low temperature systems.

Refrigerant	Boiling Temp. atmospheric pressure °C	Pressure at 30°C and –15°C (bar)	Specific Volume at –15° (m³/kg)	Latent Heat at –15°C kJ/kg	Compressor displacment per ton of refrigeration	kW per ton of (refrigration –15°C to 30°C)	C.O.P. as percentage of cornot cycle as standard condition
NH_3	–33.3	11.9 and 2.4	0.51	1316.7	5.83	0.83	82.9
F_{-12}	–29.8	7.58 and 1.86	0.093	162.5	9.85	0.86	80
F_{-22}	–41.3	12.26 and 3.03	0.0778	218.4	5.85	0.86	82
CO_2	–78.5	73.34 and 23.34	0.017	274.3	1.03	1.54	44.6
SO_2	–10.0	47.1 and 0.83	0.406	395.4	–	0.83	82.5

Standard condition is defined as 30° condenser and –15°C evaporator temperature. Carnot COP at standard condition is 5.74.

4.20 PSYCHROMETRY

Psychrometry is the study of the properties of air and water vapour mixtures. Similarly atmospheric air is a mixture of many gases plus water vapour and no. of pollutants. Where ever dry air contains nitrogen (78% by volume), oxygen (21% by volume 8.1% like carbon dioxide, hydrogen, helium, neon and oxygen etc. The pollutants are fillered before processing the air. Air mixture is known as moist air.

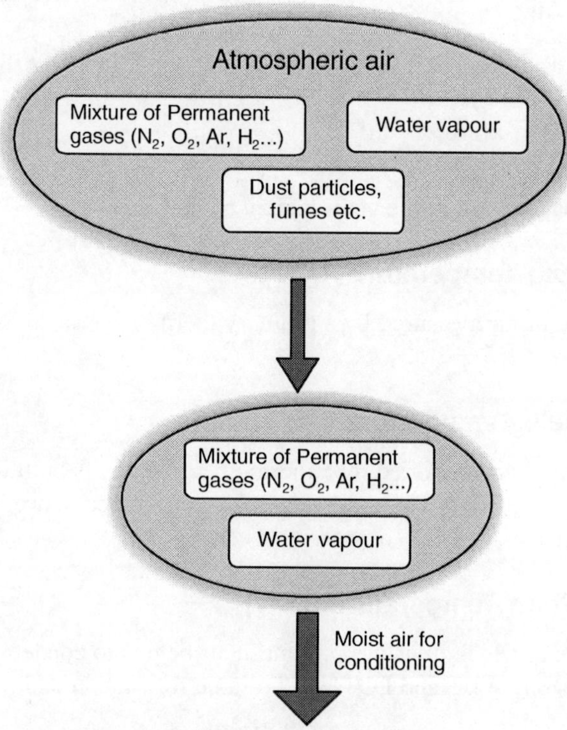

Fig. 4.13 Atmospheric air

Composition of standard air

Constituent	Molecular weight	Mol fraction
Oxygen	32.000	0.2095
Nitrogen	28.016	0.7809
Argon	39.944	0.0093
Carbon dioxide	44.010	0.0003

4.21 PSYCHROMETRIC PROPERTIES

Psychrometry is the study of the properties of moist air and the terms used in psychrometry are known as psychrometric properties.

4.21.1 Dry Air

The air which does not contain water vapour is known as dry air.

4.21.2 Moist Air

It is a mixture of dry air and water vapour. Maximum quantity of water vapour present in air at particular air temperature is known as saturation capacity of air.

4.21.3 Moisture

The water vapour present in the air is known as moisture.

4.21.4 Dry Bulb Temperature (DBT)

The temperature of air measured by a ordinary thermometer is known as dry bulb temperature of air.

4.21.5 Wet Bulb Temperature

The temperature of air measured by a thermometer when its bulb is covered with wet cloth is known as wet bulb temperature. The difference between dry bulb and wet bulb is known as wet bulb depression.

4.21.6 Dew Point Temperature (DPT)

The temperature at which moisture present in air begins to condense when the air is cooled. The DBT, WBT and DPT will be same for saturated air.

4.21.7 Specific Humidity

It is the mass of water vapour present with one kg of dry air. It is represented in grams/kg of air.

4.21.8 Relative Humidity

It is equal to the ratio of actual mass of water vapour in a given volume to the mass of water vapour. If the same volume of air is saturated at the same temperature.

4.21.9 Sensible Heat of Air

The quantity of heat which can be measured by measuring the dry bulb temperature of air is known as sensible heat.

4.21.10 Total Heat of Air

It is the sum of sensible heat of dry air and sensible heat and latent heat of water vapour associated with dry air.

4.21.11 Humidity Ratio (W)

The humidity ratio (or specific humidity W, is the mass of water vapour (m_v) to the mass of dry air (m_a) in any given volume of air.

$$W = \frac{m_v}{m_a} = \frac{\text{Mass of water vapour}}{\text{Mass of dry air}}$$

$$= \frac{P_v V / R_v T}{P_a V / R_a T} = \frac{P_v / R_v}{P_a / R_a} = .22 \frac{P_v}{P_t P_v}$$

Substituting the values of gas constants of water vapour and air R_v (0.4615 kJ/kg K) and R_a (0.287kJ/kg K).

4.21.12 Degree of Saturation (μ)

The degree of saturation is the ratio of the humidity ratio (*W*) to the humidity ratio of a saturated mixture W_s at the same temperature and pressure, *i.e.*,

$$\mu = \left| \frac{W}{W_s} \right|_{t, p}$$

where W = Humidity ratio = $\dfrac{\text{Mass of water vapour}}{\text{Mass of dry air}}$

W_s = Water vapour in the same mass and pressure of dry air when it is saturated at the same temperature.

4.21.13 Enthalpy

The enthalpy of moist air is the sum of the enthalpy of the dry air and the enthalpy of water vapour. It is expressed in kJ/Kg of dry air.

The enthalpy of moist air is given by

$$h = h_a + W_{kg}$$

$$= c_p t + W(h_{fg} + c_{pw} t)$$

where c_p = Specific heat of dry air at constant pressure, kJ/kg K *i.e.*, 1.005 kJ/kg K

c_{pw} = Specific heat of water vapour, kJ/kg K *i.e.* 1.88 kJ/kg K

t = Dry bulb temperature of air vapour mixture, 0°C

W = Humidity ratio, kg of water vapour/kg of dry air

h_a = Enthalpy of dry air at temperature *t*, kJ/kg

h_g = Enthalpy of water vapour at temperature ±, kJ/kg

h_{fg} = Latent heat of vaporisation at 0°C, kJ/kg

The unit of h is kJ/kg of dry air. Substituting the approximate values of c_p and h_g, we obtain:

$$h = 1.005t + W(2501 + 1.88)$$

Humid Specific Heat: From the equation for enthalpy of moist air, the humid specific heat of moist air can be written as:

$$c_{pm} = c_p + WC_{pW}$$

where c_{pm} = Humid specific heat, kJ/kg K *i.e.* 1.022 kJ/Kg dry air per

c_p = Specific heat of dry air, kJ/kg.K

c_{pw} = Specific heat of water vapour, kJ/kg

W = Humidity ratio, kg of water vapour/kg of dry air.

Since the second term in the above equation ($W.c_{pw}$) is very small compared to the first term, for all practical purposes, the humid specific heat of moist air, c_{pm} can be taken as 1.0216 kJ/kg dry air K.

4.22 PSYCHROMETRIC CHART

A psychrometric chart is a graphical representation of various thermodynamic properties of moist air and this chart helps to calculate readily the properties of air directly from the chart. In this psychometric chart dry bulb temperature DBT is taken as abscissa and moisture content (gm/kg) is taken as ordinate.

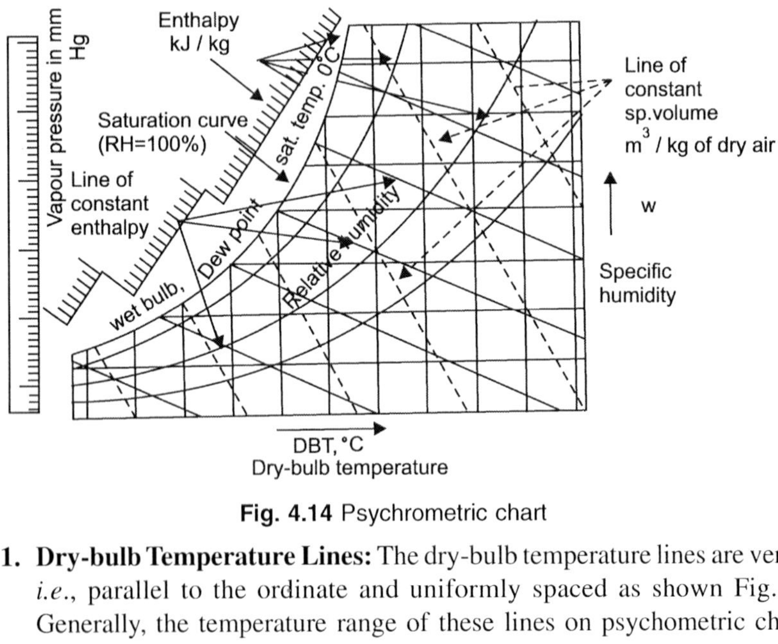

Fig. 4.14 Psychrometric chart

1. **Dry-bulb Temperature Lines:** The dry-bulb temperature lines are vertical, *i.e.*, parallel to the ordinate and uniformly spaced as shown Fig. 4.15 Generally, the temperature range of these lines on psychometric chart is from –6° to 45 °C.

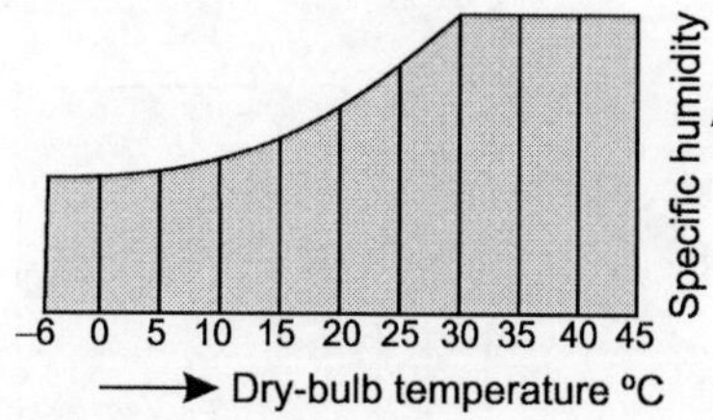

Fig. 4.15 Dry-bulb temperature line

2. **Specific Humidity:** The specific humidity (moisture content) are horizontal, *i.e.*, parallel to the abscissa and are also uniformly spaced as shown in Fig. 4.16. Generally, moisture content range of these lines on psychrometric chart is from 0 to 0.030 kg/kg of dry air.

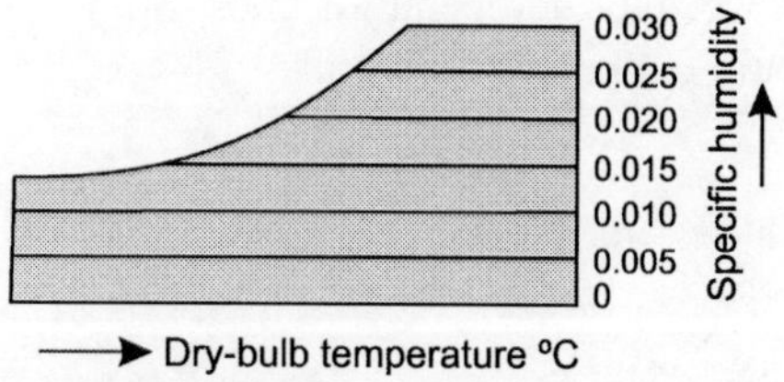

Fig. 4.16 Specific humidity lines

3. **Dew Point Temperature Lines:** The dew point temperature lines are horizontal, *i.e.* parallel to the abscissa and non-uniformly spaced as shown in the Fig. 4.17.At any point on the saturation curve, the dry bulb and dew point temperatures are equal.

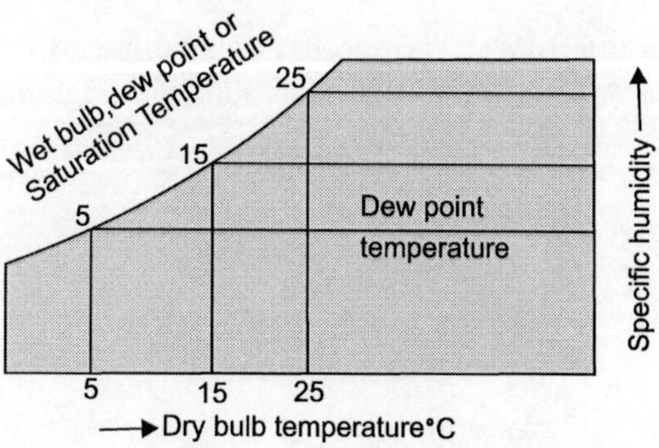

Fig. 4.17 Dry-bulb Temperature ºC

4. **Wet Bulb Temperature Lines:** The wet bulb temperature lines are inclined straight lines and non-uniformaly spaced as shwon in Fig. 4.18. At any point on the saturation curve, the dry bulb and wet bulb temperature are equal.

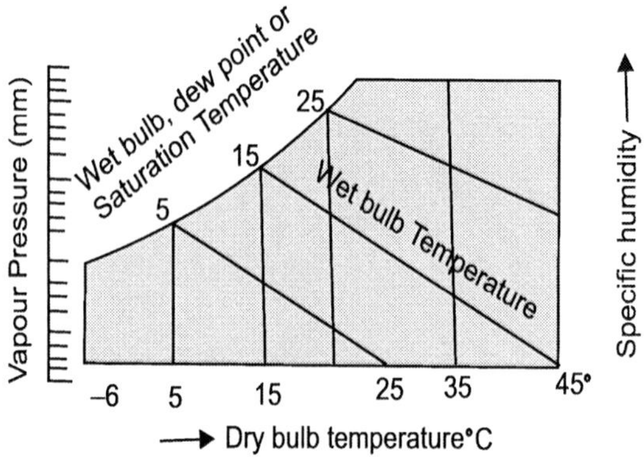

Fig. 4.18 Dry bulb temperature ºC

5. **Enthalpy (Total heat) Lines:** The enthaply (total heat) lines are inclined straight lines and uniformly spaced as shown in Fig. 4.15. These lines are parallel to the wet bulb temperature lines. Some of these lines coincide with the wet bulb temperature lines also.

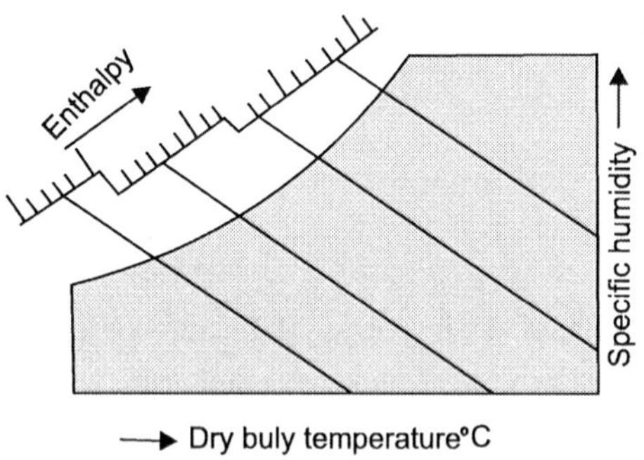

Fig.4.19 Enthaply lines

6. **Specific Volume Lines:** The specific volume lines are obliquely inclined straight lines and uniformly spaced as shown in Fig. 4.20.

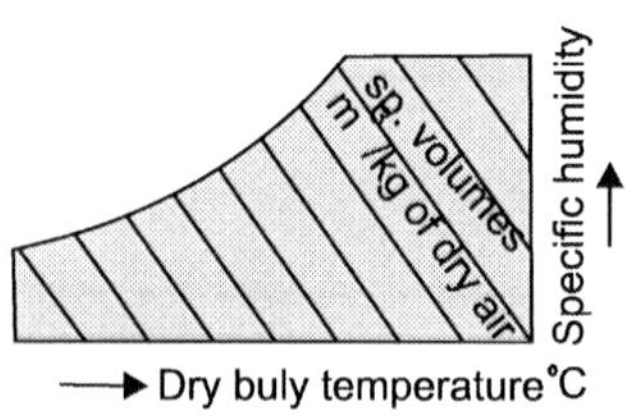

Fig. 4.20 Specific volume lines

7. **Vapour Pressure Lines:** The vapour pressure lines are horizontal and uniformly spaced. Generally, the vapour pressure lines are not drawn in the main chart. But a scale showing vapour pressure in mm of Hg is given on the extreme left side of the chart as shown in Fig. 4.21.

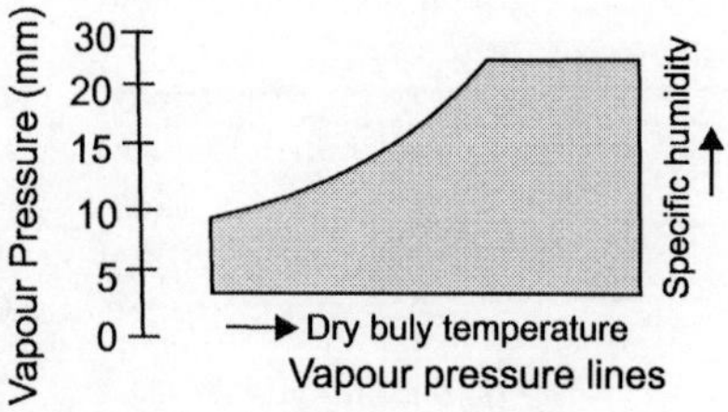

Fig. 4.21 Vapour pressure lines

8. **Relative Humidity Lines:** The relative humidity lines are curved lines and follow the saturation curve. Generally, these lines are drawn with values 10%, 20% etc. and upto 100%. The saturation curve represents 100% relative humidity as shown in Fig. 4.22.

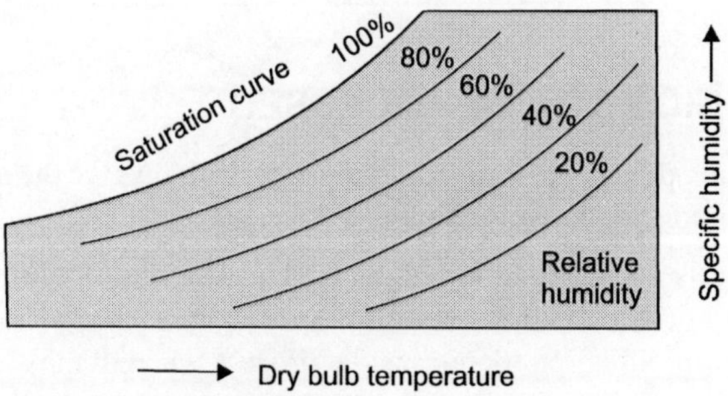

Fig. 4.22 Relative humidity lines

4.23 PSYCHROMETER

It is an instrument which measure both the dry bulb and the wet bulb temperatures of air. Psychrometer comprises of two thermometers with the bulb of one covered by moist wick.

The sling Psychrometer shown in Fig. 4.23 is widely used for measurements involving room air or other applications where the air velocity inside the room is small. The sling psychrometer consists of two thermometers mounted side by side and fitted in a frame with a handle for whirling the device through air. The required air circulation (= 3 to 5 m/s) over the sensing bulbs is obtained by whirling the psychrometer $\simeq$ 300 RPM. Readings are recorded when both the thermometers show steady-state readings.

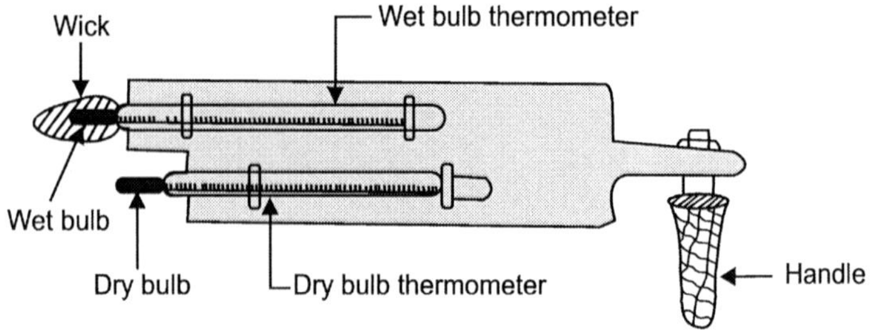

Fig. 4.23 Sling Psychrometer

The function of the wick on the wet-bulb thermometer is to provide a thin film of water on the sensing bulb. To prevent errors, there should be a continuous film of water on the wick. The wicks made of cotton or cloth should be replaced frequently, and only distilled water should be used for wetting it. The wick should extend beyond the bulb by 1 or 2 cms to minimize the heat conduction effects along the stem.

4.24 PSYCHROMETRIC PROCESSES

The most common processes in air-conditioning which involve the air water vapour mixture are mentioned below as shown in the chart.

The initial state is represented by point O. Heating and cooling processes are represented by horizontal lines while humidification and dehumidification by vertical lines. Humidification represents the process wherein the moisture is added to air but its dry bulb temperature is maintained constant. In dehumidification process. The moisture is removed from air without changing its dry bulb temperature.

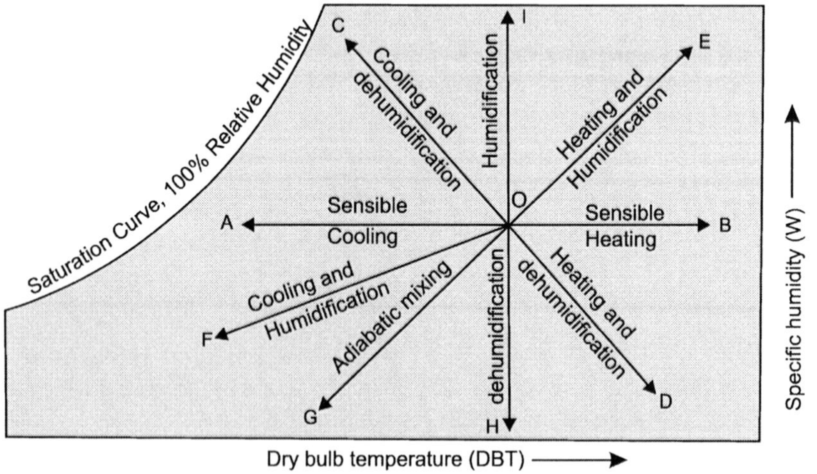

Fig. 4.24 Psychrometric Process

4.24.1 Sensible Cooling (Process O-A)

During this process, the moisture content of air remains constant but its temperature decreases as it flows over a cooling coil. For moisture content to remain constant, the surface of the cooling coil should be dry and its surface temperature should be greater than the dew point temperature of air. If the cooling coil is 100% effective, then the exit temperature of the air will be equal to the coil temperature. However, in practice, the exit air temperature will be higher than the cooling coil temperature. Fig. 4.25 shows the sensible cooling process *O-A* on a psychrometric chart. The heat transfer rate during this process is given by:

$$Qc = m_a (h_O - h_A) = m_a c_{pm} (T_O - T_A)$$

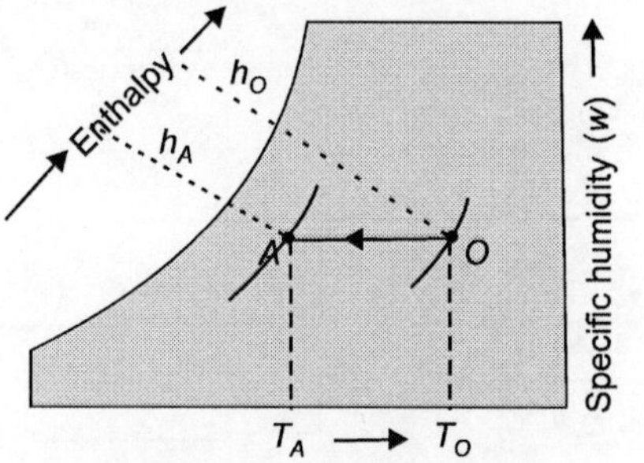

Fig. 4.25 Sensible cooling process

4.24.2 Sensible Heating (Process O-B)

During this process, the moisture content of air remains constant and its temperature increases as it flows over a heating coil. The heat transfer rate during this process is given by:

$$Qh = m_a (h_B - h_O) = m_a C_{pm} (T_B - T_O)$$

Where *c* is the humid specific heat (= 1.0216 kJ/kg dry air) and m_a is the mass flow rate dry air (kg/s). Fig.4.26 shows the sensible heating process on a psychrometric chart.

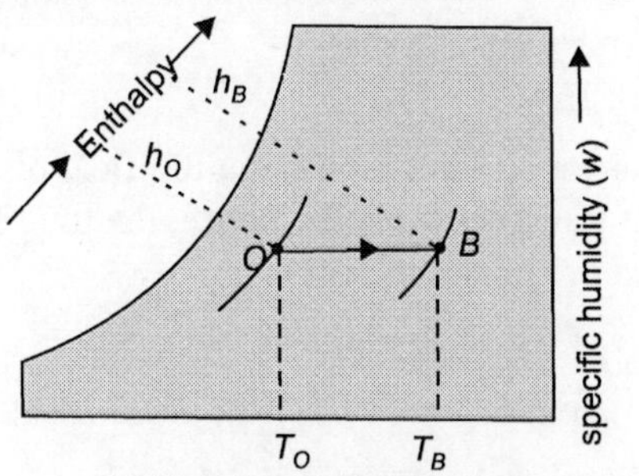

Fig. 4.26 Sensible heating process

4.24.3 Cooling and Dehumidification (Process O-C)

When moist air is cooled below its dew-point by bringing it in contact with a cold surface as shown in Fig. 4.27, some of water vapours in the air condenses and leaves the air stream as liquid, as a result both the temperature and humidity ratio of air decreases as shown. This is the process used in a typical air conditioning system. Although the actual process path will vary depending upon the type of cold surface, the surface temperature, and flow conditions, for simplicity the process line is assumed to be a straight line. The heat and mass transfer rates can be expressed in terms of the initial and final conditions by applying the conservation of mass and conservation of energy equations as given below:

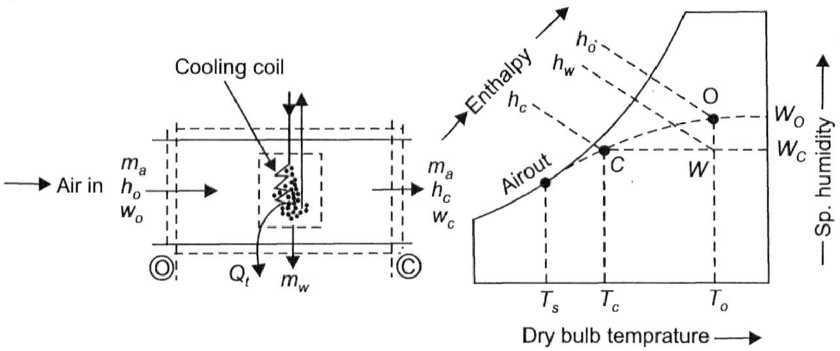

Fig. 4.27 Cooling and dehumidification process

By applying mass balance for the water:

$$m_a . W_0 = m_a W_c + m_w$$

By applying energy balance:

$$m_a h_o = Q_t + m_w h_w + m_a h_c$$

From the above equations, the load on the cooling coil, Q_t *is* given by:

$$Q_t = m_a(h_0 - h_c) - m_a(W_0 - W_c)h_w,$$

The (2) terms on the RHS of the above equation is normally small compared to the other terms, so it can be neglected. Hence,

$$Q_t = m_a(h_o - h_c)$$

It can be observed that the cooling and dehumidification process involves both latent and sensible heat transfer processes, hence, the total latent and sensible heat transfer rates (Q_t, Q_l and Q_s) can be written as:

$$Q_t = Q_l + Q_s$$

where Q_l = Latent heat $= m_a (h_0 - h_w) = m_a h_{fg} (W_o - W_c)$

Q_s = Sensible heat $= m_a (h_w - h_c) = m_a C_{pm} (T_o - T_c)$

By separating the total heat transfer rate from the cooling coil into sensible and latent heat transfer rates, a useful parameter called Sensible Heat Factor (SHF) defined (SHF) is defined as the ratio of sensible heat to total heat transfer rate, *i.e.*

$$\text{SHF} = \frac{\text{Sensible heat}}{\text{Total heat}} = \frac{\text{Sensible heat}}{\text{Sensible heat} + \text{Latent heat}}$$

$$= \frac{Q_s}{Q_s + Q_l} = \frac{Q_s}{Q_t}$$

From the above equation, one can deduce that a SHF of 1.0 corresponds to no latent heat transfer and a SHF of 0 corresponds to no sensible heat transfer. A SHF of 0.75 to 0.80 is quite common in air conditioning systems in a normal dry-climate. A lower value of SHF, say 0.6, implies a high latent heat load such as that occurs in a humid climate.

4.24.4 Heating and Humidification (Process O-E)

During winter it is essential to heat and humidify the room air for comfort. As shown in Fig. 4.28. This is normally done by first sensibly heating the air and then adding water vapour to the air stream through steam nozzles as shown in Fig. 4.29.

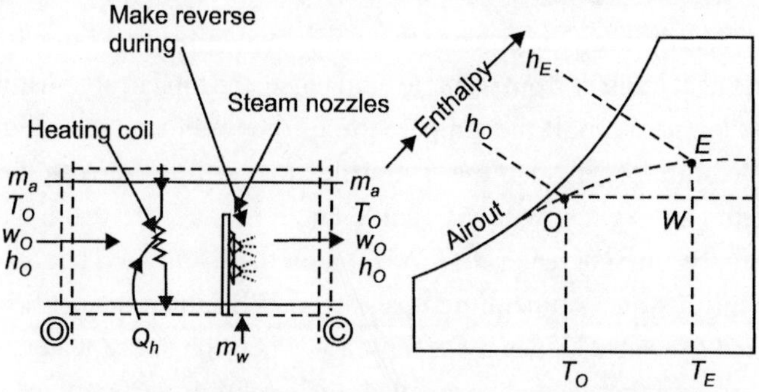

Fig. 4.28 Heating and humidification process (O- D)

Mass balance of water vapour for the control volume yields the rate at which steam has to be added, *i.e.*, m_u

$$m_v = m_a(W_E - W_O)$$

Where m_a is the mass flow rate of dry air.

From energy balance:

$$Q_h = m_a(h_E - h_O) - m_w h_w$$

Where Q_h is the heat supplied through the heating coil and h_w is the enthalpy of steam. Since this process also involves simultaneous heat and mass transfer, we can define a sensible heat factor for the process in a way similar to that of a cooling and dehumidification process.

4.24.5 Cooling and Humidification (Process O-F)

As the name implies, during the process the air temperature drops and its humidity increases. This process is shown in Fig.4.29 As shown in the figure, this can be achieved by spraying cold water in the air stream. The temperature of water should be lower than the dry-bulb temperature of air but higher than its dew-point temperature to avoid condensation.

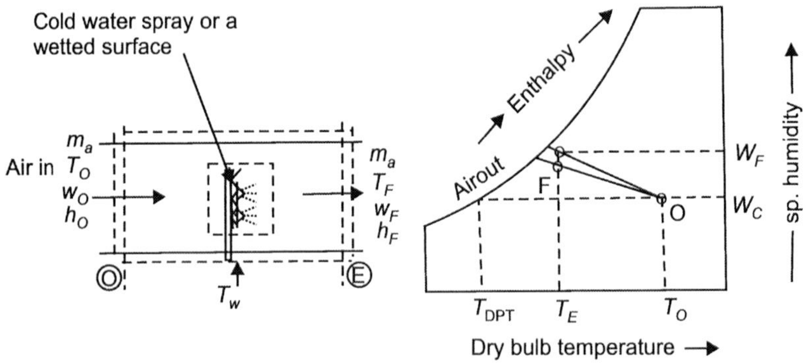

Fig. 4.29 Cooling and humidification

It can be seen that during this process there is sensible heat transfer from air to water and latent heat transfer from water to air. Hence, the total heat transfer depends upon the water temperature. If the temperature of the water sprayed is equal to the wet-bulb temperature of air, then the n_e transfer rate will be zero as the sensible heat transfer from air to water will be equal to latent heat transfer from water to air. If the water temperature is greater than WBT, then there will be *a* net heat transfer from water to air. If water temperature is less than WBT, then the net heat transfer will be from air to water. Under a special case when the spray water is entirely recirculated and is neither heated nor cooled, the system is perfectly insulated and the make-up water is supplied at WBT, then at steady-state, the air undergoes an adiabatic saturation process, during which its WBT remains constant. This is the process of adiabatic saturation. The process of cooling and humidification is encountered in a wide variety of devices such as evaporative coolers, cooling towers etc.

4.24.6 Heating and Dehumidification (Process O-G)

This process is mainly used in industrial air conditioning. This process can be achieved by using a hygroscopic material, which absorbs or adsorbs the water vapour from the moisture. If this process is thermally isolated, then the enthalpy of air remains constant, as a result the temperature of air increases as its moisture content decreases as shown in Fig. 4.30 This hygroscopic material can be a solid or a liquid. In general, the absorption of water by the hygroscopic material is an exothermic reaction, as a result heat is released during this process, which is transferred to air and the enthalpy of air increases.

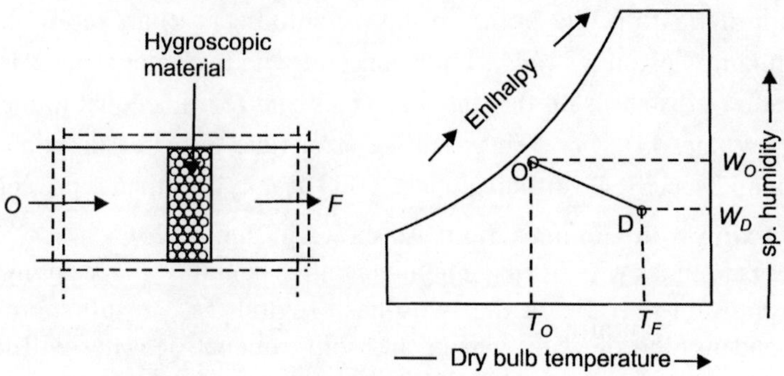

Fig. 4.30 Heating and dehumidification Process

4.24.7 Mixing of Air Streams

Mixing of air streams at different states is commonly encountered in many processes, including in air conditioning. Depending upon the state of the individual streams, the mixing process can take place with or without condensation of moisture.

(*a*) **Without Condensation:** Fig. 4.31 shows an adiabatic mixing of two moist air streams during which no condensation of moisture takes place. As shown in the figure, when two air streams at state points 1 and 2 mix, the resulting mixture condition 3 can be obtained from mass and energy balance.

From the mass balance of dry air and water vapour:

$$m_{a,1}W_1 + m_{a,2}W_2 = m_{a,3}W_3 = (m_{a,1} + m_{a,2})\, W_3$$

From energy balance:

$$m_{a,1}h_1 + m_{a,2}h_2 = m_{a,3}h_3 = (m_{a,1} + m_{a,2})\, h_3$$

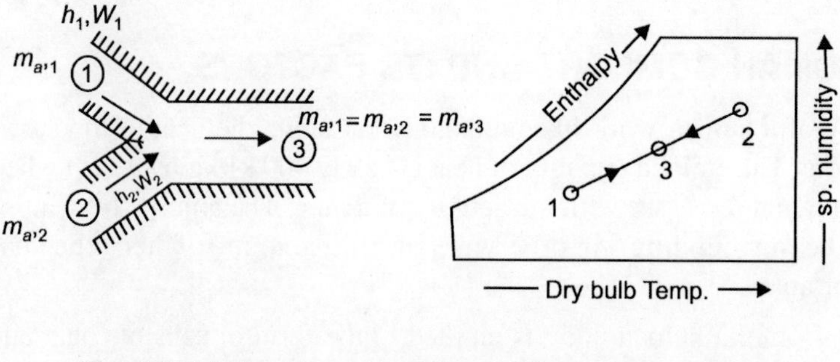

Fig. 4.31 Mixing of two air streams without condensation

From the above equations, it can be observed that the final enthalpy and humidity ratio of mixture are weighted averages of inlet enthalpies and humidity ratios. A generally valid approximation is that the final temperature of mixture is the weighted average of the inlet temperatures. With this approximation, the point on the psychrometric chart representing the mixture lies on a straight line connecting the two inlet states. Hence, the ratio of distances on the line, *i.e.*, (1–3) and (2–3) is equal to the ratio of flow rates $m_{a,2}/m_{a,1}$. The resulting error (due to the assumption that the humid specific heats being constant) is usually less than 1 per cent.

(*b*) **Mixing with Condensation:** As shown in Fig. 4.32 when very cold and dry air mixes with warm air at high relative humidity, the resulting mixture condition may lie in the two-phase region, as a result there will be condensation of water vapour and some amount of water will leave the system as liquid water. Due to this, the humidity ratio of the resulting mixture (point 3) will be less than that at point 4. Corresponding to this will be an increase in temperature of air due to the release of latent heat of condensation. This process rarely occurs in an air conditioning system, but this is the phenomenon which results in the formation of fog of frost (if the mixture temperature is below 0°C). This happens in winter when the cold air near the earth mixes with the humid and warm air, which develops towards the evening or after rains.

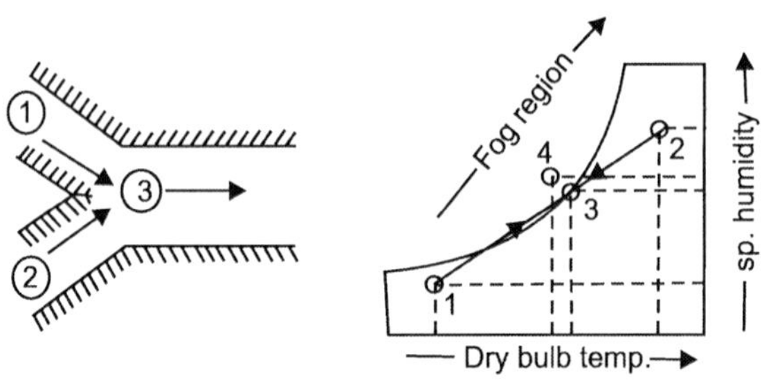

Fig. 4.32 Mixing of two air streams with condensation

4.25 HUMAN COMFORT AND ITS FACTORS

The body of a human work like an engine that emits the heat to the surrounding atmosphere. The emitted quantity of heat is nearly 400 kJ/kg by a man when at rest and the amount increases with the activity of a man. The amount of heat is carried out by the surrounding air otherwise on this amount of heat the man feels uncomfortable.

The heat from the human body is emitted in the form of sensible heat and latent heat. The sensible heat when taken out by air from human body. The temperature of air should be lower than the human body temperature about (36.4°C) and the latent heat carried out from the body in the form of sweating, which makes the body suffer, therefore it is necessary that 30% of the total heat from the body should be carried out in the form of latent heat and remaining 70% in the form of sensible heat. If this balance will not be maintained the human bengs will feel uncomfortable.

To maintain this balance the heat carried away by the air, certain DBT and RH of the air are absolutely essential in addition to DBT and RH the air movement over the body is equally important for the human body comfort.

In general, the range of temperature and humidity condition maintain in the air conditioned space during summer are 23.5° to 25.5°C (74° to 78°F) and 55 - 65% RH.

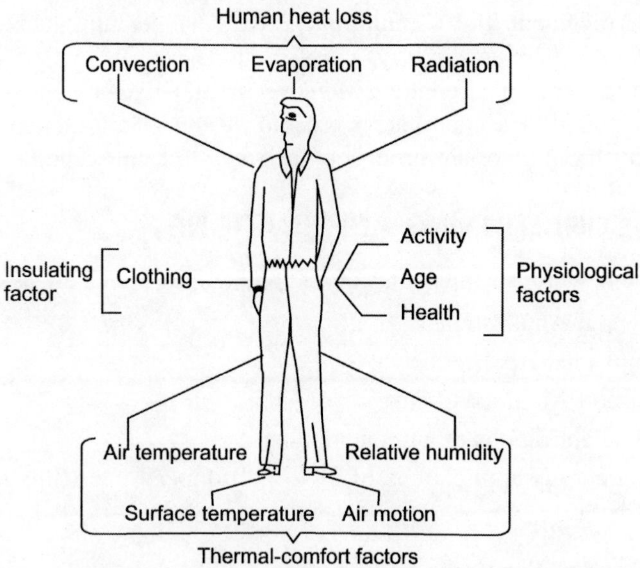

Fig. 4.33 Thermal interchange with environment

There are four important factors which affect human comfort. The important factors are as follows:

1. Temperature of air 2. Humidity of air 3. Air purify
4. Air movement and circulation

1. **Temperature of Air:** It is very important factor which affects human comfort to a great extent. The range of temperature for human being feels comfortable is 21°C with 56% relative humidity.

   In air conditioning the control of temperature means the maintenance of any desired temperature within an enclosed space even though the temperature of the outside air is above or below the desired room temperature. This is accomplished either by the addition or removal of heat from the enclosed space as and when demanded.

2. **Humidity of Air:** The control of humidity of air means the increasing or decreasing of moisture contents of air during summer or winter respectively. In order to produce comfortable and healthy conditions, the control of humidity is not only necessary for human comfort but it also increases the efficiency of the work.

   In general, for summer air conditioning, the relative humidity should not be less than 60°C whereas for winter air-conditioning it should not be more than 40%.

3. **Air Purify:** People do not feel comfortable when breathing contaminated air even if it is within acceptable temperature and humidity ranges. So proper filteration, cleaning and purification of air is necessary to keep it free from dust, dirt and other impurities.

4. **Air Movement and Circulation:** Even if temperature, humidity and air purification are satisfactory, a certain amount of air motion seems to be necessary for human comfort. Moreover the air movement and its circulation are required to be controlled in order to ensure ?. Uniform temperature and equidistribution of air throughout the space to be air-conditioned.

4.26 AIR-CONDITIONING APPLICATIONS

There are the numerous applications of air conditioning which are as follows:

1. Industrial Applications
2. Commercial Applications
3. Transport Air Conditioning
4. Special aplications of air conditioning

There are numerous applications of air-conditioning. A few applications are summarised here:

1. Industrial Applications

In most of the air-conditioning applications for industry, the common problem is to control the temperature, humidity and air motion for maintaining the quality of the product. The function of air-conditioning is to control the temperature and humidity in chemical and biochemical reactions.

(*a*) ***Photographic Industry*:** Many processes in photographic industries require accurate control of temperature and humidity conditions both in manufacturing of photographic materials and preserving the quality of manufactured materials.

(*b*) ***Textile Industry*:** Successful application of air-conditioning in textile industry has helped in the production of quality textile materials. Maintenance of specific relative humidity is required in cotton, nylon, silk and rayon industries. The humidity of air has pronounced effect on the strength, quality and rate of processing.

(*c*) ***Printing Industry*:** Specific temperature and humidity are required in paper printing and cloth printing factories. Paper becomes too dry in low humidity conditions and curling occurs. The paper swells in high humidity conditions and causes non-uniform printing. In cloth printing factories, low humidity gives faint prints and high humidity causes blots. The temperature and humidity requirements for these industries are 24° to 27°C DBT, and 20-50% RH.

(*d*) ***Machine Tool Industry*:** There are some machining processes which require accurate control of temperature and humidity. These include the manufacture of bearings, scientific instruments, electronic devices, test gauges and many others where close tolerances of dimensions are required. The measuring instruments and gauges are always preserved in controlled atmosphere for maintaining the dimensional stability.

2. Commercial Applications

The comfort of the human beings is the prime consideration in commercial air-conditioning systems. The degree of comfort maintained also differs from application to application. Better comfort condition is necessary in offices or houses compared with shops, banks and theatres as their stay time is more.

(*a*) ***Theatre:*** The main load on the air-conditioning system in theatres consists of occupancy and outdoor air for the night shows. The load during the day time is much more as solar radiation and heat load due to heat transfer from the outside through the walls and roof play important role in considering the total heat load. The common comfort conditions maintained in theatres are 26°C DBT and 56% R.H.

(*b*) ***Departmental Stores and Banks:*** Departmental stores vary in size, type and area so that the design of air-conditioning system differs from one to the other. The comfort conditions maintained are 26°C DBT and 50% R.H. at maximum outdoor summer design conditions. One of the major heat loads is the lighting load and it should be considered independently of out-door conditions. Special attention is required for air distribution system because of the location of elevators, escalators and service areas.

3. Transport Air Conditioning

The use of air-conditioning in moving vehicles has been introduced very recently. The major problems encountered in design of air-conditioning plants used in moving vehicles are the irregular variation of load due to change in speed. The lightness and compactness of the equipment required are also important considerations.

(*a*) ***Automobile Air-Conditioning:*** The equipments used for air-conditioning in an automobile, are similar to other stationary air-conditioning systems. The compressor is mounted above the engine and is belt driven. The condenser is mounted ahead of the car radiator and condensing of refrigerant is due to the Low of air over it. The cooling coil and blower are located below or behind the dash.

Air distribution system for automobile air-conditioning offers several unusual problems. The space is confined and the seats present a restriction to the air-flow and low roof creates the possibility of draft. Air-distribution system ducts are installed between the ceiling and the outer shell of the body. The grills are placed at frequent intervals along these ducts to supply conditioned air. An air-conditioning unit used for a 6-seated car requires 3 tonnes of refrigerating capacity at a vehicle speed of 80 km/hr. Air-conditioning units used for a passenger about 500 kg and require about 10 kW to provide a cooling.

(*b*) ***Air-conditioning of Trains:*** Year round system can be successfully used for train air-conditioning. Heat is supplied to each coach using steam from the boiler in steam locomotives and hot gases or heated water in diesel locomotives.

Summer air-conditioning can be achieved by using steam jet refrigeration in steam locomotives or by using vapour compression system. A normal refrigerating capacity required for each coach is about 8 tonnes of refrigeration requiring 10 kW. The quantity of air circulated is 40 m^3/min for each coach of 80 seating capacity. The condensers are generally air-cooled. A separate air-conditioning circuit is used for each coach.

(*c*) ***Aircraft Air-conditioning:*** The air-conditioning in aircraft is more difficult as compared to vehicles moving on the ground because of extreme cold conditions at high altitudes and rapid fluctuations in air-temperature. The air-conditioning system used for air-planes requires conditioning and pressurising of the cabin. Air-cycle cooling is commonly used as pressurised air is readily available and the weight of the unit (kg/ton) is considerably reduced. Electronic control systems are commonly used for aircraft air-conditioning.

(*d*) ***Air-conditioning of Ships:*** The use of air-conditioning for passenger ships as well as for cargo ships has been recently developed. The ship air-conditioning has more problems in design and construction due to some special problems encountered. The requirements of ship air-conditioning are different as it is used in sleeping rooms, theatres, offices and food store rooms. The corrosion (due to sea water) and vibration (due to tolling and pitching) of air ducts, water piping and refrigerant piping are main problems in the design of air-conditioning systems used on ships. These problems are more severe in war ships. Earlier CO_2 and NH_3 were used as refrigerant for air-conditioning plants. The use of CO_2 stopped due to high temperature at the condensing surface. The use of NH_3 created lot of problems due to its tendency of leak even under weak vibration conditions. The leaking of

NH_3 is unsafe and also spoils food stuffs. Nowadays Freon-11 and Freon-12 are successfully used for ship air-conditioning It is absolutely necessary to provide proper vibration dampers to avoid the failure of ducts and pipe Joints.

4. Some Special Applications of Air-conditioning

(*a*) ***Air-conditioning for Television Centre:*** Television and radio studios have to be sealed against all external noises, making air-conditioning essential for protecting sensitive equipments against the effect of temperature and humidity variations. Generally 20°C DBT is maintained inside the studios.

(*b*) ***Air-conditioning for Computer Centre:*** The computers are playing a dominant role in progress of present technology. One basic requirement for its successful application is the controlled environment. The heat generation rate of computer element is considerably high and localised. The lives of electron components are shortened when they are operated at elevated temperatures and also introduce errors into the system, The temperature to be maintained lies between 15 to 20°C. Relative humidity also must be controlled within certain prescribed limits for the successful operation of the computer. At high relative humidities, electrical charges leak between components and low relative humidity encourages static discharges and introduce errors into computer calculations. The relative humidity to be maintained lies between 30 to 65%.

(*c*) ***Air-conditioning for Automatic Telephone Exchange Building.*** Controlled atmosphere in telephone exchange building is essential when machinery operations are made automatic. Dry atmosphere will cause the trouble of making the contacts of the closing points and too humid atmosphere may cause point contact irrespective of call. High temperature may cause excessive expansion and low temperature may cause excessive contraction. To avoid these difficulties, air-conditioning is an essential requirement of a telephone exchange building.

(*d*) ***Hospital Air-conditioning:*** The applications of air-conditioning to hospital patients imposes many problems not encountered in the usual comfort air-conditioning systems. The major problem in the hospital air-conditioning is the rapid infection. Therefore, it is necessary to maintain sufficient air circulation to remove the bacteria. The requirements of the different sections of the hospital as operating theatres, nurseries, laboratories. X-ray department and service departments are varied in nature. These sections are usually served by individual systems because of the different requirements of temperature and humidity and it is not desirable to mix air from some of these areas to avoid infection.

In pathology department, the air distribution system should be designed to carry chemical fumes, odour and heat from the equipments. The ventilation rate of 10 times air changes per hour is recommended.

The control of temperature and humidity in nursery department is essential to take care of the new born. An adequate ventilation is necessary to remove odour and to reduce bacterial contamination. About 12 times air-changes per hour and 27°C DBT and 60% R.H. are recommended for the nursery department.

Surgical department requires a more careful control of the septic condition of the environment. The air supplied to the operation room should contain minimum bacteria. Minimum 20 times air changes per hour and 24°C DBT and 50% R.H. are recommended for the operation room. Positive air pressure (2.5 ton of water) is to be maintained to avoid the leakage of outside air into the rooms. The requirements of emergency department are the same as surgical department.

The service department includes kitchen, dining, dish washing, store and many others. The cooking equipments such as boilers, ovens and kettles-are grouped together for efficient utilization, therefore, heat load in kitchen is considerably large compared with other departments. The kitchen should be provided with proper hoods to remove heat, odour and vapour. The dish washing room should be completely enclosed as it gives out specific odour of food and ventilated with the same rote as kitchen. Minimum 10 air changes are recommended for all the service departments mentioned above.

HIGHLIGHTS OF THE CHAPTER

1. Refrigeration may be defined as it is a method of introducing and maintaining the low temperature in an enclosed space below the surrounding temperature.
2. Air conditioning means condition of air for maintaining low temperature humidity, purity, air motion and human being comfort within the enclosed space.
3. The cooling effect produced by a refrigerating machine in a given time known as refrigerating effect. It is expressed as kJ/minor kJ/sec.
4. The rating of the refrigeration machine is given by unit of refrigeration which is known as tone of refrigeration.

5. Tonne of refrigeration is defined as the amount of heat to be removed in order to form one tonne of ice at 0°C in 24 hours

$$\text{Relative } wp = \frac{\text{Actual C.O.P.}}{\text{Theoretical C.O.P.}}$$

6. C.O.P is defined as the ratio of heat extracted at low temperature from the refrigerator to the work done of the refrigerant.

$$\text{C.O.P} = \frac{\text{Refrigerating effect}}{\text{Work done}}$$

7. Refrigerator is a machine which is used to remove the heat contagiously from the enclosed space and maintain the temperature below the surrounding temperature and rejects the heat the surrounding at higher temperature.

8. Heat Pump is a device which is used to supply the heat continuously to a space for maintaining the temperature above surrounding atmospheric temperature by absorbing the heat from the surrounding at low temperature.

9. C.O.P of reversed Carnot cycle $= \dfrac{T_2}{T_1 - T_2}$

10. The following precession of a simple vapour compression cycle are as follows.

(*i*) Compression (*ii*) Condensation

(*iii*) Expansion (*iv*) Vaporization

11. Principal parts of simple vapour compression cycle are as follows:

(*i*) Evaporator (*ii*) Suction line

(*iii*) Compressor (*iv*) Discourage line

(*v*) Condenser (*vi*) Receiver Tank

(*vii*) Liquid line (*viii*) Expansion vase

12. Refrigerant is a heat carrying medium used in refrigerating cycle to absorb heat from a low temperature system and discard the heat so absorbed to another system at higher temperature.

13. Dry bulb temperature is the temperature of air measured by ordinary thermometer is known as dry bulb temperature of air.

14. The wet wick thermometer reading is called the wet bulb temperature.

15. It is defined as the ratio of actual mass of water vapours in a given volome to the mass of water vapour.

16. Humidity ratio is difined as the ratio of water vapour to the mass of dry air.
17. Psychometric chart is a graphical representation of various thermodynamic properties of moist air and helps to calculate properties of air directly from the chart.
18. Psychometers is an instrument which measure both dry bulb and wet bulb temperature of air.
19. Psychometric processes are as follows
    (*i*) Sensible cooling
    (*ii*) Sensible heating
    (*iii*) Cooling and dehumidification
    (*iv*) Heating and humidification
    (*v*) Cooling and humidification
    (*vi*) Heating and dehumidification
    (*vii*) Mixing of air streams.
20. Human comfort is the condition of mind in which human feels comfort and satisfaction with the thermal environment.
21. The factor which affect the human comfort are
    (*i*) Temperature
    (*ii*) Humidity
    (*iii*) Air purity
    (*iv*) Air movement and circulation.

REVIEW QUESTIONS

1. What do you mean by refrigeration and air conditioning?
2. State the difference between refrigerator and heat pump. How these machines satisfy the 2nd law of thermodynamics?
3. What do you mean by rating of refrigeration machine?
4. Define:
   (*i*) Tonne of refrigeration
   (*ii*) Refrigeration
   (*iii*) Refrigerating effect
5. What do you mean by Co-efficient of Performance (C.O.P.)?
6. Set up a relation for the C.O.P. of a heat pump and the refrigeration and prove that:
   $(\text{C.O.P.})_{\text{heat pump}} = 1 + (\text{C.O.P.})_{\text{refrigeraton}}$
7. Explain the working of a bell coleman cycle with neat sketch.
8. Describe with neat sketch the working of a simple vapour compression refrigeration cycle with P-V and T.S. diagrams.

9. State the function of different parts of a simple vapour compression refrigeration system.

10. Give the comparison between air refrigeration system and vapour compression system.

11. Give the comparison between reverse carnot cycle and vapour compression refrigeration system.

12. What are the properties of an ideal refrigerant?

13. Define a refrigerant and name few widely used refrigerants.

14. Write the different methods of refrigeration and explain in brief?

15. Establish the expression for specific humidity and prove:

$$w = 0.622 \times \frac{P_v}{P_t - P_v}$$

16. Establish the following expression for air vapour mixture specific humidity

$$w = 0.622 \times \frac{P_v}{P_b - P_v}$$

17. Explain the following psychometric forms

(*i*) Dew point temperature

(*ii*) Dry bulb temperature

(*iii*) Wet bulb temperature

(*iv*) Relative humidity

(*v*) Relative humidity

(*vi*) Specific humidity

18. Explain the following processes on a psychrometric chart separately:

(*i*) heating and humidification

(*ii*) Adiabatic mixing of two strams

(*iii*) Sensible cooling

(*iv*) Cooling and dehumidifiction

19. Write various applications of refrigerators.

20. What do you mean by human comfort? Discuss its important factors which are responsible for human confort.

21. What is a simple psychometer explain with neat sketch.

22. Distinguish between primary and secondary refrigerants.

23. Describe a simple vapour compression cycle with neat sketch.

24. State the functions of the following parts used in a simple vapour compression system:

(*i*) Compressor

(*ii*) Condenser

(*iii*) Expansion valve

(*iv*) Evaporator

25. State the properties and applications of the following refrigerant:

ammonia, carbon dioxide, freon-12, freon-22 times air.

26. Mention the various applications of refrigeration.

■■■

5

Hydraulic Turbines and Pumps

5.1 INTRODUCTION

The word Hydraulic means (Hydro + licks) where Hydro means fluid and licks means flow through pipe. Hence, the fluid flow through the pipe is known as hydraulics.

Hydraulic machines are those in which the fluid (water, oil etc.) is in continuous motion and constitute a medium for energy conversion. The machine may convert the hydraulic energy into mechanical energy or *vice versa* for expample, turbines, pumps, hydraulic coupling etc.

In hydraulic turbine the working fluid is water because water is cheap and easily available. In water turbine water jet acquires high velocity while flowing over the converging passage. When this high velocity jet strikes the vanes or blades fixes around the periphery of rotor, provide a rotary motion by centrifugal forces which results change in momentum and a hydrodynamic force is exerted which starts rotating the turbine shaft. This shaft is attached with alternator or generator by coupling system and thus electric power is generated.

Hydraulic Turbine: Hydraulic turbine is a device which converts hydraulic energy *i.e.* (P.E. + K.E.) into mechanical energy and later on with the addition of alternator or generator this mechanical energy is converted into electrical energy known as hydraulic turbine.

5.2 IMPULSE TURBINE

Impulse turbine is that in which whole pressure energy of water is converted into **Kinetic energy** by allowing water to pass through the nozzle and impingings with high speed on the twin split hemispherical cap shaped blade of the pelton runner which produces thrust on the runner and causing change in direction to rotate the runner is this way hydraulic energy is converted into mechanical energy.

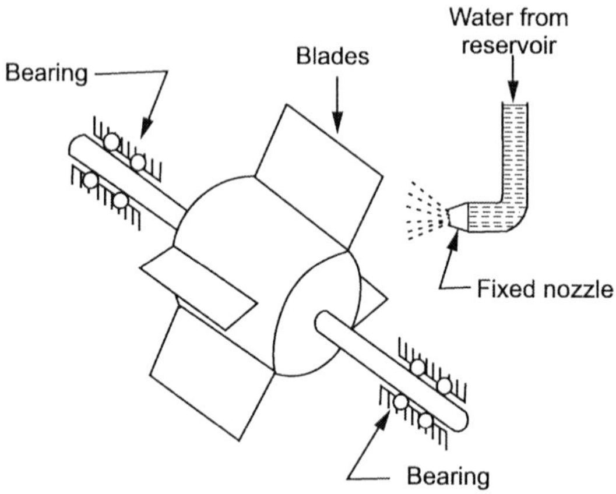

Fig. 5.1 Principle of an impulse turbine.

5.3 REACTION TURBINE

In reaction turbine water is partially converted in to pressure energy and partially in to kinetic energy. The pressure of water at inlet of turbine and at outlet of turbine is called reaction pressure. This pressure acting on the blades exerting a thrusts on the blades which are attached to the rotor and it causes the runner to rotate. Hence, we get mechanical power. Here the pressure difference is responsible to rotate the runner. Therefore, whole unit must be air tight casing to maintain the pressure difference in reaction turbine.

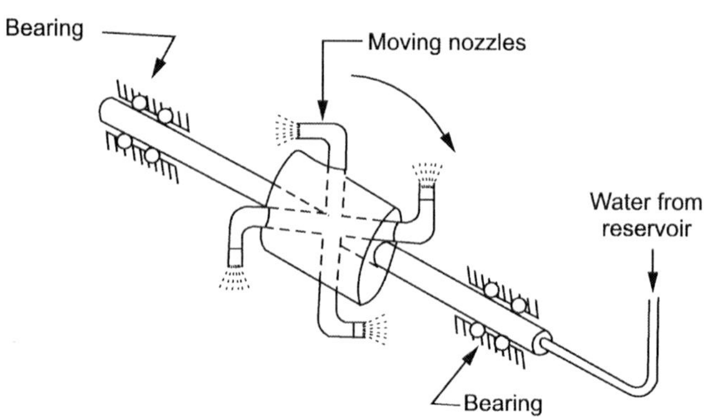

Fig. 5.2 Principle of a reaction turbine

5.4 SCHEMATIC DIAGRAM OF A HYDRAULIC POWER PLANT

Fig. 5.3 shows a general diagram of a hydroelectric power plant, consists of:

Dams: A dam is constructed across a river to store water.

Penstock: Pipes, made of steel or reinforced concrete, of large diameter called penstocks, which carry water under pressure from the storage reservoir to the turbine.

Vane: Turbine having different types of vanes where fitted to the wheels.

Tail race: Tail race which is a channel which carries water away from the turbines after the water has worked on the turbines. The surface of water in the tail race is also known as tail race.

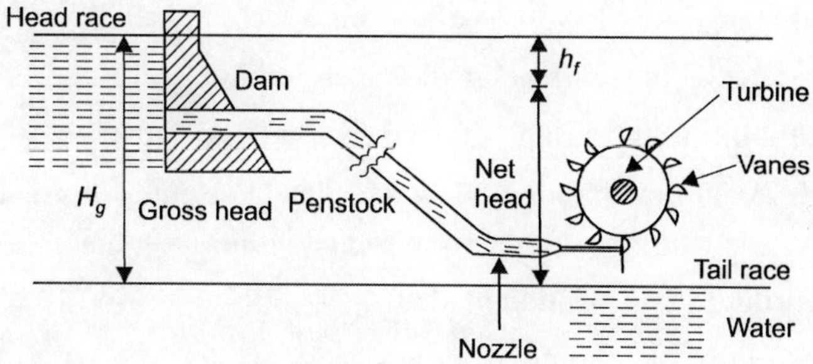

Fig. 5.3 Diagram of a hydraulic power plant

Gross Head: The difference between head race level and tail race level when no water is flowing is known as Gross Head. It is denoted by H_g.

Net Head: Head available at the inltet of the turbine

$$H = H_g - h_f$$

where $$h_f = \frac{4 \times f \times L \times V^2}{D \times 2g}$$

V = Velocity of flow in penstock

L = Length of penstock

D = Diameter of penstock.

5.5 CLASSIFICATION OF TURBINES

The turbines may be classified according to:

1. **The head and quantity of water available.**
   (*a*) High heat and low discharge turbine - Pelton wheel impulse turbine.
   (*b*) Medium head and medium discharge turbine - Francis turbine.
   (*c*) Low head and large discharge turbine - Kaplan or propeller turbine.

2. According to action of water on moving blades.

(*a*) Impulse turbine - Pelton wheel

(*b*) Reaction turbine - Francis, Kaplan

3. According to the inventor or orignator

(*a*) Pelton turbine - Laster Allen Pelton - California (USA)

(*b*) Francis turbine - Jammes Bichens Francis France

(*c*) Kaplan turbine - Dr. Victor Kaplan

4. According to the direction of flow of water along the blades

(*a*) Tangential flow turbine - Pelton wheel turbine

(*b*) Radial flow turbine - Old Francis turbine

(*c*) Mixed flow turbine - Redial entry and axial exit

(*d*) Axial flow turbine - Flow direction (1) parallel to axis of the shaft (Kaplan and propeller turbine)

5. According to disposition of shaft

(*a*) Horizontal turbine - Pelton wheel (axis of shaft is horizontal in small Francis turbine.)

(*b*) Vertical turbine - Kaplan, propeller turbine, large Francis (Axis of shaft is vertical)

6. According to specific speed of turbine:

	Specific speed		
	low	**medium**	**high**
(*a*) Slow specific speed turbine - Pelton wheel	9-17	17-25	25-30
(*b*) Medium specific speed turbine - Francis	50-100	100-150	150-200
(*c*) High specific speed turbine - Kaplan	50-100	100-150	200-250

$$NS = \frac{N\sqrt{P}}{H^{3/4}}$$

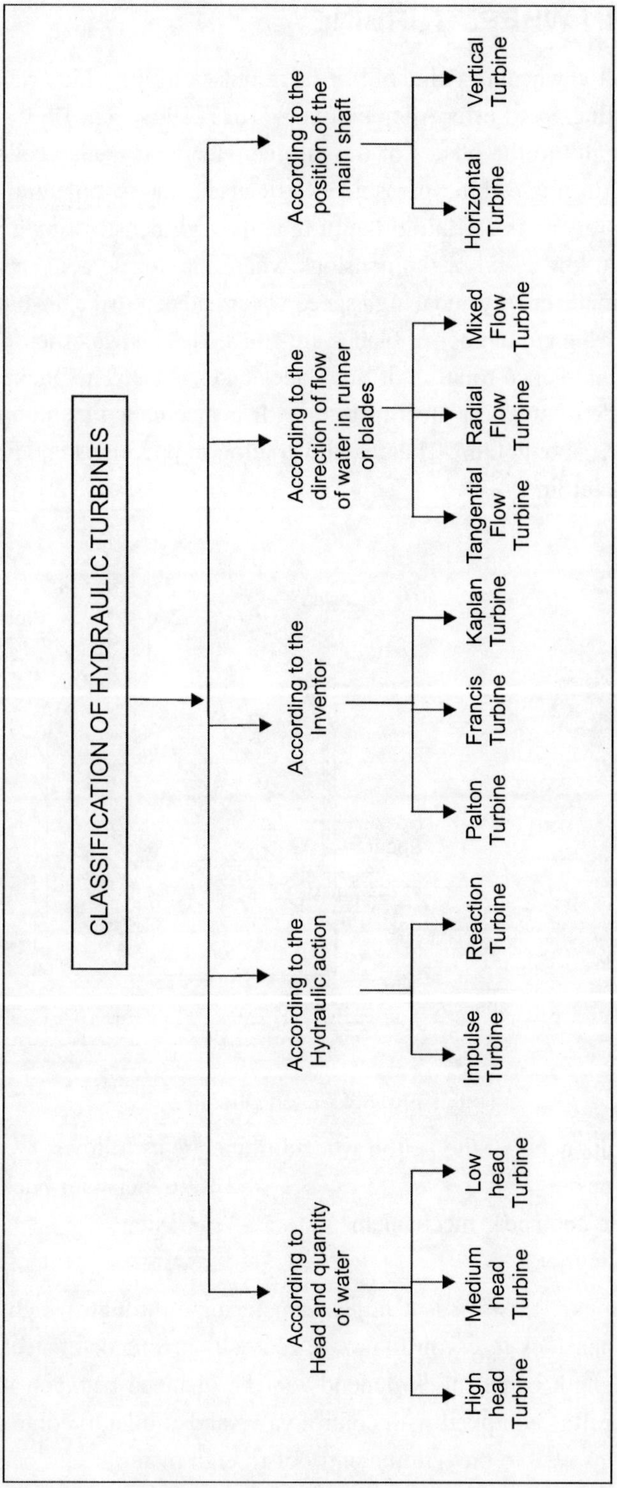
CLASSIFICATION OF HYDRAULIC TURBINES
According to Head and quantity of water
High head Turbine
Medium head Turbine
Low head Turbine
According to the Hydraulic action
Impulse Turbine
Reaction Turbine
According to the Inventor
Palton Turbine
Francis Turbine
Kaplan Turbine
According to the direction of flow of water in runner or blades
Tangential Flow Turbine
Radial Flow Turbine
Mixed Flow Turbine
According to the position of the main shaft
Horizontal Turbine
Vertical Turbine

5.6 PELTON WHEEL TURBINE

Working: Pelton wheel is a kind of free jet impulse turbine which was invented by an American Engineer Lester Allen Pelton in (1829–1908). The flow of water takes place tangentially to the blades of the turbine. Hence, it is also called tangential flow turbine. In this turbine pressure head caused due to potential energy. The water from reservoir is carried to the turbine through penstock pipe and a nozzle attached to the lower end of the penstock where the nozzle converts all pressure energy into kinetic energy and a high speed water jet head strike the outer periphery of runner on which a number of blades are attached. This water jet induced thrust on the blade and hence rotation of the blades takes place. The buckets are spliter form and made of stainless steel or bronze. It is also called as high head turbine whose height is about 400m. There is no variation of pressure so no need of strong casing for the turbine.

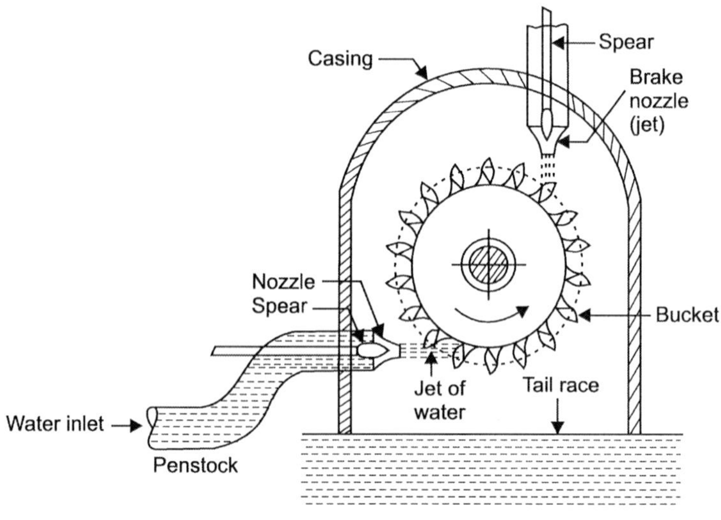

Fig. 5.4 Pelton wheel

The main components of the pelton wheel turbine are as follows:

1. Penstock
2. Runner with buckets
3. Nozzle and spear mechanism
4. Casing
5. Breaking jet

1. **Penstock:** Penstock is a large diameter pipe through which water flows from penstock reservoir to nozzle generally it is made of steel or concrete. The length of penstock depends on the distance between reservoir and runner. It is equipped with control valve and at inlet to control the flow of water as well to prevent the entry of foreign material into penstock.

2. **Runner with Buckets:** The rotor of the turbine is called runner. It consists of a circular disc, on the periphery of which a number of buckets are evenly spaced. The bucket is divided into two equal parts which is known as splitter. A notch is also provided to ensure the proper position of striking jet when the emerging jet strike the splitter a change of momentum takes place due to which the jet moves in opposite direction and it creates a thrust on the bucket due to which motion of the runner takes place. The buckets are made of cast iron, bronze, cast steel or stainless steel depending on the availability of head.

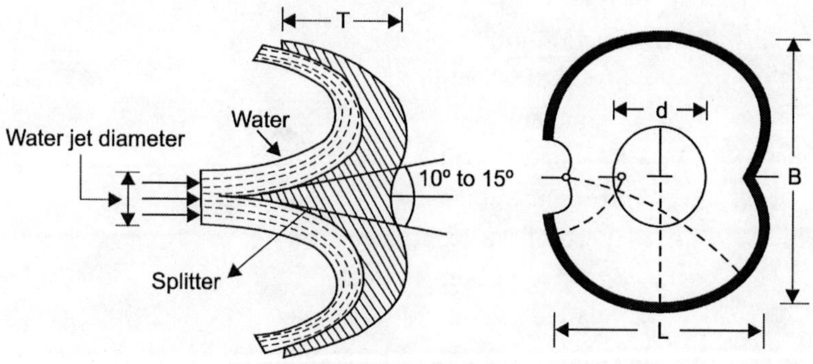

Fig. 5.5 The bucket dimensions.

3. **Nozzle and Spear Mechanism:** The nozzle is provided at the lower end of the penstock. It converts the pressure energy into kinetic energy and a high velocity jet coming out of nozzle is proportional to the load on the turbine. A needle spear mechanism is used to regulate the supply of water through the nozzle and provide the smooth flow. The movement of spear is controlled either by hand or by automatically or by mechanism through a servomotor to keep the constant load. Usually Pelton wheel is a horizontal shaft turbine and number of nozzles depends upon specific speed. The material of nozzle is cast iron or cast steel while the spear tip is of stainless steel.

4. **Casing:** The casing of a Pelton wheel does not perform any hydraulic function. It is used just to prevent splashing of water, supporting the thrust bearing in housing as well as to discharge the water into tail race.

5. **Breaking Jet:** A small nozzle is provided to direct the jet of water on the back of the vanes to stop the runner, in case of emergency or for the maintenance purpose. The nozzle is completely closed by moving the spear in forward direction. Hence, the breaking jet is used to stop the motion of turbine by creating a jet in opposite direction.

5.7 VELOCITY TRIANGLE, WORKDONE AND EFFICIENCY OF A PELTON WHEEL

When the jet of water strikes the bucket at the splitter, it splits up into two parts. Each part of the jet glides over the inner surface of the bucket and leaves at outer edge. The inlet velocity triangle is drawn at the splitter and outlet velocity triangle is drawn at the outer edge of the bucket shown in Fig.5.6

Let

H = Net head acting on the pelton wheel

$H = H_g - h_f$

H_g = Gross head

h_f = Friction loss in pipe

N = Speed of wheel in rpm.

d = Diameter of jet

V_1 = Absolute velocity of jet at inlet

V_2 = Absolute velocity of jet at outlet

u_1 = Speed of bucket at inlet

u_2 = Speed of bucket at outlet

V_{r_1} = Relative velocity of jet at inlet

V_{r_2} = Relative velocity of jet at outlet

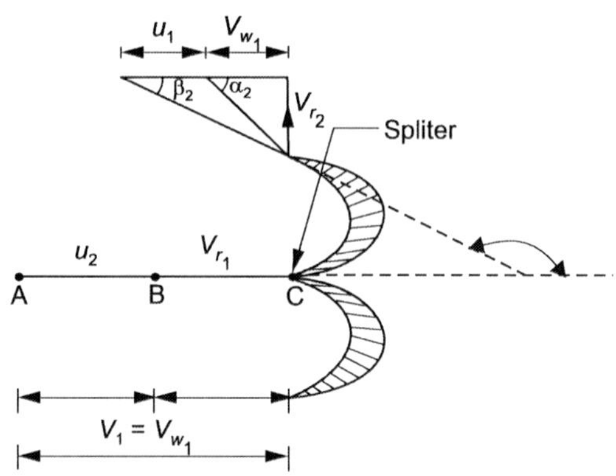

Velocity triangle for Pelton wheel

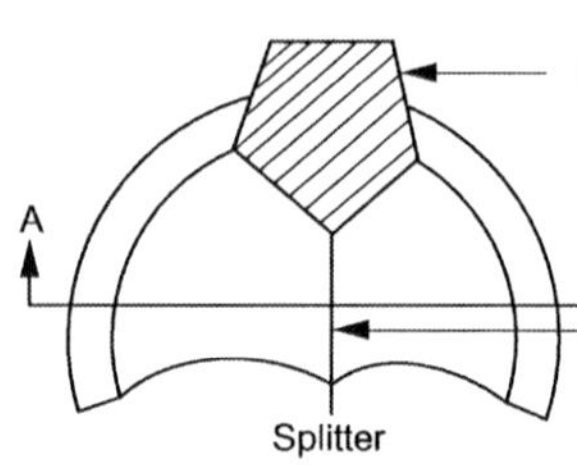

Fig. 5.6 Velocity triangle and work done efficiency of a Pelton wheel

V_{f_1} = Velocity of flow at inlet
V_{f_2} = Velocity of flow at outlet
V_{w_1} = Velocity of whirl at inlet
V_{w_2} = Velocity of whirl at outlet
α_1 = Angle between direction of jet and direction of motion of the plane.
β_1 = Angle made by relative velocity Vr_1, with the direction of motion at inlet.
α_2 = Angle made by velocity V_2, with the direction of motion of the bucket at outlet.
β_2 = Angle made by relative velocity Vr_2, with the direction of motion of the bucket at outlet.

$$a = \text{Area of jet} = \frac{\pi d^2}{4}$$

ρ = mass density.

Now,

Velocity of jet,

$$V_1 = \sqrt{2gH} \qquad \ldots(i)$$

at inlet

$$u = u_1 = u_2 = \frac{\pi DN}{60} \qquad \ldots(ii)$$

The velocity triangle at inlet will be a straight line. Therefore, we have

$$V_{r_1} = V_1 - u_1 \qquad \ldots(iii)$$
$$= V_1 = u$$

Also $V_{w_1} = V_1$

and $\alpha_1 = \beta_1 = 0$

From velocity triangle at outlet we have

$$V_{r_2} = V_{r_1} \qquad \ldots(iv)$$
$$V_{w_2} = V_{r_2} \cos \beta_2 - u_2 \qquad \ldots(v)$$

The force exerted by jet of water in the direction of motion is gives as

$$F = \rho a V_1 (V_{w_1} + V_{w_2}) \qquad \ldots(vi)$$

Now work done by the jet on the runner per second = $F \times u$.

$$\rho a V_1 (V_{w_1} + V_{w_2}) \times u \text{ Nm/s}$$

or

$$= \frac{\rho a V_1 (V_{w_1} + V_{w_2}) u}{100} kW \qquad \ldots(vii)$$

Work done per second per unit weight of water striking

$$\frac{\rho a V_1 (V_{w_1} + V_{w_2}) u}{\text{Weight of water striking}}$$

$$= \frac{\rho a V_1 (V_{w_1} + V_{w_2}) u}{\rho a V_1 \times g}$$

$$= \frac{1}{g}(V_{w_1} + V_{w_2}) u \qquad \text{...(viii)}$$

The energy supplied to the jet at inlet is in the form of kinetic energy and is equal to $\frac{1}{2} m V_1^2$.

$\therefore$ Kinetic energy supplied by jet per second $= \frac{1}{2}(\rho a V_1) V_1^2$

$$\therefore \text{ Hydraulic efficiency, } \eta_h = \frac{\text{Work done per second}}{\text{Kinetic energy supplied per second}}$$

$$\eta_h = \frac{\rho a V_1 (V_{w_1} + V_{w_2}) u}{\frac{1}{2}(\rho a V_1) V_1^2}$$

$$\eta_h = \frac{2(V_{w_1} + V_{w_2}) u}{V_1^2} \qquad \text{...(ix)}$$

from inlet and outlet velocity triangles, we have V_w, $= V$,

$$\begin{aligned} V_{w_1} &= V_1 \\ V_{r_1} &= V_1 - u_1 = V_1 - u \\ V_{w_2} &= V_{r_2} \cos \beta_2 - u_2 \\ &= V_{r_1} \cos \beta_2 - u_2 \qquad (u = u_1 = u_2) \\ &= (V_1 - u) \cos \beta_2 - u \end{aligned}$$

Substituting the values of V_{w_1} and V_{w_2} in Equation (*ix*), we get

$$\eta_h = \frac{2[V_1 + (V_1 - u)\cos\beta_2 - u]u}{V_1^2}$$

$$= \frac{2[V_1 - u + (V_1 - u)\cos\beta_2]u}{V_1^2}$$

$$\eta_h = \frac{2(V_1 - u)[1 + \cos\beta_2]u}{V_1^2} \quad \text{...(x)}$$

Maximum Hydraulic Efficiency: The efficiency will be maximum for a given value of V_1 when

$$\frac{d\eta_b}{du} = 0 \quad \text{...(xi)}$$

i.e.
$$\frac{d}{du}\left[\frac{2(V_1 - u)(1 + \cos\beta_2)u}{V_1^2}\right] = 0$$

or
$$\frac{(1 + \cos\beta_2)}{V_1^2}\frac{d}{du}(2uV_1 - 2u^2) = 0$$

or
$$\frac{d}{du}(2uV_1 - 2u^2) = 0 \qquad \left[\because \frac{1 + \cos\beta_2}{V_1^2} \neq 0\right]$$

or
$$2V_1 - 4u = 0$$

or
$$V_1 - 2u = 0$$

or
$$u = \frac{V_1}{2} \quad \text{...(xii)}$$

So the hydraulic efficiency of a pellon turbine is maximum when the velocity of the wheel is half the velocity of jet of water at inlet. The maximum efficiency can be obtained by substituting the value of $u = \frac{V_1}{2}$ in Equation (*x*).

$$\eta_{h\,max} = \frac{2\left(V_1 - \frac{V_1}{2}\right)(1 + \cos\beta_2)\frac{V_1}{2}}{V_1^2}$$

$$= \frac{2\left(\frac{V_1}{2}\right)(1 + \cos\beta_2)\frac{V_1}{2}}{V_1^2}$$

$$\eta_{h\,max} = \frac{(1 + \cos\beta_2)}{2}$$

5.8 FRANCIS TURBINE

The Francis turbine is an inward mixed flow reaction turbine named after an American Engineer James Bichens Francis (1815–1892). The flow of water takes place inward along the periphery (radially) to the centre of the runner and coming out axially. But in modern Francis turbine flow is mixal flow in which water enters the runner at its outer periphery and leave axially at its center. Such flow is against the direction of centrifugal force when the load on the turbine reduces the runner speed tends to increase, but at the same time increase in centrifugal force which records the fluid motion.

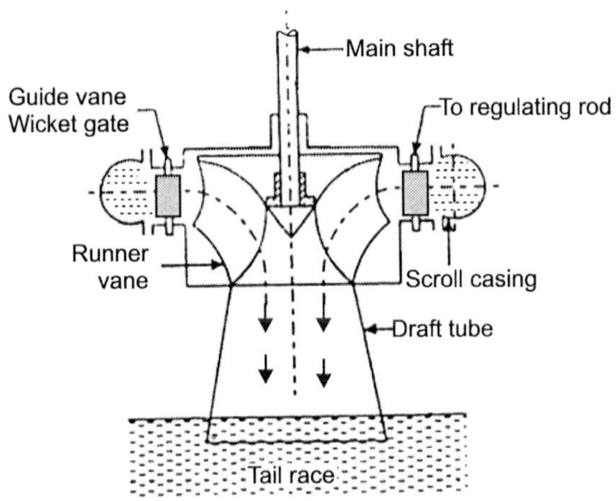

(a)

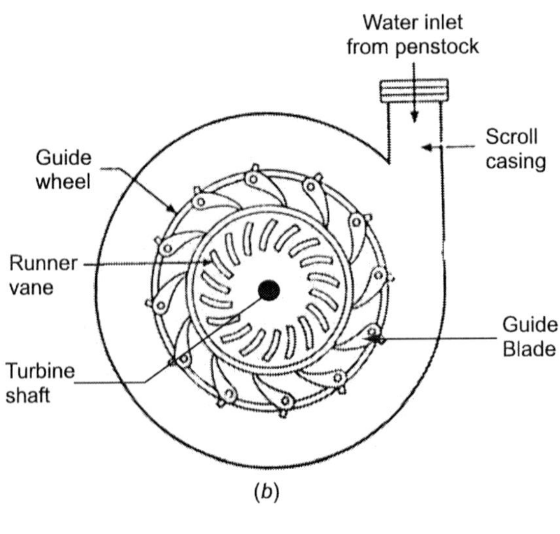

(b)

Fig. 5.7 Elements of a Francis turbine

The Francis turbine is a type of reaction turbine so its casing is closed and area of X-section of duct goes on gradually decreasing. It is a spiral type so it is called scroll casting. The number of these blades varies between 16 to 25, whereas the free end of the draft tube is submerged into tail race. Thus, the partial conversion of static (pressure) head to kinetic energy from the intel to the penstock and remains continue till the water enters the draft tube where again if kinetic energy convert into static head.

A number of guide blades are arranged within casing around the runner which gives the proper direction to water jet in such a way that the jet does not strike the runner ones in opposite direction the guide blades are adjustable about the hinged point with the help of a governing mechanism but do not rotate with the runner.

Around the runner periphery a no. of runner vanes are fixed which rotate the runner. The driving force on the runner is both due to impulse and reaction. The impulsive forces due to change in magnitude and direction of velocity whereas the reactive force is due to change in pressure.

The main components of the Francis turbine are as follows:

1. Penstock
2. Scroll casing
3. Guide mechanism
4. Runner
5. Draft tube

1. **Penstock:** Penstock is a large diameter pipe through which water flows from reservoir to scroll casing.

2. **Scroll Casing:** It is a casing in spiral form. The cross-sectional area gradually decreases around the circumference of the wheel. So that the water enters the runners at constant head and constant velocity at all points. These are made of generally from reinforced concerete or steel. The quantity of water to be allowed to runner is much more in these turbines. Control valve and screens at intel are also provided to prevent the entry of debris and foreign material in to penstock.

3. **Guide Mechanism:** The guide mechanism consists of guide vanes fixed on a stationary circular wheel around the runner of the turbine. It regulates the discharge according to the load on turbine. The guide vanes are of airfoil section. These allow the water to strike the vanes of runner without shock at inlet. These vanes act as fixed nozzles and a part of pressure energy is converted into kinetic energy as water passes the vanes. Water leaves the guide vanes in tangential direction of the runner blades. The angle of guide vanes can be adjusted by retorting a stator ring. Guide vanes are made up of cast steel.

4. **Runner:** The runner of a Francis turbines is a circular wheel on which a series of radial curved vanes are fixed. So that the water enters leaves the runner without shock. The water enters the runner blades tangentially along its periphery and proceeds radially inward. The water pressure decreases in radial inward direction along the moving blades. The runner is coupled to the turbine shaft, which may be horizontal or vertical. The number of runner blades varies between 16 to 24. These are generally made up of cast steel, stainless steel and bronze.
5. **Draft Tube:** It is a tube of enlarging cross-sectional area, which is fixed at the outlet of the francis turbine to take the water into the tail race. The static pressure of water decreases gradually as water passes over the guide vanes and runner blades and thus water coming out with large kinetic energy and a pressure, which is less than atmospheric pressure, so the water can not be discharged directly to the tail race. This water passes a draft tube, which is directing duct and act as a diffuser, and thus it recovers partially static pressure of water by lowering the speed of water.

   It performs the following functions:

   1. It provides a mechanical structure for installation of turbine and reduces the loss of head.
   2. It improves the efficiency of the turbine.
   3. The draft tube prevents the splashing of water coming out of runner and guides the water to the tail race.

5.9 VELOCITY TRIANGLE, WORKDONE AND EFFICIENCY OF A FRANCIS TURBINE

Fig. 5.8 shows the runner and the velocity diagram (inlet and outlet) for an inward flow reaction turbine.

When the water strikes on radial curved vanes fixed on a runner, it glides over the vanes and leaves at other edge. So to calculate the work output, inlet and outlet velocity triangles are drawn as given below.

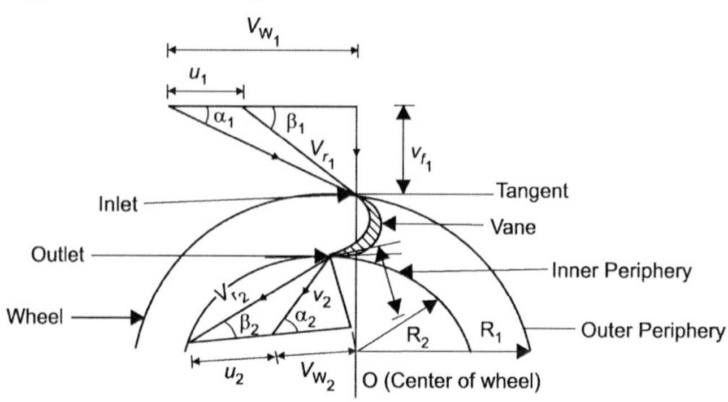

Fig. 5.8 Velocity diagram for an inward flow reaction turbine.

(*i*) The work done per second on the runner by water is

$$= \rho a V_1 \, [V_{w_1} u_1 \pm V_{w_2} \, u_2]$$

$$= \rho Q \, [V_{w_1} u_1 \pm V_{w_2} \, u_2]$$

Where Q = Discharge through the runner, m^3/s = $a.V_1$

(*ii*) Work done per second per unit weight of water striking

$$= \frac{\rho a V_1 [V_{w_1} u_1 + V_{w_2} u_2]}{\text{Weight of water striking}}$$

$$= \frac{\rho a V_1 [V_{w_1} u_1 + V_{w_2} u_2]}{\rho a V_1 \times g}$$

$$= \frac{1}{g} [V_{w_1} u_1 + V_{w_2} u_2]$$

For maximum output under given conditions is obtained when $Vw_2 = 0$

Thus the maximum work done is:

for work done per second on the runner by water is

$$= PQ[V_{w_1} u_1]$$

for work done per second per unit weight of water striking

$$= \frac{1}{g} [V_{w_1} u_1]$$

The discharge is radial, so for the radial discharge, the absolute velocity at exit is radial.

Hydraulic efficiency,

$$\eta_h = \frac{\text{Power produced by the runner}}{\text{Power supplied by water to turbine}}$$

$$= \frac{\rho a V_1 (V_{w_1} u_1)}{\rho Q g H} = \frac{\rho Q (V_{w_1} u_1)}{\rho Q g H}$$

$$\eta_h = \frac{(V_{w_1} u_1)}{gH}$$

If velocity of whirl, Vw_2 is not zero then

$$\eta_h = \frac{V_{w_1} u_1 + V_{w_2} u_2}{gH}$$

Mechanical efficiency,

$$\eta_h = \frac{\text{Power available at the shaft}}{\text{Power produced by the runner}}$$

Overall efficiency,

$$\eta_h = \frac{\text{Power available at the shaft}}{\text{Power supplied by water to turbine}}$$

5.10 ADVANTAGES AND DISADVANTAGES OF A FRANCIS TURBINE OVER A PELTON WHEEL TURBINE

Advantages and disadvantages of a Francis turbine over a Pelton wheel

Advantages:

The Francis turbine claims the following advantages over Pelton wheel:

1. In Francis turbine the variation in the operating head can be more easily controlled.
2. In Francis turbine the ratio of maximum and minimum operating heads can be even two.
3. The operating head can be utilized even when the variation in the tail water level is relatively large when compared to the total head.
4. The mechanical efficiency of Pelton wheel decreases faster with wear than Francis turbine.
5. The size of the runner, generator and power house required is small and economical if the Francis turbine is used instead of Pelton wheel for same power generation.

Disadvantages/Drawbacks :

As compared with Pelton wheel, the Francis turbine has the following drawbacks/ shortcomings:

1. Water which is not clean can cause very rapid wear in high head Francis turbine.
2. The overhaul and inspection is much more difficult comparatively.
3. Cavitation is an ever-present danger.
4. The water hammer effect is more troublesome with Francis turbine.
5. If Francis turbine is run below 50 per cent head for a long period it will not only lose its efficiency but also the cavitation danger will become more serious.

5.11 KAPLAN TURBINE

Working: Kaplan turbine is a reaction turbine and it is suitable for low head and high discharge. This turbine was developed by an Australlian Engineer Victor Kaplan. It runs fulley means closed conduct from inlet to tail race. Kaplan turbine has scroll casing, guide blade and guide mechanism similar to francis turbine. The water enters at right angle between the guide vanes and runner and then flow parallel or along the axis of the turbine shaft. So it is also called axial flow turbine. The minimum number of blades varies usually (4) four to (8) eight without shroud ring. In Kaplan turbine, the guide blades angle as well as vanes angles are adjustable and thereby have an efficient control flow. These turbines are suitable for low head and high specific speed. The efficiency of this turbine is more than the other type of turbines.

The runner is in the form of a boss which is a extension of bottom end of the shaft into a bigger diameter. It has a less frictional resistance due to the less number of blades in this turbine. The fixed blade propeller turbine is installed only at a sites where the head and load are constant. It has adjustable runner vanes hence the named variable pitch propeller turbine.

The Kaplan turbine has double regulation which comprises the movement of guide vanes and rotation of runner blades. The mechanism employs two servomotors one control the guide vanes and the second operates on the runner vanes. Both the servomotors, are synchronised. They are actuated simultaneously and a high efficiency is maintained at all loads.

Kaplan turbine blades are hollow and are attached to the hub in such a way that they are able to spin on their axis and the velocity components at the inlet and out let remain constant. The velocity of blades being directly proportional to radius and tangential component (whirl velocity) being inversely proportional to radius, hence the angles depend upon the radius of rotor. A little number in the blades helps in to avoid cavitation. Kaplan turbines are provided with draft tubes from exist of runner to the tail race.

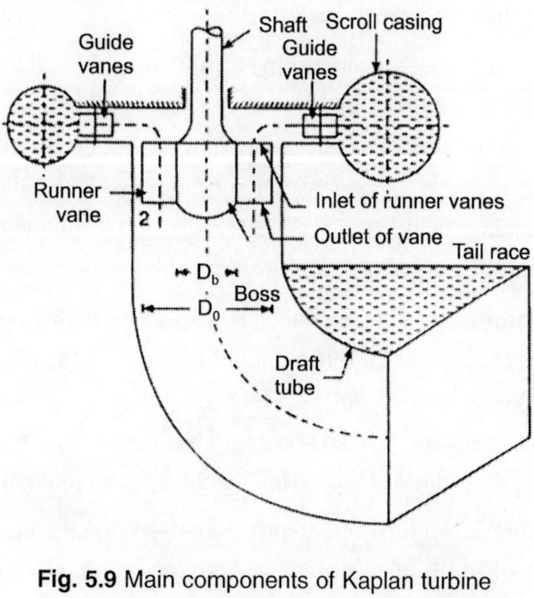

Fig. 5.9 Main components of Kaplan turbine

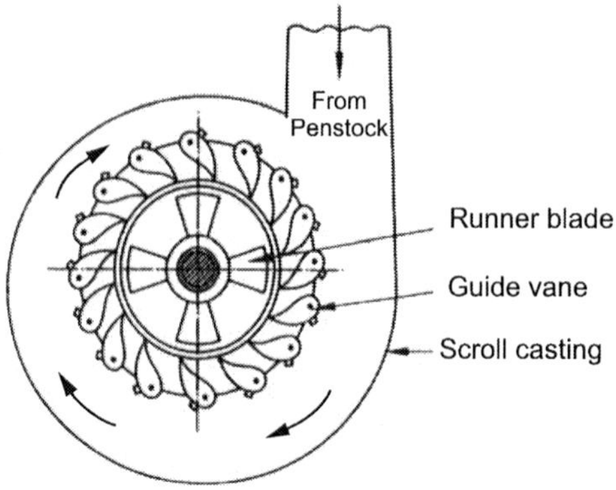

Fig. 5.10 Runner with scroll casing

The main components of Kaplan turbine are as follows:

1. **Runner:** In Kaplan turbine runner is in the shape of boss on which runner blades are adjustable, which can be rotated about the pivots fixed to the boss. The blades can be automatically adjusted with the help of servo mechanism. It is made up of stainless steel which is free from rusting.
2. **Guide Vanes:** Guide vanes are used to impact a tangential and a radial inward velocity to the liquid. It has the double regulation *i.e.* the movement of guide venes and rotation of runner blades. Two servomotors are used for this purpose which are kept in a hollow shaft of turbine runner which guide the vanes through proper angles.
3. **Scroll Casing:** The Kaplan turbine is high discharge and low head turbine, so it needs a scroll casing in order to increase the velocity and keep it constant for a fixed load. The water first enters the spiral casing in which area of cross-section decreases continuously. The reduction in area is proportional to the decreasing volume of water. So it maintains a cast iron velocity and is made up of cast iron or rolled steel.
4. **Whirl Chamber:** Whirl chamber is a space provided between the guide vanes and the runner. In this chamber the flow is turned at right angle or 90° and moves as free vortex. Here the radial component changed in to axial component due to fixed housing. The tangential component of velocity increase as the radius is reduced but angular momentum remains unchanged.
5. **Draft Tube:** It is a tube of enlarging cross-sectional area which is fixed at the outlet of the Kaplan turbine to take the water into tail race. Here the static pressure decreases and water comes out with large kinetic energy

and a pressure. This tube helps in preventing the atmospheric resistance and act as a diffuser and recovers the partially static pressure. Which helps in reduce the head loss, splashing of water as well as to improve the efficiency of the turbine.

5.12 DRAFT TUBE

It is a tube of enlarging cross-sectional area which is fixed at the outlet of the turbine and other end of the draft tube is submarged into the tail race.

It is an integral part of the low head turbine with large flow *i.e.* for reaction turbine like Francis and Kaplan. The unit consists of an airtight diverging conduit with cross-sectional area increasing along its length are end of this diverging tube is connected to the runner exit and other is below the level of tail race.

It has the following functions to perform:

1. Decrease the pressure at the runner exit to a valve less than atmospheric pressure and thereby increase the working head.
2. To recover the portion of exit kinetic energy which otherwise goes waste to the tail race.
3. It helps in increasing the efficiency of the turbine.
4. To prevent the splashing of water from runner and guide it to the tail race.

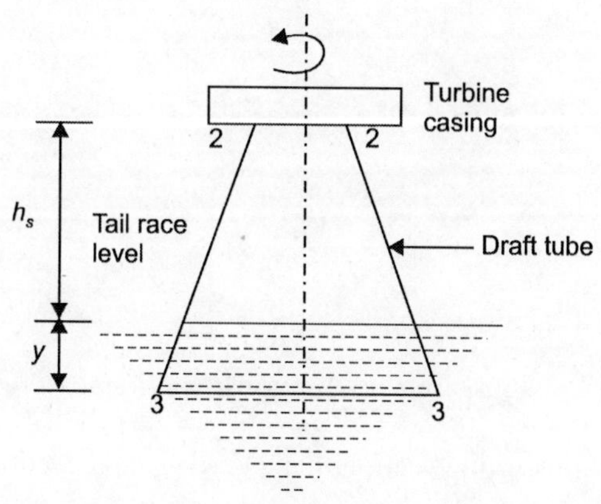

Fig. 5.11 Draft tube

The water from runner exit and freely discharges into tail race.

Let pressure P_2 at runner exit and equal to atm. pressure Pa and K.E. $\frac{V^2}{2g}$ goes waste to tail race.

Now atmospheric pressure at section 3-3

Applying Bounoulids equation between 2-2 and 3-3.

$$\frac{P_3}{w} = \frac{Pa}{w} + y, \frac{P_2}{w} + \frac{V_2^2}{2g} + y_2 = \frac{P_3}{w} + \frac{V_3^2}{2g} + 0 + h_f$$

Where h_f - hydraulic loss due to friction

$$\frac{P_2}{w} = \frac{P_3}{w} - y_2 - \left(\frac{V_2^2 - V_3^2}{2g}\right) - h_f$$

Now substituting the values of

$$\frac{P_3}{w} = \frac{Pa}{w} + y$$

$$\frac{P_2}{w} = \frac{Pa}{w} + (y - y_2) - \left[\frac{V_2^2 - V_3^2}{2g} - h_f\right]$$

Where $(y_2 - y)$ - suction head

$$\frac{P_2}{w} = \frac{Pa}{w} + hs - \left[\frac{V_2^2 - V_3^2}{2g}\right] - h_f$$

Whenre $\frac{(V_2^2 - V_3^2)}{2g}$ - dynamic head

$$\frac{P_2}{w} = \frac{Pa}{w} + hs - \left[\frac{V_2^2 - V_3^2}{2g} - hs\right]$$

Pressure at exit < atmospheric pressure so $\frac{P_2}{w}$ is (–ve)

If the draft tube is of diverging section then $(V_3 < V_2)$ and suction head h_s is (+ve).

No recovering kinetic head due to velocity changes in draft tube $\frac{(V_2^2 - V_3^2)}{2g}$ and if

h_f is friction loss than pressure head $= \dfrac{(V_2^2 - V_3^2)}{2g} - h_f$

$$\eta_d = \frac{\text{Actual conversion of Kinetic Head}}{\text{Pressure head}}$$

$$\eta_d = \frac{(1-K)\,(V_2^2 - V_3^2)/2g}{V_2^2/2g} = \frac{(1-K)\,(V_2^2 - V_3^2)/2g}{(V_2^2 - V_3^2)/2g} = 1 - K$$

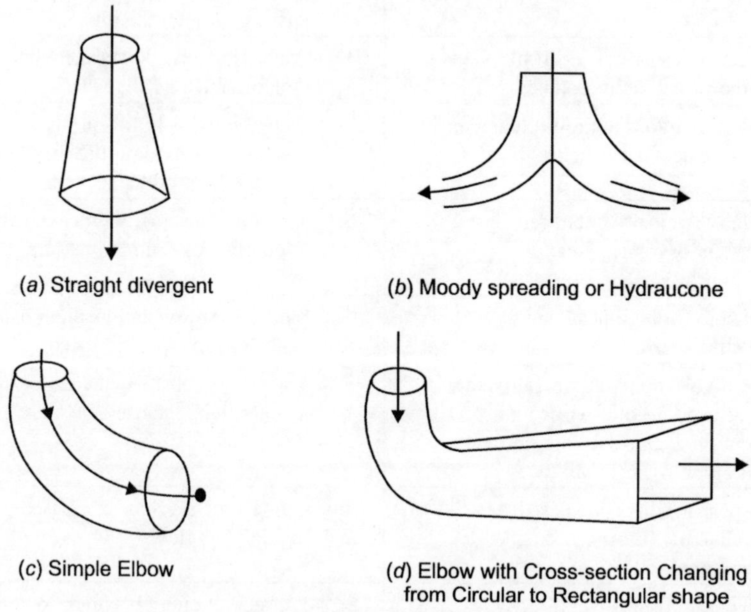

Fig. 5.12 Different forms of Draft Tube.

Straight Divergent Tube:

(*a*) The straight divergent tube is used to prevent possibility of flow separation and its efficiency is about 90%.

(*b*) Moody's spreading draft tube has a solid central core of conical shape which minimizes whirling action of discharged water and its efficiency is around 82%. This tube is used when the water leaves the runner with a whirl component.

(*c*) Simple elbow is used where the length of the shaft is kept minimum. As it has curved channel so there is loss of energy and efficiency is 60%.

(*d*) Elbow type tube is of circular section at inlet and rectangular section at outlet as shown in figure with efficiency around 80%.

5.13 COMPARISON OF IMPULSE AND REACTION TURBINES

Impulse Turbine	Reaction Turbine
1. The static head is converted into kinetic energy in stationary nozzles.	1. The static head is converted in Kinetic energy partially in fixed guide vanes and partially in moving blades.
2. Pressure remains constant through-out the action of water on the runner.	2. Water enters the runner with high pressure and then pressure is gradually converted into kinetic energy while passing through the runner.
3. Water may enter as a part of whole of the wheel circumference.	3. Water is admitted over the whole circumference of the wheel.
4. The wheel does not run full and air has free access to bucket.	4. Water completely fills all the passages between the blades and between inlet-outlet sections.
5. Relative velocity either remains constant or reduced due to friction loss.	5. The relative velocity processes due to continuous drop in pressure.
6. Suitable for high head and low specific speed.	6. Suitable for low and medium head as well as for specific speed.
7. Casing has no hydraulic purpose so there is no need of water tight. It only prevents splashing of water and guide water to tail race.	7. The runner must be enclosed within a water tight enclosed.
8. A spear mechanism is used to regulate the flow according to speed required.	8. Guide vane mechanism is used to regulate the flow.
9. Water can flow down directly to tail race.	9. Pressure at outlet becomes low due to high K.E., so a diverging draft tube is required for efficient flow to tail race.
10. It is always installed above tail race.	10. It can be installed above or below the tail race with the use of a draft tube.

5.14 COMPARISON BETWEEN INWARD AND OUTWARD FLOW TURBINES

Inward flow turbine	Outward flow turbine
1. It is quite suitable for low or medium heads.	1. Quite suitable for medium and high heads.
2. Centrifugal force developed on water due to rotation of water in the direction of flow.	2. Due to decrease in load there is increase in runner speed and quantity of water get reduced over the runner hence power will also reduce. Hence turbine try to self-regulate.

Inward flow turbine	Outward flow turbine
3. The entry of water creates shock as it is admitted with maximum velocity at the slow moving part and leaving it at points where tangential velocity is maximum.	3. The water enters without shock as it is admitted with its maximum velocity at the outer periphery and flows inward towards the center.
4. If the flow is divergent than chances increase for eddy formation and flow separation resulting a loss of energy.	4. The flow is convergent and there is a no chance of eddy formation and flow separation.
5. Guide vanes are at the inner periphery of runner, so they are not easily accessible, so the regulation flow is not easy and accurate.	5. The guide vanes are at outer periphery of the runner so they are easily accessible. Thus the regulation of flow is easy and accurate.

5.15 COMPARISON BETWEEN PELTON WHEEL AND FRANCIS TURBINE

Pelton Wheel	Francis Turbine
1. It is a high head and low discharge turbine more than 1000 meters	1. It is medium head and medium discharge turbine (60m - 250m).
2. It is a tangential flow turbine.	2. It is a mixed flow turbine (Radial entry and axial exit).
3. Pressure is atmospheric in turbine.	3. Pressure is high at inlet and low at outlet.
4. There is no need of draft tube.	4. Draft tube is essential to increase efficiency and decrease kinetic Energy loss.
5. It is a low specific speed turbine.	5. Specific speed varies from 50 to 250.
6. Casing has no hydraulic function to perform. It is used only to protect the runner and collect the splashed water and direct it to tail race.	6. To maintain difference of pressure at various parts of runner which is completely closed and it distributes water all along periphery and equal pressure at exit.
7. Unit is installed above tail race.	7. Unit is submerged into tail race.

5.16 COMPARISON BETWEEN FRANCIS TURBINE AND KAPLAN TURBINE

Francis Turbine	Kaplan Turbine
1. It is a readily inward or mixed flow turbine.	1. It is a purely axial flow turbine.
2. It is a medium head and medium discharge turbine ranging from (60m-250m).	2. It is a low head and large discharge turbine range up to 30 m.

Francis Turbine	Kaplan Turbine
3. Specific speed ranges 50 to 250.	2. Specific speed ranges from 250 to 850.
4. Large no. of runner blades 10 to 36.	4. Number of blades are 3 to 8.
5. Runner blades are fixed.	5. Runner blades are adjustable.
6. Large frictional resistance due to more vanes. Because the contact area is more.	6. Less resistance due to less number of blades having a less contact area.
7. Disposition of shaft may be horizontal or vertical.	7. Disposition of shaft is only vertical
8. Ordinary governor or servomotor is used in large size.	8. Heavy duty governor for double regulation.

5.17 SPECIFIC SPEED N_S

Specific speed of turbine is defined as the speed of a geometrical similar turbine identical in dimension, blade, angle and gate opening which produces a unit power (1 k. watt) under a unit head (1m). It is dennoted by N_S.

$$N_S = \frac{N\sqrt{P}}{H^{5/4}}$$

where N = Revolution per minute r.p.m.

P = Power developed

H = Net head

According to the definition.

Overall efficiency

$$\eta_0 = \frac{\text{shaft power}}{\text{water power}}$$

$$\eta_0 = \frac{\text{Power available at the shaft of turbine}}{\text{Power supplied at the inlet of turbine}}$$

$$\eta_0 = \frac{Pt}{\frac{WQH}{1000}} \qquad \ldots(i)$$

where Pt = shaft power

Q = discharge

w = wt. density = pg

H = head on which turbine is working

ρ = density of water

g = acceleration due to gravity

$$Pt = \eta_0 \times \frac{WQH}{1000}$$

$$P = QH$$

$$P \propto QH \qquad \ldots(ii) \; (\because \eta_0 \text{ and } W \text{ are constants})$$

Let D = diameter of actual turbine

N = speed of actual turbine

μ = tangential velocity of the turbine

V = absolute velocity

N_s = specific speed of turbine

The absolute velocity, tangential velocity and head on the turbine are related as

$$u \propto V \qquad \ldots(iii)$$

$$V \propto \sqrt{H} \qquad \ldots(iv)$$

$$\because \quad V = Ku\sqrt{2gH}$$

$$u = v,$$

$$V = \sqrt{H},$$

$$u = \sqrt{H}$$

from equation (*iii*) and equation (*iv*)

$$u \propto \sqrt{H} \qquad \ldots(v)$$

$$u = \frac{\pi DN}{60} \qquad \text{where } D\text{- diameter of wheel} \quad \ldots(vi)$$

$$\sqrt{H} = DN$$

$$\sqrt{H} \propto DN$$

$$D = \frac{\sqrt{H}}{N}$$

$$D \propto \frac{\sqrt{H}}{N} \qquad \ldots(vii)$$

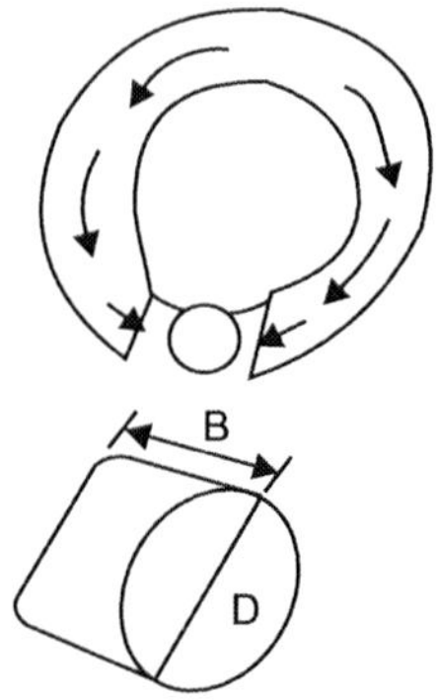

Fig. 5.13

Discharge through turbine is given by

Discharge = Area × Velocity

$Q = A \times V$ ...(*viii*)

where Area = $(B \times D)B$ - Width of blade

But $B \propto D$

Hence Area = D^2 ...(*ix*)

so $Q = A \times V$

$= D^2 \times \sqrt{H}$ ...(*x*)

In case of Francis turbine

$A = \pi D^2 B$

$A \propto D^2$

A = Arc of flow

$B = D$

$= \frac{\pi}{4} d^2$

$A = d^2$

$A \propto d^2$

By putting the valve of equation (*vii*) in equation (*x*)

$$Q = \left(\frac{\sqrt{H}}{N}\right)^2 \times \sqrt{H}$$

$$Q = \frac{H^{3/2}}{N^2}$$

$$Q \propto \frac{H^{3/2}}{N^2} \quad \text{...(xi)}$$

From equation (*vii*) and (*xi*)

$$P \propto \frac{H^{3/2}}{N^2} \times H \qquad P \propto QH$$

$$P \propto \frac{H^{3/2}}{N^2} \qquad P = QH$$

$$P = K.\frac{H^{3/2}}{N^2} \qquad \ldots(xii)$$

where K is constant of proportionality.

By definition

we know $\quad P = 1,\ H = 1$ and $N = N_s$

$H = 1$

$N = N_S$

By substituting the valves of P, H, and N in equation (*xii*)

$$P = K.\frac{H^{3/2}}{N^2}$$

$$1 = K.\frac{1^{3/2}}{N_S^2}$$

$$N_S^2 = \text{K} \qquad \ldots(xiii)$$

From equation (*xii*) and (*xiii*)

$$P = N_S^2.\frac{H^{5/2}}{N^2}$$

$$N_S^2 = \frac{P \times N^2}{H^{5/2}}$$

$$N_S = \frac{N\sqrt{P}}{H^{5/4}}$$

Hence, specific speed

$$N_S = \frac{N\sqrt{P}}{H^{5/4}}$$

5.18 SELECTION OF TURBINES

Selection of turbine is made on the basis of following consideration of discharge:

1. **Specific Speed:** The specific speed of a turbine is selected corresponding to the synchronous speed of the generator and turbine is selected on the basis of specific speed.

The specific speed should be high when available head is low and out put is high even then high rotational speed can be obtained in medium capacity plants for low specific speeds.

Specific speed (N_s)	Pelton turbine	Francis turbine	Kaplan turbine
	8 – 30 – 50 ↑ Single jet ↑ multijet	50 – 340	250 – 950

2. **Head:** The head available also plays an important role in the selection of the turbine.

Head	Pelton turbine	Francis turbine	Kaplan turbine
	150 - 350 m	60 - 150 m	30 - 60 m

3. **Maximum Power**

Maxi power M watt.	105	600	125

4. **Part Load Operation:** The load on turbine varies with power element very frequently. The efficiency decreases at part load condition for efficiency fixed blades are not suitable hence variable vanes are used for part load and over load Deriaz and Kaplan turbines can be used.

5. **Turbine Efficiency:**

$\eta_{turbine}$	Pelton turbine	Francis turbine	Kaplan turbine
	85 - 90%	80 - 95%	90 - 95%

6. **Number of Units:** It is better to go in for a large unit as far as possible. But there must be at least two units at any particular site so that one unit is always available for power production. The number of units will depend upon the discharge available at a particular head.

7. **Cavitation:** Cavitation is likely to occur when the pressure at the runner outlet equals vapour pressure. This means that a turbine should be installed closer to the tail race level. Francis turbine cannot be used for very high heads because of cavitation. Pelton turbines are free from cavitation because the pressure at runner outlet is atmospheric.

8. **Position of Turbine Shaft:** On the position of shaft turbines are classified as:

    1. Pelton turbine - horizontal shaft

    2. Francis turbien - horizontal vertical

    3. Kaplan turbine - vertical shaft

5.19 HEAD LOSSES AND EFFICIENCIES OF THE TURBINE

HEAD: The head is the water height above certain reference level. It is measured in meters and characterised as:

(*a*) **Gross Head:** The gross (total head) is defined as the difference in water level at headrace and tail race is denoted by H_g.

(*b*) **Net Head or Effective Head:** It is the effective head available which is available at the turbine inlet. It is denoted by H. When water flows through penstock pipes from reservoir to water turbine so there is a certain loss due to friction.

So Net head obtained by substracting the head loss from gross head.

$$H = H_g - h_f - h.$$

where

H = net head

H_g = gross head

h_f = friction head loss

h = height on nozzle above tail race.

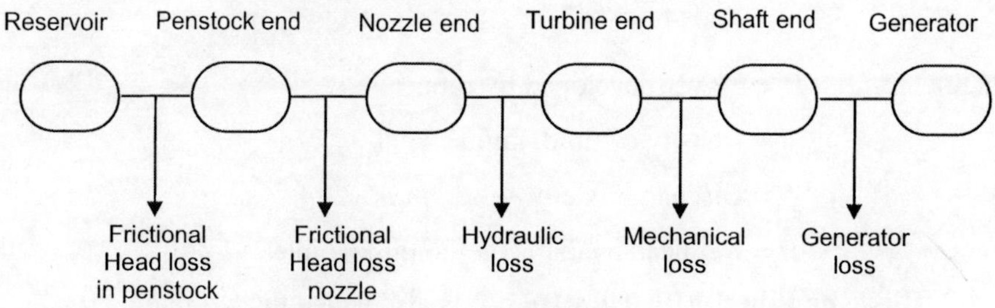

Fig. 5.14 Various types of losses

Losses : Various types of losses in power plant are :

1. *Head loss in the penstock:* This is the friction loss in the penstock pipe.
2. *Head loss in the nozzle:* This is the friction loss in the nozzle.
3. *Hydraulic loss:* This loss is due to blade friction, eddy formation and kinetic energy of the leaving water and friction in draft tubes.
4. *Mechanical losses:* Power loss in bearing friction is a mechanical loss. So power produced by the runner is not available as useful work of the shaft.
5. *Generator losses:* Due to generator losses power produced by generator is lesser than the power obtained at the shaft output.

Efficiency: In general term the efficiency of the any machine can be defined as ratio of useful energy delivered to the energy supplied.

(*a*) **Manometric Efficiency (η_{mano}):** It is defined as the ratio of manometric head developed by pump to the head imported by the impeller to the liquid.

$$\eta_{mano} = \frac{\text{Manometric head}}{\text{Head imparted by impeller to liquid}}$$

$$= \frac{\eta_{mano}}{\frac{V_{w_2} v_2}{g}} = \frac{g\,\eta_{mano}}{V_{w_2} v_2}$$

(*b*) **Hydraulic Efficiency (η_h):** It is the ratio of the power produced by turbine runner to the power supplied by the water jet at the turbine inlet.

$$\eta_{hydraulic} = \eta_h = \frac{\text{Power produced by runner}}{\text{Power supplied by water to turbine at inlet}}$$

$$\eta_h = \frac{P_r}{PQgh}$$

where P_r = Power developed by runner

P = Density of fluid, kg/m^3

Q = Discharge rate of water, m^3/s

H = Net head available at the turbine inlet.

(*c*) **Blade or Wheel Efficiency (η_b):** It is also called the hydraulic efficiency of the wheel. It is defined as the ratio of power produced by the wheel (runner) to the energy input (K.E.) of jet to the wheel:

$$\eta_b = \frac{\text{Power produced by runner}}{\text{Kinetic energy of the jet}}$$

$$= PQ\frac{(V_{w_1} + V_{w_2})\mu}{\frac{1}{2}PQV_1^2}$$

$$\eta_b = \frac{2\mu(V_{w_1} + V_{w_2})}{V_1^2}$$

(*d*) **Volumetric Efficiency (η_v):** It is defined as the ratio of volume of water actually flowing over the runner blades to the total volume of water supplied to the turbine.

$$\eta_{\text{volumetric}} = \frac{\text{Volume of water actually flowing over the runner blade}}{\text{Total volume of water supplied to turbine}}$$

$$(\eta_v) = \frac{Q}{Q + \delta Q}$$

(*e*) **Mechanical efficiency:** It is defined as the ratio of the useful power available at the turbine shaft to the power produced by the runner:

$$\eta_{\text{mech}} = \frac{\text{Power available at the shaft turbine}}{\text{Power produced by the runner}} = \frac{P}{P_r}$$

$$= \frac{P}{wQH}$$

(*f*) **Overall Efficiency:** It is defined as the ratio of useful power available at the turbine shaft to power supplied by water at the turbine inlet. It is also called the turbine efficiency:

$$\eta_{\text{overall}} = \frac{\text{Useful power available at the turbine shaft}}{\text{Power supplied by water at turbine inlet}}$$

$$= \frac{P}{PQgH_m}$$

It can also be expressed as

$$\eta_{\text{turbine}} = \eta_{\text{mech}} \times \eta_{\text{hydraulic}}$$

5.20 CHARACTERISTICS CURVES OF HYDRAULIC TURBINES

Characteristic curves of a hydraulic turbine are the curves, with the help of which the exact behaviour and performance of the turbine under different conditions can be known. These curves are plotted from the research of the test performed on the turbine under the different working conditions. The important parameters which are varied during a test on turbine are:

1. Speed (N)
2. Head (H)
3. Discharge (Q)
4. Power (P)
5. Gate opening
6. Overall efficiency (h_0)

Out of the above six paremeters, three parameters namely speed (N), Head (H) and discharge (Q) are independent parameters.

The variation of all other quantities with respect to any one of three gives a curve or a characteristic of the turbine. These curves are known as characteristic curves.

Types of Characteristic Curves

1. **Main Characteristic Curve (Constant Speed Curves):** In this case, the head is kept constant and the speed is varied by changing the load on the turbine. The governing mechanism is disconnected from the system so that the experiments are performed at constant gate opening.

   Different set of curves like speed-discharge, speed-power and speed-efficiency are drawn. Figure shows the main characteristic curves for a Pelton wheel. Figure shows the main characteristic curves for Francis turbine and Figure shows the main characteristic curves for Kaplan turbines.

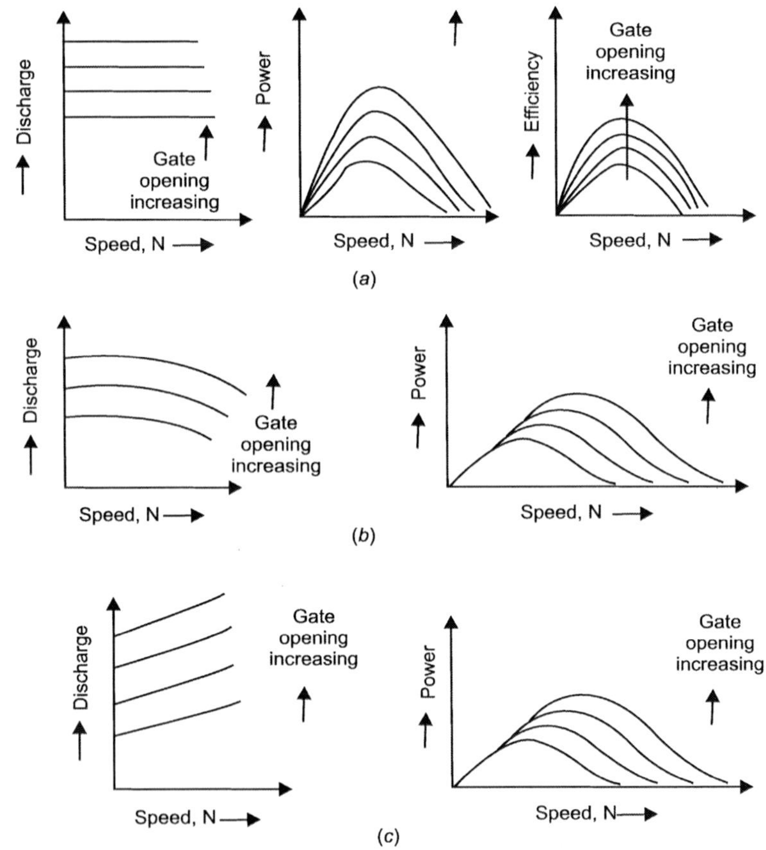

Fig.5.15 (*a*) Main characteristics of Pelten turbine (*b*) Francis turbine (*c*) for Kaplan turbine

2. **Operating Characteristics (Constant Speed Curves):** In hydroelectric power stations, turbines are coupled with electrical alternators for producing electricity at constant frequency and therefore they must run at constant speed The head and discharge depends upon their availability in the storage reservoir. The power output depends upon the demand from the locality, and the power output-efficiency, discharge power and efficiency curves are important and have been shown in Fig.5.16.

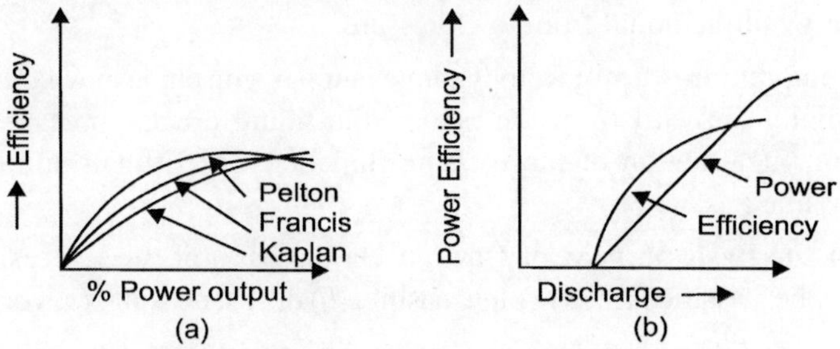

Fig. 5.16 Operating characteristic curves

3. **Muschel Curves (Constant Efficiency Curve):** It is clear from speed-efficiency and speed-power curves that there are two values of speed for the same power output and efficiency, except for tne point of maximum efficiency occurring at the designed speed. Thus, we should locate the region of constant efficiency so that the turbines are operated with maximum efficiency. The efficiency curves are plotted in Fig. 5.17
   (*a*) Power output-efficiency curve
   (*b*) Discharge-power and efficiency curve

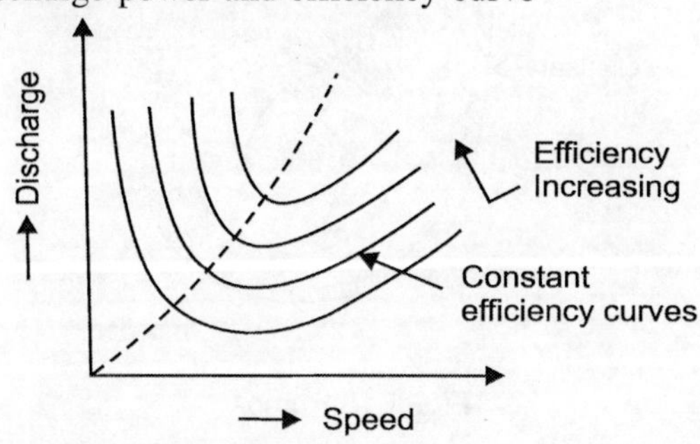

Fig. 5.17 Constant efficiency curves for a reaction turbine

5.21 WATER PUMP

A water pump is a device which lifts the water from lower level and discharges it to higher levels at the expense of mechanical energy pump in a device which converts the mechanical energy into hydraulic energy. The pump is a power absorbing device and is used to increase the pressure of liquids.

The pump creates the lower pressure (less than atmospheric pressure) at inlet and sucks the liquid from low level reservoir to pump and create a higher pressure by the impeller to deliver the water at outlet.

5.22 CLASSIFICATION OF PUMP

Pumps are generally classified in two categories: 1. Reciprocating pump, 2. Rotodynamic pump.

1. **Reciprocating Pump:** Reciprocating pump is a pump in which piston reciprocates inside the cylinder to suck the water in suction stroke and exerts thrust on liquid during return stroke and this increases the pressure energy of the liquid from low pressure.
2. **Rotodynamic Pump:** Rotodynamic pump is a machine in which fluid flow radially outward from the center, the liquid creates pressure load by centrifugal action of the rotating impeller. Centrifugal pumps may be classified as:

1. **On the Basis of Type of Casing:** On the basis of type of casing, pump may be classified as: (*a*) volute casing, (*b*) diffuser casing, (*c*) vortex casing
   (*a*) *Volute casing:* In volute casing the impeller is surrounded by a spiral casing, where X-section area increases throughout from the tounge to the delivery pipe. The increasing X-section area accommodates more and move quantity of water coming out the impeller vanes and reduce the absolute velocity of water in casing due to its volute shape and that results in increase in pressure head.

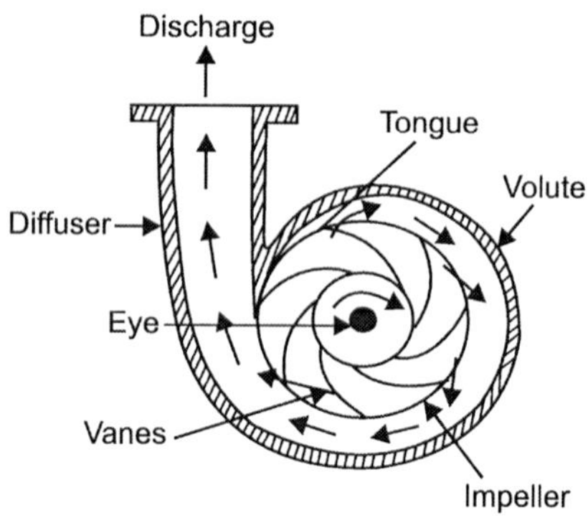

Fig. 5.18 Pump

   (*b*) *Diffuser casing:* In diffuser type casing, the impeller is surrounded by the stationary guide vanes, which provides gradually increasing X-section area to effect a gradual reduction of velocity the water coming out of impeller vanes flows in the guide vanes where its K.E. It is converted into pressure energy. The guide vane at inlet consides with the direction of the velocity of water coming out the impeller. It resembles a reverse reaction turbine and that it is also called the turbine pump.

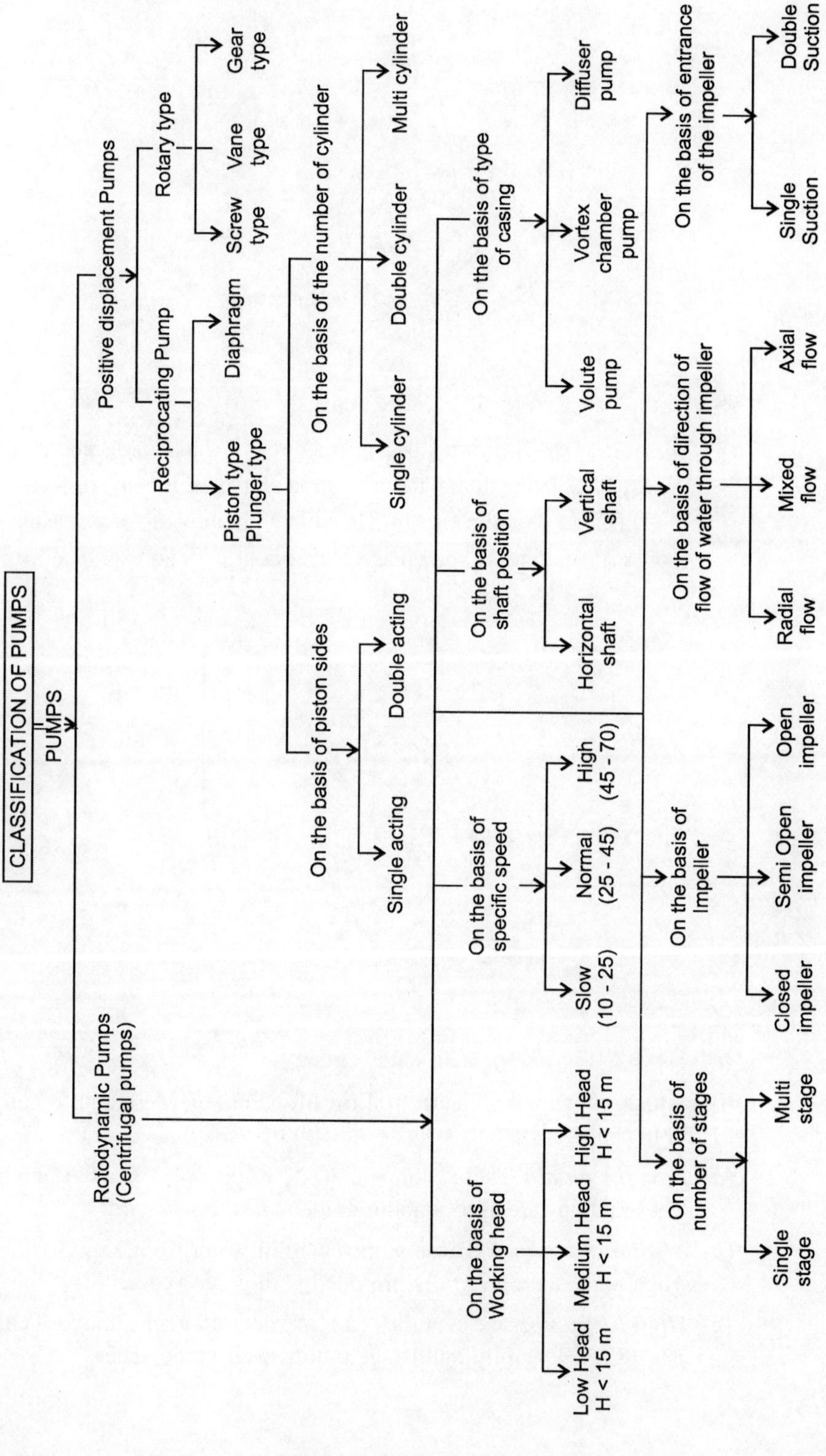
CLASSIFICATION OF PUMPS
PUMPS
Positive displacement Pumps
Reciprocating Pump
Rotary type
Piston type
Plunger type
Diaphragm
Screw type
Vane type
Gear type
On the basis of piston sides
Single acting
Double acting
On the basis of the number of cylinder
Single cylinder
Double cylinder
Multi cylinder
Rotodynamic Pumps (Centrifugal pumps)
On the basis of Working head
Low Head H < 15 m
Medium Head H < 15 m
High Head H < 15 m
On the basis of specific speed
Slow (10 - 25)
Normal (25 - 45)
High (45 - 70)
On the basis of shaft position
Horizontal shaft
Vertical shaft
On the basis of type of casing
Volute pump
Vortex chamber pump
Diffuser pump
On the basis of number of stages
Single stage
Multi stage
On the basis of Impeller
Closed impeller
Semi Open impeller
Open impeller
On the basis of direction of flow of water through impeller
Radial flow
Mixed flow
Axial flow
On the basis of entrance of the impeller
Single Suction
Double Suction

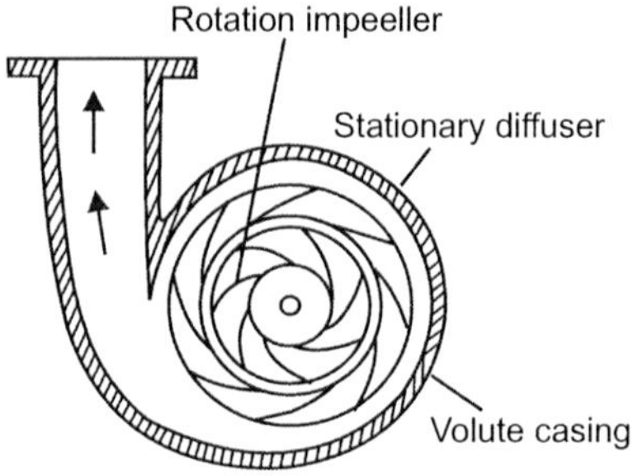

Fig. 5.19 Diffuser casing

(*c*) *Vortex casing:* The vortex casing is a circular chamber. It reduces the energy loses by avoiding, the formation of addies. It converts some of the K.E. into pressure energy. The volute pump with vortex casing is more efficient and is known as volute pump with vortex chamber.

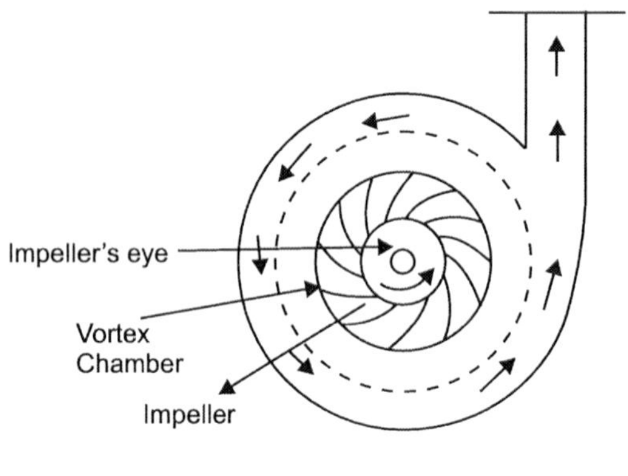

Fig. 5.20 Vortex casing

2. **On the Basis of Working Head:** (*a*) Low lift pump, (*b*) Medium lift pump, (*c*) High lift pump.

   (*a*) ***Low lift pump:*** These pumps can lift water upto 15 m in height. These pumps are used without guide vanes.

   (*b*) ***Medium lift pump:*** These pumps can lift water from 15 m to 45m. Then pumps are generally provided with guide vanes.

   (*c*) ***High lift pump:*** These pumps can lift water 45 m and above. These pumps are generally multistage pumps with guide vanes.

3. **On the Basis of Specific Speed:** The specific speed on a pump is defined as the speed of a geometrically similar pump that delivers unit discharge under unit head.

Thus, the specific speed, N_s is given as $N_s = \dfrac{N\sqrt{Q}}{N^{3/4}}$

where

N is the speed of the pump in r.p.m.

Q is discharge in m^3/s

H is head in metres.

Table 5 gives the values of specific speed of different types of pump.

Table 5. Specific speed of centrifugal pump

S.No.	Pump	Specific speed
1.	Radial flow	10-30 slow speed 30-50 medium speed 50-80 high speed
2.	Mixed flow	80-160
3.	Axial flow	160-425

4. **On the Basis of Direction of Flow of Water by Impeller:** On the basis of direction of flow of water, pumps may be classified as: (*a*) Radial flow, (*b*) Mixed flow, (*c*) Axial flow.

(*a*) *Radial flow*: In radial flow pump, liquid flows through the impeller in the radial direction only.

This type of pump is most suitable for high head and low discharge. Ordinarliy all the centrifugal pumps are manufactured with radial flow impeller. A radial flow impeller is shown in Fig.5.21 (*a*)

(*b*) *Mixed flow*: In mixed flow pump, liquid flows through the impeller axially as well as radially that is there is a combination of radial and axial flows. These pumps are most suitable for irrigation purpose where a medium quantity of water is required at a medium head. A mixed flow impeller is shown in Fig.5.21 (*b*).

(*c*) *Axial flow:* In axial flow pumps, liquid flows through the impeller in the axial direction only. The axial flow pump is not a certrifugal force pump because there is no any centrifugal force in their operation. These pumps are usually designed to deliver very large quantity of liquid at low head.

An axial flow impeller is shown in Fig. 5.21(*c*).

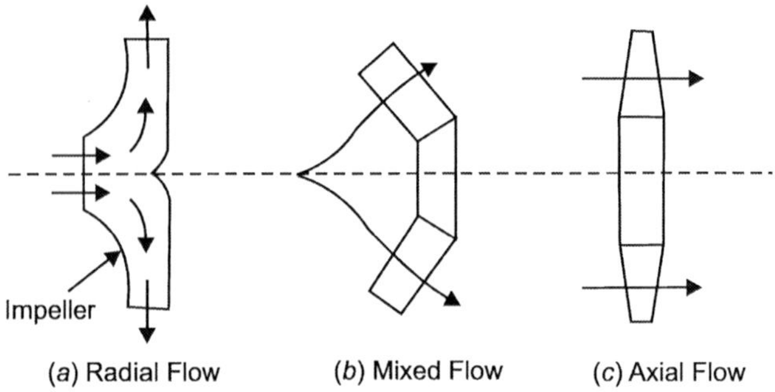

Fig. 5.21 Different impeller shapes.

5. **On the Basis of Entrance of Impeller:** On the basis of entrance of impeller pumps may be classified as: (*a*) Single suction pump, (*b*) Double suction pump.

   (*a*) *Single suction pump:* Pumps which have suction pipe only on one side of the impeller are called single suction pump. They are shown in Fig. 5.22(*a*).

   (*b*) *Double suction pump:* In double suction pumps, the suction is made from both sides of the impeller. It is shown in Fig. 5.22(*b*). The double suction pumps are suitable for pumping large quantities of liquid and have the added advantages that there is no axial thrust.

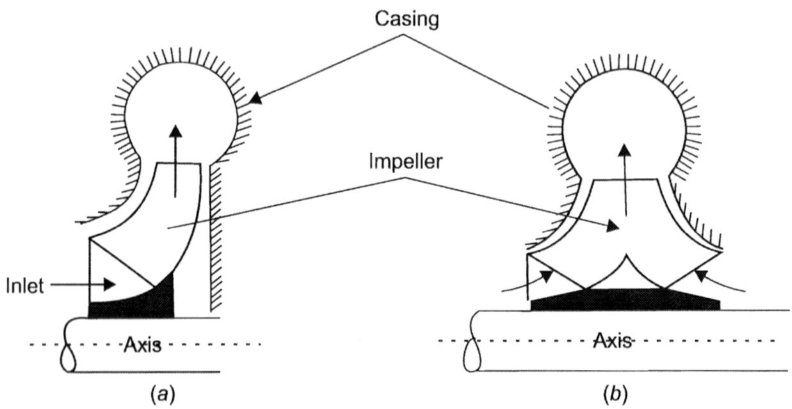

Fig. 5.22 Single and double entry impeller.

6. **On the Basis of Impeller:** On the basis of shaft position, pumps may be classified as: (*a*) Horizontal shaft, (*b*) Vertical shaft
   (*a*) *Horizontal shaft:* Centrifugal pumps are used with horizontal shafts.
   (*b*) *Vertical shaft:* Vertical shaft pumps are used where there is space limitation *i.e.*, deep well mines etc.
7. **On the Basis of Number of Stages:** On the basis of number of stages, pumps may be classified as: (*a*) Single stage pump, (*b*) Multistage pump.
   (*a*) *Single stage pump:* It has one impeller keyed to the shaft.
   (*b*) *Multistage pump:* It has two or more impellers keyed to a single shaft enclosed in the same casing. Pressure is buit up in steps.
8. **On the Basis of Shaft Position:** On the basis of impeller, pumps may be classified as : (*a*) Closed impeller, (*b*) Semi-open impeller, (*c*) Open impeller.

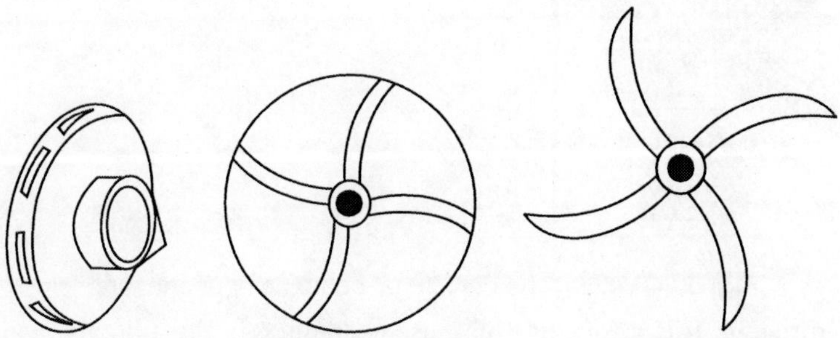

(*a*) Closed impeller (*b*) Semi-open impeller (*c*) Open impeller

Fig. 5.23 Pumps on the basis of impellers

5.23 CENTRIFUGAL PUMP

Centifugal pump is a rotadynamic machine, in which fluid flow radially outward from the center, the liquid creates pressure head by centrifugal action of the rotating impeller. Centrifugal pump may be classified as:

The main components of the centrifugal pump are:

1. Impeller
2. Casing
3. Suction pipe with foot valve and a strainer,
4. Delivery pipe.

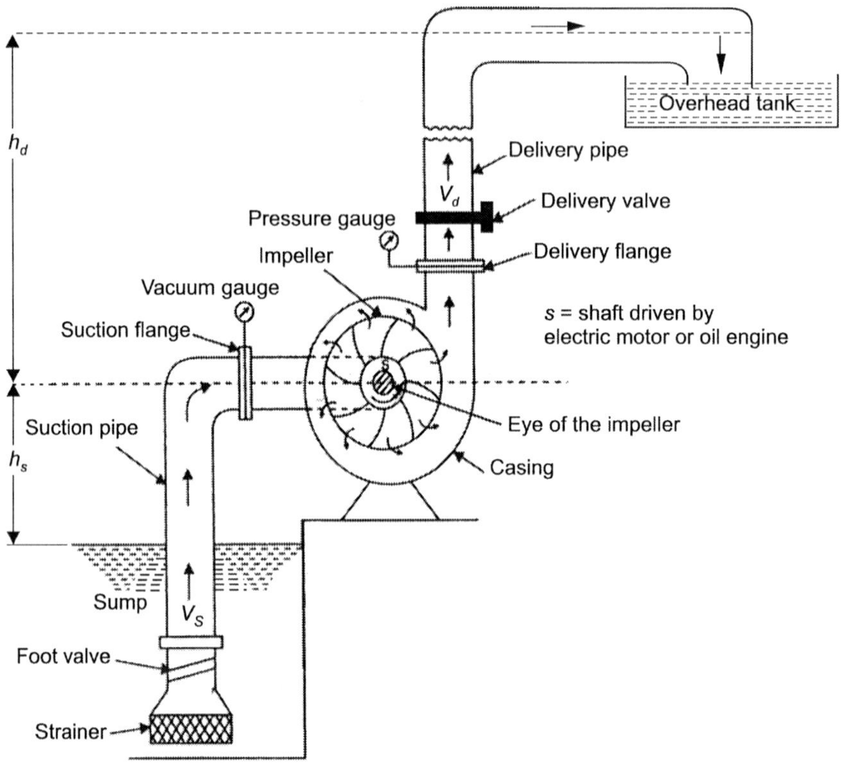

Fig. 5.24 Volute type centrifugal pump component parts.

1. **Impeller:** It is a rotating solid disc containing backward curve vanes. The fluid enters the impeller at its centre known as eye and discharges into the casing. The impeller is mounted on a shaft, which is coupled with prime mover.

   Mainly impellers are of three types: (*a*) Shrouded or closed impeller. (*b*) Semi-open impeller, (*c*) Open impeller

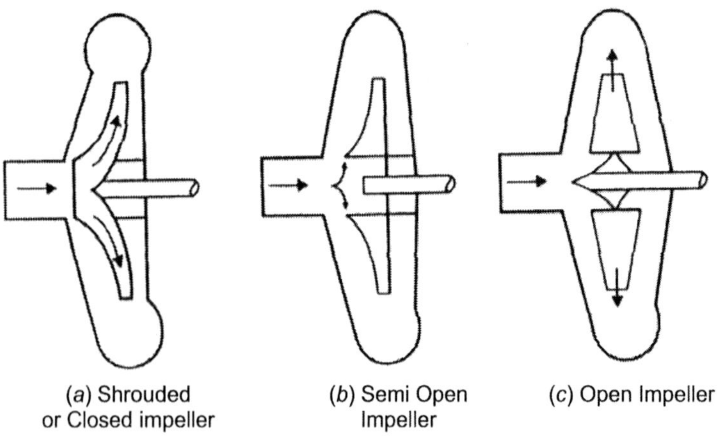

Fig. 5.25 Types of Impellers

2. **Casing:** The casing of the pump forms the housing for the impeller and passage for liquid these casing are of three types: (*a*) Volute casing, (*b*) diffuser casing, (*c*) Vortex casing.
3. **Suction Pipe with Foot Valve and a Strainer:** A pipe where one end is connected to the inlet of the pump and other end dips into water in a sump is called as suction pipe. The end of suction pipe submerged into water fitted with a valve called foot valve which does not allow water to flow back. A strainer is also fitted at the lower end of suction pipe, which do not allow to enter foreign material into pipe and hence prevent it from being chocked.
4. **Delivery Pipe:** A pipe whose one end is connected to the outlet of the pump and other end to a place where the water is required to deliver is called delivery pipe. The diameter of the pipe is always less than suction pipe.

Working of Pump: The pump is initially primed in suction pipe, casing and portion of the delivery pipe upto delivery valve are completely filled with the liquid to be pumped. The rapid motion imparts a centrifugal force on the impeller which throws the liquid towards the periphery. Thus, a partial vacuum is created at the centre of impeller eye, and then water is sucked into impeller eye through the suction pipe. The sucked water is forced out by rolating vanes and thus discharged with increased pressure and velocity into the casing. The diffuser blades converts the K.E. of water in to pressure head and thus water is discharged with greater pressure.

5.24 MULTISTAGE CENTRIFUGAL PUMP

When two or more identical impellers are mounted on the same shaft or on different shafts is called multistage centrifugal pump.

Centrifugal pumps can be made to run either in series or in parallel.

Pumps in Series: When pumps are placed in series, the outlet of the first impeller becomes the inlet to the second impeller and so on. This type of multistaging is done to obtain high heads.

Pumps in Parallel: When the pumps are placed in parallel, two or more pumps are discharging water through the same delivery pipe and so on.

This type of multistaging is done to increase the total discharge.

Figure 5.26 shows pumps in series and parallel.

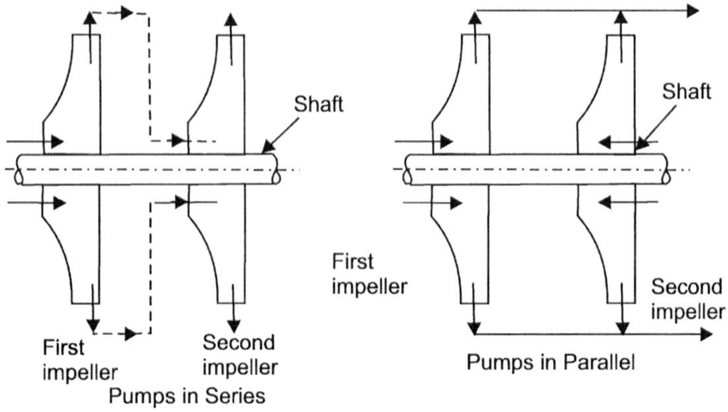

Fig. 5.26 Multistage Centrifugal pump

5.25 HEADS OF A CENTRIFUGAL PUMP

1. Suction Head: Distance between centre line of pump and water level of supply is called suction head. It is denoted by H_s.

2. Delivery Head: Distance between centre line of pump and level of discharge is called delivery head. It is denoted by H_d.

3. Static Head: It is the sum of suction and delivery head. If H_s and H_d be the suction and delivery heads respectively, it makes or it is called the static head.

$$H_{static} = H_s + H_d$$

Manometric head: It is the head measure across the pump inlet and outlet flanges. It is to express the increase in pressure energy per unit weight of liquid handled by the impeller

Let H_m = Manometric head of the pump,

H_{ms} = Reading of the pressure gauge on the suction side, and

H_{md} = Reading of the pressure gauge on the delivery side

Then.. $H_m = H_{md} - H_{mx}$

So manometric head is the difference between the delivery and suction pressure head.

Total Head: This is the actual head against which the pump has to work. It is equal to the static head plus all the head losses occurring to flow before, through and after the impeller. This is the actual head against which the pump is working and is practically equal to the manometric head, because the kinetic head at suction and delivery pipes are equal (since the diameter of the suction and delivery pipes are equal).

Let H = Total head

h_s = Suction head

h_d = Delivery head

h_{fs} = Losses in suction pipe

h_{fd} = Losses in delivery pipe

hf = Total friction loss in pipes

$= h_{fs} + h_{fd}$

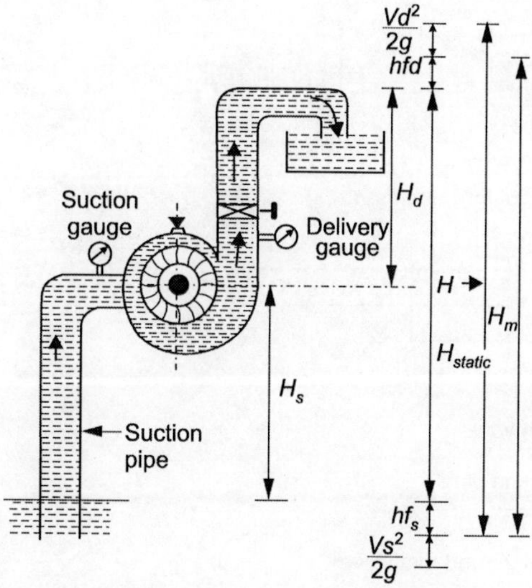

Fig. 5.27 Heads of a pump

V_d = Velocity of liquid in delivery pipe

Then
$$H = H_s + H_d + H_{fs} + H_{fd} + \frac{V_d^2}{2g}$$

$$\text{Total Head} = H_{\text{static}} + h_f + \frac{V_d^2}{2g}$$

Total head is also known as gross head, effective head or dynamic head.

5.26 RECIPROCATING OR POSITIVE DISPLACEMENT PUMP

In reciprocating pump, the piston reciprocates inside the cylinder to suck the water in section stroke and exert thrust on liquid during return stroke and thus increases the pressure energy of the liquid from low pressure to higher pressure.

Construction: A reciprocating pump consists of a piston or plunger having a reciprocating motion inside a cylinder with the help of a connecting rod and crank as shown in the figure. The crank is rotated by an electric motor and the cylinder is connected by pipe to the sump *i.e.* called suction pipe and to reservoir by the delivery pipe. The bottom end of the suction pipe has a strainer and a foot valve (non return type). The foot valve allows water to flow only in upward direction through suction pipe. There are the two valves - suction and delivery valves located near the cylinder.

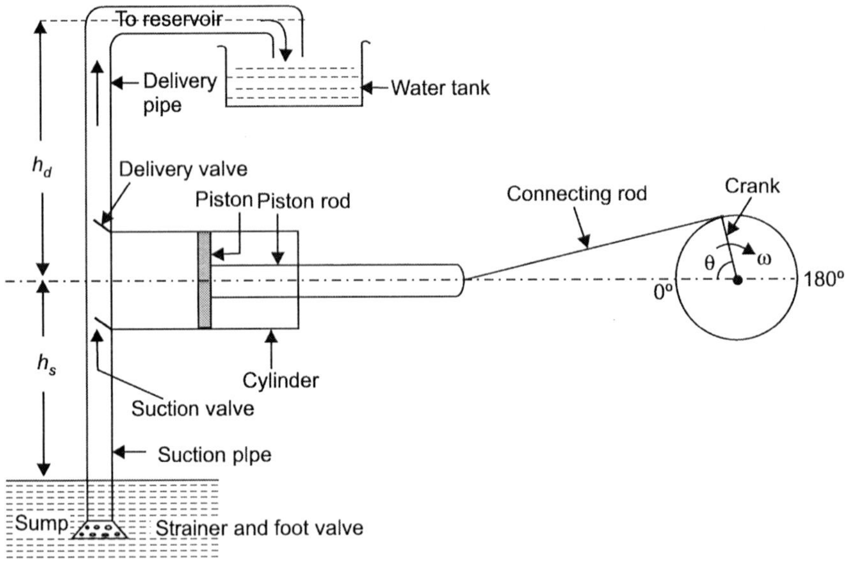

Fig. 5.28 Schemetic diagram of a single acting reciprocating pump.

Main parts of a single acting reciprocating pump.

1. A cylinder with a piston, piston rod, connecting rod and a crank.
2. Suction pipe
3. Delivery pipe
4. Suction valve
5. Delivery valve

Working: When the crank starts rotating from (0–180°) means from TDC to BDC, the piston moves outward and it creates vaccum inside the cylinder thus the suction valve opens due to pressure difference. The atmospheric pressure on the water in the sump forces the water to enter in the cylinder through the suction pipe till the piston reaches to BDC *i.e.* outermost positon (180°).

When the crank rotates from 180° to 360°, the piston moves from BDC to TDC inward in the cylinder at first due to pressure the suction valve closes and the water pressure opens the delivery valve and water is pushed upward in the delivery pipe to deliver the water at outlet.

The height difference between water in sump and pump centre line is called the suction head, 'h_s' and the vertical distance between centre line of the pump and the delivery pipe is called delivery head h_d the quantity of water lifted through static head $H = h_s + h_d$, during one revolution of the crank.

The rate of water discharge/sec $= Q = \dfrac{ALN}{60}$

where A = Cross - sectional area of the cylinder (m^2)

L = Stroke length (m)

N = Crank speed

Q = Discharge/sec.

weight of water/sec

$$W = WQ = W\frac{ALN}{60}$$

Workdone/sec = Weight of water lifted/sec × height

$$= W(h_s + h_d) = W\frac{ALN}{60}\ (h_s + h_d)$$

Hence the power required to drive pump $= \dfrac{WALN}{60}\ (h_s + h_d)\ K$ watt

where $W = \text{N/m}^3$

Working of a Double-acting Reciprocating Pump:

Refer Fig. 5.29. In a double-acting reciprocating pump, suction and delivery *strokes occur simultaneously.* When the crank rotates from *I.D.C.* in the clockwise direction, a vaccum is created on the left side of piston and the liquid is sucked in from the sump through valve S_r. At the same time, the liquid on the right side of the piston is pressed and a high pressure causes the delivery valve D_2 to open and the liquid is passed on to the discharge tank. This operation continues till the crank reaches *O.D.C.*

With further rotation of the crank, the liquid is sucked in from the sump through the suction valve S_2 and is delivered to the discharge tank through the delivery valve D_r When the crank reaches *I.D.C.*, the piston is in the extreme left position. Thus, one cycle is completed and as the crank further rotates, cycles are repeated.

Because of continuous delivery strokes, a double-acting reciprocating pump gives more uniform discharge as compared to a single-acting pump which pumps the liquid intermittently). To get a still more uniform feed, invariably a multi-cylinder arrangement having two or more cylinders is employed.

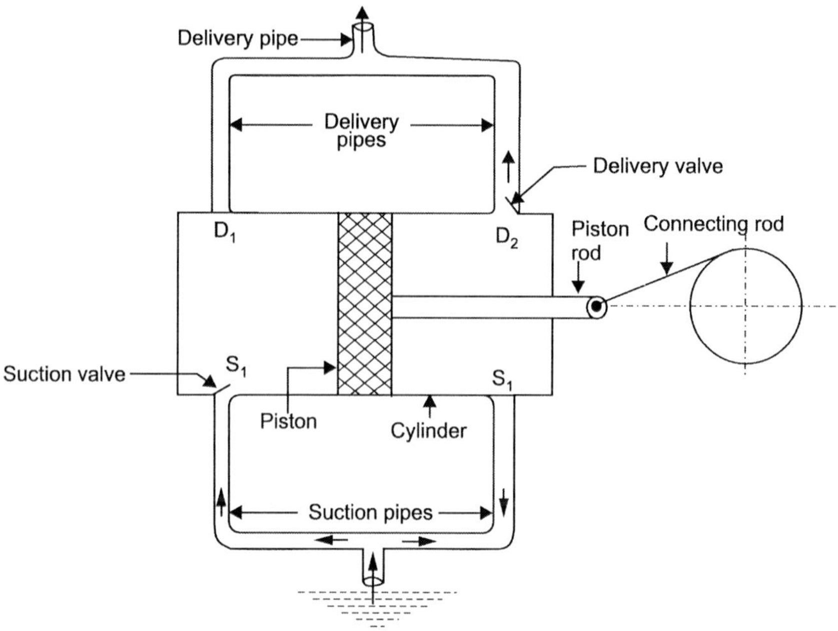

Fig. 5.29 Double acting reciprocating pump

Figures 5.30 and 31 show the variations of discharge through delivery pipe (Q_d) with crank angle (θ) for single-acting and double-acting pumps respectively.

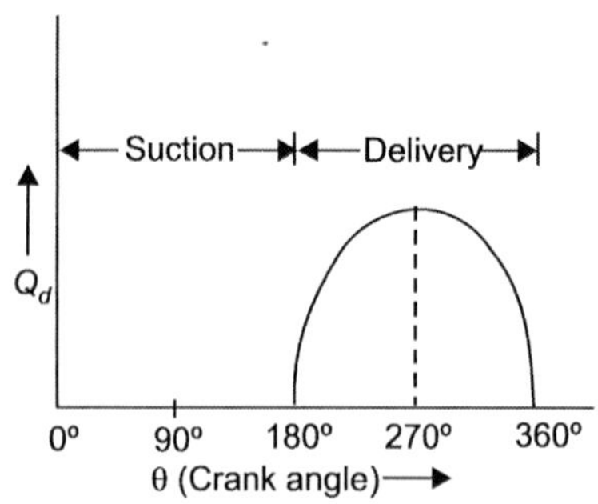

Fig. 5.30 Dischage crank angle ($Q_{d,}$ vs θ) variations for single-acting pump

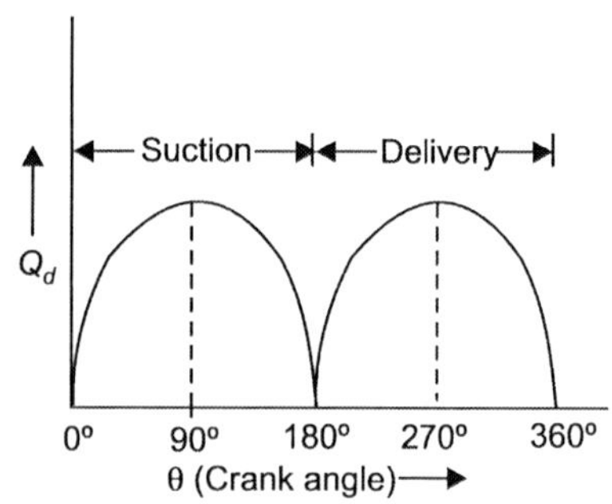

Fig. 5.31 Discharge crank angle ($Q_{d,}$ vs θ) variations for double-acting pump

5.27 COMPARISON OF RECIPROCATING AND CENTRIFUGAL PUMP

Reciprocating	Centrifugal Pump
1. Mechanical energy is directly converted into pressure energy of water.	1. It requires diffusion action of casing for conversion of velocity of water into pressure energy.
2. Mechanical power input is converted into piston force by means of crank and connecting rod, so losses in energy transfer.	2. The pump impeller is directly coupled to the prime mover shaft so losses in energy transfer are negligible.
3. It is positive displacement type, it ensures water delivery for each delivery stroke of piston.	3. It is dynamic action type pump, and the centrifugal action given to the water is converted to pressure head.
4. Presence of any gas do not effect suction to large extent.	4. It requires diffusion action of casing for conversion of velocity of water into pressure energy.
5. Primming is not required.	5. Priming is necessary, or the casing should always be full of water.
6. Reciprocating pump can handle low viscosity fluid free from impurties.	6. Centrifugal pump can easily handle high viscous fluid like oil, sugar, molasses, paper, pulp, etc.
7. Cost is more.	7. Cost is less.
8. Maintenance is more and costly.	8. Maintenance is less and cheap.
9. They are usesd in automobile workshops, cycle pump, kerosene pump, hydraulic jack etc.	9. These are used for domestic as well as industrial application ranging from 0.5 H.P. to 1000. H.P.

Advantages of Centrifugal Pump Over Displacement (Reciprocating) Pump

The centrifugal pump claims the following *advantages* with reference to a positive displacement (reciprocating) pump.

1. The cost of a centrifugal pump is less as it has fewer parts.
2. Installation and maintenance are easier and cheaper.
3. Its discharging capacity is much greater than that of a reciprocating pump.
4. It is compact and has smaller size and weight for the same capacity and energy transfer.
5. Its performance characteristics are superior.
6. It can be employed for lifting highly viscous liquid such as paper pulp, muddy and sewage water, oil sugar molasses etc.
7. It can be operated at very high speeds without any danger of separation and cavitation.

8. It can be directly coupled to an electric motor or an oil engine.
9. The torque on the power source is uniform, the output from the pump is also uniform.

However, because of higher efficiency the reciprocating pumps are still employed for high heads and small discharges. A reciprocating pump can build up very high pressures (as high as 700 bar or even more) and as such these pumps are used for lifting oils from very deep oil wells

5.28 HYDRAULIC DEVICES

There are many hydraulic devices/machines which are based on the principles' of fluid statics and fluid kinematics and are used for either storing the hydraulic energy and then transmitting when needed or magnifying the hydraulic energy several times and transmitting the same. In all such machines power is transmitted with the help of a fluid which may be a liquid (water or oil). The various hydraulic devices/machines in use are :

1. Hydraulic lift,
2. Hydraulic press,
3. Hydraulic ram,
4. Hydraulic crane,
5. Hydraulic coupling,
6. Hydraulic torque converter,
7. Hydraulic accumulator,
8. Hydraulic intensifiers
9. Air lift pump
10. Jet pump.

Here only Hydraulic press and Hydraulic lift will be disscussed.

5.28.1 Hydraulic Lift

Hydraulic lift is a device used for carrying persons and loads from one floor to another, in a multi-storeyed building. The hydraulic lifts are of the following two types :

1. Direct acting hydraulic lift
2. Suspended hydraulic lift.

1. Direct Acting Hydraulic Lift: Refering to Fig.5.35. It consists of a ram sliding in a cylinder. A platform or a cage is fitted to the top end of ram on which goods may be placed. As the liquid under pressure is admitted to the cylinder, the ram moves in the downward direction when the liquid from the fixed cylinder is removed.

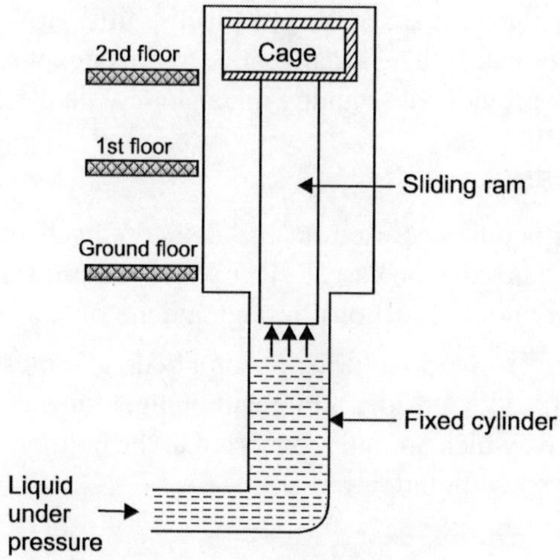

Fig. 5.32 Direct acting hydraulic lift.

2. Suspended Hydraulic Lift: The suspended hydraulic lift is a modified form of the direct acting hydraulic lift. It is fitted with a jigger. The cage is suspended by ropes. It runs between guides of hard wood or round steel. In order to balance the weight of the cage, sliding balance weights are provided.

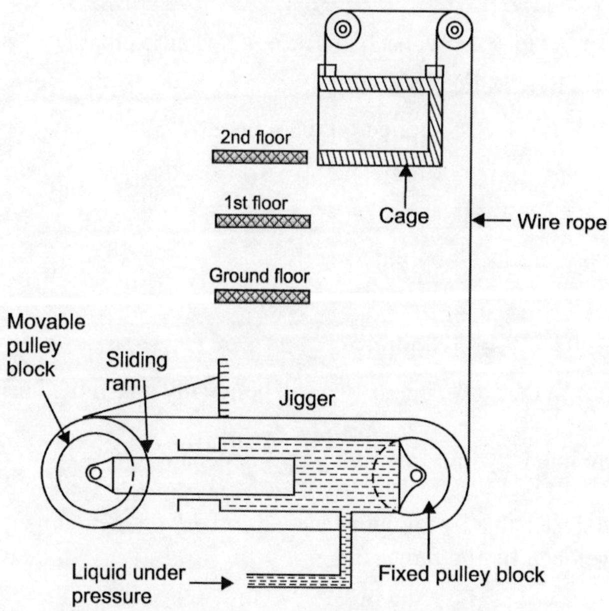

Fig. 5.33 Suspended hydraulic lift.

Modem lifts are of suspended type and have a high velocity ratio ; the lifting speeds range from 100 to 120 m per minute

The hydraulic lifts have been suspended by electric lifts. However, hydraulic lifts are preferred in places where there is danger due to fire or explosions. The hydraulic lifts are also usually provided as standby units along with electric lifts.

2. Hydraulic Press

The hydraulic press is a device used far lifting heavy loads by the application of much smaller force. It is based on Pascal's law, which states that intensity of pressure is transmitted equally in all directions through a mass of fluid at rest.

Working Principle: The working principle of a hydraulic press may be explained with the help of Fig.5.34. Consider a ram and plunger, operating in two cylinders of different diameters, which are interconnected at the bottom, through a chamber, which is filled with some liouid.

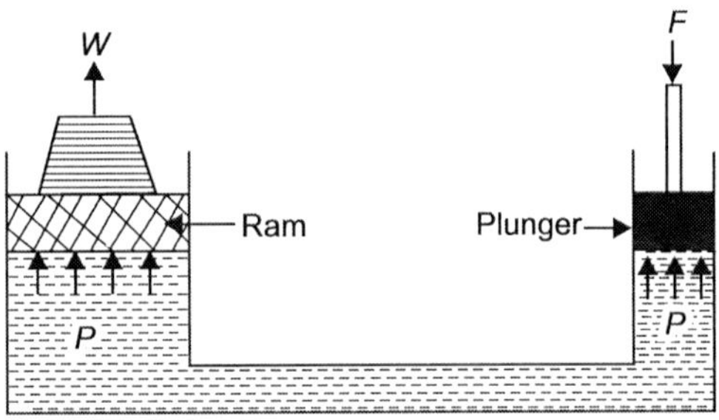

Fig. 5.34 Working principle of hydraulic press.

Let W = Weight to be lifted,

F = Force applied on the plunger,

A = Area of ram, and

a = Area of plunger.

Pressure intensity produced by the force F,

$$P = \frac{F}{\text{Area of plunger}} = \frac{F}{a}$$

As per Pascal's law, the above intensity p will be equally transmitted in all directions.

$$\therefore \text{ The pressure intensity on ram } = P = \frac{F}{a} = \frac{W}{A} \text{ or } W = F \times \frac{A}{a} \quad ...(i)$$

Equation (i) indicates that by applying a small force F on the plunger, a large force W may be developed by the ram.

$$\text{Mechanical advantage of press} = \frac{A}{a}$$

If the force in the plunger is applied by a lever which has a mechanical advantage $\frac{L}{l}$ (Fig.5.35) then total mechanical advantage of machine $= \frac{L}{l} \times \frac{A}{a}$

The ratio $\frac{L}{l}$ is known as *leverage of press.*

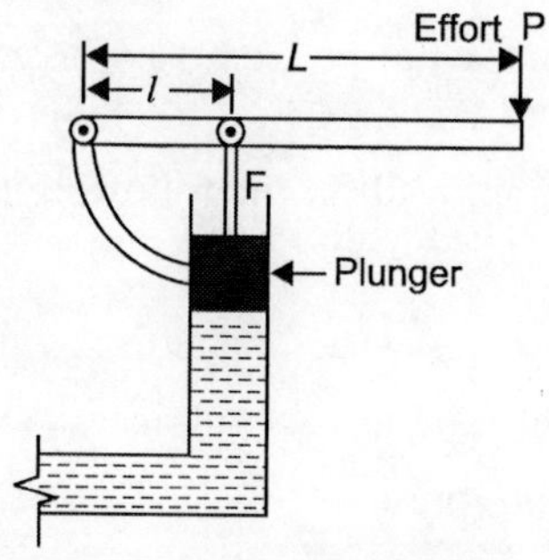

Fig. 5.35 Plunger.

Fig. 5.36 shows an 'Elementary Inverted Hydraulic Press'.

- It consists of a cylinder (fixed) in which the ram slides. The lower end of the ram carries a movable platen which moves up and down with the ram. The upper and lower stationary platens are joined by columns.
- When liquid under high pressure is supplied to the cylinder, the ram moves downward and applies tremendous pressure (equal to the product of intensity of pressure supplied and area of the ram) upon any material placed between the movable platen and lower stationary platen.

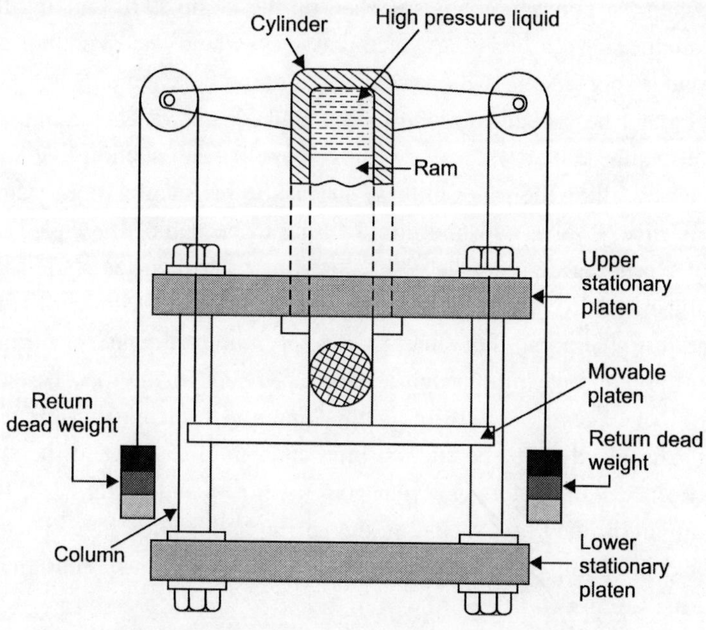

Fig. 5.36 Elementary inverted hydraulic press.

For bringing the ram back in position, the liquid from the cylinder is taken out, subsequently the ram (along with movable platen) moves up by the action of return weights.

It may be noted that in some large presses it is possible to produce total thrust ranging from about 50 MN to 100 MM.

Hydraulic presses may be employed for the following jobs :

(*i*) Forging press.

(*ii*) Metal press work (to press sheet metal to any required shape).

(*iii*) Bending and straightening any metal piece.

(*iv*) Autoclave vulcanising press.

(*v*) Plate press etc.

(*vi*) Drawing and pushing rods.

(*vii*) To prepare moulds and casting of bakelite (Bakelite press).

(*viii*) Cotton press.

(*ix*) Packing press.

5.29 PRIMING OF A PUMP

When centrifugal pump is not running for some time, the water present in the pump casing and suction pipe flows back to the sump and these spaces get filled with air, when the pump motor is switched on the pump starts running the head develop equals $H = Vu_2u_2/gm$ of air since $\rho aw << \rho$ water. The Head thus genrated cannot produce spontaneously the vaccum required to start pumping action. Hence the water cannot be sucked in along the suction pipe to reach the impeller. For making the pump deliver water, the casing, impeller and suction line should be free from air by filling the water in these spaces. So the suction pipe, casing and a portion of delivery valve with the liquid filling to be pumped is called priming. After proper priming delivery valve is closed and the pump is started the head genrated and then delivery valve is gradually opened and thus a contineous discharge takes place from the pump. For small pumps, primming is done by pouring water directly in the pump casing through a funnel provided at the top of casing. The flowing out of excess water through this opening indicate that air has been completely displaced from the suction pipe and pump casing and the system is thoroughly filled with water. The vent cock is closed and pump is switched on larger pump are primed by evacuating the casing and suction pipe. This work is due by a vacuum pump or steam ejector, and the water get sucked into the suction pipe from the sump.

5.30 CAVITATION

The formation, growth, and collapse of vapour filled cavities or bubbles in a flowing liquid due to local fall in fluid pressure is called cavitation. When the pressure at any point, in a flow field equals the vapour pressure of the liquid at that temperature vapour cavities (bubbles of vapour) begin to appear. It is presumed that a vapour cavity is formed around a dust nuclei which is in the liquid (The vapour pressure values of water at 15° and 20°C are 1.74 m and 2.38 m of water column absolute). The cavities thus formed, due to motion of liquid, are carried to high pressure regions where the vapour condenses and they suddenly collapse. The adjoining liquid rushes with a very great velocity (and hence with very great force) to occupy the empty spaces thus created, causes series of violent, irregular, spherical shock waves. When these irregular implosions occur on the metallic surface, they produce noise and vibration.

When the cavities collapse (the collapsing pressure is of the order of 100 times the atmospheric pressure) on the surface of a body, due to repeated 'hammering' action, the metal particle gives way ultimately due to fatigue and indentations are formed; this erosion of material is called **pitting** (Fig. 5.37).

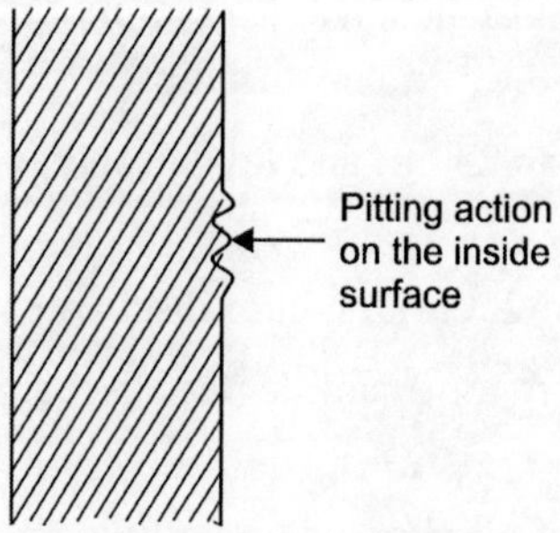

Fig. 5.37 Pitting action on the inside surface (shown on large scale).

In reaction turbines the cavitation may occur at the runner exit or the draft tube inlet where the pressure is negative. The hydraulic machinery is affected by the cavitation in the following three ways :

1. Roughening of the surface takes place due to loss of Fig.5.37. Pitting action on the inside material caused by pitting.surface (shown on large scale).
2. Vibration of parts is caused due to irregular collapse of cavities.
3. The, actual volume of liquid flowing through the machine is reduced (since the volume of cavities is many times more than the volume of water from which they are formed) causing sudden drop in output and efficiency.

HIGHLIGHTS OF THE CHAPTER

1. Hydraulic turbine is a device which convert hydraulilc energy (PE + KE) into mechanical energy.

2. In an impulse turbine the pressure energy of water is converted into kinetic energy when passed through the nozzle and forms the high velocity jet of water which drive the wheel. The Pelton wheel is a high head and tangential flow impulse turbines.

3. Reaction turbine is that in which water enters at the inlet and is partially converted into pressure energy and partially converted into kinetic energy known as reaction turbine.

4. If the water flows through the runner along the direction parallel to the axis of rotation of the runner. The turbine is called axial flow turbine *e.g.*, Kaplans and propeller turbine.

5. If the water flows in radial direction through the runner, turbine is called radial flow turbine.

6. If the water enters through the runner in the radial direction but leaves in the direction parallel to the axis of rotation of the runner, turbine is called mixed flow turbine *e.g.*, Francis turbine.

7. High head (more than 250m) and low discharge turbine *e.g.*, Petton wheel turbine.

8. Medium Head (45m-250m) medium discharge *e.g.*, fronas turbo.

9. Low Head (less than 45m) and degree discharge turbine, *e.g.*, Kaplan or propeller turbine.

10. Main components of a Pelton wheel turbine are as follows.

(*i*) Pen Stock
(*ii*) Ranner with buchet
(*iii*) Guide Mechanism
(*iv*) Runner
(*v*) Draft Tube

11. Main components of a frontal turbine are as follows.

(*i*) Pen Stock
(*ii*) Scroll Casing
(*iii*) Guide Mechanism
(*iv*) Runner
(*v*) Draft Tube

12. Main components of a leaplen turbine are as follows

(*i*) Runner
(*ii*) Guide Vane
(*iii*) Scrolling Casing
(*iv*) Whirl Chamber
(*v*) Draft Tube

13. Draft tube is an enlarging cross-sectional area tube which is fixed at the outlet of the turbine and other end is submerged into water *i.e.,* (tail race). It is also an integral part of low head turbine.
14. Types of draft tube are as follows:
    (*a*) Straight divergent
    (*b*) Moody sepronng tube
    (*c*) Simple elbow
    (*d*) Elbow of cross section from circular to rectangular shape.
15. Specific speed of a turbine is defined as the speed of a geometrical similar turbine which produce a unit power when working order a unit head.
16. Pump is a device which lifts the water from lower level and discharges it to higher level at the expanse of mechanical energy into hydraulic energy.
17. A centrifugal pump is a rotodynamic pump in which fluid flows radically outward from the center, the liquid creates pressure head by centrifugal action of the rotoimpeller.
18. A positive displacement pump, in which liquid is sucked inside the pump and forced out of the liquid from low pressure to higher pressure.
19. Hydraulic lift is a device used for carrying persons and load from one floor to another floor.
20. Hydraulic press is a device used for lifting heavy loads by the application of much smaller force. It is based on Pascal's law.

REVIEW QUESTIONS

1. Define Hydraulic turbine. How turbines can be classified?
2. Differentiate between impulse turbine and reaction turbines.
3. Explain the construction and working of Pelton wheel turbine with a neat sketch.
4. Draw a neat diagram of the layout of a Pelton wheel unit. Name the different parts and state the purpose by each part.
5. Describe with neat sketches, the constructions and working of a Kaplan turbine. Why a draft tube is used in reaction turbine?
6. Describe with neat sketches, the construction and working of a Francis turbine. Why a draft tube is used in such turbines.

7. State the advantages and disadvantages of Francis turbine over a Pelton wheel.
8. Differentiate between inward and outward flow turbine.
9. Differentiate between pelton wheel and Francis turbine.
10. Differentiate between Francis turbine and Kaplan turbine.
11. Draw a neat diagram of the layout of a Kaplan turbine unit. Name the different parts and state the purpose served by each part.
12. Describe, with neat sketch the construction and working of Francis turbine hydropower unit.
13. Define and drive an expression for the specific speed of a hydraulic turbine:

$$N_s = \frac{N\sqrt{P}}{H^{5/4}}$$

14. Drive an expression for the specific speed of a hydraulic turbine and point out how the classification of hydraulic turbine is based on specific speed.
15. What are the factors which govern the selection of water turbine?
16. What are the consideration for selection of turbine?
17. What is a draft tube also describe its types?
18. What is the difference between turbine pump?
19. How would you classify a pump?
20. What is priming in centrifugal pumps why is it necessary?
[Dec., 2003 MDU).
21. What is a centifugal pump? Differentiate between the volute casing and vortex casing for the centrifugal pump. [May, 2004 MDU].
22. Explain how does a centrifugal pump operate. [Dec., 2004 MDU].
23. How do you classify the water pumps? Explain. [May, 2004 MDU]
24. How are centrifugal pumps classified? State the advantages of a centrifugal pump over reciprocating pump?
25. Explain the working principal of reciprocating pumps with the help of a sketch and name all the main parts.
26. What are the rotary pumps and how they are classified?
27. Explain the heads of a centrifugal pump?
28. Differentiate between centrifugal and reciprocating pump.

29. Differentiate between closed, semiclosed and open impellers of a centrifugal pump.
30. Describe with neat sketch, the working of a reciprocating pump.
31. What do you mean by multi stage centrifugal pumps?
32. What are the characteristic curves?
33. Explain the following centrifugal hydraulic devices.
    (*a*) Hydraulic left (*b*) Hydraulic press
34. What do you mean by priming of a pump and why it is done ?
35. What is a phenomenon of a cavitation?

■■■

6

Power Transmission Methods and Devices

6.1 INTRODUCTION

A mechanical drive is a mechanism used to transmit power from the prime mover to the machine and usually involve in changing velocities forces and torques etc. The power transmission from shaft to another at a considerable distance is obtained by fixable connection drives such as belt, rope and chain drives. Belt and rope drives are used to transmit power between two parallel shafts by means of friction similarity chain and gear drives are used to transmit power with positive or direct contact.

6.2 MECHANICAL DRIVE

Mechanical arrangement for power transmission by prime mover with its application is known as mechanical drive.

Mechanical Drive

↓

Flexible drive

(*a*) Belt drive
- (*i*) Flat belt
- (*ii*) V-belt
- (*iii*) Circular (rope) drive
- (*v*) Chain drive

Friction drive

(*a*) Clutch plate drive

Direct drive

(*a*) Gear drive
(*b*) Cam drive
(*c*) Chain drive

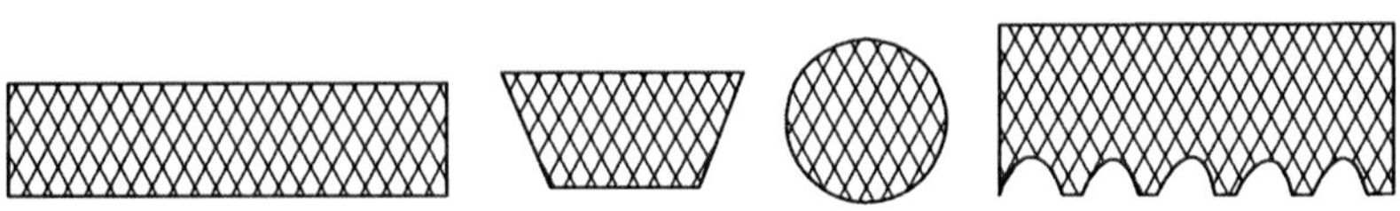

(*i*) (*ii*) (*iii*) (*iv*)

Fig. 6.1 Mechanical drive

6.3 TYPES OF DRIVE SYSTEMS

There are two types of drive systems:

1. Individual drive system
2. Group drive system

6.3.1 Individual Drive System

When the power is transmitted by a prime nover to the single machine known as individual drive system. *e.g.*, An electric motor drive on air compressor.

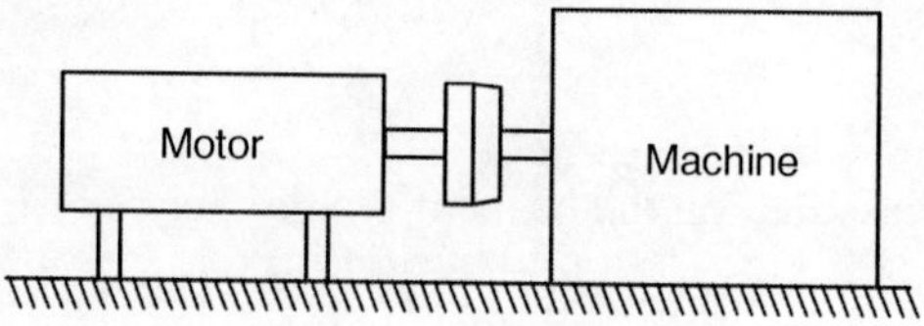

Fig. 6.2 Individual drive.

6.3.2 Group Drive System

When the power is transmitted by prime mover to a group of machines known as group drive system.

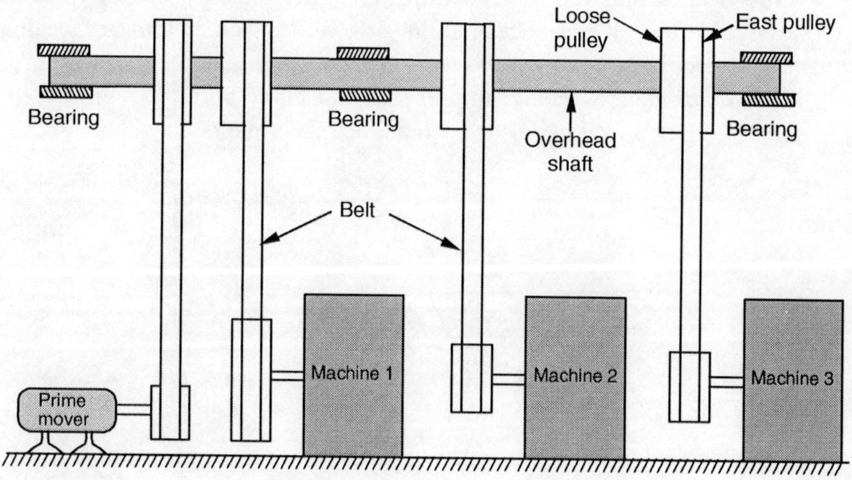

Fig. 6.3 Schematic of a group drive

6.4 BELT DRIVES

The transmission of power in industries from one rotating shaft to another shaft at a considerable distance is achieved by belt. Belts are used to transmit power between two parallel shafts consisting of driving and driven pulleys by means of friction. The grip between the belt and pulley is obtained by friction.

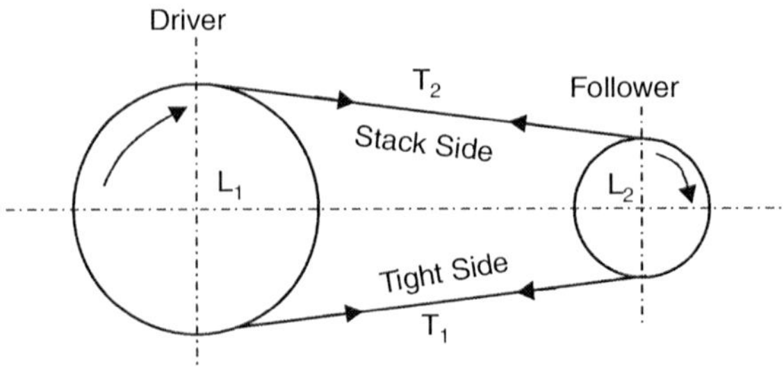

Fig. 6.4 Belt drives

6.4.1 Types of Belts

Belt drives can be classified on the basis of type of belt.

1. **Flat Belt:** It is mostly used for high speeds. They are used to transmit the power from one shaft another through a number of pulley when the centre to centre distance between the pulley is large. Those are rectangular in section and made of leather, canvas, batala and steel whose two ends are connected together with butt or strap, joint.
2. **V-belt:** These are used when two pulleys are nearer to each other having a shorter distance. These are less efficient then flat belts. It is trapezoidal in cross-section and slip is almost negligible in case of V belt drive to the wedge action belt and sides of pulley groove. The belt bottom surface does not contact with bottom of groove on the pulley. It is also a noiseless drive. These are made of rubber impregnated on fabric with an angle 30-40°. Hence, more power can be transmitted for the same belt tension.

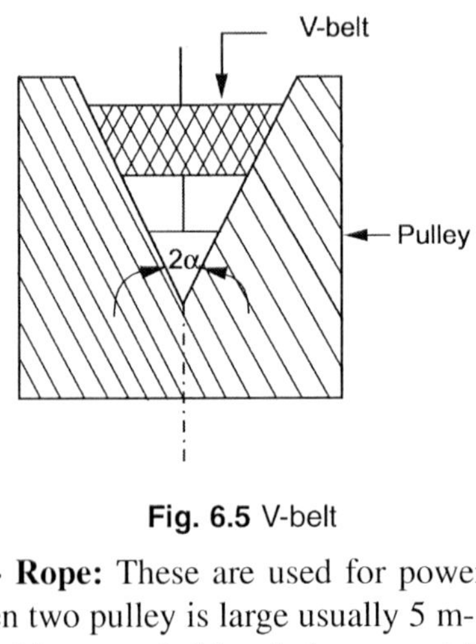

Fig. 6.5 V-belt

3. **Circular Belt - Rope:** These are used for power transmission when the distance between two pulley is large usually 5 m- 30 m. These are used in machines like table type clothing industry and in household appliances. They may be made of leather, canvas and rubber and fabric.

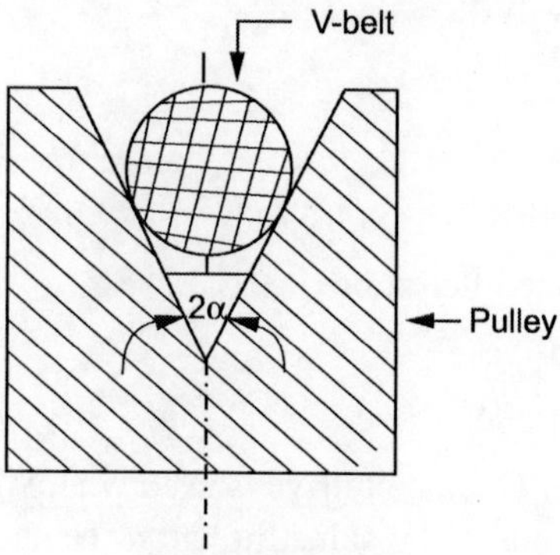

Fig. 6.6 Rope

The ropes are of two types: (*i*) Fibre rope, (*ii*) Wire rope

(*i*) **Fibre Rope:** These are made of fibre material like hemp cotton. These are used in hand operated hoisting machinery and used for transmission of light power.

(*ii*) **Wire Rope:** These are used to transmit power over a long distance from one pulley to another and used for elevators, mine host cranes, conveyors and for the suspension bridges. Ropes are made from cold drawn wires in order to increase its strength and durability.

6.4.2 Advantages of V-belt Drive

The salient features of V-belt are as follows :

1. V-belts are suited for small centre distances.
2. The drive is capable in absorbing high shock.
3. V-belts offer more positive drive because of reduced slippage.
4. Multiple V-belt can be employed for greater power out put.
5. Standardization gives better initial installation.
6. The gripping action results in lower belt tension.
7. No possibility of belt coming out of grooves.
8. In case of multiple belt drive, it one belt fails, the machine does not come to stop.
9. Large reduction in speed are possible in a single drive by using V-belts over small pulleys.
10. V-belt may be fitted at any angle according to requirement.

6.4.3 Types of Belt Drives

(*a*) Open belt drive

(*b*) Cross belt drive

(*c*) Quarter turn belt drive

(*d*) Belt drive with one idler pulley

(*e*) Compound belt drive.

(*f*) Stepped or cone pulley drive

(*a*) **Open Belt Drive:** Open belt drive is used where the shafts are arranged in parallel and rotating in the same direction. Hence the belt is subjected to tension and bending. The lower side of the belt is known as tight side and upper side of the belt is known as slack side.

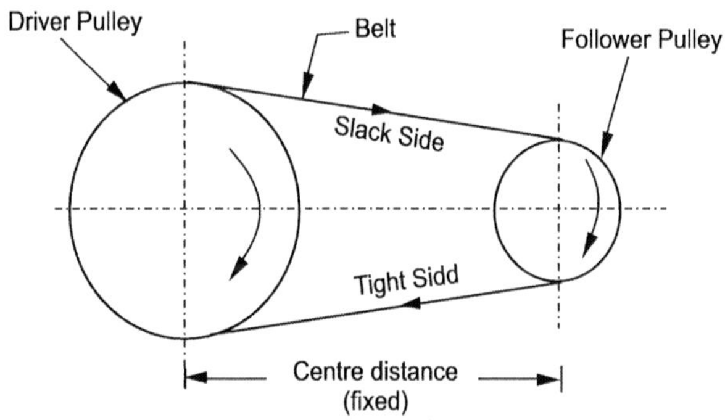

Fig. 6.7 Open belt drive

(*b*) **Cross Belt Drive:** Cross belt drive system is used for where the shaft axis are parallel to each other and rotating in opposite directions since angle of contact between pulley and belt is more, It can transmit more power than open belt drive. However there is more wear and tear takes place in the belt due to rubbing of belt with each other.

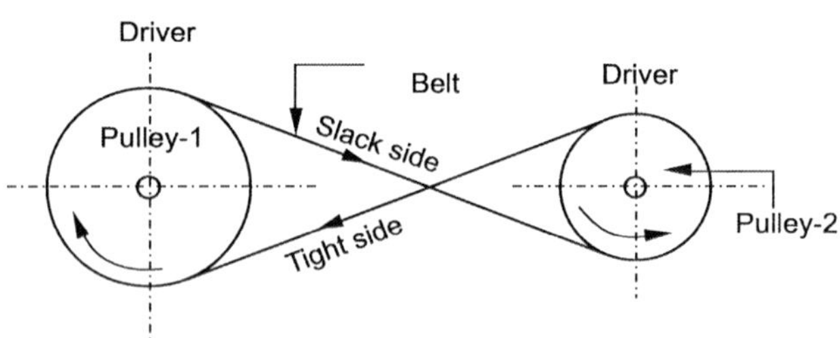

Fig. 6.8 Cross belt drive

(*c*) **Quarter Twist Belt Drive:** It is used when the shafts are at right angles and rotation is required in one definite direction.

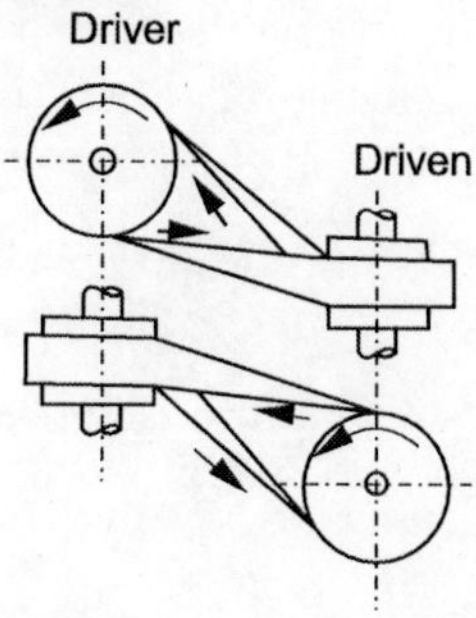

Fig.6.9 Quarter-twist belt drive

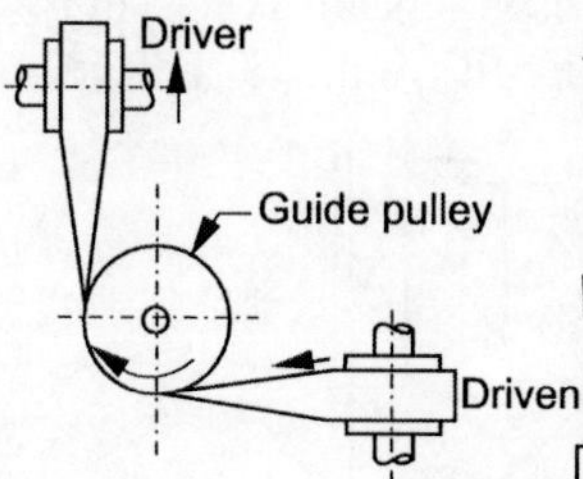

Fig. 6.10 Quarter-twist belt drive with guide pulley

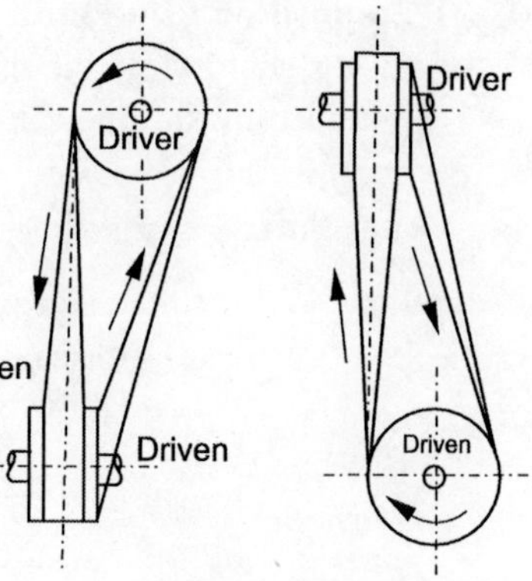

Fig. 6.11 Quarter turn belt

(*d*) **Belt Drive with Idler Pulley:** It is used when the open belt drive cannot be employed because of small arc contact over the smaller pulley.

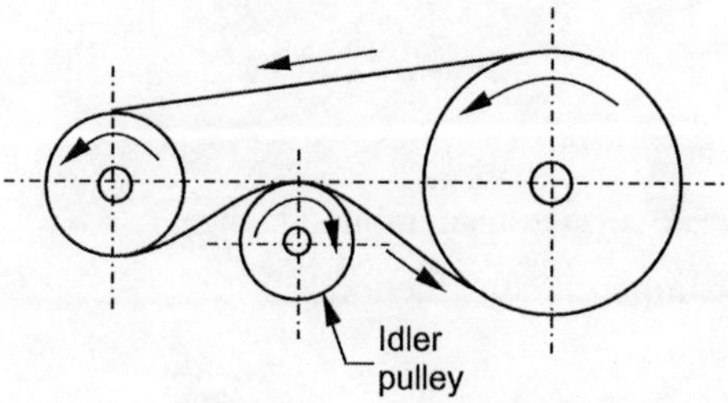

Fig. 6.12 Belt drive with an idler pulley

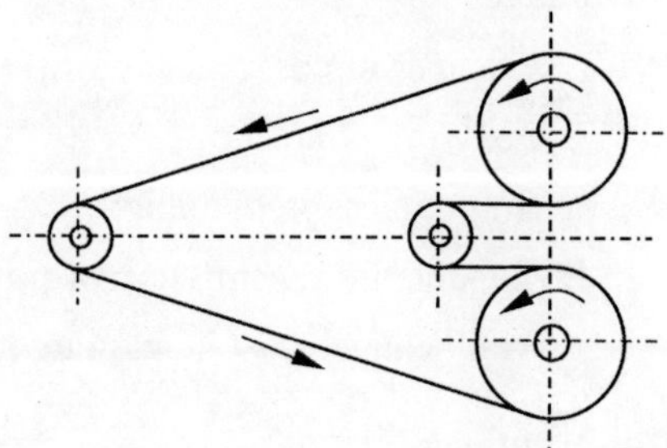

Fig. 6.13 Belt drive with many pulleys

(*e*) **Compound Belt Drive:** It is used to transmit power from one shaft to several shafts arranged in parallel.

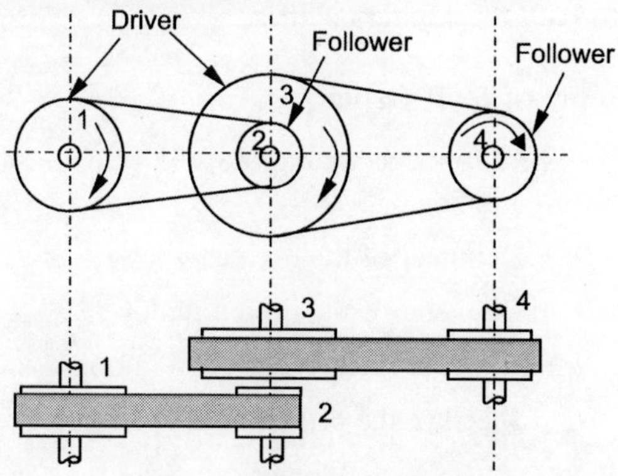

Fig. 6.14 Compound belt drive

(*f*) **Stepped or Cone Pulley Drive:** It is used to obtain different speeds on drives shaft while the driving shaft is moving with same speed. Speed is obtained by shifting the belt from one shaft to another.

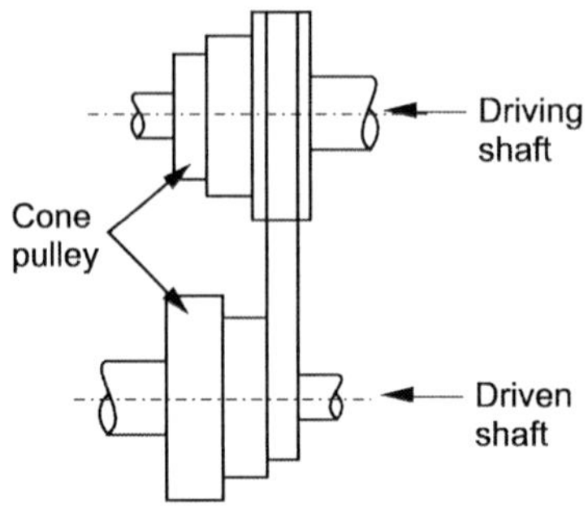

Fig. 6.15 Stepped or cone pulley drive

6.4.4 Advantages and Disadvantages of Belt Drive

Advantages

1. Cheaper in cost and easily available.
2. Easy in maintance
3. Smoothness in operation.
4. Capable in transmission of power over a considerable distance.
5. Maintenance cost is less due to simple in construction.

Disadvantages

1. Less velocity ratio
2. Stretching of belt loads to slip and creep.
3. Larger in size.
4. Short service life in high speed drives.
5. Necessity for belt tensioning devices.

6.4.5 Velocity Ratio of Belt Drive

The velocity ratio may be defined as the ratio of the velocities of the drive pulley and the follower or driven pulley.

Let D_1 = Diameter of the driver pulley

D_2 = Diameter of the driven pulley

N_1 = Speed of the driver pulley in rpm

N_2 = Speed of the driven pulley in rpm.

ω_1 = Angular velocity of the driver pulley

ω_2 = Angular velocity of the driven pulley

$\therefore$ Length of the belt that passes over the driver pulley in one minute = $\pi D_1 N_1$

Similarly, length of belt that passes over the driven pulley in one minute = $\pi D_2 N_2$

Assuming thickness of belt to be negligible and no slip between the belt and pulley.

$$\omega_1 \times \frac{D_1}{2} = \omega_2 \times \frac{D_2}{2}$$

Hence, velocity of the belt $v = \omega_1 R_1 = \omega_2 R_2 =$

$$\frac{\omega_2}{\omega_1} = \frac{R_1}{R_2} = \frac{D_1}{D_2}$$

or
$$\frac{2\pi N_2}{2\pi N_1} = \frac{D_1}{D_2} \qquad \therefore \omega = 2\pi N$$

$$\frac{N_2}{N_1} = \frac{D_1}{D_2}$$

or
$$\frac{\text{Speed of the driven pulley}}{\text{Speed of the driver pulley}} = \frac{\text{Diameter of driver pulley}}{\text{Diameter of driven pulley}}$$

Thus, in a belt, speed of a pulley is inversely preportional to its diameter.

The ratio $\frac{N_2}{N_1}$ is called velocity ratio of the drive. If thickness of the belt is taken into account then

$$v = \omega_1 \times \frac{D_1 + t}{2} = \omega_2 \times \frac{D_2 + t}{2}$$

$$\frac{\omega_2}{\omega_1} = \frac{D_1 + t}{D_2 + t} \text{ or } \frac{2\pi N_2}{2\pi N_1} = \frac{D_1 + t}{D_2 + t}$$

$\therefore$ Velocity ratio $\frac{N_2}{N_1} = \frac{D_1 + t}{D_2 + t}$

6.4.6 Slip of the Belt

Considering slip between belt and the pulley. (Thickness is assumed to be as negligible). Slip is the relative motion between the pulley and belt. The different between the linear speeds of the pulley rim and belt is the measure of slip. Belt drive is not a positive drive because of the insufficient friction between pulley and belt. The slip is also expressed in percentage (%).

the result of belt slipping is to reduce the velocity ratio of the system.

Let S_1 = Percentage slip between driver and belt

S_2 = Percentage slip between belt and follower.

Linear velocity of driving pulley $v_1 = \omega_1 \times \frac{D_1}{2} = \omega_1 R_1$ $\qquad \omega = 2\pi n$

$$\text{Speed of the belt} = \omega_1 r_1 \left(\frac{100 - S_1}{100}\right)$$

Peripherial speed of the follower = $\omega_2 R_2$

$$= \left(\frac{100 - S_2}{100}\right) \times \text{Speed of the belt.}$$

$$\omega_2 R_2 = \left(\frac{100 - S_2}{100}\right) \times \omega_1 R_1 \left(\frac{100 - S_1}{100}\right)$$

$$\frac{\omega_2}{\omega_1} = \frac{N_2}{N_1}\left(\frac{100 - S_2}{100}\right) \times \left(\frac{100 - S_1}{100}\right) \times \frac{R_1}{R_2}$$

$$\frac{2\pi N_2}{2\pi N_1} = \left(\frac{100 - S_2}{100}\right) \times \left(\frac{100 - S_1}{100}\right) \times \frac{D_1}{D_2}$$

If S = Total slip between the driver and follower.

$$\frac{N_2}{N_1} = \left(\frac{100 - S}{100}\right) \times \frac{D_1}{D_2}$$

$$\frac{N_2}{N_1} = \left(\frac{100 - S}{100}\right) \times \frac{D_1}{D_2}$$

Considering slip as well as thickness of the belt

$$\boxed{\frac{N_2}{N_1} = \left(\frac{100 - S}{100}\right) \times \left\{\frac{D_1 + t}{D_2 + t}\right\}}$$

6.4.7 Creep of Belt

Creep : When the belt passes from the Slack side to tight side a certain portion of the belt extends, and when the belt passes from tight side to slack side of belt contracts. Due to these changes in length, there is a relative motion between belt and pulley surfaces. This Phenomena is termed as creep.

Example 6.1: An engine shaft running at 120 rpm. is required to drive a machine shaft by means of a belt. The pulley on the engine shaft is of 2 m diameter and that of machine shaft is of 1 m diameter of the belt thickness is 5 mm find the speed of the machine shaft when.

(*i*) There is no slip

(*ii*) There is a slip of 3 %

Solution :

Given Speed of the shaft N_1 = 120 rpm.

Diameter of the pulley on the engine shaft D_1 = 2 m

Diameter of the pulley on the machine shaft D_2 = 1 m

Thickness of the belt = 5 mm

Slip = 3 %

(*i*) Speed of the machine shaft

$$N_2 = N_1 \times \frac{D_1 + t}{D_2 + t}$$

$$= 120 \times \frac{2 + 0.005}{1 + 0.005}$$

$$= \mathbf{239.4\ rpm.}$$

(*ii*) Speed of the machine shaft when there is a slip of 3 %.

$$N_2 = N_1 \times \frac{D_1 + t}{D_2 + t}\left(1 - \frac{S}{100}\right)$$

$$= 120 \times \frac{2 + 0.005}{1 + .005}\left(\frac{1 - 3}{100}\right)$$

$$= \mathbf{232\ rpm.}$$

6.4.8 Length of an Open Belt Drive

Let O_1, O_2 = Centres of two pulleys

r_1, r_2 = Radius of larger and smaller pulley.

l = Distance between O_1 and O_2

$2d$ = Angle subtended.

The belt leaves the larger pulley at E and G the smaller pulley at F and H as shown in Fig. 6.16 given below. Through O_2 draw O_2 M parallel to EF we find that from the Fig. O_2M is perpendicular to O_1E.

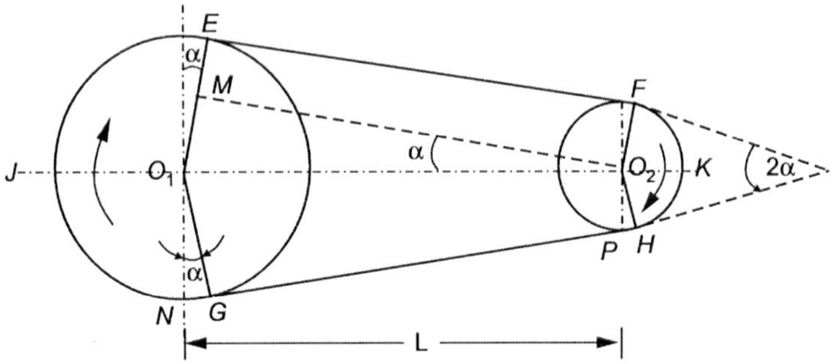

Fig. 6.16 Length of an open drive

$$\angle NO_1 G = \angle PO_2H = \angle O_1O_2M = \alpha$$

$\therefore$ $$\sin \alpha = \frac{r_1 - r_2}{l} \qquad ...(i)$$

Where α is the angle in radians. Which is very small

$$\text{Arc } JE = r_1\left(\frac{\pi}{2} + \alpha\right) \qquad ...(ii)$$

$$\text{Similarly Arc } FK = r_2\left(\frac{\pi}{2} - \alpha\right) \qquad ...(iii)$$

$$\text{Length } EF = \text{Length } MO_2 = \sqrt{l_2^2 - (r_1 - r_2)^2} = l\sqrt{I - \frac{(r_1 - r_2)^2}{l}}$$

Since $\left(\frac{r_1 - r_2}{l}\right)^2$ is very small,

$$\text{Then through Binomial expression } EF = l\left[1 - \frac{1}{2}\left(\frac{r_1 - r_2}{l}\right)^2 + ...\right]$$

$$= l - \frac{(r_1 - r_2)^2}{2l} \qquad ...(iv)$$

Hence, the length of belt

$$L = \text{Arc } GJE + EF + \text{Arc } FKH + HG = 2(\text{Arc } JE + EF + \text{Arc } FK)$$

Substituting the values of Arc JE from equation (*ii*) and value of FK from equation (*iii*) and values EF from equation (*iv*).

$$L = 2\left[r_1\left(\frac{\pi}{2} + \alpha\right) + l - \frac{(r_1 - r_2)^2}{2l} + r_2\left(\frac{\pi}{2} - \alpha\right)\right]$$

$$= 2\left[r_1\frac{\pi}{2} + r_1\alpha + l - \frac{(r_1 - r_2)^2}{2l} + r_2\frac{\pi}{2} - r_2\alpha\right]$$

$$= 2\left[\frac{\pi}{2}(r_1 + r_2) + \alpha(r_1 - r_2) + l - \frac{(r_1 - r_2)^2}{2l}\right]$$

$$= \left[\pi(r_1 + r_2) + 2\alpha(r_1 - r_2) + 2l - \frac{(r_1 - r_2)^2}{2}\right]$$

Substituting the value of $\alpha = \frac{(r_1 - r_2)}{l}$ from equation (*i*)

$$L = \left[\pi(r_1 + r_2) + 2\times\frac{(r_1 - r_2)}{l}(r_1 - r_2) + 2l - \frac{(r_1 - r_2)^2}{l}\right]$$

$$= \left[\pi(r_1 + r_2) + 2\frac{(r_1 - r_2)^2}{l} + \frac{2l - (r_1 - r_2)^2}{2}\right]$$

$$= \left[\pi(r_1 + r_2) + 2l + \frac{(r_1 - r_2)^2}{l}\right]$$

$$L = \left[\pi(r_1 + r_2) + 2l + \frac{(r_1 - r_2)^2}{l}\right]$$

Hence, the Length of the open belt

$$L = \pi(r_1 + r_2) + \frac{(r_1 - r_2)^2}{l} + 2l$$

Length of Cross Belt Drive

In case of cross belt drive both the pulleys rotate in opposite direction and angle of contact between belt and pulley is also large.

Let O_1 and O_2 = Centre of the two pulleys

r_1 and r_2 = Radius of the larger and smaller pulleys

l = Centre distance between O_1 and O_2

The belt leaves the larger pulley at *E* and *G* and the smaller pulley at *F* and *H* as shown in the figure given below. Through O_2 draw O_2M Parallel to *FF* and from Fig.6.17 O_2M is perpendicular to O_1E.

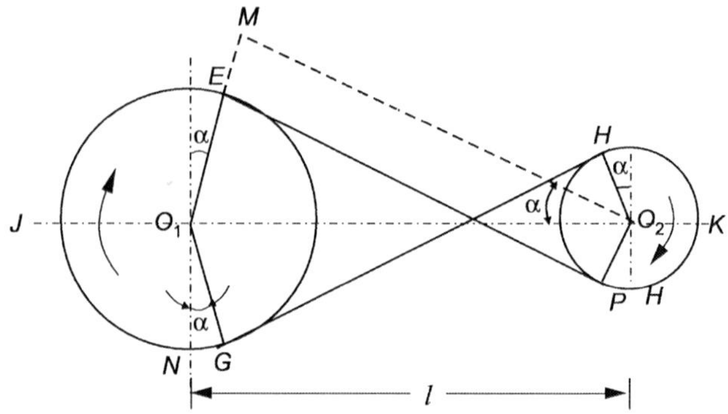

Fig.6.17 Length of a cross belt drive

$\therefore \quad \sin\alpha = \dfrac{r_1 + r_2}{l}$

Since the angle α is very small $\quad \therefore \sin\alpha = \alpha$

Hence $\quad \sin\alpha = \dfrac{r_1 + r_2}{l} \quad$ ...(*i*)

$$\text{Arc } JE = r_1\left(\frac{\pi}{2} + \alpha\right) \quad \text{...(ii)}$$

Simillarly $\text{Arc } FK = r_2\left(\dfrac{\pi}{2} + \alpha\right) \quad$ ...(*iii*)

$$EF = MO_2 = \sqrt{l^2 - (r_1 + r_2)^2} = l\sqrt{1 - \frac{(r_1 - r_2)}{l}}$$

Since $\dfrac{(r_1 - r_2)^2}{l}$ is very small, then through Binomial expression

$$\text{EF} = l\left[1 - \frac{1}{2}\frac{(r_1 + r_2)^2}{l} + \ldots\right] = l - \frac{(r_1 + r_2)^2}{2l}$$

$$= l - \frac{(r_1 + r_2)^2}{2l} \quad \text{...(iv)}$$

Hence, length of belt L = Arc *GJE* + *EF* + Arc *FKH* + *HG*

= 2(Arc *JE* + *EF* + Arc *FK*)

Substituting the values of Arc *JE* from equation (*ii*)

Arc *FK* from equation (*iii*) and *EF* from equation (*iv*) then

So $$L = 2\left[r_1\left(\frac{\pi}{2}+\alpha\right)+l-\frac{(r_1+r_2)^2}{2l}+r_2\left(\frac{\pi}{2}+\alpha\right)\right]$$

$$= 2\left[r_1\frac{\pi}{2}+r_1\alpha+l-\frac{(r_1+r_2)^2}{2l}+r_2\frac{\pi}{2}+r_2\alpha\right]$$

$$= 2\left[\frac{\pi}{2}(r_1+r_2)+\alpha(r_1+r_2)+l-\frac{(r_1+r_2)^2}{2l}\right]$$

$$= \left[\pi(r_1+r_2)+2\alpha(r_1+r_2)+2l-\frac{(r_1+r_2)^2}{l}\right]$$

Now substituting the values of $\alpha = \left(\frac{r_1+r_2}{l}\right)$ from equation (*i*)

$$L = \left[\pi(r_1+r_2)+2\frac{(r_1+r_2)}{l}(r_1+r_2)+2l-\frac{(r_1+r_2)^2}{l}\right]$$

$$= \left[\pi(r_1+r_2)+2\frac{(r_1+r_2)^2}{l}+2l-\frac{(r_1+r_2)^2}{l}\right]$$

$$= \left[\pi(r_1+r_2)+\frac{(r_1+r_2)^2}{l}+2l\right]$$

$$L = \left[\pi(r_1+r_2)+\frac{(r_1+r_2)^2}{l}+2l\right]$$

6.4.9 Power Transmitted by a Belt

In the arrangement of driver and driven pulley, the driving pulley pulls the belt from one side and delivers the same to the other side.

Hence, the tension in tight side will be more than the tension in slack side.

2et T_1 = Tension in Tight side

T_2 = Tension in slack side

v = Velocity of the belt, m/s

Effective Turning force = $T_1 - T_2$

∴ workdone = Force × distance

$= (T_1 - T_2) \times v \quad N - \text{m/s} \qquad V = \text{Distance/time}$

Hence $\quad \text{Power} = \dfrac{(T_1 - T_2)}{1000} \times v \quad$ K.watt

Determination of Angle of Contact

When the two pulleys are connected by a belt of unequal size, the angle of must be taken as minimun angle of lap or contact.

For open belt = $\cos \theta/2 = \dfrac{r_2 - r_1}{l}$

where θ = Angle of lap of smaller pulley = π–2α

$$\theta = 2\cos^{-}\left[\frac{(r_2 - r_1)}{l}\right]$$

For cross belt = $\cos(\pi - \theta/2) = \dfrac{r_1 + r_2}{l}$

$$\theta = 2\left[\pi - \cos^{-1}\frac{(r_1 + r_2)}{l}\right]$$

where θ = angle of lap (in radian)

Hence $\theta = (\pi + 2\alpha)$

6.4.10 Ratio of Tensions

Let us consider a belt resting over the flat rim of a pulley which is not allowed to rotate. The tensions T_1 and T_2 are such that the motion is impending between the belt and pulley. Consider a driven pulley rotates in a clockwise direction, in this case, tension T_1 is greater than T_2 on account of frictional resistance between belt and rim. Hence it is required to establish a relation between T_1, T_2, μ_1 and θ.

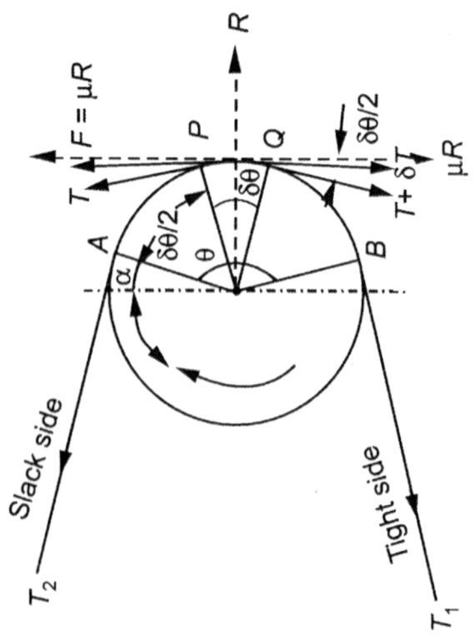

Fig. 6.18 Ratio of tensions

Let T_1 = Tension in the belt on tight side

T_2 = Tension in the belt on slack side.

θ = Angle of contact in radiane (with belt and pulley)

μ = Coefficient of friction between belt and pulley

Considering a small portion of belt *PQ* which makes an angled θ motion as the centre of pulley when *T* and $(T + \delta T)$ at the tensions at the extremities *PQ*.

The belt portion *A* is in equibrium under the action of following forces:

1. Tension in the belt at *P*
2. Tension $T + \delta T$ in the belt at θ
3. Normal Reaction *R*
4. Frictional Force $F = \mu R$

Resolving the forces horizontally.

$$R = (T + \delta T) \sin \frac{\delta\theta}{2} + T \sin \frac{\delta\theta}{2} \qquad ...(i)$$

$$R = \mathrm{T} \sin \frac{\delta\theta}{2} + \delta T \sin \frac{\delta\theta}{2} + T \sin \frac{\delta\theta}{2} \quad 2T \sin \delta\theta/2$$

$$\sin\left[\frac{\alpha\theta}{2} = \frac{\alpha\theta}{2}\right]$$

$$R = (T + \delta T) \frac{\delta\theta}{2} + T \frac{\delta\theta}{2}$$

$$= \frac{T.\delta\theta}{2} + \frac{\delta T.\delta\theta}{2} + \frac{T.\delta\theta}{2} \qquad \text{Neglecting } \delta T\ \delta\theta/2$$

$$= T.\delta\theta \qquad ...(ii)$$

$$\mu R = (T + \delta T) \cos \frac{\delta\theta}{2} \quad T \cos \frac{\delta\theta}{2} \qquad ...(iii)$$

Since the angle $\delta\theta$ is very small

Hence $\cos \frac{\delta\theta}{2} = 1$

$$\mu R = T + \delta T - T = \delta T \qquad ...(iv)$$

$$\boxed{R = \frac{\delta T}{\mu}}$$

Now equating the value of R in equations (*iv*) and (*ii*)

$$T.\,\delta\theta = \frac{\delta T}{\mu}$$

$$T\,.\,\delta\theta = \frac{\delta T}{\mu}$$

$$\delta T = \mu\,\delta\theta$$

On integrating we get

$$\int_{T_2}^{T_1} \mu \frac{\delta T}{T} = \mu \int_0^{\theta} \delta\theta$$

$$\text{loge}\left(\frac{T_1}{T_2}\right) = \mu\theta$$

$$\frac{T_1}{T_2} = e^{\mu\theta}$$

$$\boxed{\frac{T_1}{T_2} = e^{\mu\theta}}$$

6.4.11 Centrifugal Tension

Whenever belt rotates in a circular path around the pulleys, because of its having mass, and speed, it is subjected to centrifugal force, which tends to lift the lift from the rim by reducing the normal reaction. It causes the reduction in driving power capacity. The centrifugal force produces additional tension which is known as centrifugal tension.

Consider small portion L.M. of the belt as shown in Fig. 6.19.

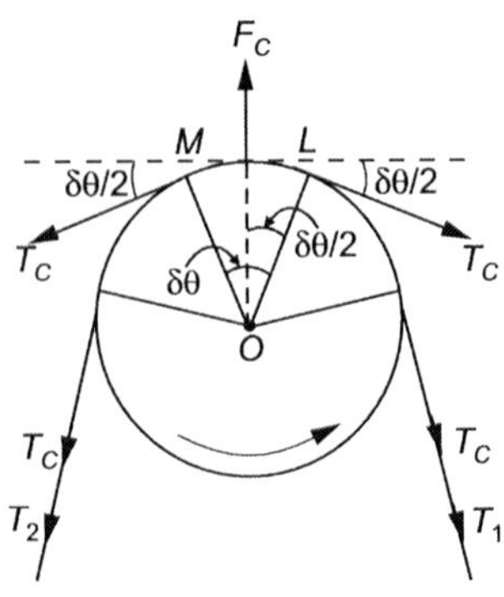

Fig. 6.19 Centrifugal tension

Let m = Mass of the belt per unit length

v = Linear velocity of the belt .

r = Radius of the pulley over which the belt runs.

T_c = Centrifugal tension acting tangentially at L and m

$\delta\theta$ = Angle subtended by the belt LM at the centre of the pulley.

$\therefore$ Length of belt $LM = r.\delta\theta$

and mass $m = m.r.\delta\theta$

We know that centrifugal force of the belt LM

$$F_c = \frac{m.v^2}{r} = \frac{(m.r.\delta\theta)}{r_2}v^2 = m.\delta\theta.v^2$$

Now Resolving the forces (*i.e.*, F_c and T_c) vertically and equating the same.

$$2T_c \sin\left(\frac{\delta\theta}{2}\right) = m.\delta\theta.v^2$$

Since $\delta\theta$ is very small, therefore substituting

$$\sin\left(\frac{\delta\theta}{2}\right) = \frac{\delta\theta}{2}$$

Putting the value in above equation.

$$2T_c\left[\frac{\delta\theta}{2}\right] = m.\delta\theta.v^2$$

$$T_c = mv^2$$

From the above expression. The centrifugal tension in the elementary strip is independent of the angle of lap, orignal tension. Hence, it is good for the entire belt.

When centifugal tension taken in account

Total tension in slack side = $T_1 + T_c$

Total tension in slack side = $T_2 + T_c$

Power transmitted $[(T_1 + T_c) - (T_2 + T_c)] \times v = (T_1 - T_2)\, v$

6.4.12 Condition for Transmission of Maximum Power

From the relation $T_c = mv^2$ and $(T_1 - T_2)v$. It is evident that both the centrifugal tension and power transmitted increase with the increase in belt speed but the value of T_1 and consequently the value of $(T_1 - T_2)$ diminishes. Hence the power transmitted increases with the increase of speed upto a certain value of velocity and after which it decreases.

Let T = Maximum permissible tension in a given belt.

T_1 = Tension on the tight side of the belt.

T_2 = Tension on the slack side of the belt.

T_c = Centrifugal tension.

Power Transmitted $P = (T_1 - T_2)v = T_1$ $\qquad T_1\left[1 - \frac{T_2}{T_1}\right]v$

$$= T_1\left[1 - \frac{1}{e^{-\mu\theta}}\right]v \qquad \left[\because \frac{\pi}{T_2} = e^{\mu\theta} \quad \frac{T_2}{\pi} = \frac{1}{e^{\mu\theta}}\right]$$

$$T = T_1 + T_c$$

$$T_1 = T - T_c$$

$$P = (T - T_c)\left[1 - \frac{1}{e^{\mu\theta}}\right]$$

$$= (T - mv^2)\left(1\frac{1}{e^{\mu\theta}}\right)v$$

$$= (Tv - mv^3)\left[1 - \frac{1}{e^{\mu\theta}}\right]$$

$$= (Tv - mv^3) \times K \qquad \text{where } K = \left[1 - \frac{1}{e^{\mu\theta}}\right]$$

$$= (Tv - mv^3) \times K$$

Differentiating p wrt. v and equating to zero

$$\frac{\delta p}{\delta v} = (T - 3mv^2) = 0 \qquad \because T_c = mv^2$$

$$T = 3mv^2 = 3T_c$$

$$v = \sqrt{\frac{T}{3m}}$$

Hence, the power transmission shall be maximum

When

1. The centrifugal tension is $\frac{1}{3}$rd of the belt strength

2. The belt is run at the velocity $\sqrt{\frac{T}{3m}}$.

6.5 V BELT DRIVE

When a *V* belt runs in a groove pulley.

Let 2α = Angle of groove

N = Normal reaction between belt and sides of the *V* groove pulley.

μ = Co-efficient of friction between belt and pulley.

R = Total Reaction in the plane of groove.

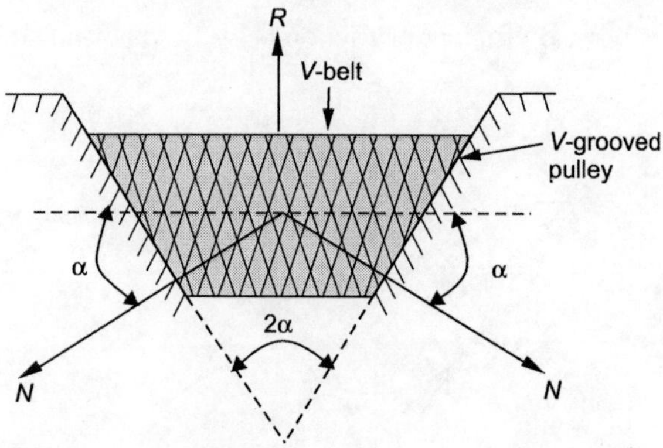

Fig. 6.20 *V* belt drive

Now resolving the forces vertically.

$$R = N \sin \alpha + \text{N} \sin \alpha$$

$$= 2\text{N} \sin \alpha$$

$$\therefore \quad N = \frac{R}{2 \sin \alpha} = \frac{R}{2} \text{ cosec } \alpha$$

$\therefore$ Frictional resistance $= \mu N + \mu N = 2\mu N$

$$= 2\mu + \frac{R}{2} \text{ cosec } \alpha$$

$$= \mu R \text{ cosec } \alpha \qquad \left[\because N = \frac{R}{2} \text{cosec } \alpha\right]$$

In case of flat belt ratio of tension in belt

$$\frac{T_1}{T_2} = e^{\mu\theta} \text{ and frictional force} = \mu R.$$

But in case of *V* belt the frictional force = (μ cosec $\alpha \times R$)

$$\therefore \quad \text{For } V \text{ belt } \frac{T_1}{T_2} \; e^{(\mu \text{ cosec } \alpha) \times \theta}$$

Hence,

$$\boxed{\frac{T_1}{T_2} = e^{\mu\theta \text{ cosec } \alpha}}$$

6.6 ROPE DRIVE

Rope drive is generally used when the distance between the driving shaft and driven shaft is large. Ropes are generally of circular section and the frictional grip between the pulley and rope is more than the belt grip.

Let N = Normal reaction between rope and side of the groove.

R^2 = Total reaction in the plane.

2α = Angle of groove.

μ = Co-efficient of friction between rope and sides of groove.

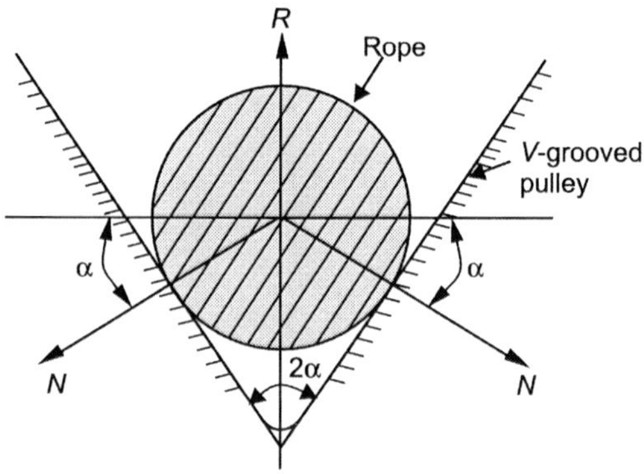

Fig. 6.21 Rope drive

Now resolving to forces vertically.

$$R = N \sin \alpha + N \sin \alpha$$

$$= 2 N \sin \alpha$$

$$N = \frac{R}{2 \sin \alpha} = \frac{R}{2} \text{ cosec } \alpha$$

$\therefore$ Frictional resistance $= \mu N + \mu N = 2\mu N$

$$= 2\mu + \frac{R}{2} \text{ cosec } \alpha$$

$$= \mu R \text{ cosec } \alpha \qquad \because N = \frac{R}{2} \text{ cosec } \alpha$$

Hence, ratio of tension in this case will also be same as in case of V belt.

$$\frac{T_1}{T_2} = e^{\mu\theta \text{ cosec } \alpha}$$

6.6.1 Advantages of Rope Drive

1. The rope drive is particularly suitable, when the distance between the shafts is large.
2. The friction grip in case of rope drive is more than that in belt drive.
3. The net driving tension in case of rope drive is more than the belt drive.

6.7 CHAIN DRIVE

Chain drive is a medium of power transmission having links connected by joints in order to product flexibility for wrapping driving and driven wheels. The chain has many special features like constant velocity ratio, creep free or no slippage (*i.e.*, long life and can drive a number of shafts from single source of power). It is generally used in combination with sprocketwheel.

Uses:

1. Power transmission for medium centre to centre distance.
2. To drive cycle, motor cycle and automobiles.
3. To drive agriculture machinery
4. Heavy Earth moving machinery.
5. Oil well drilling rigs.

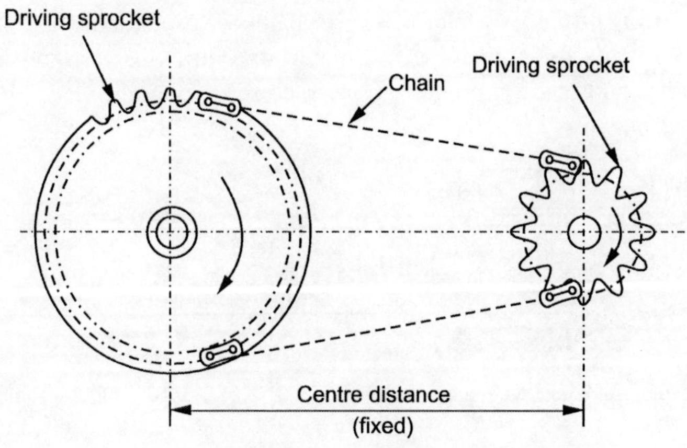

Fig. 6.22 Chain drive

6.7.1 Classification of Chains

The chains are classified according to their uses :

1. Power transmission chains
2. Hosting and having chains
3. Elevating or convey chains

1. **Power Transmission Chain:** These chains are used for high speed power transmission from shaft to another shaft upto 13 m/s. These are strong and simple in construction as well required little lubrication.

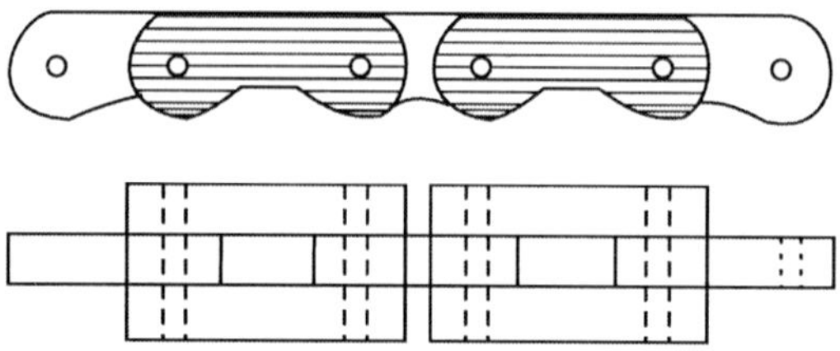

Fig.6.23 Block chain

2. **Hosting and Having Chains:** These chains are used for low speeds having a oval or square in shape with low manufacturing cost are used for hoisting and marine work.

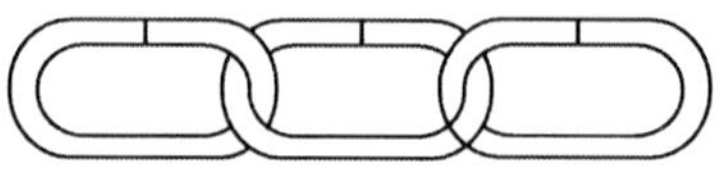

(*a*) Chain with oval links

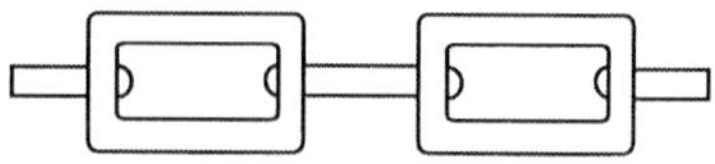

(*b*) Chain with square link

Fig.6.24 Hosting and having chains

3. **Elevating or Convey Chain:** These chains are used for average slow speeds of 1.75 m/s for conveying the material continuously from one point to another and made from hook type or closed type joint.

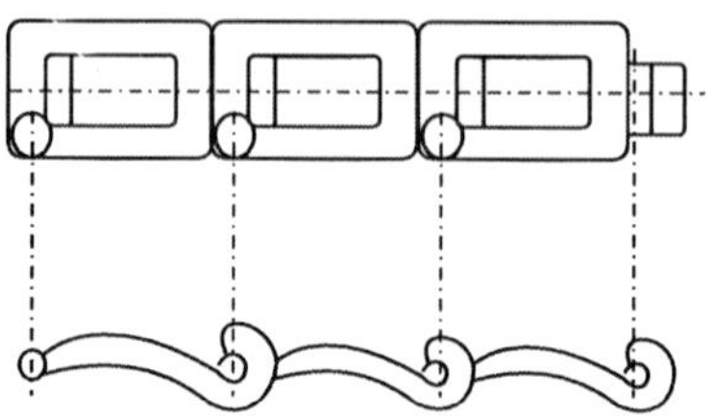

(*a*) Detachable or hook joint type chain

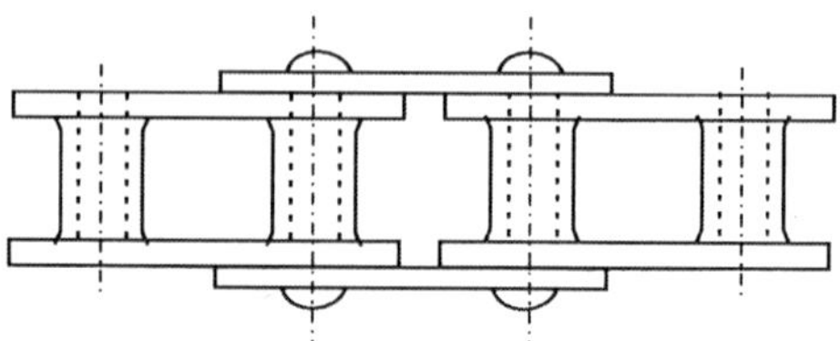

(*b*) Closed joint type chain

Fig.6.25 Elevating or convey chain

6.7.2 Advantages and Disadvantages of Chain Drive

Advantages

1. It maintains a constant velocity ratio.
2. Heigher transmission efficiency
3. It is more durable.

4. Effective transmission of power
5. Compact as compare to belt drive
6. It can carry heavy loads.
7. Drive is positive.
8. Smaller size than a belt drive.
9. It can be used when the disatnace between the shafts is less.
10. It can be operated at higher temparature and adverse atmospheric conditions.

Disadvantages

1. Operation is noisy.
2. Cost is comparatively high.
3. It requires housing.
4. It is complicated in design.
5. Need accurate mounting and careful maintenance.
6. The chain gets stretched and requires a tensioning device.

6.7.3 Comparision between Chain and Belt Drive

Chain drive	Belt drive
1. It is used for shorter distances.	1. Used for longer distance
2. Velocity ratio is constant.	2. Velocity ratio is variable due to slippage.
3. Lubrication is required.	3. Lubrication is avoided.
4. Flexible connection, mesh drive.	4. Flexible connection, friction drive
5. It requires precise alignment of shaft.	5. Slight misalignment is allowed.
6. Relatively high cost.	6. Relatively cheaper.

6.8 GEAR DRIVE

Gear is a toothed wheel which meshes with the another wheel to give the positive transmission of motion from one component to another either they are parallel intersecting or angular in position. These are also widely used to transmit power when the change in speed or change in direction is desired. They are also used to transmit moderate or large power over shorter distances without slippage as a positive drive with constant velocity ratio. The gears are always operate in pairs, the smaller gear is called pinion and larger one is called gear and combination of two called gear drive.

Uses:

In automobile, differentials, transmission gear boxes, Metal cutting machine tools, rolling mills, twisting machinery, marine engine and construction machinery etc.

6.8.1 Definitions

1. **Pitch Circle:** It is an imaginary circle which would transmit the same motion as the actual gear, by pure rolling action.
2. **Addendum Circle:** A circle concentric with the pitch circle and bounding the outer end to the teeth is called an addendum circle.

   The diameter of the addendum circle is known as addendum circle diameter.
3. **Addendum:** It is radial distance between the pitch circle and addendum circle.
4. **Dedendum (or root) Circle:** It is a circle concentric with the pitch circle and bounding the bottom of the tooth.
5. **Dedendum:** It is the radial distance between the pitch circle and the dedendum circle.
6. **Clearance:** The difference between the dedendum of (one gear) and addendum (of the mating gear) is called as clearance.

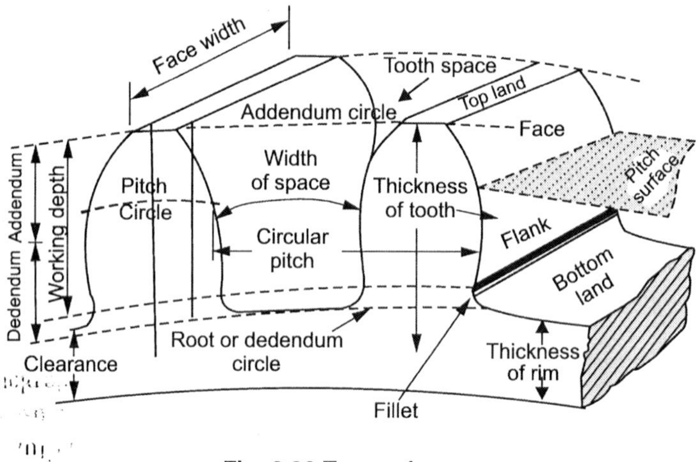

Fig. 6.26 Terms of gears

7. **Working Depth:** It is the sum of the addenda of the two mating gears.
8. **Circular Thickness (or Thickness of Tooth):** The length of arc between the sides of a gear tooth, measured on the pitch circle is known as circular thickness or thickness of tooth.
9. **Tooth Space:** It is the width of the recess between two adjacent tooth measured along pitch circle.
10. **Backlash:** It is the difference between the tooth space and the tooth thickness.
11. **Face:** It the action or working surface of the addendum
12. **Flank:** The working face of the dedendum is called the flank.
13. **Top Land:** It is the surface of the top of the tooth.

14. **Bottom Land:** It is the surface of the bottom of the tooth space.
15. **Whole Depth:** It is the total depth of the tooth space, equal to addendum plus dedendum; also it is equal to the working depth plus clearance.
16. **Tooth Fillet:** It is the radius which connects the root circle to the tooth profile.
17. **Circular Pitch:** The distance measured along the pitch circle from a point on one tooth to the corresponding point on an adjacent tooth is called circular pitch. It is represented by P_c.

$$P_c = \frac{\pi D}{T}$$

where D = pitch diameter, T = number of teeth.

18. **Pitch Diameter:** It is the diameter of a pitch circle. It is usually represented by d_p or d_f for pinion and gear respectively.
19. **Diameteral Pitch:** Number of teeth on a wheel per unit of its pitch diameter is called the diameteral pitch. It is denoted by

$$P_d = \frac{T}{D}$$

From equation of circular pitch and diameteral pitch we have

$$P_c P_d = \pi$$

20. **Module:** It is the reverse of the diameteral pitch. Ratio between the pitch diameter and the number of teeth is known module, it is denoted by m.

$$m = \frac{D}{T}$$

6.8.2 Advantages and Disadvantages of Gear Drive

Advantages:

1. Positive drive with constant velocity ratio.
2. It can operate at high speeds as well as low speed.
3. They can transmit large power beyond the range of chain, belt and rope drive.
4. Long service life.
5. More compact and reliable.
6. High efficiency.
7. Can be used where precise timing is required.
8. High torque can be transmitted.

Disadvantages:

1. Noisy in operation at considerable speeds.
2. High manufacturing cost due to special equipments
3. Costly drive.
4. Higher maintenance cost.
5. They require precise alignment of shaft.
6. They require more lubrication and cleanliness.

6.8.3 Types of Gears

1. Spur gear
2. Helical gear
3. Bevel gear
4. Spiral gear
5. Worm gear
6. Rack and pinion
7. Internal and external gear

Spur Gear

Spur gears are cylindrical in shape and has its teeth parallel to the axis of the shaft. They are used to transmit motion between two shafts whose axis are parallel and coplaner. A high efficiency upto 98% can be achived in power transmission. Spur gears are more noisy in operation, wear out at faster rate and due to which develop backlash in system. They can be used at any speed.

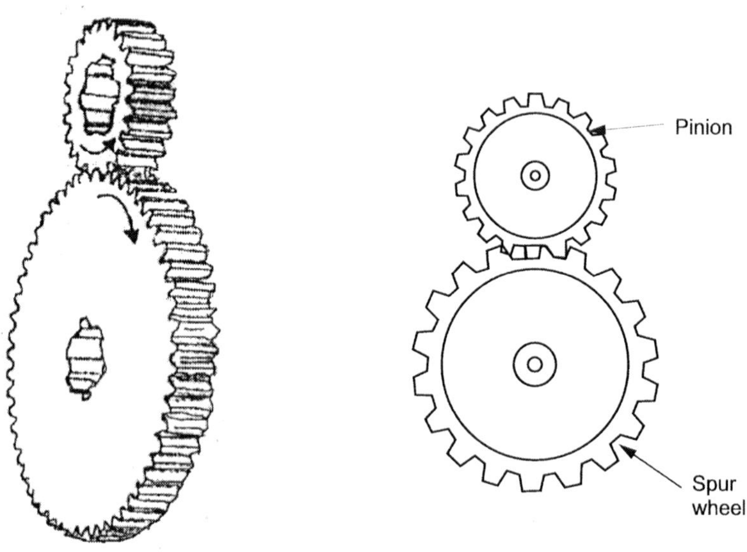

Fig. 6.27 Spur gear

Helical Gear

These are cylindrical gears whose tooth trace are straight helix and teeth are inclined at an angle to the gear axis. These gears are as usually known as high speed gears and can take higher loads than the spur gears. The motion of drive is smooth and quick.

Uses : Gear boxes, crane, transmission system, rolling mill etc.

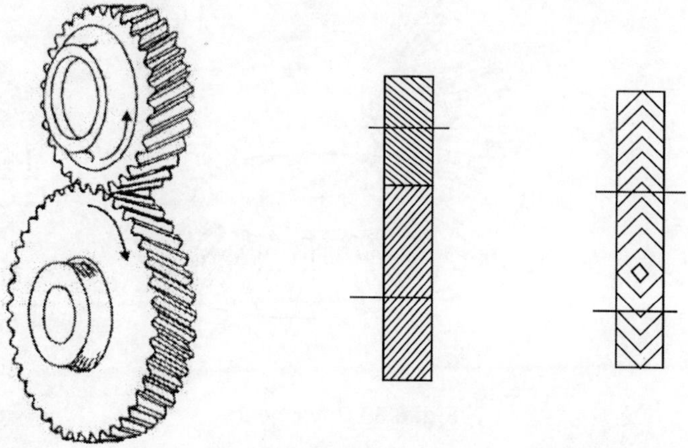

Fig. 6.28 Helical gear

Herringbone Gear

These are also known as double helical gears. The two different teeth may be continuous or separated by a gap. They are formed by joining two helical gears of identical pitch and diameter of opposite hand on the same shaft. Lateral thrust can be eliminated by using double helical gears.

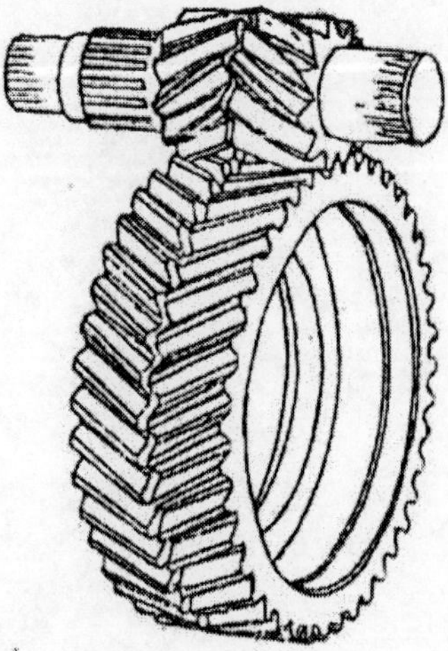

Fig. 6.29 Herringbone gear

Bevel Gear

These gears are used to transmit motion between two intersecting and coplaner shafts and inclination may be at any angle from 0° - 180° of the shafts are at right angle and wheels in equal size called mitre gear and if it is inclined to the face then it is known as helical bevel gear.

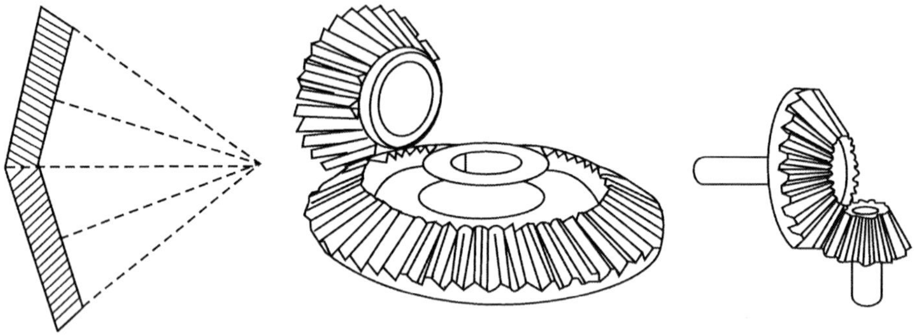

Fig. 6.30 Bevel gear

Spiral Gear

Spiral gears are also called cross helical gears. These gears have a point contact rather than a line contact. These are used to transmit power between non-parallel and non-intersecting shafts. They cannot transmit heavy loads.

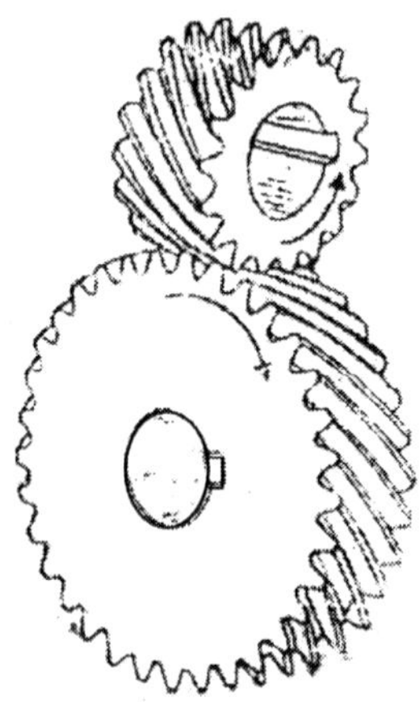

Fig. 6.31 Sprial gear

Worm Gear

A worm is basically a part of screw worm gear connect two non-parallel, non-intersecting shafts which are usually at right angles.

This gearing system is smooth and quiet in operation and provide a high gear ratio. This system of gears reecommended when speed reduces is more than 10 : 1

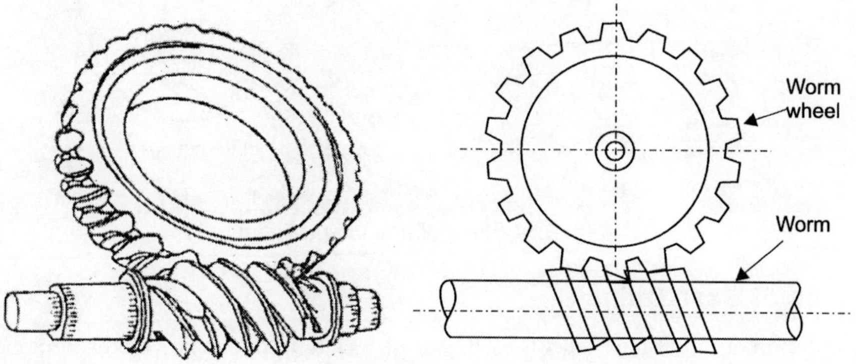

Fig. 6.32 Worm and worm wheel

Rack and Pinion

Rack is a spur gear of infinite diameter. It meshes with the both internally and externally geares. That is called pinion. Rack and pinon arrangement use to convert rotary motion into rotary or rectilinear motion.

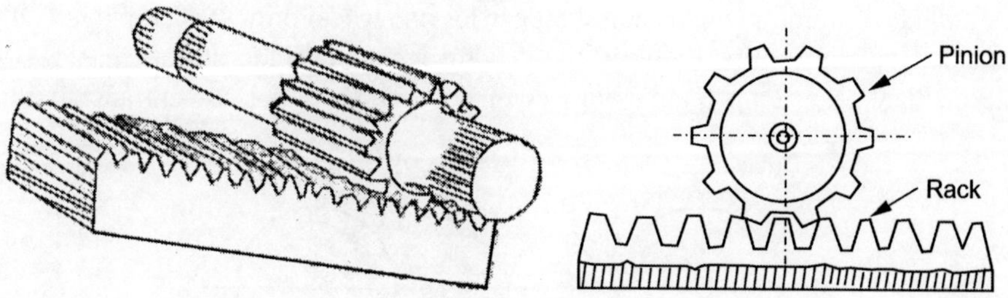

Fig. 6.33 Rack and pinion

Internal and External Gearing

Set of two toothed wheel is on parallel shafts may have gearing arrangement either externally or internally.

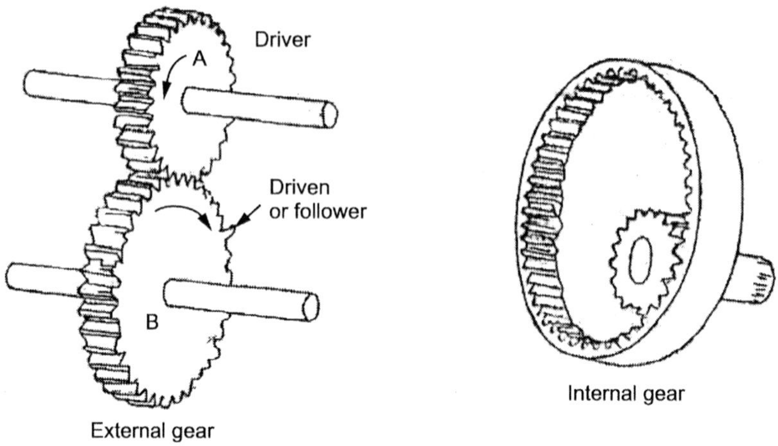

Fig. 6.34 External and Internal gear

Types of Gear Trains

The combination of gear wheels by means of which motion is powered and motion is transmitted from one shaft to another shaft called gear train.

Various trains are as follows:

1. Simple gear train
2. Compound gear train
3. Epicyclic gear train
4. Reverted gear train

6.9 SIMPLE GEAR TRAIN

A simple gear train in which each shaft carries one wheel only. It is employed only for small velocity ratio. Between the driving gear 1 and the driven gear 4, two intermediate gears 2 and 3 are employed on co-axial shafts for power transmission.

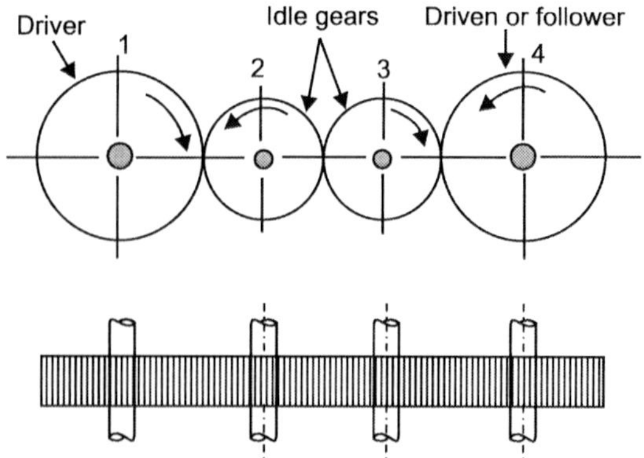

Fig. 6.35 Simple gear train

Let N_1 = Speed of the driver in r.p.m.

N_2 = Speed of the idle gear 2 in r.p.m.

N_3 = Speed of the idle gear 3 in r.p.m.

N_4 = Speed of the driven gear in r.p.m.

Similarly T_1, T_2, T_3 and T_4 be the number of teeth on gears 1, 2, 3, 4 respectively. and D_1, D_2, D_3 and D_4 be the pitch diameter of gears 1, 2, 3 and 4 respectively. The driving gear 1 meshes with the intermeadiate gear 2.

Hence $\pi D_1 N_1 = \pi D_2 N_2$

$$\frac{N_1}{N_2} = \frac{D_2}{D_1}$$

diameteral pitch of gear 1 = diameteral pitch of gear 2

Hence, the speed ratio

$$\text{for gears 1 and 2} = \frac{N_1}{N_2} = \frac{T_2}{T_1} \qquad ...(i)$$

Simillarly the speed ratio for (*i.e.* velocity ratio) gears 2 and 3.

$$\frac{N_2}{N_3} = \frac{T_3}{T_2} \qquad ...(ii)$$

Further the speed ratio (velocity ratio) for gears 3 and 4.

$$\frac{N_3}{N_4} = \frac{T_4}{T_3} \qquad ...(iii)$$

Therefore from equations (*i*), (*ii*) and (*iii*) we get speed ratio or

$$\text{Velocity ratio} = \frac{N_1}{N_2} \times \frac{N_2}{N_3} \times \frac{N_3}{N_4} = \frac{T_2}{T_1} \times \frac{T_2}{T_2} \times \frac{T_4}{T_3}$$

$$\frac{N_1}{N_4} = \frac{T_4}{T_1}$$

$$\text{Velocity ratio} = \frac{\text{Speed of driving wheel}}{\text{Speed of driven wheel}} = \frac{\text{No. of teeth on driven wheel}}{\text{No. of teeth on driving wheel}}$$

Train value = It is the reciprocal of velocity ratio.

$$= \frac{\text{Speed of driven}}{\text{Speed of driven}} = \frac{\text{No. of teeth on drives}}{\text{No. of teeth on driven}}$$

Note: The velocity ratio is independent of the intermediate gears 2 and 3. The gears 2 and 3 are called idlers.

If the gears have external teeth. The rotation of first and last gear will be the same (in case of intermediate odd gear)

The rotation of the driver and follower will be in opposite direction (in case of idlers even gear).

Compound Gear Train

A compound gear train is that in which two or more gear pairs arranged in series, at least one shaft should have two gear mounted on it. These gear trains are used for high velocity ratio. In this gear train gear is the driver for gear 2 and gear 3, and 5 are drivers for gears 4 and 6 respectively.

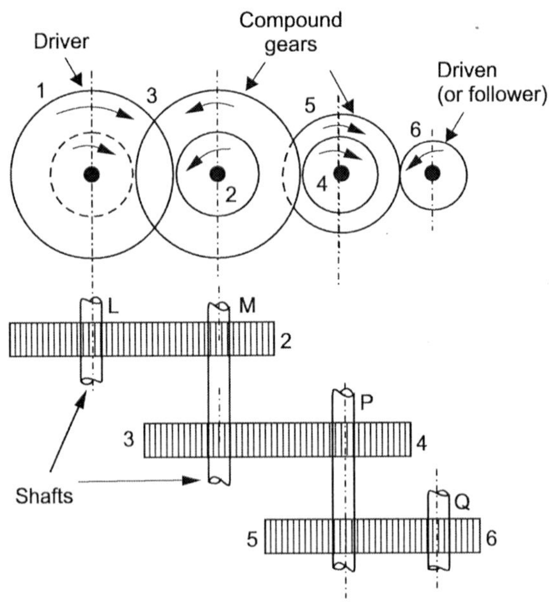

Fig. 6.36 Compound gear train

Let

$N_1, N_2, N_3, N_4, N_5, N_6$ be the speed of the respective gear.

and $T_1, T_2, T_3, T_4, T_5, T_6$ are the number of teeth on respective gears.

$$\text{Velocity ratio of gear 1 and gear 2} = \frac{N_1}{N_2} = \frac{T_2}{T_1} \quad ...(i)$$

$$\text{Similarly velocity ratio for gear 3 and gear 4} = \frac{N_3}{N_4} = \frac{T_4}{T_3} \quad ...(ii)$$

$$\text{Velocity ratio for gear 5 and gear 6} = \frac{N_5}{N_6} = \frac{T_6}{T_5} \quad ...(iii)$$

Therefore from equations (*i*), (*ii*) and (*iii*)

$$\frac{N_1}{N_2} \times \frac{N_3}{N_4} \times \frac{N_5}{N_6} = \frac{T_2}{T_1} \times \frac{T_4}{T_3} \times \frac{T_6}{T_5}$$

$$\text{Hence, velocity ratio} = \frac{\text{Speed of the first driver}}{\text{Speed of the last follower}} = \frac{\text{No. of teeth on drivens}}{\text{No. of teeth on drivers}}$$

Hence, a compound gear train helps to obtain large speed reduction from first to the last shaft with an assembly of small gear. Compound gear train is employed when the ratio 8 : 1 for speed reduction is required.

Epicyclic Gear Train

In a epicyclic gear train one gear rotates over another gear is called epicyclic path in the epicylictra in the axis of rotation of one or more of the wheel is carried on an arm and this arm is free to rotate about the axis of rotation of one or other wheel in train. The basic component of an epicyclic gear train is shown in Fig.6.37 as given below. It is generally useful for transmitting very high velocity ratio with gears of moderate size. It is used in workshop and electrical machines gear boxes for automobile.

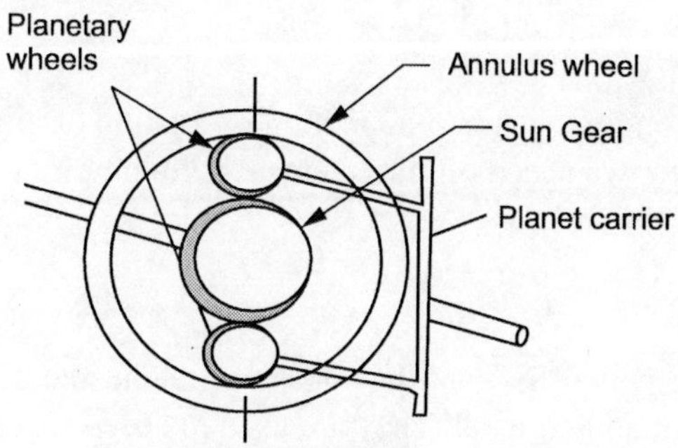

Epicyclic gear train

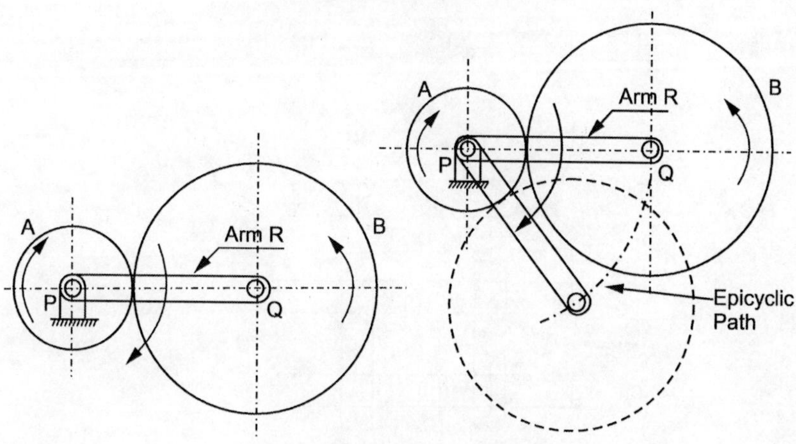

Fig. 6.37 Epicyclic gear train

A gear A has fixed axis P and gear A and Arm R both rotate about axis P. The gear B rotates about an axis and about arm R. This causes an epicyclic path. For velocity ratio of epicyclic train.

Tabulated Method

S.N.	Condition	Relation of elements		
		Arm R	**Wheel A**	**Wheel B**
1.	Arm fixed and when A rotates through + 1 rev.	0	+1	$-\frac{TA}{TB}$
2.	Arm fixed and wheel A rotates through + x rev	0	+ x	$-x\frac{TA}{TB}$
3.	Add y revolutions to all elements	+ y	+ y	$+y$
4.	Total motions (2 + 3)	y	$x + y$	$y - x.\frac{TA}{TB}$

$$N_R = +y$$

$$N_A = x + y$$

$$N_B = y - x\,\frac{TA}{TB}$$

when there is two conditions about the motion of rotation of any two elements are known then the unknown speed of third element is found by substituting the given data in 3rd column and 3rd row.

Reverted Gear Train

It is a gear train in which first and last gear turn about the same axis. The axis should be coincident and co-axial. They are used in speed reversing in head stock of lathe and in clocks.

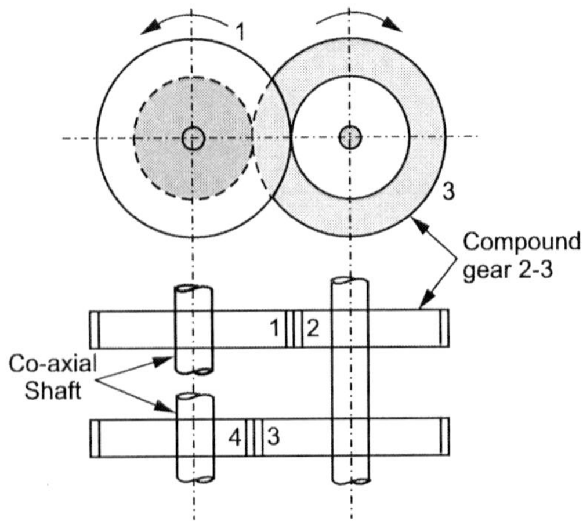

Fig. 6.38 Epicyclic gear train

Let T_1 = No. of teeth on wheel 1

T_2 = No. of teeth on wheel 2

T_3 = No. of teeth on wheel 3

T_4 = No. of teeth on wheel 4

Simillary r_1, r_2, r_3, and r_4 are radius for wheel 1, 2, 3, and 4

The distance between shaft centers of gear 1 and 2 and between gear 3 and 4 is same.

Hence, $$r_1 + r_2 = r_3 + r_4$$

Since the no. of teeth are proportional to their pitch circle diameters.

So $$T_1 + T_2 = T_3 + T_4$$

Speed ratio $$\frac{N_1}{N_4} = \frac{\text{No. of teeth on driver}}{\text{No. of teeth on driver}} = \frac{T_2}{T_1} \times \frac{T_4}{T_3}$$

6.10 CLUTCH

Clutch is a device which is used to engage and disengage the driving and driven shafts generally the clutches are used in between engine shaft and gear box shaft to transmit motion in automobile without stopping the prime mover. These are operated without slipping under, normal load.

6.10.1 Functions of Clutch

1. It can connect and disconnect the shafts when desired.
2. It helps to reduce the shocks transmission between machine shafts
3. It helps to start and stop a rotating element without any jerk.
4. It helps to eliminate noise produced in transmission.
5. It helps to maintain constant speed, torsional moment and power.

Material Used

1. Asbestos
2. Fabric
3. Ferodo
4. Reybestos
5. Leather

6.10.2 Working Principle of a Clutch

Clutch is used to engage and disengage the engine shaft to the gear shaft by the three main components the driving plate or flywheel, clutch plate or driven plate and a pressure plate basically it consist of a steel plate whose both sides are lined with a friction material. The pressure is applied through a pedal on the pressure plate and pressed together with the clutch plate and flywheel to transmit power from driving shaft to driven shaft due to physical contact.

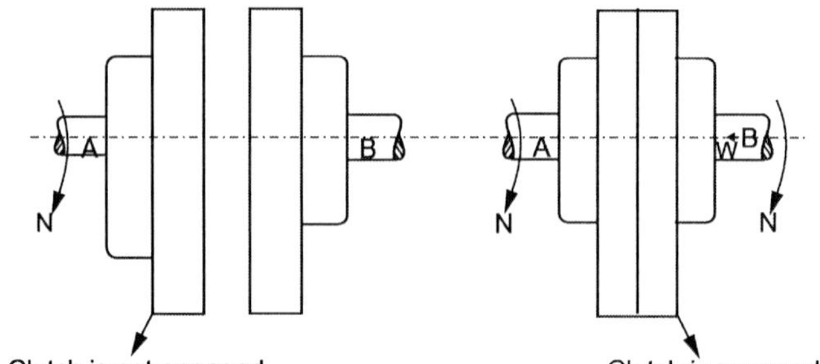

Fig. 6.39 Principle of friction clutch

6.10.3 Types of Clutches

The clutches are mainly classified as :

1. Positive clutch
2. Friction clutch

The positive type of clutches are used where positive drive is required. In this type of clutch one flage consists of Jaws and other locked recesses. These Jaws locked with recesses to make positive drive. They are suitable for use between the engine and gear box.

The friction clutches are used in transmission of power between shafts when machines are started and stopped frequently. These clutches are operated due to the friction between two contact surfaces. These are also known as positive drive type due to their better performance.

Types of Friction Clutches

The friction clutches of following types:

1. Disc type
   (*a*) Single plate clutch
   (*b*) Multiplate type
2. Cone clutch
   (*a*) External clutch
   (*b*) Internal clutch
3. Centrifugal clutch
4. Dog clutch
5. Hydraulic clutch
6. Vaccum clutch
7. Electromagnetic clutch

6.10.4 Single Plate Clutch

A single plate clutch is fitted to the driving shaft by means of a key. The driven flange is free to slide axially on driven shaft and rotate with it. The driven shaft flange is covered with the friction lining. It includes a friction plate held between the flywheel and pressure plate. Wheres the spring provided axial pressure to keep the clutch in engaged position. It is mounted on a splined hub so that it can slide over the gear box. A pedal is provided to disengage the clutch against the spring force whenever required. When pedal is released a thrust bearing moves back due to the action of spring.

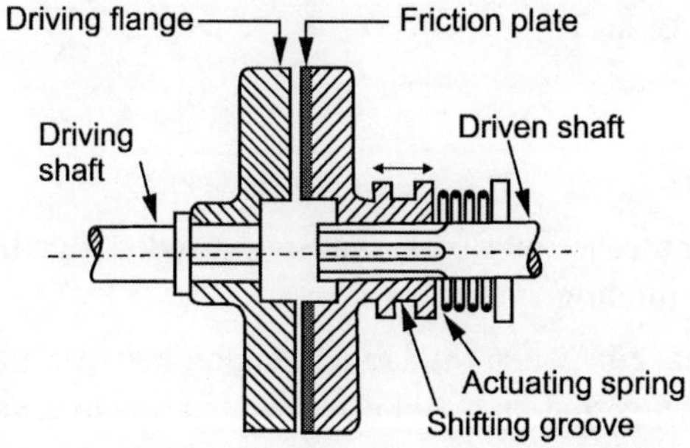

Fig. 6.40 Single plate clutch

6.10.5 Friction Disk

The friction disk has cushion springs and dampenning springs hub and plate.

The cushion springs are slightly waved or curled. The cushion springs are attched to the plate and the friction facings are attached to the cushion springs. When the clutch is engaged, the cushion springs compress slightly to take up the shock of engagement.

The dampening springs or torsional springs are heavy coil springs set in a circle around the hub. The hub is driven through these springs. They help to reduce the torsional vibration (The power pluses from the engine) so that the power flow to the transmission is smooth.

There are grooves in both sides of the friction disk facing. These grooves prevent the facing from sticking to the flywheel face and pressure plate when clutch is disengaged.

The groove breaks any vacuum that might cause the facings to stick to the flywheel or pressure plate.

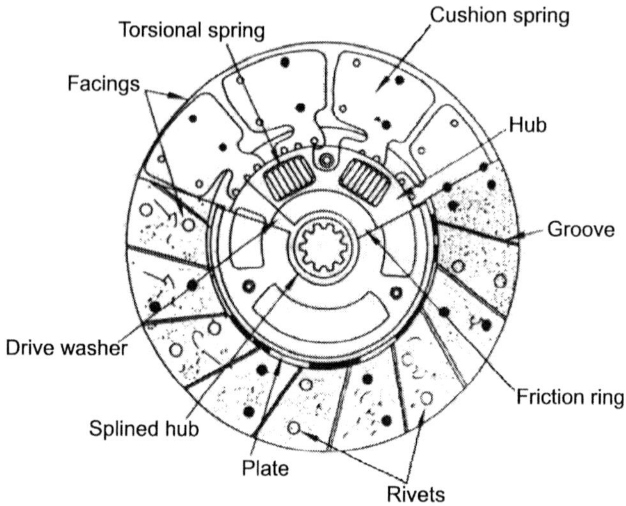

Fig. 6.41 Friction disk

Material: Cotton and asbestos fibers woven or molded together and impregnated with resins or other binding agent.

Connection takes place between the lab and clutch between the drum and driving discs. Due to large frictional area and effective heat dissipation makes the design successful.

6.10.7 Cone Clutch

The cone clutches are generally simillar to disc clutch but the contact area in the form of inner cone and outer cone. The outer cone is keyed to the driving shaft while the inner cone is free to slide axially on the driven shaft due to splines. A axial thrust force keep the clutch engaged for transmission of torque from engine via flywheel by a helical compression spring. The fork in the collar is used to disengage the clutch through the clutch pedal. These clutches are used in auto mobile. The large contact area of conical working surface results in normal pressure and friction to serve the purpose.

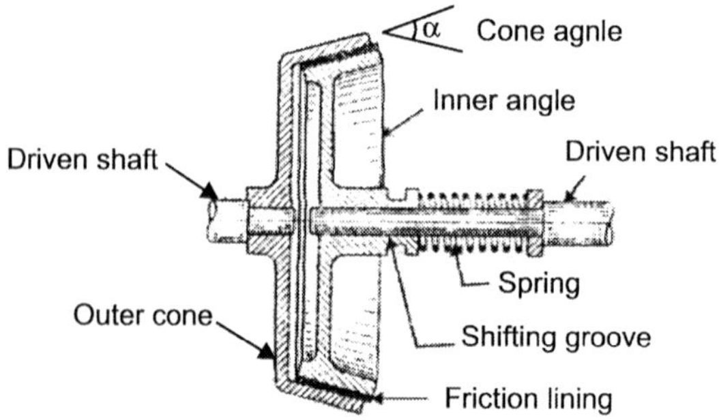

Fig. 6.42 Cone clutch

6.10.8 Centrifugal Clutch

The centrifugal clutch works on the principle of centrifugal force. The spring force is completely eliminated and only the centrifugal force is utilised to apply the required pressure needed for engaging the clutch to transmit the derived torque.

There is no separate pedal required to operate the clutch. The driver shaft carries the spider sliding shoes (mass) in guides and springs. While the dirven shaft carries pulley. A friction linear is also used in addition to the shoes. When the driver shaft rotates a centrifugal force acts on shoes to overcome the spring force, the shoes (mass) fly out radially and comes in contact with drum thus the power is transmitted due to frictional force between the shoe lining and inner surface of the drum.

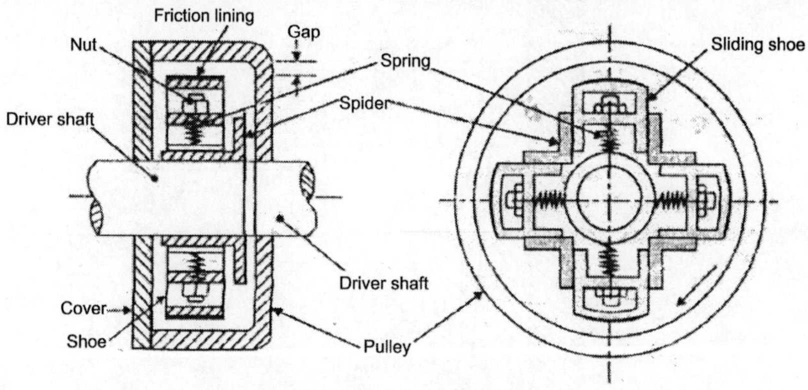

Fig. 6.43 Contrifugal clutch

6.10.9 Semi-centrifugal Clutch

The semi-centrifugal clutch is used for high torque transfer. Both centrifugal as well spring force are used together for engeged position. Here spring force used for low speed and centrifugal forces for high speed. It is almost similar to the single plate clutch.

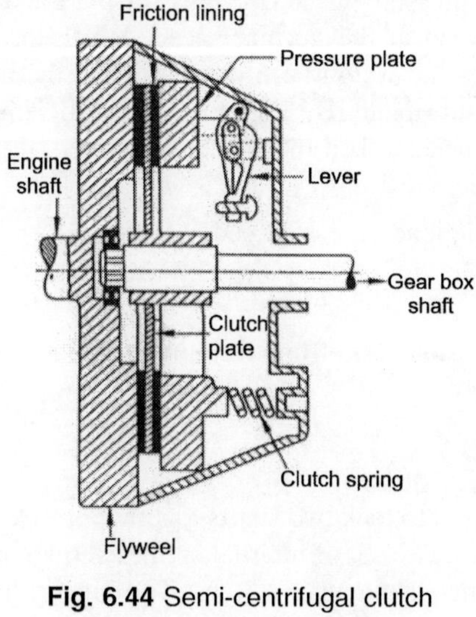

Fig. 6.44 Semi-centrifugal clutch

There are three levers which act due to centrifugal force. Fig.6.45 shows the diagram of a lever. The flywheel, the clutch plate and the pressure plate are arranged in the same manner as single plate clutch. Springs are also used to keep the clutch partially engaged. The levers are hinged on the pressure plate. The outer arm is called the weight arm. A weight is put on this arm. There is an adjusting screw at the end of the inner arm of the lever. This screw is used to adjust the amount of pressure being applied on the pressure plate.

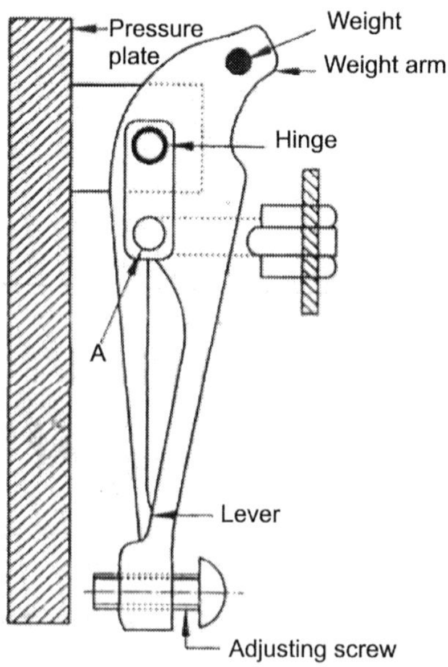

Fig. 6.45 Level of Semi-centrifugal clutch

When the clutch pedal is pressed, the clutch is disengaged. After releasing the clutch ped[illegible] the spring pressure pushed the pressure plate and the clutch is engaged. The cl[illegible]arts working at the normal speed. When the speed increases, the centri[illegible]e moves the weight arm outwards and the inner arm of the lever start p[illegible]he pressure plate. By this way, the sufficient amount of torque is transm[illegible] higher speed. When the speed is lowered, the weight arm starts to fall and i[illegible]r arm is retracted.

So after applying a little load on clutch pedal, the clutch is disengaged.

Advantages

For high torque transmission, less effort is required for clutch operation as compared to single plate clutch.

6.10.10 Dog Clutch

This type of clutch is used to lock two shafts together or to lock a gear to a shaft. It consists of sleeve having two set of internal splines shown in Fig.6.44 It slides on a splined shaft with smaller diameter splines. The bigger diameter splines match with the external dog clutch teeth on driving shaft. When the sleeve is made to slide on the splined shaft, it teeth match with the dog clutch teeth of the driving shaft. Thus the sleeve turns the splined shaft with the driving shaft. The clutch is said to be engaged for disengaging the clutch, the sleeve is moved back on the splined shaft to have no contact with the driving shaft.

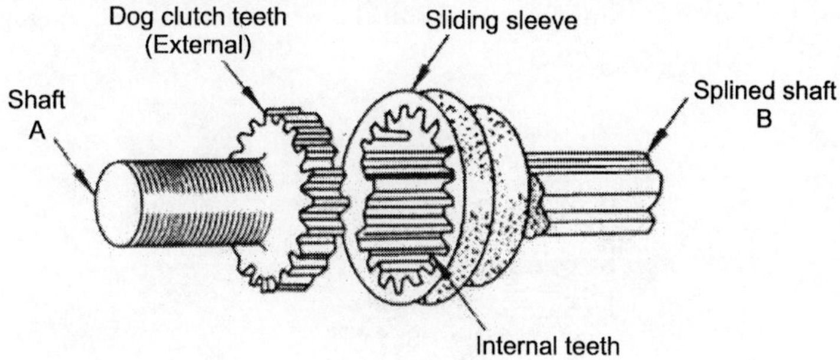

Fig. 6.46 Dog clutch

6.10.11 Hydraulic Clutch

A hydraulic clutch is used where a heavy duty machinery is operated a fluid pressure is used for the engagement of the clutch. It is suitable where a clutch pedal is away from the clutch. Generally, these types of clutches are used for earth moving machinery. In a hydraulic clutch a master cylinder remains connected to the clutch pedal. As the pedal released the clutch becomes engage. The pressure of the master cylinder keeps the plunger in its back position and valve shank contacts the spring retainer. Due to pressure the valve seat lifted from its seat and fluid flow through it to maintain the clutch in engaged position.

When the clutch pedal is pressed the clutch becomes disengage due to the cut off fluid from reversed to cylinder and the plunger displaces fluid to cylinder. The spring maintains a back pressure on release fork so that thrust bearing can make contact with the release leavers.

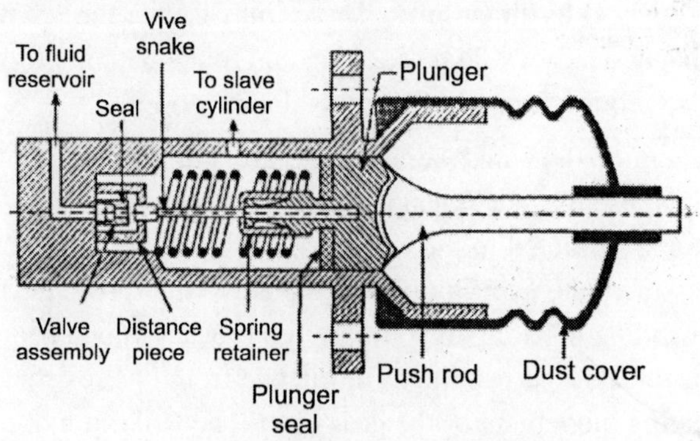

Fig. 6.47 Hydraulic clutch

6.10.12 Pneumatic Clutch (Tyre Type)

Pneumatic clutch consists of a rim connected to driving shaft and a tyre with shoes on the side facing rim. The drum is mounted on the driven shaft these are oval in X-section. The friction shoes are attached to the type by means of smooth pins. When the compressed air is passed it expands and forces the shoes on rim. Hence due to the friction between shoes and rim the driving shaft gets engaged with driven shaft to deliver the power.

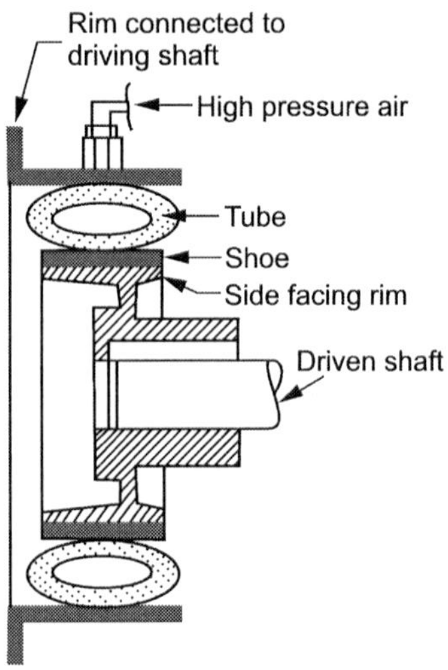

Fig. 6.48 Pneumatic clutch

6.10.13 Electromagnetic Clutch

Electromagnetic clutch is that in which force is provided by the electromagnet instead of springs. Where a splined shaft is carried in a spigot bearing at the left and gear box in right. A pressure plate is bolted to the armature disc which is secured to the flywheel by three links. These links isolate the fly wheel from the armature disc magnetically while positioning and rotate with the flywheel. An axis movement is also required for engaging the clutch.

There is an air gap of 0.45 mm between the armature and face of the flywheel when the clutch is disengaged. When the winding is energised. The armature with the pressure plate is attracted to the flywheel, so that the dirven plate is squeezed between the pressure plate and rigidly bolted plate to the flywheel. The current for energising the winding is led in and out through brushes which bears on slip rings connected to the ends of the winding. The change from one resistance to the other is made by selecting movement of the gear lever. The striking motion produces an interruption of the current to the clutch coil and gives the momentary clutch disengagement to enable a gear change to be made. A cam actuated by the accelerator pedal also opens a switch when the pedal is in idling and starting position. This ensuriny disengagement of the clutch at the starting and during idling.

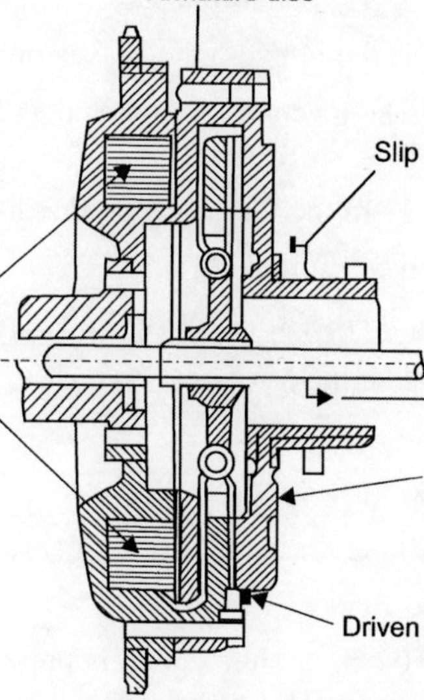

Fig. 6.49 Electromagnetic clutch

6.13 BRAKES

6.13.1 Introduction

A brake is a device by means of which artificial frictional resistance is applied to a moving machine member, in order to retard or stop the motion of a machine.

The brake, in its operation, absorbs either kinetic energy of moving member or potential energy given up by objects being lowered by hoists, elevators etc. The energy absorbed by brakes is dissipated in the form of hunt in the surrounding air or water which is circulated through the passages in the brake drum. By doing so excessive heating of the brake lining does not take place. The materials are commonly used for facing or lining of brakes :

1. Cast iron on cast iron
2. Bronze
3. Steel
4. Wooden
5. Fibre on moto
6. Cork
7. Leather
8. Wire asbestos
9. Asbestos blocks on metal.

6.13.2. Types of Brakes

Depending upon the means employed for transforming the energy by the braking elements, the brakes are classified as follows :

1. Mechanical brakes 2. Hydraulic brakes 3. Electric brakes

We shall discuss here mechanical and hydraulic brakes only.

1. **Mechanical Brakes.** The mechanical brakes may be further classified as follows :

(*i*) ***Radial brakes:*** Here the force acts on the drum in radial direction,

(*a*) External brakes (*b*) Internal brakes

These blocks may be *block* or *shoe brakes* and hand brakes.

(*ii*) ***Axial brakes:*** Here the force acts on the drum in

— Disc brakes

— Cone brakes

The mechanical brakes are discussed briefly below :

Single Brake or Shoe Brake

- It consists of a block or shoe which is pressed against the rim of a revolving brake drum. The material of the block is softer than that of the rim of the wheel.
- The block is pressed against the wheel by a force applied to one end of a lever to which the block is rigidly fixed as shown in Fig. 6.50. The other end of the lever is pivoted on a fixed fulcrum.
- Due to the frictional force between the block and the wheel, a tangential force comes into existence which acts on the wheel and retard the rotation of the wheel? "this type of brake is commonly used on railway trains and tram cars.

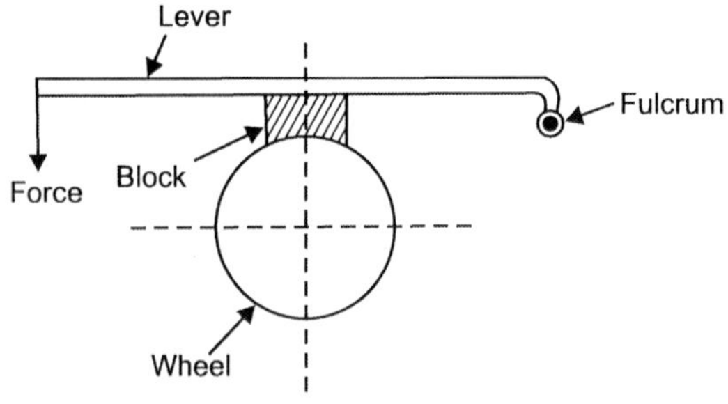

Fig. 6.50 Single or shoe brake

Pivot Block or Shoe Brake :

In this type of brake the block or shoe is not rigidly attached to the lever. This pives uniform wear of the brake lining in the direction of the applied force.

Double Block or Shoe Brake :

This type of brake overcomes the drawback of a single block brake in which an additional load is thrown on the shaft bearings (due to the normal force), which causes bending of the shaft.

It consists of two brake blocks applied at the opposite ends of a diameter of the wheel which eliminates or reduces the unbalanced force on the shaft. The brake is set by a spring which pulls the upper ends of the brake arms together. When a force is applied to the bell crank lever, the spring is compressed and the brake is released

In this type of brake, the braking action is doubled by the use of two blocks and these blocks may be operated practically by the force which operate one. A double block brake is often used on electric cranes.

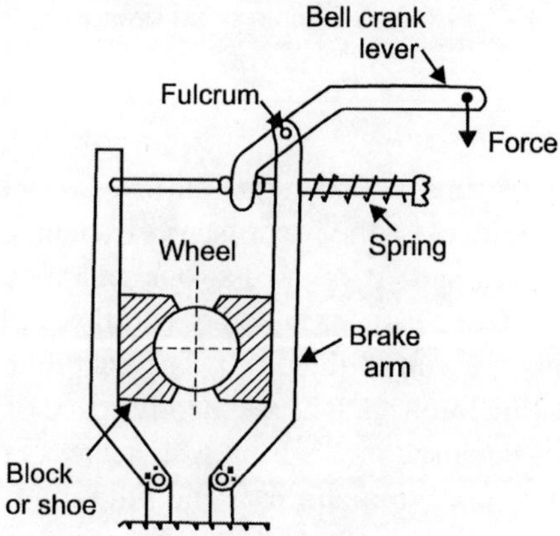

Fig. 6.51 Double block or shoe brake

Simple Band Brake :

It consists of a flexible band of leather, one or more ropes, or a steel lined with friction material, which embraces a part of the circumference of the drum.

Here, one end of the band is attached to a fixed pin or fulcrum of the lever while other end is attached to the lever at A as shown in the Fig. 6.52

When a force is applied to the lever at B, the lever turns about the fulcrum pin 0 and tightens the band on the drum and hence the brakes are applied. The necessary braking force is provided by the friction between the band and the drum the band brake is lined with blocks of wood or other material and the friction between the blocks and drum provides braking action, the brake is termed as band and block brake.

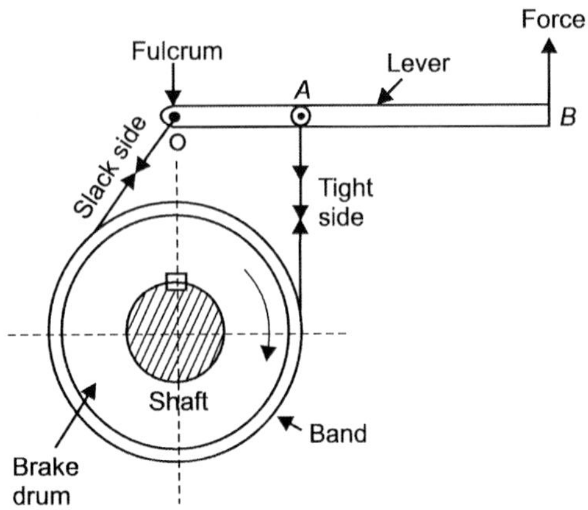

Fig. 6.52 Simple band brake

Internal Expanding Brake :

It consists of two shoes S_1 and S_2. The outer surfaces of the shoes are lined with some friction material, to increase the frictional coefficient and to prevent wearing away from the metal. Each shoe is pivoted at one end about a fixed fulcrum (O_1 and O_2) and made to contact a cam, at the other end the cam rotates, the shoes are pushed outwards against the rim of the drum. The friction between the shoes and drum produces the braking torque and consequently speed of the drum reduces,and the entire mechanism is enclosed in the drum which checks the ingress of dust and moisture. It is commonly used in motor cars and light trucks.

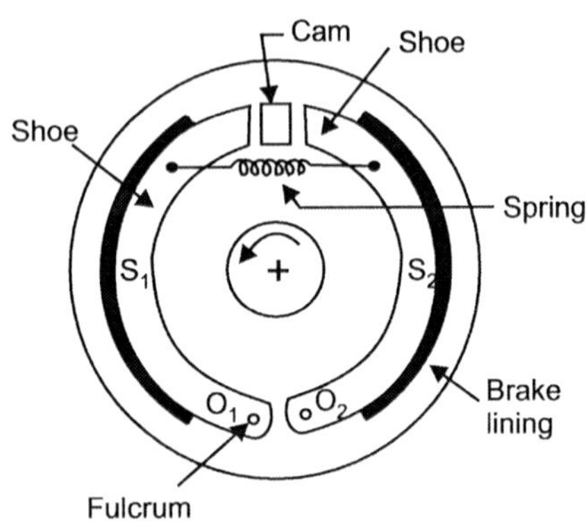

Fig. 6.53 Internal expanding brake

6.13.3 Hydraulic Brakes

In a hydraulic brake system, force is applied to a piston in a master cylinder that corresponds to the central cylinder in Fig. 6.52. The brake pedal operates the piston by a linkage as shown in Fig. 6.53. Each wheel brake is provided with a cylinder fitted with opposed piston within the master cylinder, thus forcing the brake liquid of fluid, as it is commonly called from the master cylinder through tubing and flexible hose into the four wheel cylinders.

The brake fluid enters each of the wheel cylinders between opposed pistons, making the pistons move the brake shoes outward against the brake drum. As pressure on pedal is increased, greater hydraulic pressure is built up within the wheel cylinders, which ensures effective braking of the moving wheel.

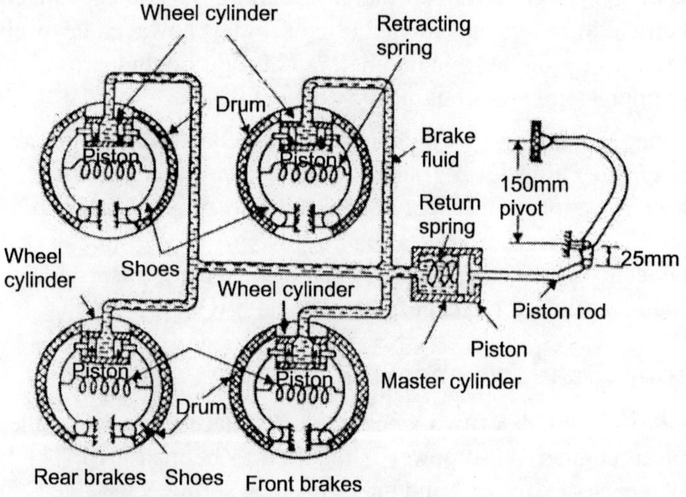

Fig. 6.54 Hydraulic brake system used on a motor vehicle

6.14 DYNAMOMETERS

A dynamometer is a brake but in addition it has a device to measure the frictional resistance. Knowing the frictional resistance, the torque transmitted and hence the power of the engine can be determined.

6.14.1 Types of Dynamometers

Following types of dynamometers are need for measuring the brake power (B.P.) of an engine :

1. **Absorption Dynamometers**

   (*i*) Prony brake dynamometer

   (*ii*) Rope brake dynamometer

   In which the energy is converted into heat by friction whilst being measured. The heat is dissipated to the surroundings where it generally serves no useful purpose. Absorption dynamometers are used when the test-machine is a power generator such as an engine, turbine and an electric motor. The types commonly used include prony brakes, hydraulic or fluid friction brakes, fan brakes and eddy current dynamometers.

2. **Transmission Dynamometers**
   (*i*) Epicyclic train dynamometer
   (*ii*) Belt transmission dynamometer
   (*iii*) Torsion dynamometer.

   Transmission dynamometers which the energy being transmitted either to or from dynamometers is not absorbed or dissipated. After measurement, the energy is conveyed to the surroundings in a useful mechanical or electrical form. A small amount of power may, however, be lost by friction or the joints of the dynamometer. The type includes torsion and belt dynamometers, and strain gauge dynamometers.

3. Driving dynamometer which may be coupled to either power absorbing or power generating devices; since it may operate either a motor or a generator. These instruments measure power and also supply energy to operate the tested devices. They are essentially useful in determining performance characteristics of such machines as pumps and compressors. Electric cradled dynamometer is a typical example of the driving dynamometer.

6.14.2 Prony Brake Dynamometer

Construction: It consists of two wooden blocks placed around a pulley fixed to the shaft of an engine whose power is required to be measured. The blocks are clamped by means of two bolts and nuts. A helical spring is provided between the nut and the upper block to adjust the pressure on the pulley to control its speed. The upper block has a long lever attached to it and carries a weight W at its outer end. A counter weight is placed at the other end of the lever which balances the brake when unloaded. In order to limit the motion of the lever, two stops are provided as shown in Fig. 6.55

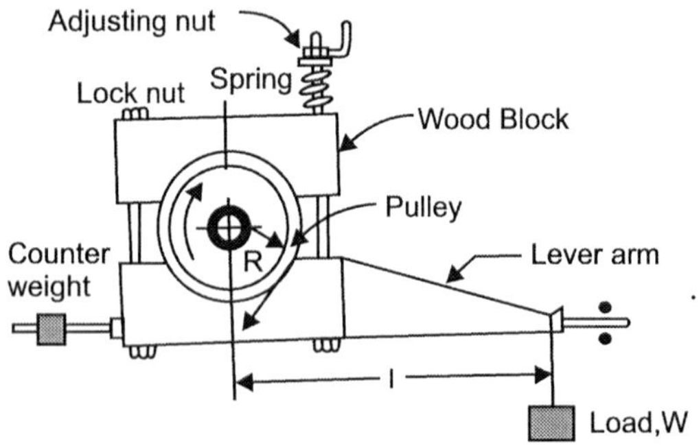

Fig. 6.55 Prony brake dynamometer.

Working: When the brake is to be operated, the long end of the lever is loaded with suitable weight *W* and nuts are tightened until the engine shaft runs at a constant speed and the lever is in horizontal position. Under these conditions, the moment due to the weight *W* must balance the moment of frictional resistance between the blocks and the pulley.

Let W = Weight at the outer end of the lever N (newtons),

L = Horizontal distance of the weight W from the centre of the pulley, m and

F = Frictional resistance between the blocks and, the pulley, N

R = Radius of the pulley, m, and

N = Speed of the shaft, *r.p.m.*

Now, the moment of the frictional resistance or torque of the shaft,

$$T = W \times L = F \times \text{R Nm}$$

Work done in one revolution

$$T = T \times \theta \text{ (angle turned)} = T \times 2\pi \text{ Nm}$$

∴ Work done per minute = $T \times 2\pi$ N Nm

∴ Brake power of the engine = $T \times 2\pi$ N Nm

$$\text{B.P.} = \frac{T \times 2\pi N}{60} = \frac{(W \times L) \times 2\pi N}{60}$$

When the driving torque on the shaft is not uniform, this dynamometer is subject to severe oscillations.

6.14.3. Rope Brake Dynamometer

This type of absorption dynamometer is most commonly used. Its schematic arrangement is shown in Fig. 6.56.

Construction: It consists of one, two or more ropes wound around the flywheel or rim of a pulley fixed rigidly to the shaft of the engine. The upper end of the ropes is attached to a spring balance while the lower end of tru; rope is kept in position by applying a dead weight as shown in Fig. 6.56. In order to prevent the slipping of the rope over the flywheel, wooden blocks are placed at intervals around the circumference of the flywheel.

Working: In the operation of the brake operates, the engine is made to run at a constant speed. The frictional torque, due to rope, must be equal to the torque being transmitted by the engine

Let W = Weight at the end of the rope N

S = Spring balance reading, N

N = Engine speed, *r.p.m,*

D = Diameter of the brake wheel, m

d = Diameter of the rope, m, and

$(D + d)$ = Effective diameter of the brake wheel

Then, Work/revolution = Torque × Angle turned per revolution

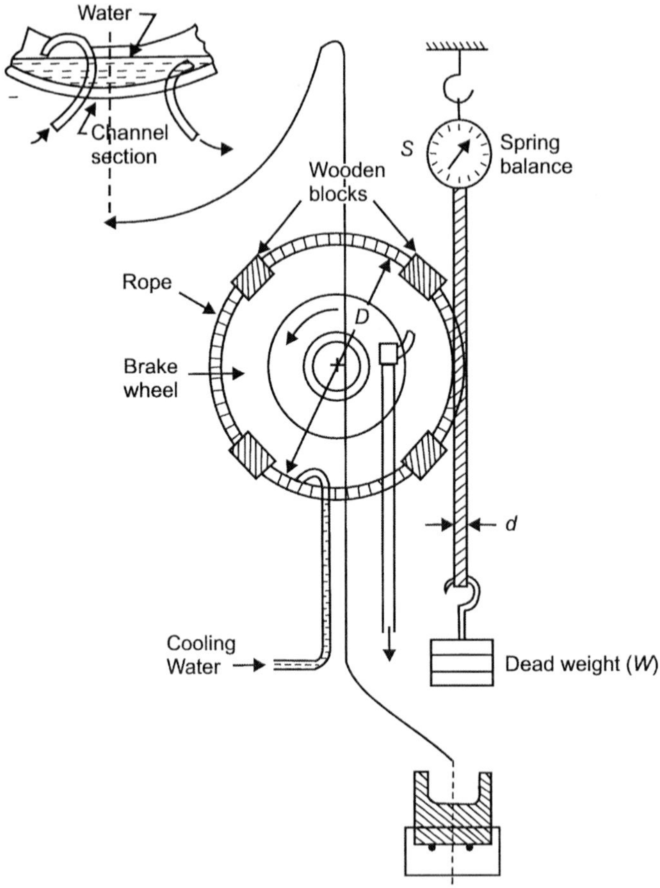

Fig. 6.56 Rope brake dynamometer

$$= (W - S) \times \left(\frac{D + d}{2}\right) \times 2\pi$$

$$= (W - S) \times (D + d) \times \pi$$

Work done/min. $= (W - S)\pi\,(D + d)N$

Work done/sec. $= \dfrac{(W - S)\pi(D + d)N}{60 - 1000}$

$$\text{B.P.} = \frac{(W - S)\pi(D + d)N}{60 - 1000}\,\text{kW}$$

$$= \frac{(W - S)\pi DN}{60 - 1000}\,\text{kW} \qquad \text{... if } d \text{ is neglected}$$

$$\text{B.P.} = \frac{T\pi DN}{60-1000}\,\text{kW}$$

Since the energy produced by the engine is absorbed by the frictional resistors of the brake and is converted into heat. So cooling arrangement is necessary for this purpose.

6.14.4 Fluid-friction Hydraulic Dynamometer

Construction: A hydraulic dynamometer uses fluid-friction rather than dry friction for dissipating the input energy. The unit consists essentially of two elements namely a rotating disk and a stationary casing. The rotating disk is keyed to the driving shaft of the prime-mover and it revolves inside the stationary casing. The casing is mounted on antifriction bearings and has a brake arm and a balance system attached to it. Such bearings allow the casing to rot; the freely except for the restraint imposed by the brake arm. Further, the casing is in two-holes; one of which is placed on either side of the rotating disk. Semi-elliptical recesses in the casing match with the corresponding grooves inside the rotating disk to form chambers through which a stream of water flow is maintained.

When the brake is operating, the water rall > as a helical path in the chamber.' Vertices and eddy-currents are set-up in the water and the e tend to turn the dynamometer casing in the direction of rotation of the engine shaft. This tendency is resisted by the brake arm and balance system that measure the torque.

$$\text{Brake Power} = \frac{WN}{K}$$

where W is weight as lever arm; N is speed in revolutions per minute and K is dynamometer constant.

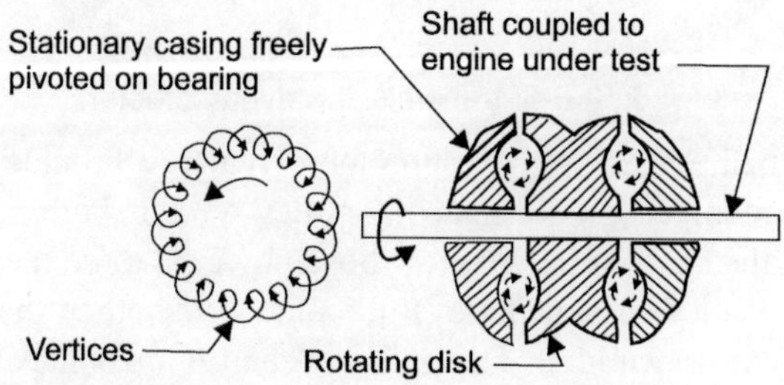

Fig. 6.57 Water vortex dynamometer

The amount of braking action can be controlled by varying either the amount of water and its pressure, or the distance between the rotating disk and the stationary casing. Maximum absorption is obtained with the casing full, and minimum absorption with the minimum amount of water required to keep the shaft seals lubricated. The amount of water circulated must be sufficient to prevent the formation of steam, since such action would cause momentary unloading. Power absorption by a hydraulic dynamometer varies approximately as the cube of rotational speed and as the fifth power of rotor diameter.

- high absorption capacity in a small space and at low cost
- constant supply of water provides cooling effect in addition to braking action.
- protection from hunting effects made possible by providing a dashpot — damper system.

Approximate speed limit = 10,000 rpm

Usual power limit = 20,000 kW

6.15 BELT TRANSMISSION DYNAMOMETER

When power is being transmitted by the belt from one pulley to another, the tangential effort on the driven pulley is equal to the difference between the tensions in the tight and slack sides of the belt. A belt dynamometer is introduced to measure directly the difference between the tensions of the belt, while it is running.

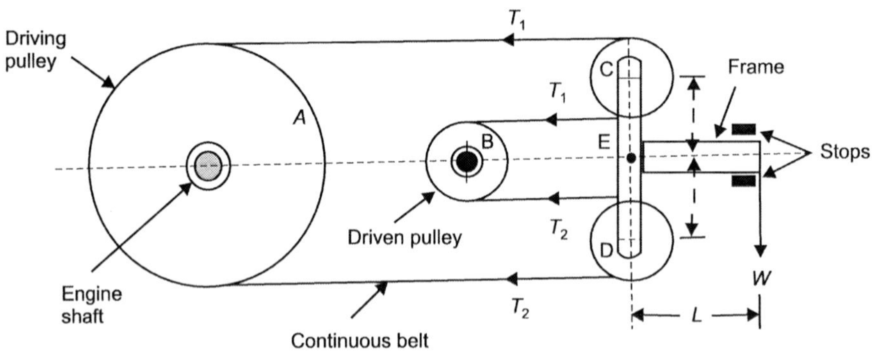

Fig. 6.58 Belt transmission dynamometer

Fig. 6.58 shows a belt transmission dynamometer (called a Froucle or Throneycraft transmission dynamometer). It consists of a pulley A (called driving pulley) which is rigidly fixed to the shaft of an engine whose power is to be measured. There is another pulley *B* (called driven pulley) mounted on another shaft to which the power from pulley is transmitted. The pulleys A and *B* are connected by means of a continuous belt passing round two pulleys *C* and *D* which are mounted on a T"-shaped frame. The frame is pivoted at *E* and its movement is controlled by two stops as shown in Fig. 6.58.

Since the tension in the tight side of the belt (T_1) is greater than the tension in the slack side of the belt (T_2) therefore, the total force acting on the pulley *C* (*i.e.*$2T_2$) is greater than the total force acting on the pulley *D* (*i.e.*, $2T_2$). It is thus evident that the frame causes movement *about* E in the anticlockwise direction. In order to balance it, a weight W is applied at a distance L from E on the frame (Refer Fig. 6.58).

Taking moments about the pivot E, neglecting friction, we have

$$2T_1 \times l = 2T_2 \times l + W \times L$$

$$2l(T_1 - T_2) = W \times L$$

$$l_1 - T_2 = \frac{WL}{2l}$$

Let D = Diameter of the pulley A, m

N = Speed of engine shaft, r.p.m.

Then, work done/revolution $= (T_1 - T_2)\pi D$ Nm

Work done/min $= (T_1 - T_2)\pi DN$ Nm

$\therefore$ Brake power of the engine, $= \dfrac{(T_1 - T_2)\pi DN}{60 \times 1000}$ kW

Construction: The band type prony brake consists of an adjustable steel band which are fixed on wooden blocks in contact with the engine brake-drum. The frictional steel band and the brake drum can be adjusted by tightening or loosening the clamp. The torque is transmitted to the knife edge through the torque arm. The knife edge rests on a platform or connects with a spring balance.

Working : The force transmitted to bend type prong brake

Let F be the frictional force at the brake drum surface, W be the scale or spring balancereading, l be the effective lever arm, r be the brake drum radius, and N be the revolutions of the crank shaft. Then :

Frictional torque at the drum $= Fr$

Balancing torque $= Wl$

Under equilibrium conditions

$$T = Fr - Wl \qquad \text{in Nm}$$

$$\text{Power} = \frac{2\pi N}{60} \times T \qquad \text{in Nm /s}$$

$$= \frac{2\pi N}{60 \times 1000} \times Wl \qquad \text{in kW}$$

The prony brake is inherently unstable, its accuracy and capacity of power measurement being affected by the following factors :

(*i*) The blocks wear out and a variation results in the value of coefficient of friction between the block and pulley rim. This necessitates continual tightening of the clamp. In the long run, the system may become unservicable for measurement of large power and for long continued runs.

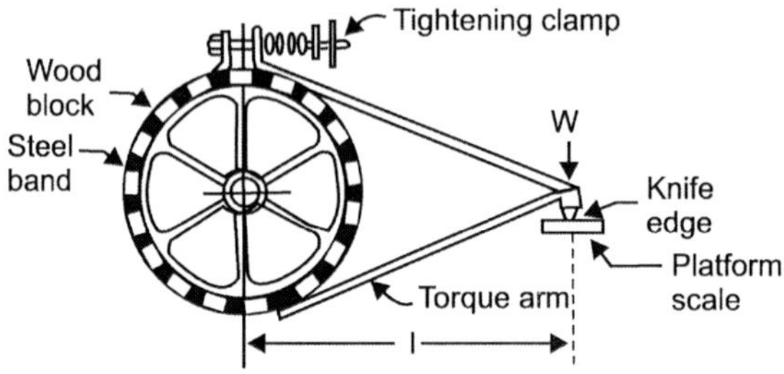

Fig. 6.59 Band type prony brake

(*ii*) The unit is subject to oscillations which may result from fluctuation in the power of the machine. This aspect makes accurate readings of the pan weight quite difficult to obtain.

(*iii*) Due to dissipation of heat, temperature rises. This results in a decrease in coefficient of friction and a consequent brake failure. This aspect is railed brake failing. The capacity of a brake ultimately depends on the balance between temperature rise and heat dissipation. To increase the brake capacity and to render the operation more efficient, the brake system requires cooling. Cooling can be achieved by making pulley rim of a hollow channel cross-section in which water can be made to flow. Lower temperature can be mantained by providing scope to take away hot water rim.

6.15.1 EPICYCLIC TRAIN DYNAMOMETER

The epicyclic train dynamometer consists of a simple epicyclic train comprising a spur gear, an annular gear (gear heaving internal teeth) and a pinion. The spur gear keyed to the engine (driving) shaft rotates in anti-clockwise direction and the annular gear keyed to the driven shaft rotates clockwise during operation. The pinion, also called the intermediate gear, meshes with both the annular and spur gears. The pinion revolves freely on a lever which is pivoted to the common axis of the driving and driven shafts. In order to keep the lever in a horizontal position between the stops S and S, there is a counter weight to balance the brake when unloaded.

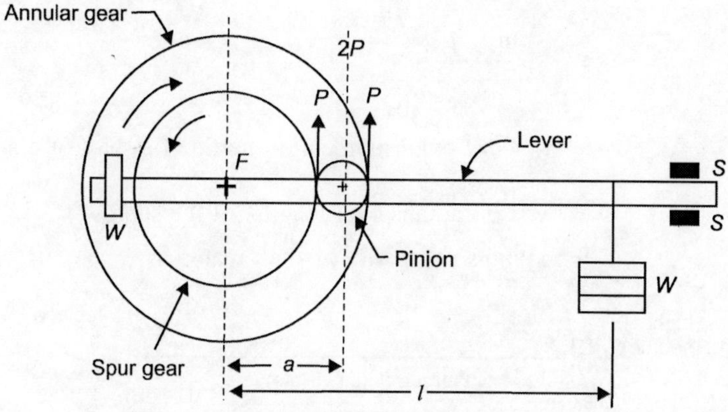

Fig. 6.60 Epicyclic gear train

Neglecting friction of the pin on which the pinion rotates, the tangential effort P exerted by the spur gear on the pinion equals the tangential reaction of the annular gear on the pinion. Since both these forces act upwards, the total upward force on the lever acting through the axis of the pinion is 2 P. This force tends to rotate the lever about its fulcrum. This tendency is prevented by adding weight W at the extremity of the lever arm so that it remains horizontal in a position of equilibrium.

Considering equilibrium of the lever, the momentum balance at the fulcrum point gives.

$$2P \times a = Wl \quad \text{or} \quad P = \frac{Wl}{2a} \text{ in newtons}$$

Let r be the pitch circle radius of the spur gear, and N be the revolutions of the engine shift per minute, then

$$\text{Torque transmitted } T = P \times r = \frac{Wl}{2} \times r \text{ in Nm}$$

$$\text{Power transmitted} = \frac{2\pi N}{60} \times T \text{ in Nm /s}$$

$$= \frac{2\pi N}{60 \times 1000} \times \left(\frac{Wl}{2a} r\right) \text{ in kW}$$

6.15.2 Bevis Gibson Flash Light Torsion Dynamometer

This torsion dynamometer is based on the fact that for a given shaft, the torque transmitted is directly proportional to the angle of twist. This twist is measured and the corresponding torque estimated from the relation

$$T = I_p \times \frac{C\theta}{l}$$

where $\quad I_p = \frac{\pi}{32} d^4 =$ polar moment of inertia of a shaft of diameter d

θ = twist in radians over length l of the shaft

C = modulus of rigidity of shaft material

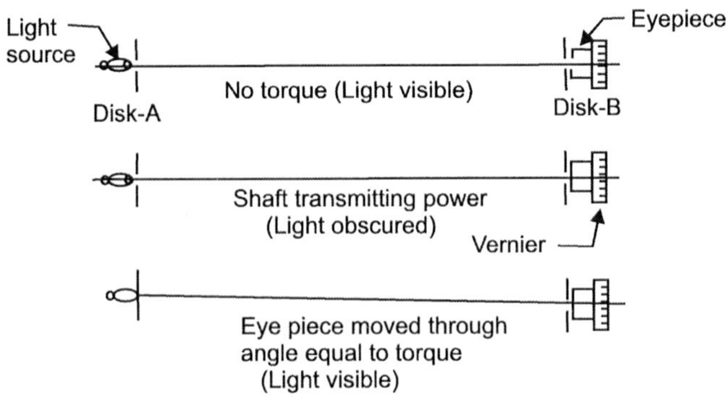

Fig. 6.61 Flash light torsion dynamometer

The essential components of the units are:

(*i*) Two similar disks A and B which are fixed to and revolve with the shaft at a known distance apart. Each disk has a narrow radial slot and the two slots are in line when no torque is being transmitted along the shaft. Thus if a light is passed through the slot of disk A, it would also pass through the slot in disk B.

(*ii*) A light source which is provided by a bright electric lamp fitted to the left of disk A. The lamp is equipped with a mask so that it throws a narrow pencil of light parallel to the axis of the shaft, and at the same distance from the axis as are the radial slots in the two disks.

(*iii*) An eye piece or torque finder fitted to the right of disk B. The eye piece can move circumferentially on a vernier.

When the shaft transmits power, the disks move relative to one another and the light gets obscured, *i.e.*, the ray of light can no longer pass through the slots. The flash is picked up by moving the eye piece circumferentially by an amount equal to the twist of the shaft. From the angular twist and the shaft constant I_p l and C, torque can be determined.

HIGHLIGHTS OF THE CHAPTER

1. The transmission of power from one rotating shaft to another shaft by means of a belt at considerable distance known as belt drive.

2. Types of belts are

(*i*) Flat belt (*ii*) V-belt

(*iii*) Circular belt

3. Types of belt drives are:

(*i*) Open belt drive (*ii*) Cross belt drive

(*iii*) Quarter turn belt drive (*iv*) Compound belt drive

(*v*) Stepped or

4. Lengths of open and cross belt drive are

$$\text{Open belt } L = \left[\pi(r_1 + r_2) + \frac{(r_1 - r_2)^2}{l} + 2l\right]$$

$$\text{Cross belt } L = \left[\pi(r_1 + r_2) + \frac{(r_1 + r_2)^2}{l} + 2l\right]$$

5. Ratio of tension is $\frac{T_1}{T_2} e^{u\theta}$

6. Centrifugal tension $T_c = mv^2$

7. Condition for maxi power transmission is $v = \sqrt{\frac{T}{3m}}$

8. Velocity ratio of belt is $\frac{N_1}{N_2} = \frac{d_2}{d_1}$

Where N_1 = Speed of driver pulley

N_2 = Speed of driven pulley

d_1 = Diameter of driver pulley

d_2 = Diameter of driven pulley

9. Velocity ratio of compound drive

$$\frac{N_1}{N_4} = \frac{d_2 \times d_4}{d_4 \times d_3}$$

$$\frac{\text{Speed of driver}}{\text{Speed of driven}} = \frac{\text{Product of diameters of drivens}}{\text{Product of diameters of drivers}}$$

10. Velocity ratio with stip

$$\frac{N_2}{N_1} = \frac{d_1}{d_2}\left[1 - \frac{5}{100}\right]$$

11. **Chain drive:** It is a medium of power transmission which consists of a endless chain running over sporkets.
12. **Gear drive:** Toothed wheel master with another wheel to give positive transmission of mobin for a shorter distance.
13. Type of gears
    - (*i*) Spur gear
    - (*ii*) Helical gear
    - (*iii*) Bevel gear
    - (*iv*) Spiral gear
    - (*v*) Worm gear
    - (*vi*) Rack and driven
    - (*vii*) Internal and external gear
14. Clutch is a device used to engage and disengage the driving and driven shafts.
15. Clutches classified
    - (*i*) Positive clutch
    - (*ii*) Cone clutch
    - (*iii*) Centrifugal clutch
    - (*iv*) Dog and spline clutch
    - (*v*) Hydraulic clutch
    - (*vi*) Vacuum clutch
    - (*vii*) Electromagnetic clutch
16. Break is a device which produces artificial friction to stop the motion of the machine.
17. Types of brakes
    - (*i*) Mechanical brakes
    - (*ii*) Hydraulic brakes
    - (*iii*) Electric brakes
18. Dynamometer is a device used to measure the frictional resistance *i.e.* how much torque is produced by the engine.
19. Types of dynamometers
    - (*i*) Absorption types
    - (*ii*) Transmission type
    - (*iii*) Driving type

REVIEW QUESTIONS

1. What do you mean by power transmission?
2. Explain the belt drives with neat sketches.
3. List the advantages of V-belt drive.
4. What are the advantages and disadvantanges of belt drive?

   [M.D.U. Dec. 2005]
5. Explain and derive an expression for the velocity ratio of belt drive.

6. What is slip? How does it affect the velocity ratio of belt drive?
7. Drive an expression for the length of flat best drive
   (*i*) Open belt drive (*ii*) Cross belt drive
8. Set up the followng expression for the flat belt drive.[M.D.U. Dec., 2005]

$$\frac{T_1}{T_2} = e^{-u\theta}$$

   Where T_1 and T_2 are the tensions on the tight side and slack sides of the belt, θ is the angle of contact and it is the coefficient of friction between the belt and pulley rpm.
9. Draw an expression for centrifugal tension of a flat belt drive.
10. Find the condition of transmission of maximum power.
11. Derive an expression for the ratio of driving tensions in rope drive assuming the angle of pulley to be 2β.
12. What do you mean by chain drive also give the classification of chain?
13. What are the advantages and disadvantages of chain drive?
14. Give the comparison between chain and belt drive.
15. Explain the terminology of a gear with neat sketch.
16. What do you mean by gear train ? Explain the type of gear train.
17. What are the advantages and disadvantages of gear drive?
    [M.D.U. Dec., 2005]
18. How are gears classified? [M.D.U. Dec., 2004]
19. Explain briefly an epicycle gear train. What are the special advantages of epicycle gear train?
20. What do you mean by reverted gear train?
21. What is the working principle of a clutch?
22. What is the function of a clutch?
23. Name the different types of clutches.
24. Explain the working of a single plate clutch with neat sketch.
25. What is clutch? Describe the working of a multiplate disc clutch with neat sketch.
26. What is a brake? How are they classified?
27. Explain briefly with a neat sketch of hydraulic brake system used in a motor car.
28. Explain the working of a double block or shoe brake.

29. Explain the working of a internal expanding brake.

30. What is a dynamometer? How does it differ from a brake?

[M.D.U. May, 2004]

31. How are dynamometers classified?

32. Explain briefly any two of the dynamometers:

    (*i*) Prony brake dynamometer.

    (*ii*) Rope brake dynamometer.

    (*iii*) Hydraulic

    (*iv*) Belt transmission type.

33. Explain working and construction of prony brake dynamometer.

[M.D.U. May, 2006]

34. Explain with neat sketch, the band brake type dynamometer.

[M.D.U. Dec. 2004]

■■■

7

Stress and Strain

7.1 INTRODUCTION

Strength of material is basically a study with the relation between external forces and internal effects 'mechanical properties' on bodies. These properties are primary related to the elastic and plastic behaviour, as well as strength and fracture characteristics of the materials some time these properties also known as engineering properties. Like tensile strength, ductility, compressive strength fatigue, creep and others. The bodies are no longer assumed to be rigid and deformation are of major interest. The deformation takes place due to external forces acting upon body to do work. Hence, it is assumed that the material is homogeneous, isotropic as well there is no internal stress present in the material before the application of load.

7.2 STRESS

When an external force is applied on a body an internal resistance is developed within the body. An internal resistance is developed to keep the body in equilibrium with the external force. Hence, external force per unit area is called stress.

Stress, $\sigma = \dfrac{P}{A}$

where P = force applied

A = cross-sectional area of the body

The SI unit of stress in N/m^2 or Pascal.

7.3 TYPES OF STRESS

The various types of stresses may be classified as:

1. Simple or direct stress

    (*a*) Tensile stress

(*b*) Compressive stress

(*c*) Shear stress

2. Indirect stresses

(*a*) Bending stress

(*b*) Torsion stress

3. Combined stress

1. Simple Stress

(*a*) **Tensile Stress:** When a body is subjected to two equal and opposite forces which tends to increase its length called tensile stress *i.e.* cross-sectional area of the body gets reduced:

$$\sigma_t = \frac{P}{A}$$

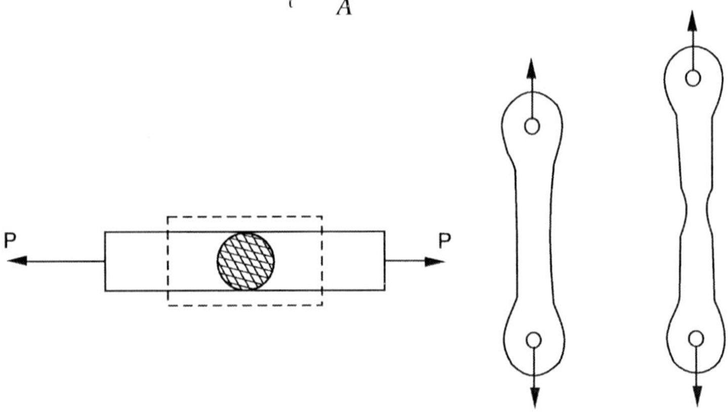

Fig.7.1 Tensile stress

(*b*) **Compressive Stress:** When a body is subjected to two equal and opposite forces which tends to decrease its length called compressive stress *i.e.* cross sectional area of the body gets increased:

$$\sigma_c = \frac{P}{A}$$

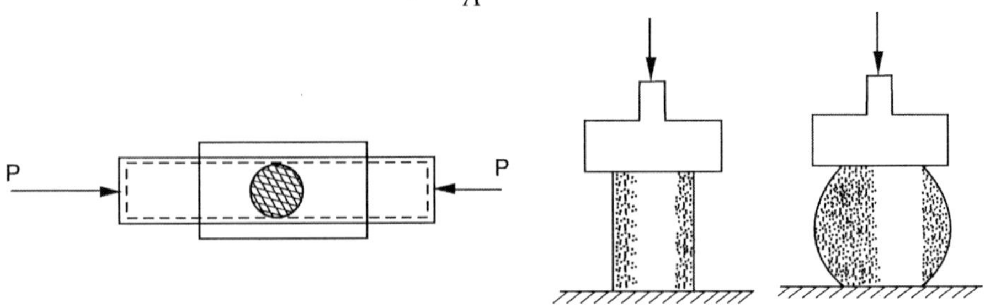

Fig.7.2 Compressive stress

(*c*) **Shear Stress:** When a body is subjected to two equal and opposite forces, acting tangentially across the resisting section and the body tends to shear off across the section called shear stress *e.g.*, shear stress occurs in rivited joint.

$$\text{shear stress } Z = \frac{P}{A}$$

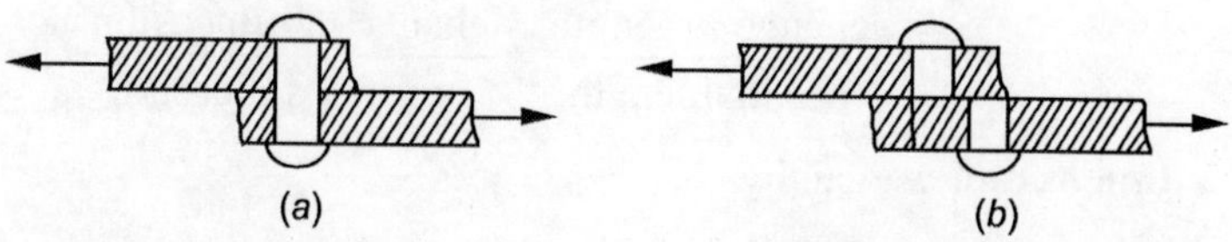

Fig.7.3 (*a*) Rivet resisting shear (*b*) Rivet failure due to shear

Consider Fig. 7.4 in which body *ABCD* is fixed with bottom. Let the force *P* be applied at the force *DC* tangencially to force *AB*. The body displaces *ABCD* to ABC_1D_1 at an angle ϕ

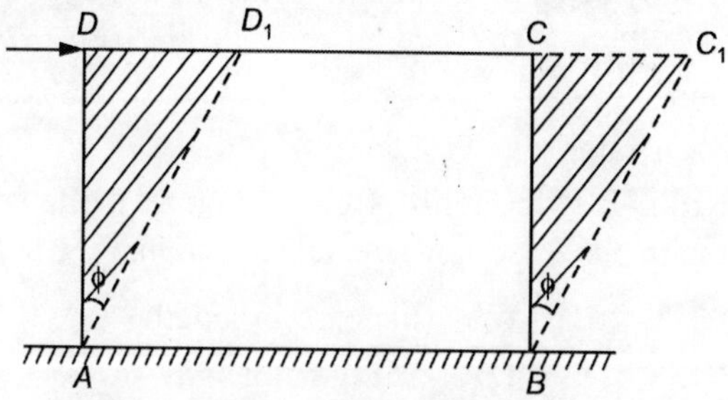

Fig.7.4 Tensile stress

$$\text{Shear stress } Z = \frac{P}{A} = \frac{\text{Shear force}}{\text{Shear area}}$$

2. Indirect Stresses

(*a*) **Bending Stress:** These are those stresses which are used to bend the structure of a component.

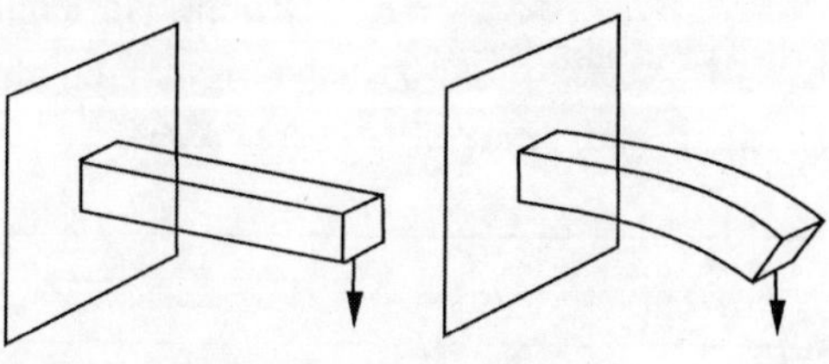

Fig.7.5 Bending sresses.

(*b*) **Torsion Strees:** Torsional stresses are those which are used to twist the position of axis of the component.

3. Combined Stress

Combined stresses are those in which tensile as well as compresive stresses take place simultaneously.

7.4 STRAIN

Strain is a measure of deformation or change in the shape of a body in application of force. Deformation per unit length is also known as strain.

$$\varepsilon = \frac{\delta l}{l} = \frac{\text{change in length}}{\text{original length}} = \frac{\text{change in dimension}}{\text{original dimension}}$$

Strain is a **dimensionless** quantity.

7.5 TYPES OF STRAINS

7.5.1. Normal Strain

Normal strains generally depending upon the type of force applied and it is divided into two categories:

(*a*) Tensile strain

(*b*) Compressive strain.

(***a***) **Tensile Strain:** It is the ratio of elongation in the length to the original length of the member is known as tensile strain.

$$\text{Tensile strain } \varepsilon_t = \frac{\delta L}{L} = \frac{\text{Increase in length}}{\text{Original length}}$$

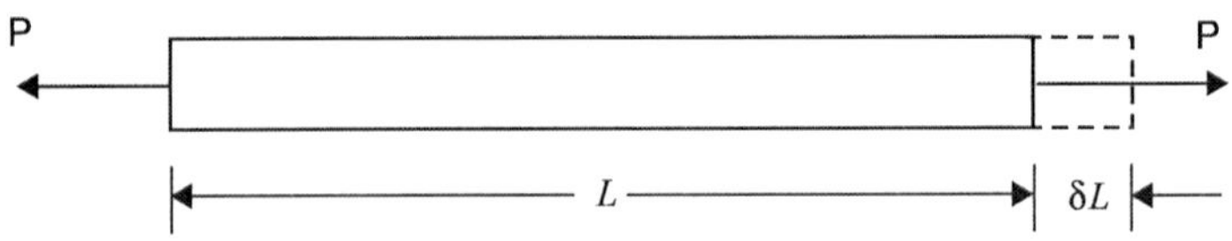

Fig.7.6 Tensile strain

(***b***) **Compressive Strain:** It is the ratio of reduced length to the original length of the member is known as compressive strain.

$$\text{Compressive strain } \varepsilon_c = \frac{\delta L}{L} = \frac{\text{Decrease in length}}{\text{Original length}}$$

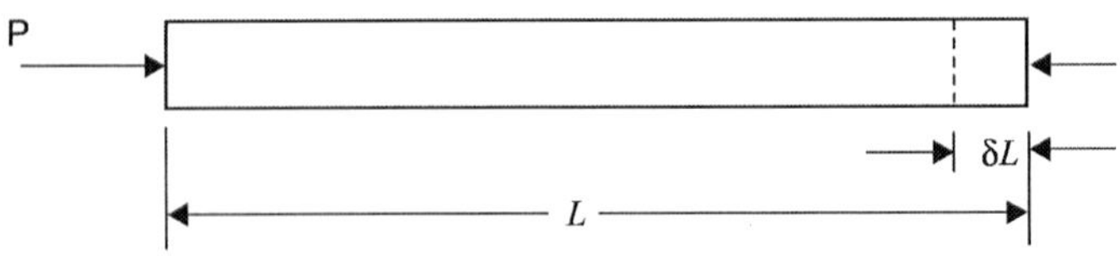

Fig.7.7 Compressive strain

7.5.2. Shear Strain

It is defined as when two equal and opposite non-collinear parallel forces tends to shear off the resisting section.

$$\text{Shear strain } \varepsilon_s = \frac{\text{Deformation}}{\text{Original length}} = \frac{\delta N^1}{\delta P} = \tan\phi = \phi \qquad (\because \phi \text{ is very small})$$

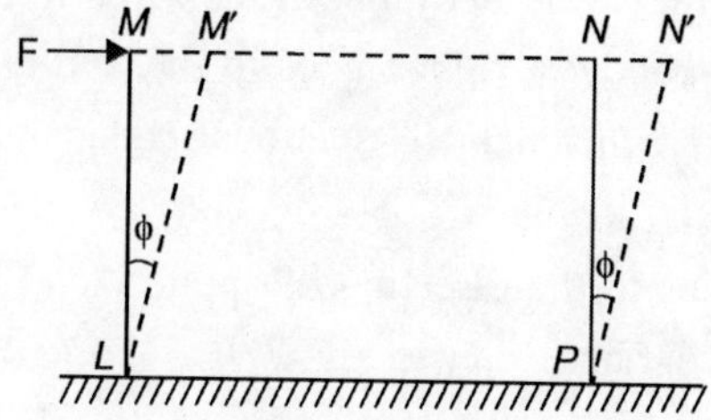

Fig.7.8 Shear strain

7.5.3. Volumetric Strain

It is defined as the ratio between change in volume and original volume of the body known as volumetric strain or bulk strain,

$$\text{Volumetric strain } \varepsilon_v = \frac{\text{Change in volume}}{\text{Original volume}} = \frac{\delta v}{v}$$

7.5.4. Longitudinal Strain

When a body is subjected to an axial tensile load, there is an axial deformation in the length of the body.

The ratio of axial deformation to the original length of the body is known as longitudinal or linear or primary strain. Refer to Fig. 7.9.

$$\text{Longitudinal strain} = \frac{\delta l}{l}$$

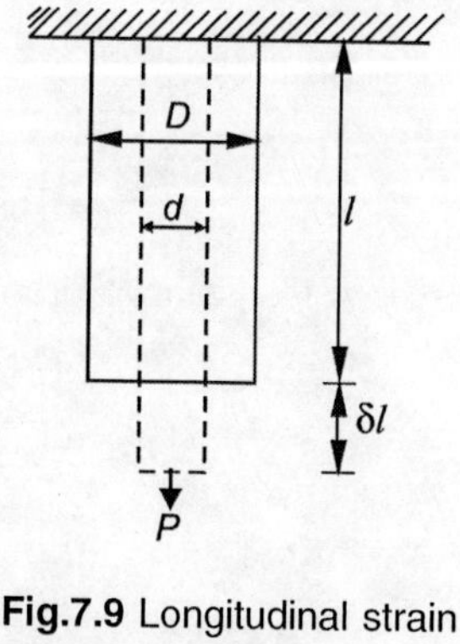

Fig.7.9 Longitudinal strain

where l = Original length of the rod

P = Tensile force acting on the rod

δl = Increase in length of the rod in the direction of P

7.5.5. Lateral Strain

When a body is subjected to an axial tensile or compressive load, there is an axial deformation in the length of the body but at the same time there is a change in other dimensions of the body at right angles to the line of action of the applied load. Thus, the body is having axial deformation at right angles to the line of action of the applied load.

The strain at right angles to the direction of applied load is known as lateral or transverse or secondary strain. Refer to Fig. 7.9.

where D = Original length of the rod

d = Reduced diameter of rod after elongation

P = Tensile force acting on the rod

Hence

$$\text{Lateral strain} = \frac{D-d}{D}$$

If longitudinal strain is tensile, the lateral strain will be compressive strain. If longitudnal strain is compressive, then the lateral strain will be tensile.

7.6 HOOK'S LAW

It state that when a material is loaded, within its elastic limit. The stress is proportional to strain. Mathematically,

$$\text{Hook's Law} = \frac{\text{Stress}}{\text{Strain}} = E = \text{Constant}$$

$$\because \quad \text{Stress} \propto \text{Strain}$$

$$\sigma \propto \varepsilon$$

$$\sigma = E.\varepsilon$$

$$\text{Stress} = \text{Constant of Proportionality} \times \text{Strain}$$

$$E = \frac{\sigma}{\varepsilon}$$

where σ = Stress

ε = Strain

E = Modulus of elasticity, const. of proportionality

Hence, Hook's law is hold good equally for tension as well as compression load.

7.7 YOUNG'S MODULUS OF ELASTICITY

The ratio of normal stress to normal strain within the elastic limit is called Young modulus of elasticity.

$$\text{Modulus of elasticity } E = \frac{\text{Normal stress}}{\text{Normal strain}}$$

$$E = \frac{\sigma}{\varepsilon}$$

In SI units the value of E is N/m^2 or Pascals.
The stress and strains may be tensile or compressive.

7.8 MODULUS OF RIGIDITY OR SHEAR MODULUS

It is defined as the ratio of shear stress to shear strain is known as modulus of rigidity. It is denoted by C or G.
Modulus of rigidity

$$G = \frac{\tau}{\phi} = \frac{\text{Shear stress}}{\text{Shear strain}}$$

or $$C = \frac{\tau}{\phi}$$

where Z is shear stress and ϕ is shear strain.

7.9 BULK MODULUS OF ELASTICITY

It is a ratio of direct or normal stress to the volumetric strain known as bulk modulus of elasticity. It is depoted by K. Mathematically,

$$\text{Bulk modulus, } K = \frac{\text{Direct stress}}{\text{Volumetric strain}} = \frac{\sigma}{\varepsilon_v}$$

7.10 EFFECT OF STRESS AS A LONGITUDINAL AND LATERAL STRAIN

The effect of longitudinal stress and lateral strain as shown by a element in Fig. 7.10

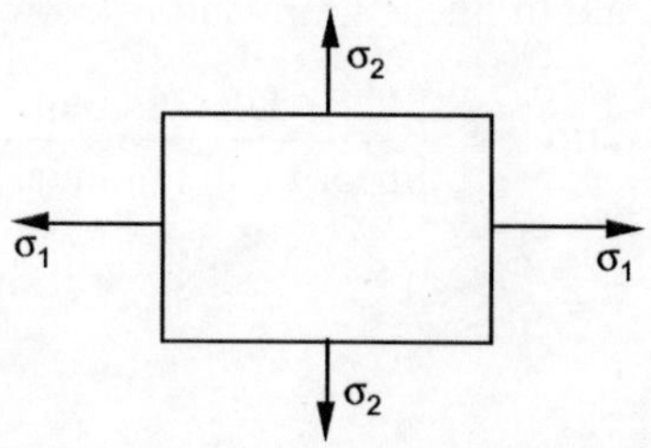

Fig.7.10 Stress as a longitudinal and lateral strain

Strain due to σ_1:

In x direction *i.e.* longitudinal strain $= \frac{\sigma_1}{E}$

$$[\because E = \frac{\sigma}{\varepsilon}, \text{ so longitudinal strain} = \frac{\sigma}{\varepsilon}]$$

In y direction *i.e.* lateral strain $= -v\frac{\sigma_1}{E}$

–ve sign shows that it is reverse, because if longitudinal strain is tensile then lateral strain is compressive.

$$[v = \frac{\text{Lateral strain}}{\text{Longitudinal strain}}, \text{ Lateral strain} = v \times \text{Longitudinal strain}]$$

Strain due to σ_2:

In x direction *i.e.* lateral strain $= -v\frac{\sigma_2}{E}$

In y direction *i.e.* longitudinal strain $= \frac{\sigma_2}{E}$

Now strain in x direction due to stress (σ_1 and σ_2)

$$\frac{\sigma_1}{E} - v\frac{\sigma_2}{E}$$

and strain in y direction due to stresses (σ_1 and σ_2)

$$\frac{\sigma_2}{E} - v\frac{\sigma_1}{E}$$

7.11 POISSON'S RATIO

When a bar is subjected to normal tensile stress, it produces tensile strain in the direction of normal stress, and the body elongates in the direction of stress. There is a contraction in cross-sectional area of the body and produces lateral strain. Hence

"It is the ratio of lateral strain to linear strain and is known as Poisson's ratio."

$$\text{Poisson's ratio } \mu = \frac{\text{lateral strain}}{\text{linear or longitudnal strain}}$$

$$\mu = \frac{1}{m}$$

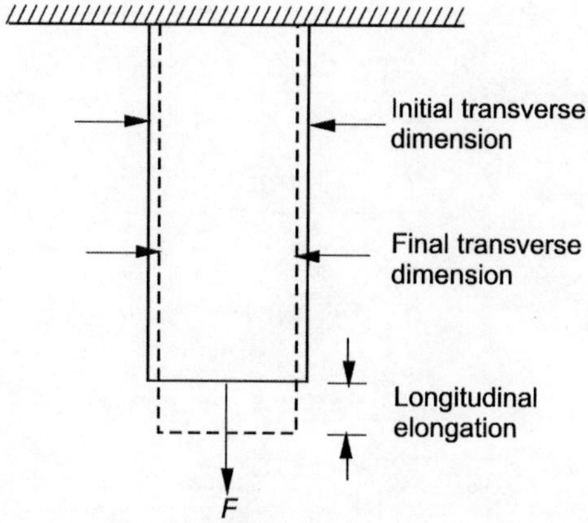

Fig.7.11 A bar

Generally the valve of μ lies between 0-25 to .35 for most of the engineering materials.

Table 7.1: Poisson's ratio for some of the common materials

S. No.	Material	Poisson's ratio
1.	Aluminium	0.330
2.	Brass	0.340
3.	Bronze	0.350
4.	Cast iron	0.270
5.	Concrete	0.200
6.	Copper	0.355
7.	Steel	0.288
8.	Stainless stell	0.305
9.	Wrought irlon	0.278

7.12 RELATIONSHIP BETWEEN ELASTIC CONSTANTS

We are using three elastic constants:

1. Modulus of elasticity (E)
2. Modulus of rigidity (G)
3. Bulk modulus (K)

7.12.1 Relation between *E* and *G* or *C*

Consider a cube of unit size acted upon by a shear stress of intensity τ as shown in figure. Due to stresses some distortion will take place. Hence, the diagonal AC will elongate and diagonal BD will shorten.

Let the shear stress τ cause shear strain ϕ so shear strain

$$\text{Shear strain } AC = \frac{AC_1 - AC}{AC} = \frac{C_1C_2}{AC}$$

$$= \frac{CC_1 \cos 45^\circ}{AB\sqrt{2}} = \frac{CC_1}{2AB}$$

$$= \frac{1}{2}\left(\frac{CC_1}{AB}\right) = \frac{1}{2}\phi \qquad (\because AB = BC)$$

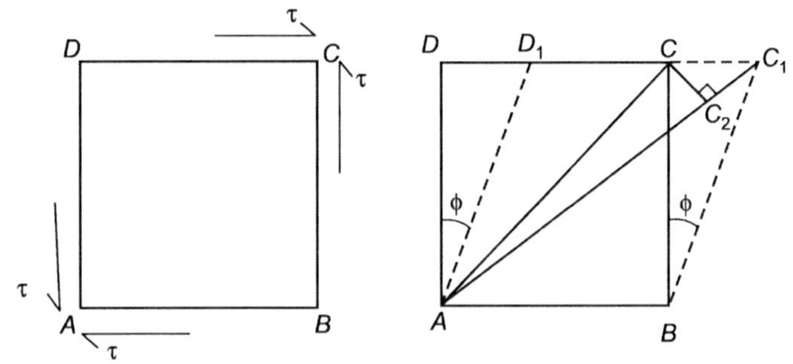

Fig.7.12 A curbe before and after distortion

Now linear strain in diagonal AC

$$= \frac{\phi}{2} = \frac{\tau}{2G} \qquad \ldots(i)$$

where τ = shear stress

G = modulus of rigidity

$$\text{Strain of } AC = \frac{\tau}{E} \qquad \left[\because E = \frac{\sigma}{\varepsilon}\right]$$

and tensile strain of diagonal AC due to compressive stress on diagonal BD

$$= V.\frac{\sigma}{E}$$

Now total strain on diagonal AC

$$= \frac{\tau}{E} + V.\frac{\tau}{E}$$

$$= \frac{\tau}{E}(1 + v) \qquad \ldots(ii)$$

Equating equations (*i*) and (*ii*)

$$\frac{\tau}{2G} = \frac{\tau}{E}(1+v)$$

$$G = \frac{E}{2(1+v)}$$

where

G = Modulus of rigidity

E = Modulus of elasticity

$$G = \frac{E}{2(1+v)} \qquad \ldots(iii)$$

7.12.2 Relation between Young's Modulus (*E*) and Bulk Modulus (*K*)

Consider a cube of volumetric stress σ which acts simultaneously along perpendicular direction *x*, *y* and *z*.

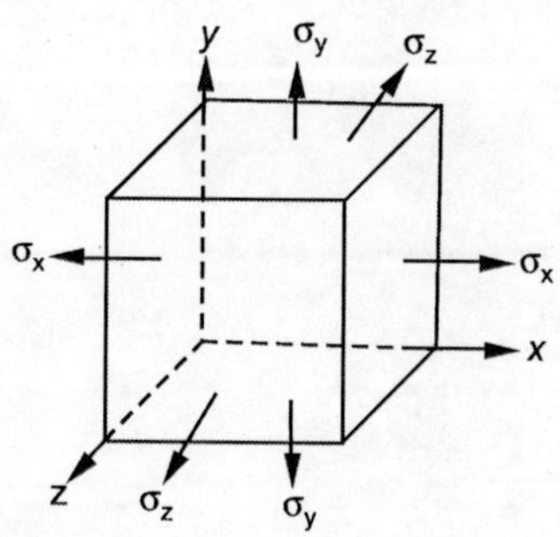

Fig.7.13 A cube of volumetric stress

$$\text{Bulk modulus } (K) = \frac{\text{Normal/Direct stress}}{\text{Volumetric strain}}$$

$$K = \frac{\sigma}{\varepsilon_v} \qquad \text{Where} \qquad \text{stress } (\sigma) = \sigma_x = \sigma_y = \sigma_z$$

$$\text{volumetric strain } \varepsilon_v = \varepsilon_n + \varepsilon_y + \varepsilon_z = \frac{3\sigma}{E}(1-2v)$$

$$K = \frac{\sigma}{\frac{3\sigma}{E}(1-2v)}$$

$$E = 3K(1-2v) \qquad \ldots(iv)$$

7.12.3 Relation between *E*, *G*, *K*

With reference to the equations (*iii*) and (*iv*)
from equation (*iii*)

$$G = \frac{E}{2(1+v)}$$

$$v = \frac{E}{2G} - 1 \qquad \ldots(v)$$

Now putting the valve of equation (*v*) in equation (*iv*)

$$E = 3K\,(1 - 2v)$$

$$= 3K\,(1 - 2)\left[\frac{E}{2G} - 1\right]$$

$$E = 3K\left[1 - \left(\frac{E}{G} - 2\right)\right] = 3K\left[3 - \frac{E}{G}\right]$$

$$E = 9K - \frac{3KE}{G}$$

$$E + \frac{3KE}{G} = 9K,$$

$$EG + 3KE = 9KG$$

$$E = \frac{9KG}{3K + G}$$

$$E = \frac{9KG}{G + 3K}$$

$$E = \frac{9KG}{G + 3K}$$

7.13 PRINCIPLE OF SUPERPOSITION

Principle of superposition states that, when number of forces acting on a body along lengthwise (sectionwise). The total resulting strain (deformation) is equal to algebraric sum of strains due to individual loads.

Resulting strain,

$$\delta l = \sum_{i=1}^{i=n}\left(\frac{PL}{AE}\right) = \left[\frac{P_1L_1}{A_1E_1} + \frac{P_2L_2}{A_2E_2} + \frac{P_3L_3}{A_3E_3}\right]$$

7.14 MECHANICAL PROPERTIES OF METALS

Mechanical properties are the most important parameters of materials from the engineering point of view, for the design purposes. Some important mechanical properties of material are as follows:

1. **Strength:** It is the ability of a material to resist externally applied load or forces without breaking. It is usually defined as tensile strength, compressive strength and proof stress, shear strength, etc.
2. **Hardness:** It is the property of the metal by virtue of which it can resist abrasion, indentation machining and scratching. It is usually expressed in relation to the hardness of other materials. Hardness decreases by heating and completely disappears on the removal of external forces.
3. **Elasticity:** The property by virtue of which certain materials return back to their original position after the removal of the external force, is called elasticity. Example: Steel, rubber, Steel is said to be more elastic than rubber.

   There is a limiting value of force within which, the deformation completely disappears, on the removal of the force: The value of stress upto which the material is perfectly elastic is known as elastic limit.
4. **Plasticity:** Plasticity is the property of the material which causes the formation of permanent deformation.

   The deformation takes place in material even without fracture, remains after removing the load, the material is called plastic. Example: Gold, lead.
5. **Ductility:** It is the property of a metal which enables it to be drawn out in sheets or wires or elongated to an appreciable extent without rupture occurs. Ductile material must be strong and plastic in nature. Ductility of a metal generally measured by the percentage of elongation and the percentage of reduction of area before rupture of a test piece.
6. **Brittleness:** It is the property of breaking of a material with little permanent distortion. Lack of ductility is brittleness.

   Metals having very little property of deformation, either elastic or plastic is called brittle. Usually the tensile strength of brittle materials is only a fraction of their compressive strength. Example: Cast Iron.
7. **Malleability:** The property of a metal by virtue of which it can be beaten or rolled into thin sheets without rupture is known as malleability. This property generally increases with increase of temperature.
8. **Toughness:** It is the property of a material to resist fracture due to high impact loads like hammer blows.

   It is treasured by the amount of energy (in mN) that a unit volume of the metal has absorbed after being stressed upto the point of fracture. Genrally toughness decreases when material is heated. Examples: Mild steel, brass etc.

7.15 DETERMINATION OF TENSILE STRENGTH AND OTHER PROPERTIES ON UNIVERSAL TESTING MACHINE

The tensile strength of material is determined by loading it in a testing machine known as universal testing machine (UTM). The test is known as tensile test and is widely used to indicate elastic strength, yield strength, ultimate strength, ductility and toughness of a metal.

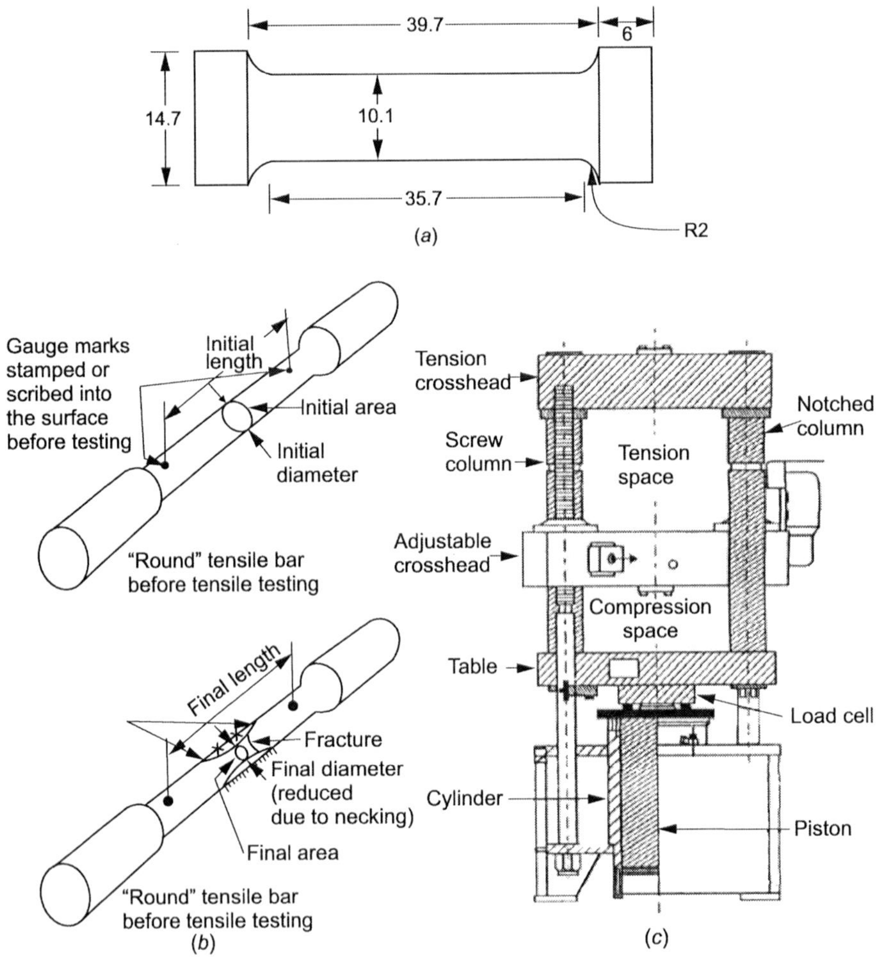

Fig. 7.14 (*a*) The dimensions of specimen for tensile testing on universal testing machine. (*b*) The diameter of specimen before and after tensile listing. It is clearly shown that diameter of specimen reduces before fracture due to necking. (*c*) Universal testing machine.

The load commands of preparation of a test specimen as per standard. The test specimen is held in machine by suitable means (*i.e.*, Jaws) and then uniformly increasing tensile load is applied on it.

As the load increases the specimen initially gets elastically elongated. On further elongation the specimen starts necking at some point when the material goes beyond the elastic range. The reduced cross-section area of the specimen would further be reduced under the force of the load and finally develop fracture when test is completed. During the test a record is maintained for the load and elongation correspondingly .

(*i*) Stress and strain are calculated as per following relations:

$$\text{Stress} = \frac{\text{Load}}{\text{Cross - sectional Area}}$$

(*ii*) $\text{Strain} = \dfrac{\text{Change in Length}}{\text{Original Length}}$

Stress, strain are calculated and stress strain curve is drawn. The shape of stress strain curve is essentially similar to the load- elongation curve.

7.15.1 Stress Strain Curve of Mild Steel

Stress strain curve of mild steel is shown in Fig. 7.15. The various properties of material are as follows.

O to P $\rightarrow$ Proportionality limit

O to E $\rightarrow$ Elasticity limit

y_1 and y_2 $\rightarrow$ Upper and lower yield point

U $\rightarrow$ Ultimate load point

F $\rightarrow$ Breaking limit

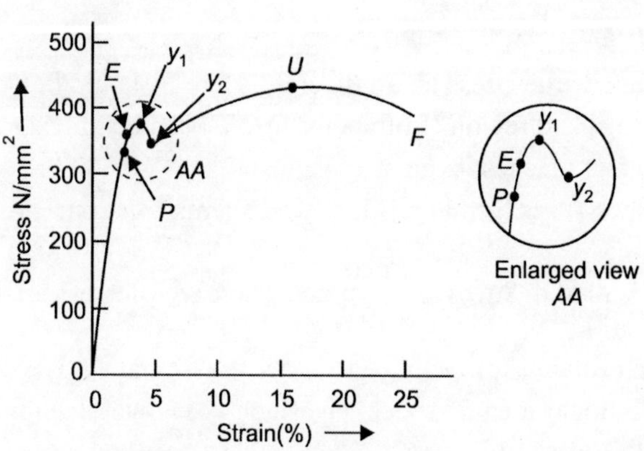

Fig. 7.15 Engineering Stress Strain Curve for Mild Steel.

Proportionality Limit (*O-P*) : From zero load point *O* to *P* the stress stain curve is geometrically a straight line which represents stress is proportional to strain. Beyond point *P* curve slightly deviates from the straight line. It is obvious that Hooke's law holds good up to point *P*. It is known as proportional limit.

Elastic Limit (*P-E*) : From point *P* to *E* Hooke's law is not obeyed although the metal is still in elastic state *i.e.* test piece will come to original position or state upon unloading, but the load extension relation is non-linear. After point *E* plastic flow or yielding starts, hence *E* is known as elastic limit for the material. For mild steel it occurs at 0.12% of strain. If the material is stressed beyond this point plastic deformation will occur *i.e.*, strain which is not recoverable if the load is removed.

Yield Point (y_1 and y_2): Beyond point *F* load increases a little after which there is momentary decrease of load as the extension increases. This upper point-y_1 just after the elastic limit is known as yield point or more precisely as upper yield point. The lower load point after y_1 is known as lower yield point y_2. Upper and lower yield points can be obtained only in a carefully controlled test otherwise the region from *E* and y_1 to y_2 merges into one curve almost horizontal showing apparently an increase of strain from 0.12% to 2% at constant stress.

Ultimate Load Point (*U*) : After y_2 point curve is smooth and much flatter. The stress strain curve continues to rise up to the point *U*. The strain in the region from (y_1, y_2) to *U* is 100 times more than in region from *O* to y_1, y_2 and is partly elastic (recoverable) but mainly plastic (permanent strain). The stress corresponding to point *U* is called ultimate strength of material.

Breaking Point (*F*) : On further leading beyond point *U*, a neck formation starts, the apparent stress decreases and material breaks at *F*. The stress corresponding to *F* is called breaking (fraction) strength of the material.

7.15.2 True Stress Strain Curve

The Fig. 7.16 shows true stress strain diagram. Here it should be pointed out that true stress in the necked region (from point *D* to *E*) is much higher than shown in Fig. 7.16 and in fact reaches its greatest value at the (Point *E*) breaking load. It is because as the neck starts forming, [Fig. 7.16(*b*)] cross-sectional area is decreased and so stress is increased. Stress $= \dfrac{\text{Load}}{\text{Area}}$ and when area decreases stress increases.

However, it is an established practice to draw stress-strain curve on the basis of original cross-sectional area of specimen which is known as engineering stress-strain curve and shown in Fig.7.16.

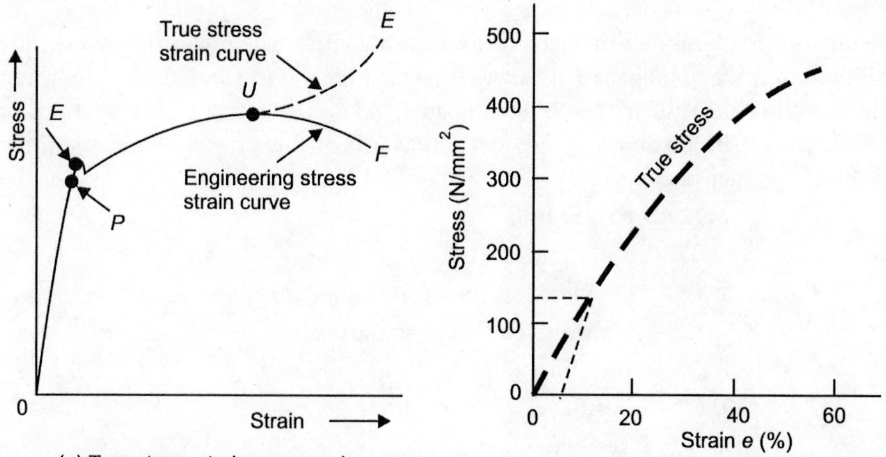

(*a*) True stress strain curve and Engineering stress strain curve

(*b*) True strain curve for copper

Fig. 7.16 True stress strain curves

7.15.3 Stress Strain Curve for Brittle Material

Fig. 7.17 shows stress strain curve for a brittle material. As it can be observed from figure that material fractures with very little or no plastic strain. If the percentage elongation is less than 5% material is considered to be brittle.

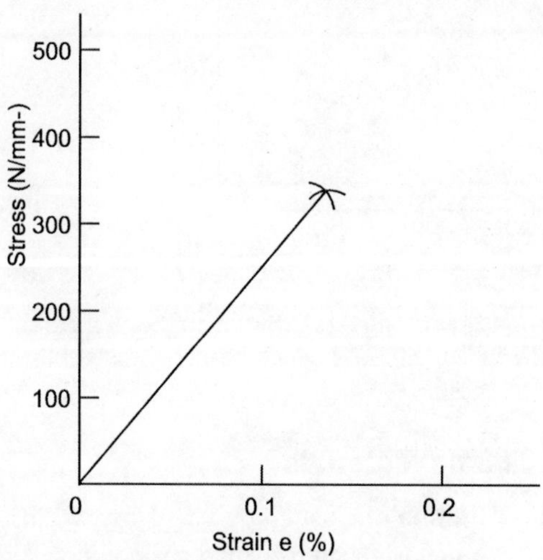

Fig. 7.17 Stress Strain Curve for Cast Iron. Some Other Brittle Materials Like Glasses. Ceramics, Some Metals and Some Plastics Shows Same Type of Stress Strain Curve.

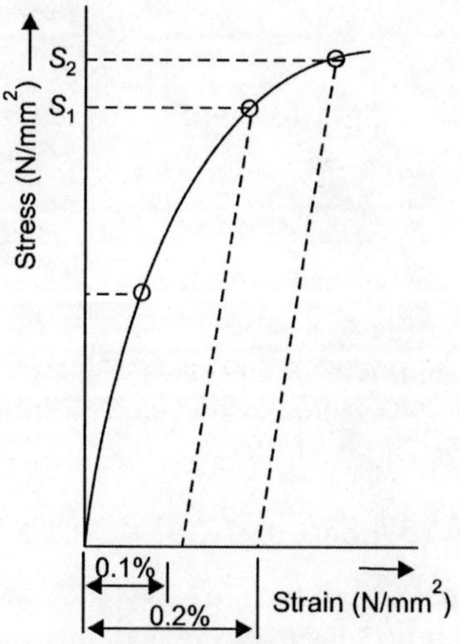

Fig. 7.18 Stress Strain Curve for Material not having Well Defined Yield Point. Offset Method is used for Determining Yield Strength. S_1 is the 0.1% Offset Yield Strength; S_2 is 0.2% Offset Yield Strength.

Resilience

Resilience is the ability of material to absorb energy when deformed elastically and return it when unloaded. The cross-hatched area *O* to *I* in Fig. 7.19 indicates the modulus of resilience for structural steel and area *O* represent the modulus of resilience for high carbon spring steel. Because of higher yield strength spring steel has greater resilience.

E → Elastic limit
Y → Yield point
U_1 → Ultimate load point for high carbon spring steel
U_2 → Ultimate load point for structural steel
F → Breaking point
OYG → Resilience of high carbon spring steel
O to I → Resiliene of structural steel

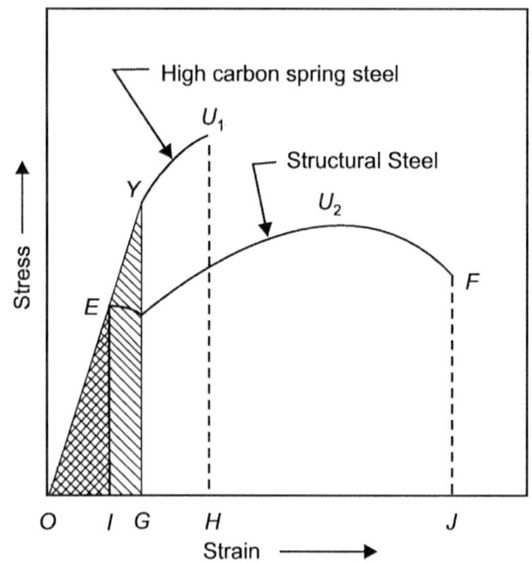

Fig. 7.19 The Resilience of High Carbon Spring Steel and Structural Steel as area *OYG* is greater than area *OEI*, it clearly indicates that high carbon spring steel has greater resilience.

7.15.4 Stress Strain Curve for Elastic Materials

In the beginning of Chapter, as per the definition of elasticity given. Now we shall discuss elasticity and plasticity formally. A material can be called perfectly elastic, if and only if:

1. While loading or unloading, the deformation and recovery are instantaneous.
2. The recovery is complete and immediate *i.e.*, no permanent set is left after unloading.

3. The load deformation curve or load-elongation curve has same shape while loading and unloading as shown in Fig. 7.20. If load elongation curve follows a straight line equation then material is called as linear elastic material, Fig. 7.20 (*a*) and if does not follow then ,we say it is non-linear elastic material Fig. 7.20 (*b*)

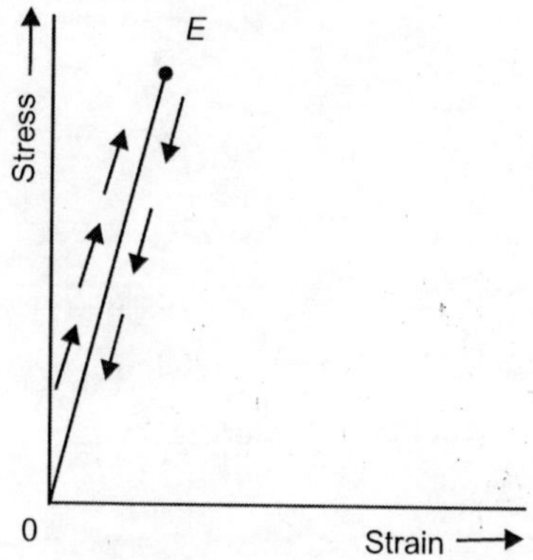

(*a*) Linear elastic material

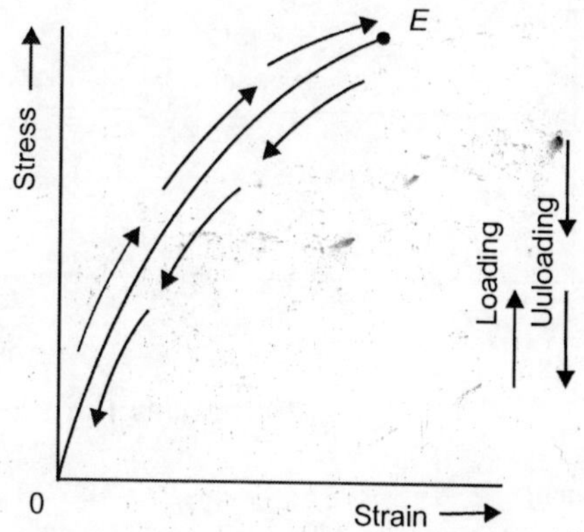

(*b*) Non-linear elastic material

Fig. 7.20 Ideal Case of Linear Plastic and Non-linear Elastic Material. These are ideal cases as loading and unloading follows same curve.

Actually there is no material which satisfies above three requirements for perfect elasticity. However, for practical purposes most of steels and other structural material satisfy these requirement up to their proof stress. Mild steel can be considered as elastic upto point E (Elastic limit) and linearly elastic upto point P (proportionality limit). Fig. 7.20(*a*) shows stress strain curve for an ideal elastic plastic material.

7.15.5 Perfectly Plastic Material

While discussing elasticity it was casually assumed that all the deformation above elastic limit are plastic. The characteristics of a perfectly plastic material are as follows:

(*i*) A perfectly plastic material shows an unlimited amount of deformation or strain at constant stress.

(*ii*) After removal of load plastic strain cannot be recovered.

(*iii*) Plastic deformations could be instantaneous.

Stress strain curve for ideal plastic material is shown in Fig. 7.21(*b*). If the elastic deformation is negligibly small, the rigid plastic idealisation is valid and for a linear strain hardening material Fig. 7.21(*c*) could be reasonable approximation. It should be kept in mind that like perfect elasticity perfect plasticity is also only an ideal concept and no material in nature is perfectly plastic. Plastic deformation are almost always time dependent. Perfect plastic idealisation can be used only when time is very small for deformation. In general all non-recoverable deformation are termed as plastic and all recoverable deformations are called elastic irrespective of the time of deformation.

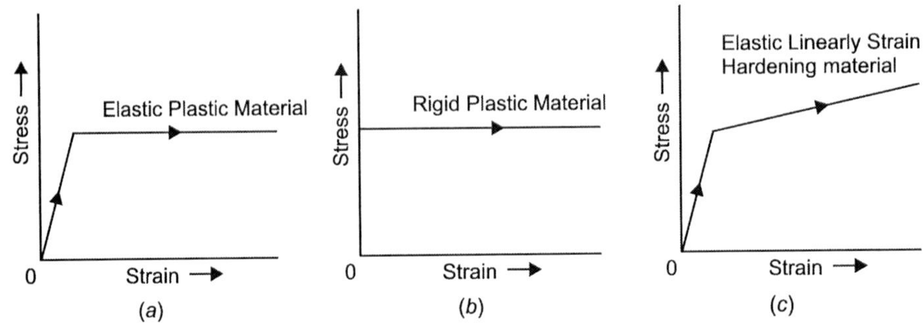

Fig. 7.21 Represents Three Ideal Cases of Plastic Materials.

Creep

While discussing perfectly elastic or plastic materials we assumed that material shows no time dependent flow. However, metal under high stress or high temperature shows this phenomenon. It can be defined as slow and progressive deformation of a material under constant stress. If a specimen is tested under the conditions of constant load and constant temperature and a graph is plotted for deformation (elongation) versus time then it would be similar to as shown in Fig. 7.22. Although the rate of elongation is small it is of great importance in design of equipment such as steam turbine, power plants and high temperature pressure vessels that operate under high loads and temperature for a long period of time.

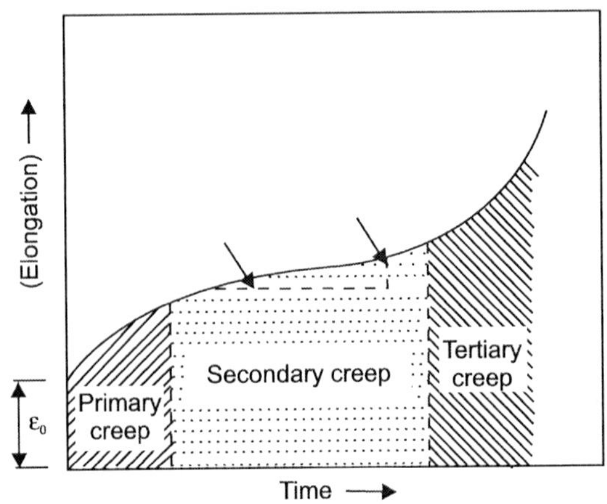

Fig. 7.22 Creep Curve of a Specimen at a Fixed Temperature and Fixed Load.

7.16 MAXIMUM OR ULTIMATE TENSILE STRESS

It is calculated by dividing the maximum load carried out by the specimen during a tension test by the original cross-sectional area of the specimen.

$$\therefore \quad \text{Ultimate stress} = \frac{\text{Maximum load}}{\text{Original cross-sectional area}}$$

7.17 WORKING STRESS

The stress used in practical design is termed as the working stress. It is also called safe stress or allowable stress. This is obtained by dividing the ultimate stress of a material by the factor of safety.

$$\therefore \quad \text{Working stress} = \frac{\text{Ultimate stress}}{\text{Factor of safety}}$$

7.18 FACTOR OF SAFETY

The ratio of ultimate stress to working stress is called factor of safety

$$\therefore \text{Factor of safety} = \frac{\text{Ultimate stress}}{\text{Working stress}}$$

It depends upon the following factors:

(*i*) Nature of loading

(*ii*) Frequency of variation of load

(*iii*) Degree of reliability required, and

(*iv*) Decrease of section by corrosion etc.

Factor of safety is decided by experience. Various values of factor of safety for different types of loading are given in Table 7.1.

Table 7.1: Factor of Safety

Material	Dead load	Live load	Shock
Cast Iron	4	0 to 10	10 to 15
Wrough Iron	3	5 to 8	9 to 13
Mild steel	3	5 to 8	9 to 13
Copper	5	6 to 9	10 to 15
Timber	5	9 to 14	14 to 20

7.19 BREAKING STRESS

It is determined by dividing the load at the time of fracture or breaking by the original cross-sectional area.

$$\therefore \quad \text{Breaking stress} = \frac{\text{Load at the breaking point}}{\text{Original cross-sectional area}}$$

7.20 PROOF STRESS

Proof stress is the stress necessary to cause a non-proportional or permanent extension equal to a defined percentage of gauge length. If the specified percentage is 0.1% of the gauge length, the corresponding proof stress is designed as 0.1% of proof stress. The stress-strain curve for most of non-ferrous metals like copper and their alloys do not have a definite yield point. In case of such metals and alloys the term proof stress is used instead of yield point.

Using 50 mm gauge length, 0.1% proof stress is the stress required to produce a permanent extension of 0.1% of 50 mm *i.e.* of 0.05 mm. This can be obtained by marking a point *A* at a distance of 0.05 mm (to a scale of extension) from the origin O and drawing a line *AB* parallel to straight line portion of load-extension curve which intersect, the curve at B. (Refer Fig. 7.23)

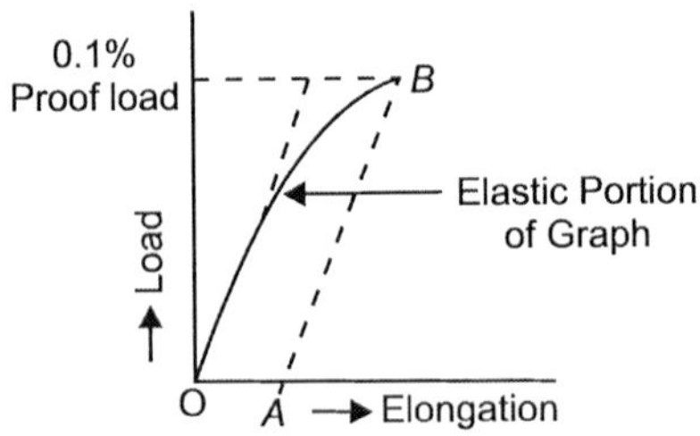

Fig. 7.23 Proof stress

The load corresponding to *B* is 0.1% proof load. The proof load divided by original cross-sectional area gives 0.1% proof stress.

7.21 PERCENTAGE ELONGATION

The percentage elongation is the percentage increase in length of the gauge length. If L_0 is the original gauge length and L_f the final gauge length (measured after fracture), then

$$\text{Percentage elongation} = \frac{L_f - L_0}{L_0} \times 100$$

Material with more than 15% elongation are usually considered as ductile. Those with less than 5% elongatior are considered as brittle. Those between 5% to 15% are of intermediate ductility.

7.22 PERCENTAGE REDUCTION IN AREA

If the original cross-sectional area is A_o and final cross-sectional area is A_f (at the plane of failure), then

$$\text{Percentage reduction in area} = \frac{A_0 - A_f}{A_0} \times 100$$

7.23 ANALYSIS OF BAR OF VARYING SECTIONS

Fig. 7.24 shows a bar which consists of three lengths l_1, l_2, and l_3 with cross-sectional area A_1, A_2 and A_3 and subjected to an axial load P

Stress for the part AB,

$$\sigma_1 = \frac{P}{A_1}$$

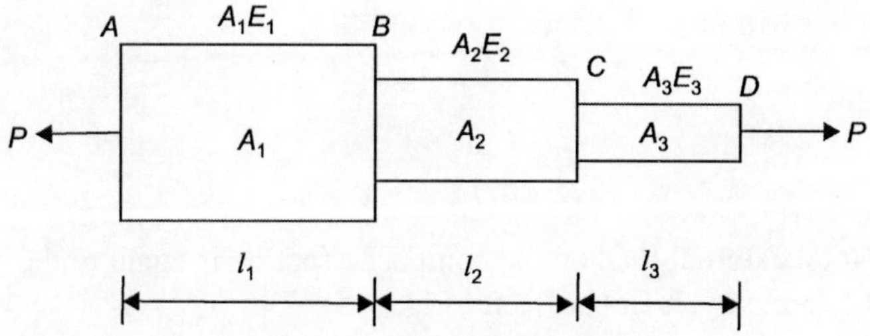

Fig. 7.24 Bar of varying sections

Similarly stress for the part BC and part CD are given as

$$\sigma_2 = \frac{P}{A_2} \text{ and } \sigma_3 = \frac{P}{A_3}$$

Change in length of the part AB, $\delta l_1 = \dfrac{\sigma l_1}{E}$ $\qquad \because = \dfrac{\sigma}{E} = \dfrac{\sigma}{\delta l / l}$

$$\delta l_1 = \frac{\sigma l_1}{E_1}$$

or $\qquad = \dfrac{Pl_1}{A_1E_1}$ $\qquad \because \sigma = \dfrac{\rho}{A}$

Similarly change in length for the part BC and part CD are given as

$$\delta l_2 = \frac{Pl_2}{A_2E_2} \text{ and } \delta l_3 = \frac{Pl_3}{A_3E_3}$$

Total change in the length of the bar, $\delta l = \delta l_1 + \delta l_2 + \delta l_3$

(Take +ve sign for increase in length and –ve sign for decrease in length)

7.24 AN ANALYSIS OF BARS OF COMPOSITE SECTIONS

If a number of bars of same or different materials are joined together in such a way that they extend or contract equally when subjected to an axial tensile or compressive loads they will form a composite section.

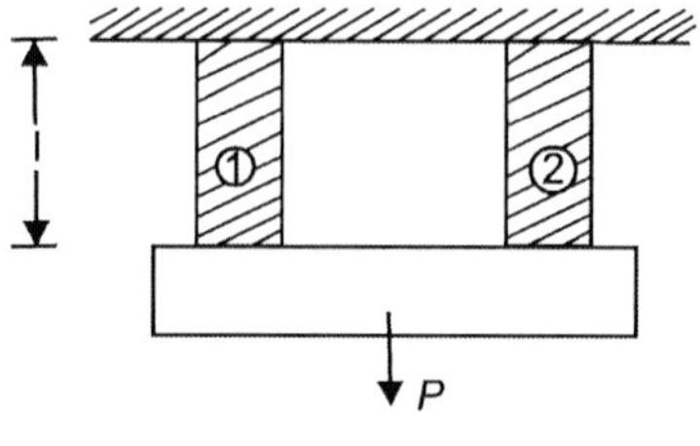

Fig. 7.25 Bar of composite section

The following two points are important for the composite section.

1. The extension or contraction in each bar is equal, so strain produced in each bar will also be equal.

$$E_1 = E_2$$

or

$$\frac{\sigma_1}{E_1} = \frac{\sigma_2}{E_2}$$

2. The total external load on the composite section is equal to the sum of the loads shared by each bar (Refer Fig.7.25)

$$\therefore \quad P = P_1 + P_2$$

$$P = \sigma_1 A_1 + \sigma_2 A_2$$

7.25 FLEXURE LOADING

7.25.1 Beam

A structural member which carries lateral or transverse force (forces at right angles to the axis of the member) is termed as a beam. Generally a beam is of moderate size and is made up of one piece. If the size is large and the beam is made of many parts joined together, than it is called girder.

7.25.2 Type of Beam

Depending upon the end conditions the various types of beams are:

1. Cantilever beam
2. Simply supported beam
3. Overhanging beam
4. Fixed beam
5. Continuous beam.

1. **Cantilever Beam:** A beam which is fixed at one end and free at the other end, is known as cantilever beam. (Fig. 7.26)

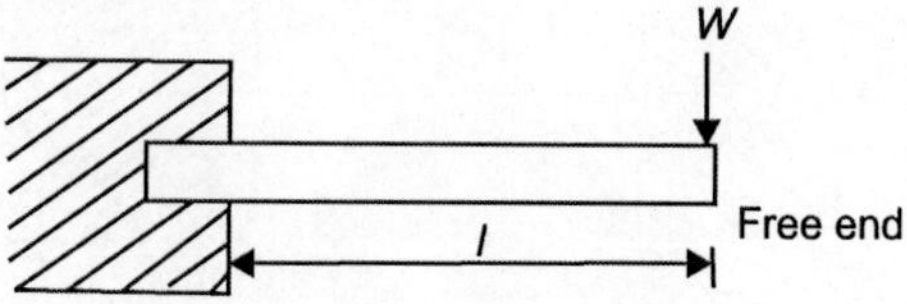

Fig. 7.26 Cantilever beam

where l = Length of Cantilever Beam.

W = load on the beam.

2. **Simply Supported Beam:** Abeam supported or resting freely on the supports at its both ends, is known as simply or freely supported beam (Fig. 7.27). The supported may be in the form of walls or columns.

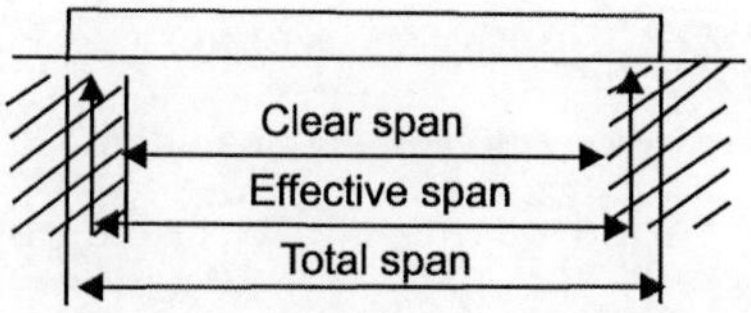

Fig. 7.27 Supported beam

The whole length of the beam is known as its total span. The clear horizontal distance between the walls is called the clear span of the beam. The horizontal distance between the centres of the end bearing is called the effective span of the beam.

3. **Overhanging Beam:** A beam in which its end portion (or portion) is extended beyond the support, is known as overhanging beam (Fig.7.28). A beam may be overhanging on one side or on both sides. (As shown in Fig. 7.28 *a*, *b* and *c*).

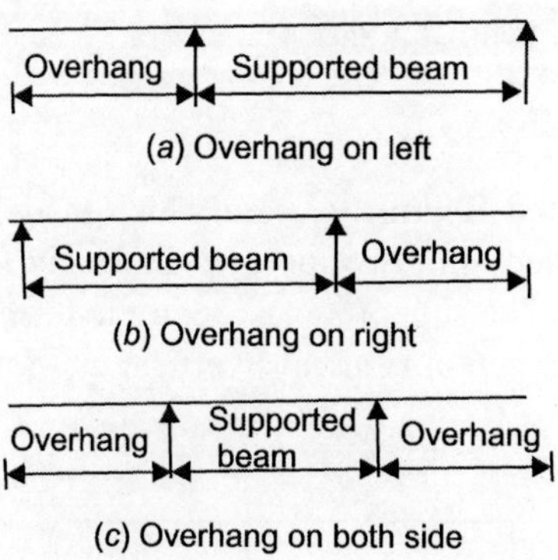

Fig. 7.28 Overhanging beam

4. **Fixed Beam:** A beam whose both ends are fixed or built-in walls, is known as fixed beam As shown in Fig.7.29

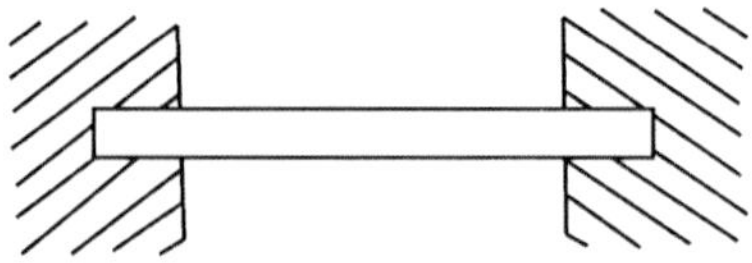

Fig. 7.29 Fixed beam

5. **Continuous Beam**: A Beam supported on more than two supports is known as continuous beam (Fig. 7.30). The extreme left and right supports are called end supports and all the remaining supports are known as intermediate supports.

(Continuous Beam)

Fig. 7.30 Continuous beam

Cantilever, simply supported beams and overhanging beams are statically determinates beams.

In such beams supports reactions can be determined by using the equations of static equilibrium. The equations of static equilibrium are

$$\Sigma H = 0, \Sigma V = 0 \ \ \Sigma M = 0$$

Fixed end beams and continuous beams are statically indeterminates beams. In such beams the supports reactions can not be determined by using the equations of static equilibrium.

7.25.3 Types of End Supports of Beam

The following are the important types of supports for the beams:

1. Simply supported beam
2. Roller supported beams, and
3. Hinged beams.
4. Fixed support

1. **Simply Supported Beams:** Abeam which rests freely on supports at its both ends is called simply supported or freely supported or knife edge supported beam. The supports may be in the form of walls or columns. In such a case, the reaction is always vertical as shown in Fig. 7.31.

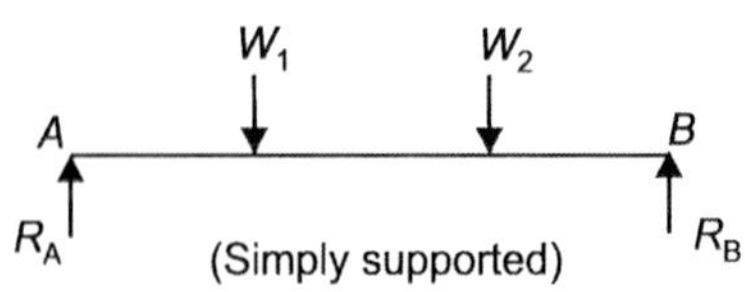

(Simply supported)

Fig. 7.31 Simply supported

2. **Roller Supported Beams:** In this case one of the end of the beam is supported on rollers in order to permit free movement in horizontal direction. Roller reaction is vertical as shown in Fig. 7.32.
3. **Hinged Beam:** In such a case, the end of a beam is hinged to the supported as shown in Fig. 7.33.

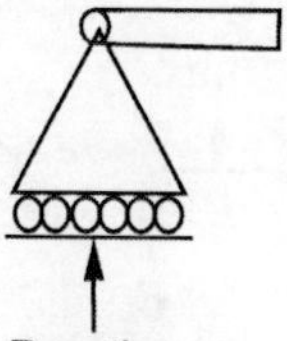

Fig. 7.32 Roller supported beam

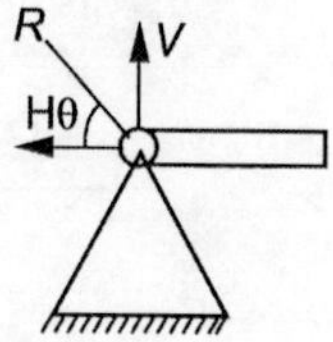

Fig. 7.33 Hinged beam

The structural members supported on hinged supported can rotate about hinge but cannot move side ways Thus the position is fixed in this case. The reaction on a hinged support may be either horizontal or vertical or inclined depending upon the types of loading on the structure.

4. **Fixed Support:** It is a type of support at which three reaction components are developed.

1. For horizontal direction - *H*, 2. For vertical direction - *V*
3. For moment - *M*

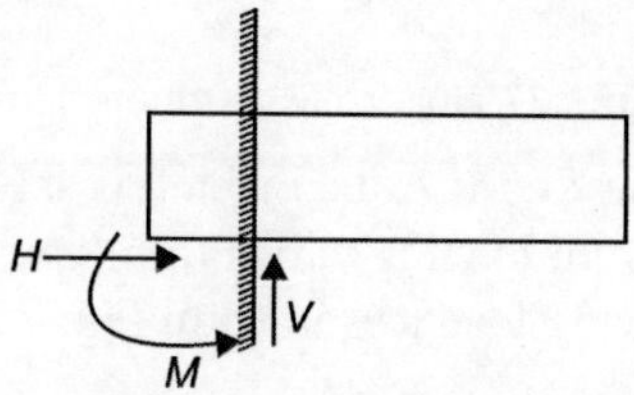

Fig. 7.34 Fixed support

7.25.4 Types of Load

The following are the important types of load from the subject point of view.

1. Concentrated or point load.
2. Distributed load
3. Uniformly distributed load, and
4. Uniformly varying load.

1. **Concentrated or Point Load:** A load acting at a point on the beam is known as a concentrated or a point load as shown in Fig. 7.35.

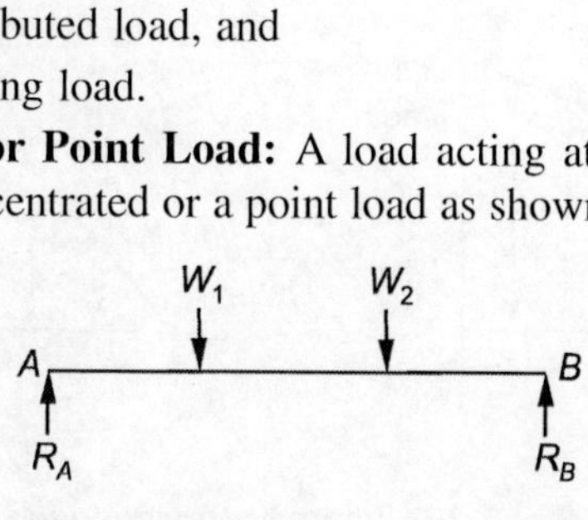

Fig. 7.35 Point load

2. **Distributed Load:** A distributed load may be uniform over the length or it may vary uniformly such loads are measured by their intensity which is expressed by the force per unit distance along the axis of the beam as shown in Fig. 7.36.

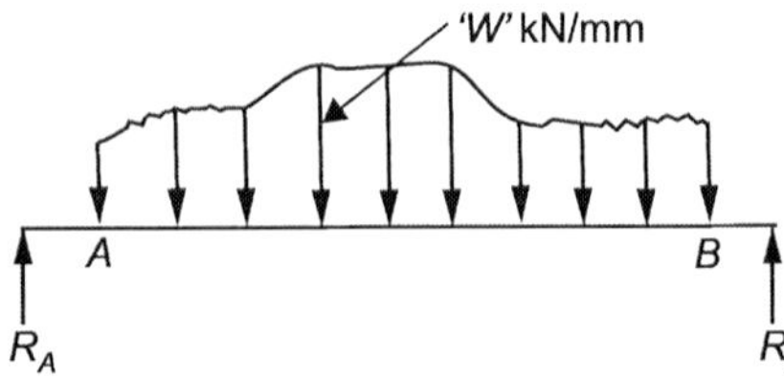

Fig. 7.36 Distributed load

3. **Uniformly Distributed Load:** A load which is spread uniformly over the entire span or small portion of the beam is known as uniformly distributed load (U.D.L) as shown in Fig. 7.37.

   This load is assumed to act at the C.G. of the load for all types of calculations.

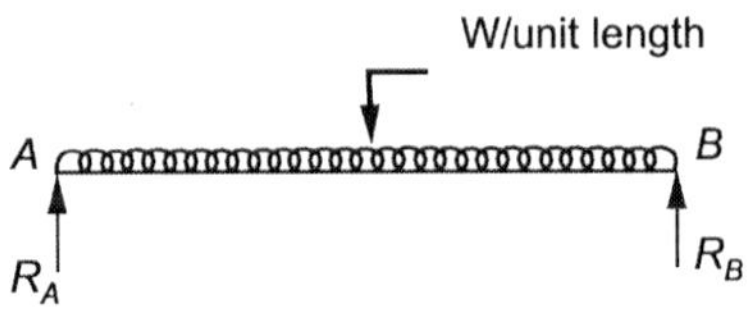

Fig. 7.37 Uniformly distributed load

4. **Uniformly Varying Load:** A load which is spread over a beam, in such a manner that the rate of loading varies from one point to another, is known as a uniformly varying load as shown in Fig. 7.38.

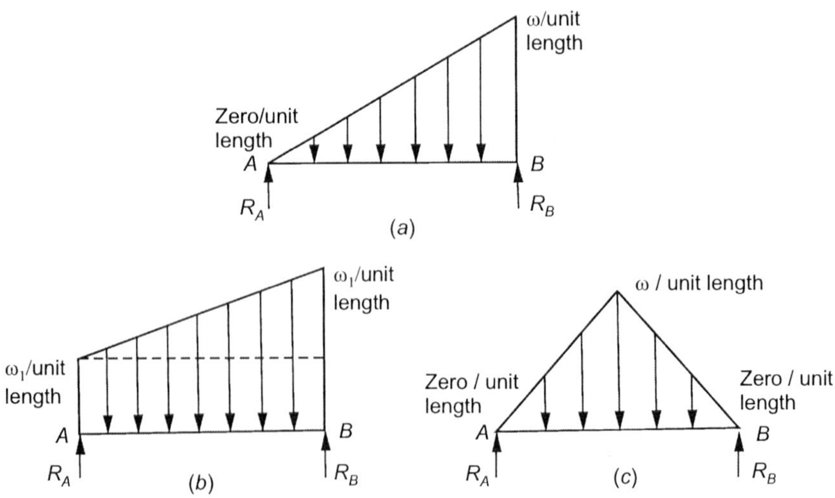

Fig. 7.38 Uniformly varying load

7.25.5 Bending Stresses

When some external load acts on a beam, the shear force and bending moments are set up at all section of the beam. Due to the shear force and bending moment, the beam undergoes certain deformation. The material of the member will offer resistance or stresses against these deformations.

The stresses produced to resist the bending moment are known as bending stresses. The stresses produced to resist the action of shearing force are known as shearing stresses.

7.25.6 Bending Stresses in Beams

A beam is a structural member whose longitudinal dimension is large compared to the transverse dimenstion. The beam is supported along its length and is acted upon by a system of loads at right angles to its axis. Due to external loads and couples, shear force and bending moment develop at any section of the beams. Pure bending refers to loading of beam in a fashion that the beam is absolutely free from shear force and is subjected only to a constant bending moment.

7.25.7 Pure Bending

Consider a simply supported beam loaded as shown in Fig. 7.39. The corresponding shear force depicted along with the beam segment CD, the shear force is zero and the bending moment has a constant value. Apparently only bending stresses will be developed in portion CD of the beam. This state of loading where the beam segment CD is absolutely free from shear force (and shear stress) and is subjected only to constant bending moment (and bending stress) is called pure bending or simple bending. The radius of bend for pure bending portion does not vary.

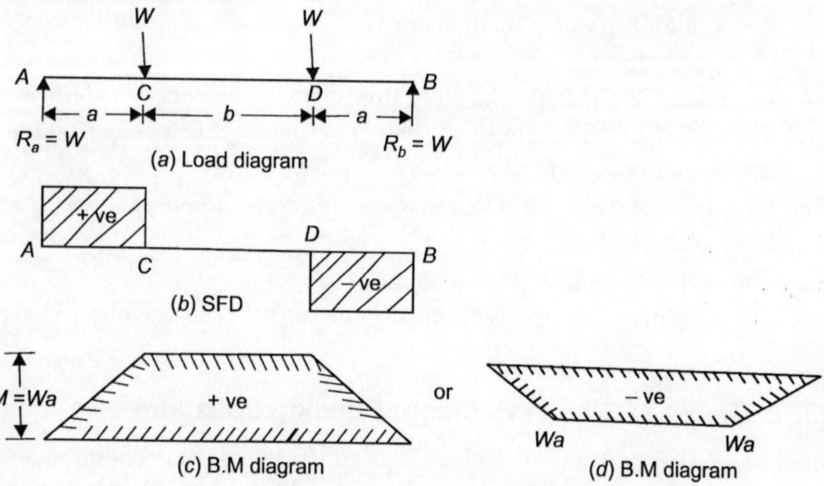

Fig.7.39 Pure bending

7.25.8 Neutral Axis

Due to external loads, bending effects are produced and these cause the beam to sag or hog as shown in Fig. 7.40.

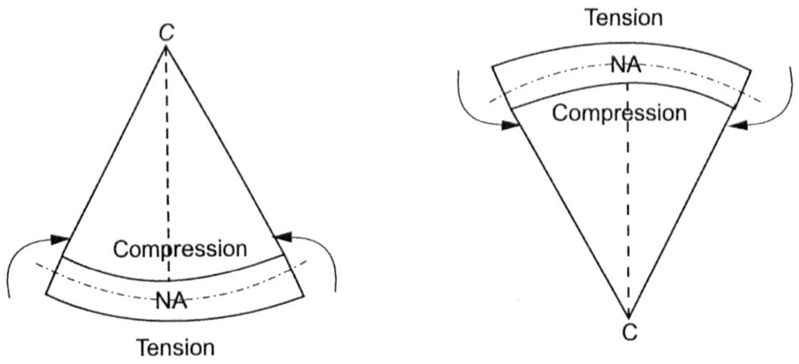

Fig. 7.40 Neutral axis.

During sagging of beam, the fibres at the upper portion of the beam section get compressed and accordingly upper side of the beam is in compression. The fibres at the lower portion of the beam get stretched and are in tension. At a particular section of the beam there exists a surface where the fibres are neither in compression nor in tension. Such a surface is called neutral surface or neutral plane. A trace of this plane on any transverse plane of the beam is called neutral axis (NA).

7.25.9 Assumptions in the Theory of Simple Bending

The following assumptions are made in the theory of simple bending:

1. The material of the beam is homogenous and isotropic.
2. A transverse section of a beam, which is plane before bending will remain a plane after bending.
3. The value of Young's modulus of elasticity is the same in tension and compression,
4. There is no resultant push or pull across the transverse section of the beam.
5. The radius of curvature of the beam before breeding is very large as compared to the transverse ection of the beam.
6. Each layer of the beam is free to expand or contract, independently, of the layer, above or below it.
7. The loads are applied in the plane of bending.
8. The beam material is stressed with in its elastic limit and thus obeys Hooke's law.

7.25.10 Theory of Simple Bending (Bending Equation)

Consider a portion (*ABCD*) of the beam which is subjected to bending moment '*M*'. Let this portion of bend into an arc of a circle with *O* as centre under the effect of applied loads.

Let R = Radius of curvature *i.e.* Radius of Neutral layer.

Consider a layer EF at a distance y from N-axis. Fibres above N-axis will be under compression below N-axis will be under tension.

After deformation EF will become $E'F'$. (Shown in Fig. 7.41).

Change in length of fibre $EF = E'F' - EF$

But $\quad GH = EF = R\theta$ [$\because$ Arc length = Radius of arc × Angle subtended]

$\therefore \quad E'F' = (R + y)\theta$

$\therefore$ Change in length of fibre $EF = (R + y)\theta - R\theta = R\theta + y\theta - R\theta = y\theta$

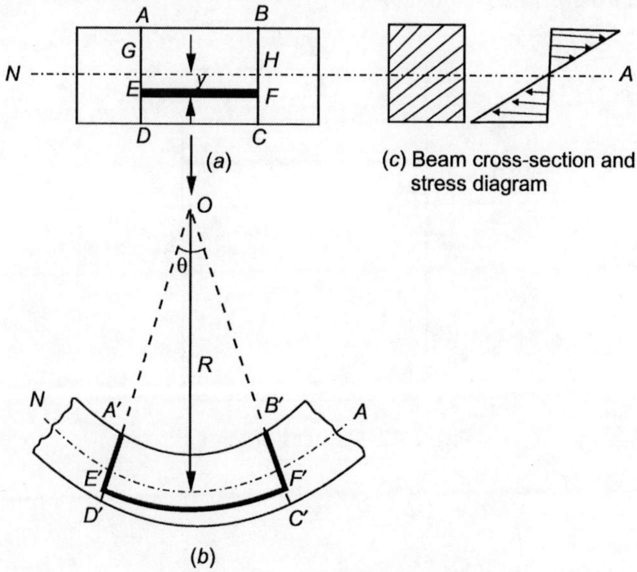

Fig. 7.41 Change in length of fibre.

Strain in layer EF;

$$E = \frac{\text{Change in length of } EF}{\text{Original length}}$$

$$E = \frac{E'F' - EF}{EF} = \frac{y\theta}{R\theta} = \frac{y}{R}$$

$$\text{But strain} = \frac{\text{Stress}}{E} = \frac{\sigma}{E}$$

$$\therefore \quad \frac{\sigma}{E} = \frac{y}{R}$$

$$\text{or} \quad \frac{\sigma}{y} = \frac{E}{R} \qquad \ldots(i)$$

E and R are constant for a given beam. Hence, stress in any fibre is proportional to the distance of the fibre from the N-axis. The maximum bending stresses of opposite nature being at maximum distance on either side of neutral axis are also known as skin stresses.

Consider elementary areas a_1, a_2, a_3, etc. at a distance of y_1, y_2, y_3 etc. respectively from N-axis (Fig.7.42)

Force on elementary area 'a_1' = $\sigma_1 \times a_1$

Force on elementary area 'a_2' = $\sigma_2 \times a_2$ and so on.

Moment of all these forces about N-axis.

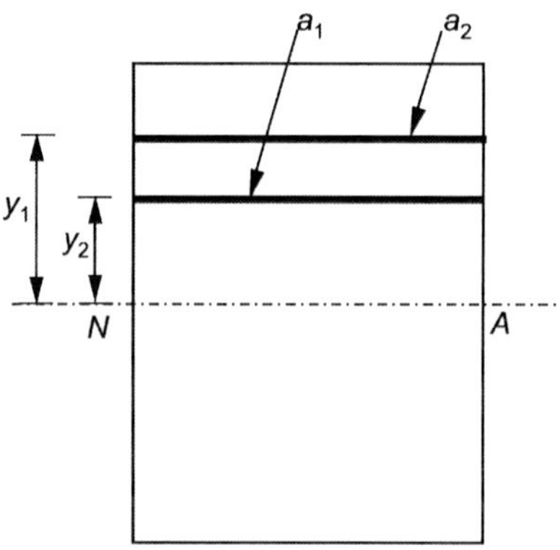

Fig. 7.42 Moment of forces.

$$M = \sigma_1 a_1 y_1 + \sigma_2 a_2 y_2 + \sigma_3 a_3 y_3 + \ldots \qquad \ldots(ii)$$

But $\quad \dfrac{\sigma}{y} = \dfrac{E}{R}$

$\therefore \quad \sigma = \dfrac{E}{R} y \qquad$ (E and R are constants)

$$\sigma_1 = \frac{E}{R} y_1$$

and $\quad \sigma_2 = \dfrac{E}{R} y_2$ and so on.

So put values of σ_1, σ_2 etc. in equation (ii), we get,

$$M = \frac{E}{R} a_1 y_1 y_1 + \frac{E}{R} a_2 y_2 y_2 + \frac{E}{R} a_3 y_3 y_3 + \ldots \text{ so on}$$

$$M = \frac{E}{R}(a_1 y_1^2) + \frac{E}{R}(a_2 y_2^2) + \frac{E}{R}(a_3 y_3^2) + \ldots \text{ so on}$$

$$M = \frac{E}{R}[a_1 y_1^2 + a_2 y_2^2 + a_3 y_3^2 + \ldots]$$

Here

$a_1y_1^2$ is moment of inertia of a_1 about N-axis

$a_2y_2^2$ is moment of inertia of a_2 about N-axis

$\therefore \quad a_1y_1^2 + a_2y_2^2 + a_3y_3^2 + ...$

Moment of inertia on entire section of the beam about N-axis *i.e.* = I

$$\therefore \quad M = \frac{E}{R} I$$

$$\therefore \quad \frac{M}{I} = \frac{E}{R}$$

$$\therefore \quad \frac{M}{I} = \frac{\sigma}{y} = \frac{E}{R} \qquad ...(iii)$$

This is called bending equation.

where

M = Bending Moment or Moment of resistance of the section in N mm

I = Moment of Inertia of the section about N.A in mm^4

σ = Bending stress at distance y from N.A. in N/mm^2

y = Distance of fibre from N-axis in mm

R = Radius of gurvature of N.A. in mm.

E = Young's modulus of Ealsticly in N/mm^2

7.25.11 Moment of Resistance

When a beam is subjected to Bending Moment. The section above and below N.A. axis develop internal forces of stresses of opposite nature. These internal stresses or forces have certain moments about N.A.

The algebraic sum of these moments is equal and opposite to the bending moment acting on the section, thus keeping the section of beam in equilibrium.

The sum of moments of the internal forces about the N.A is known as the Moment of resistance. from Bending Equation; we have

$$\frac{M}{I} = \frac{\sigma}{y}$$

or
$$M = \frac{\sigma}{y} \times I$$

or
$$M = \sigma \times Z$$

where Z is called section modulus. The Moment of resistance depends upon the moment of inertia (or section modulus) of the section.

7.26 TORSION OF CIRCULAR SHAFTS

When a structural or machine member is subjected to a moment about its longitudinal axis, the member twists and shear stress is induced in every cross-section of the member called torsion and the twisting moment is referred to as torque like door knob, screw driver, drill bits and shafts etc.

A shaft is rotating member used in workshops and factories to transmit power from a driving motor to a machine tool. Shafts are also used to transmit power from a motor to a pump or compressor, from an engine or turbine to a generator, and from an engine to axle in automobiles. During power transmission, the shaft is subjected to torque which causes twist of the shaft. The distortion of a shaft can also occur when it is fixed at one end and is subjected to twisting moment at the other end (Fig. 7.43) or acted upon by equal and opposite couples at its ends. Obviously, a shaft needs to be so designed that excessive twist is avoided and the induced shear stress is within prescribed limits.

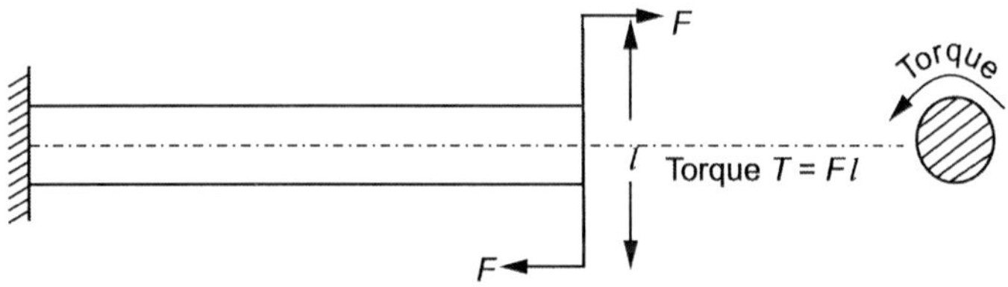

Fig. 7.43 Distortion of a shaft

The shafts are generally circular, either solid or hollow in cross-section. The material of shaft usually of mild steel or alloy steel.

Pure Torsion

Generally two types of stresses are induced in a shaft:

(*i*) torsional (shear) stresses due to transmission of torque,

(*ii*) bending stresses due to weight of pulleys, gears etc. mounted on the shaft.

A circular shaft is said to be in a state of pure torsion when it is subjected to torque only, without being acted upon by any bending moment or axial force. Such a state is rather difficult to attain because even when the shaft is transmitting only torque, some bending occurs due to self-weight of the shaft.

7.26.1 Torque

A shaft is said to be in torsion, when equal and opposite torques are applied at the two ends of the shaft.

The torque is equal to the product of the force applied (tangentially to the ends of a shaft) and radius of the shaft.

A torque causes twist or torsion in the member. Therefore, it is also known as twisting moment or turning moment. It is denoted by 'T'.

If the shaft is subjected to two opposite turning moments it is said to be in pure tension and as a result, it has a tendency to shear off at every cross-section perpendicular to the longitudinal axis. So the effect of torsion is to produce shear stress in the material of the shaft.

7.26.2 Torsion Equation

The torsion equation is based on the following assumptions:

1. The material of the shaft is uniform throughout.
2. The shaft is of uniform circular section throughout
3. Cross-section of the shaft, which are plane before twist remain plane after twist.
4. The twist along the length of shaft is uniform throughout.
5. Maximum shear stress induced in the shaft due to application for que does not exceed its elastic limit value.
6. The distance between any two normal cross-section remains the same after the application of torque.

Consider a shaft fixed at one end AA' and free at the other and BB' as shown in Fig. 7.44.

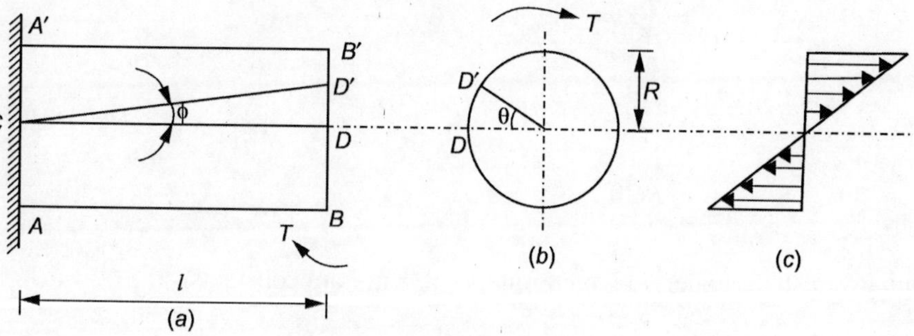

Fig. 7.44 A shaft fixed at AA′

Let

T = Torque or twisting moment

G = Modulus of rigidity of the material of the shaft.

θ = Angle of twist.

τ = Shear stress

R = Radius of shaft

l = Length of the shaft

Let a line CD on the surface of the shaft take a new position CD' after the application of torque. Thus, cross-sections will be twisted through angle θ and surface by angle ϕ.

$$\text{Shear strain} = \frac{DD'}{l} = \phi$$

$$= \frac{DD'}{CD} = \tan\phi$$

$$= \phi \qquad \text{(If } \phi \text{ is very small than } \tan\phi = \phi\text{)}$$

$$\therefore \text{ Shear strain, } \quad \phi = \frac{DD'}{l}$$

$$\text{Arc } DD' = OD \times \theta = R\theta \qquad (\because OD = R = \text{Radius of shaft})$$

$$\therefore \text{ Shear strain, } \quad \phi = \frac{R\theta}{l} \qquad \ldots(i)$$

$$\text{Also} \quad \text{Shear strain} = \frac{\text{Shear stress}}{\text{Modulus of rigidity}}$$

$$\text{or} \qquad \phi = \frac{\tau}{G} \qquad \ldots(ii)$$

from equations (*i*) and (*ii*), we get

$$\frac{\tau}{G} = \frac{R\theta}{l}$$

$$\text{or} \qquad \frac{\tau}{R} = \frac{G\theta}{l} \qquad \ldots(iii)$$

$$\tau = \frac{RG\theta}{l}$$

In a given shaft, under a given torque, G, θ, and l are constants.

$$\therefore \qquad \tau \propto R$$

$$\text{or} \qquad \frac{\tau}{R} = \text{Constant} \qquad \ldots(iv)$$

from equation (*iv*), it is clear that shear stress at any point in the shaft in proportional to the distance of the point from the axis of the shaft. Hence, the shear stress is maximum at the outer surface and shear stress is zero at the axis of the shaft.

Consider an elementary circular ring of thickness 'dr' at a distance 'r' from the centre as shown in Fig. 7.45.

Area of the ring $= 2\pi r\, dr$

Shear stress at radius

$$r = \frac{\tau}{R} \times r \qquad \left[\because \frac{\tau}{R} = \text{Constant}\right]$$

Turning force on the elementary circular ring

$$= \text{Stress} \times \text{Area}$$

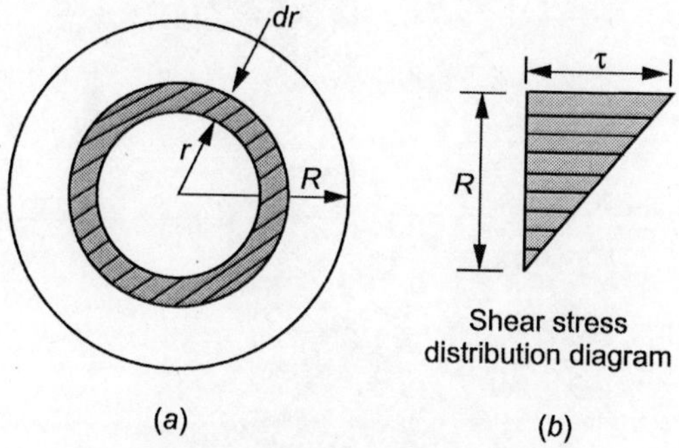

Fig. 7.45 An elemetary circular ring

$$= \frac{\tau}{R} \times r \times 2\pi r.dr$$

$$= \frac{\tau}{R} \times 2\pi r^2.dr$$

Turning moment due to turning force on the elementary ring.

$$dT = \frac{\tau}{R} \times 2\pi r^2.dr.r$$

$\therefore$ Total turning moment,

$$T = \int dT$$

$$\int_0^R = \frac{\tau}{R} \times 2\pi r^3.dr$$

$$T = \frac{\tau}{R} \times 2\pi \int_0^R r^3.dr$$

$$= \frac{\tau}{R} \times 2\pi \left| \frac{r^4}{4} \right|_0^R$$

$$= \frac{\tau}{R} \times 2\pi \times \frac{r^4}{4}$$

$$= \tau \times \frac{\pi}{2} \times R^3$$

$$= \tau \times \frac{\pi}{2} \times \frac{D^3}{8}$$

$$= \frac{\pi}{16} \tau D^3$$

$$T = \frac{\pi}{16} \tau D^3$$

$$\tau = \frac{16T}{\pi D^3}$$

Substituting the value of τ in equation (*iii*), we get.

$$\frac{16T}{\pi D^3 . R} = \frac{G\theta}{l}$$

$$\frac{16T.2}{\pi D^3 . D} = \frac{G\theta}{l}$$

$$\frac{32T}{\pi D^4} = \frac{G\theta}{l}$$

But $\frac{\pi D^4}{32}$ = J (polar moment of Inertia)

$$\therefore \quad \frac{T}{J} = \frac{G\theta}{l} \qquad \ldots(v)$$

from equations (*iii*) and (*v*), we get

$$\frac{T}{J} = \frac{G\theta}{l} = \frac{\tau}{R}$$

This equation is known as torsion equation.

7.26.3 Torsional Stiffness, Torsional Rigidity and Polar Modulus

(*i*) The parameter $\frac{T}{\theta}$, *i.e.*, the torque required per unit twist is called torsional stiffness of the shaft.

(*ii*) The parameter $\frac{T}{\theta / l}$, *i.e.*, the torque divided by the angle of twist per unit length is called torsional rigidity of the shaft.

For torsion equation,

$$\frac{T}{J}=\frac{C\theta}{l}, \text{ the torsional rigidity is given by } \frac{T}{\theta/l}=CJ$$

Where CJ represents the torque required to produce a twist of 1 radian per unit length of shaft.

(*iii*) The maximum shear stress occurs at the shaft radius R and is given by

$$\tau=\frac{T}{J}R;\ T=\tau\left(\frac{J}{R}\right)=\tau Z_P$$

The parameter $Z_p = (J/R)$ is called the *polar modulus* of the shaft section. It also replicates the strength of shaft in torsion. Hence, the polar modulus of beams subjected to bending.

$$Z_P=\frac{J}{R}=\frac{\pi R^4/2}{R}=\frac{\pi R^3}{2}=\frac{\pi}{16}d^3$$

where d is the shaft diameter.

$$\therefore \qquad T=\tau Z_P=\frac{\pi}{16}\tau d_3$$

7.26.4 Power Transmitted by a Shaft

Consider a shaft used to trasmit power from a driving motor to a machine tool, from a turbine to a generator, or from an engine to an axle.

If the shaft turns N revolutions per minute and T (Nm) is the torque applied on the shaft, then angular displacement $\omega=\frac{2\pi N}{60}$ radians/sec.

$$\text{Work done} = \text{torque} \times \text{angular displacement} = \frac{2\pi NT}{60} \text{ Nm/s}$$

$\therefore$ Power transmitted by the shaft $P=\frac{2\pi NT}{60}$ watt

Quite often the torque fluctuates and accordingly calculations for the power transmitted are made on the basis of mean torque. However, maximum value of torque is considered while calculating the stress developed in the shaft material.

Hollow Circular Shafts

The torsion equation equally holds good for hollow shafts and can be established. Consider a hollow shaft subjected to a torque T, take an elementary circular ring of thickness dr at a distance r from the centre as shown in Fig. 7.46.

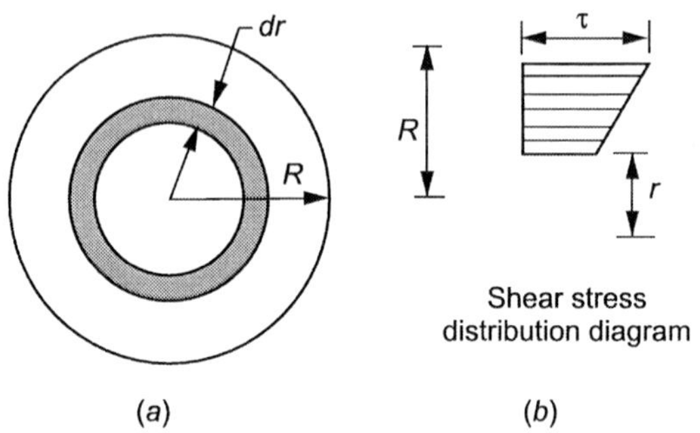

Fig.7.46 Thickness of circular ring (*dr*) at a distance *r*

Area of the ring = $2\pi r\,dr$

Shear stress at radius

$$r = \frac{\tau}{R} \times r \qquad \left(\because \frac{\tau}{R} = \text{Constant}\right)$$

Turning force on the elementary circular ring

$$= \text{Stress} \times \text{Area}$$

$$= \frac{\tau}{R} \times r \times 2\pi r\,dr$$

$$= \frac{\tau}{R} \times 2\pi r^2\,dr$$

Turning moment due to turning force on the elementary ring

$$dT = \frac{\tau}{R} \times 2\pi r^2 .dr.r$$

$$dT = \frac{\tau}{R} \times 2\pi r^3 .dr$$

Total turning moment,

$$T = \int dT$$

$$= \int_r^R \frac{\tau}{R} \times 2\pi r^3 .dr$$

$$T = \frac{\tau}{R} \times 2\pi \int_r^R r^3 dr$$

$$= \frac{\tau}{R} \times 2\pi \left| \frac{r^4}{4} \right|_r^R$$

$$T = \frac{\tau}{R} \times 2\pi \left(\frac{R^4 - r^4}{4} \right)$$

$$= \frac{\tau \times 2 \times 2\pi}{D} \left\{ \frac{\left(\frac{D}{2}\right)^4 - \left(\frac{d}{2}\right)^4}{4} \right\}$$

$$= \frac{4\pi\tau}{D} \left(\frac{D^4 - d^4}{64} \right)$$

$$T = \frac{\pi}{16} \tau \left(\frac{D^4 - d^4}{D} \right)$$

$$\tau = \frac{16T}{\pi} \frac{D}{(D^4 - d^4)}$$

Now $$\frac{\tau}{R} = \frac{G\theta}{l}$$

or $$\frac{16T.D}{R\pi(D^4 - d^4)} = \frac{G\theta}{l}$$

$$\frac{16T\,D.2}{D\pi(D^4 - d^4)} = \frac{G\theta}{l}$$

$$\frac{32T}{\pi(D^4 - d^4)} = \frac{G\theta}{l}$$

But $\frac{\pi}{32}(D^4 - d^4) = J$ (Polar moment of inertia)

$\therefore$ $$\frac{T}{J} = \frac{G\theta}{l} = \frac{\tau}{R}$$

7.27 COMPARISON BETWEEN HOLLOW AND SOLID CIRCULAR SHAFTS

Let R_o = outer raddii of the hollow shaft

R_i = inner radii of hollow shaft

R = radius of solid shaft

Further, let the hollow and solid shafts be designated by subscripts h and s respectively.

7.27.1 Comparison by Strength

Ratio of torques transmitted when the hollow and solid are of the same material, length and weight.

From the torsion equation,

$$\frac{T_h}{\frac{\pi}{2}(R_0^4 - R_1^4)} = \frac{\tau_h}{R_0}; \quad \tau_h = \frac{2T_h R_0}{\pi(R_0^4 - R_1^4)}$$

$$\frac{T_s}{\frac{\pi}{2}R^4} = \frac{\tau_s}{R}; \quad \tau_s = \frac{2T_s}{\pi R^3}$$

Since both the shafts are made of the same material, their maximum allowable shear stress will be same ($\tau_h = \tau_s$). Accordingly

$$\frac{T_h}{T_s} = \frac{R_0^4 - R_1^4}{R_0 R^3} \quad \ldots(i)$$

Again both the shafts are of same material, length and weight. That gives

$$\rho\,\pi\,(R_0^2 - R_i^2)l = \rho\,\pi\,R^2 l \quad \text{(mass = density × volume)}$$

or $$R_0^2 - R_i^2 = R^2 \quad \ldots(ii)$$

From the equations (*i*) and (*ii*), we have

$$\frac{T_h}{T_s} = \frac{R_0^2 + R_i^2}{R_0 R} = \frac{R_0}{R}\left[1 + \frac{1}{(R_0/R_i)^2}\right]$$

$$= \frac{R_0}{R}\left[1 + \frac{1}{n^2}\right] \quad \text{where } n = \frac{R_0}{R_i} \quad \ldots(iii)$$

From the equation (*ii*) $R_0^2 - R_i^2 = R^2$; $1 - \left(\frac{R_i}{R_0}\right)^2 - \left(\frac{R}{R_0}\right)^2$

or $$1 - \frac{1}{n^2} = \left(\frac{R_i}{R_0}\right)^2 \text{ or } \frac{R_0}{R} = \frac{n}{\sqrt{n^2 - 1}}$$

Substituting $\dfrac{R_0}{R} = \dfrac{n}{\sqrt{n^2 - 1}}$ in expression (*iii*), we get

$$\frac{T_h}{T_s} = \frac{n}{\sqrt{n^2-1}}\left(\frac{n^2+1}{n^2}\right) - \frac{n^2+1}{n\sqrt{n^2-1}}$$

If the ratio $\dfrac{R_0}{R_i} = n = 2$

$$\frac{T_h}{T_s} = \frac{2^2+1}{2\sqrt{2^2-1}} = \frac{5}{\sqrt{3}} = 1.44$$

Thus, a hollow shaft with $R_0/R_i = 2$ will transmit 44% more torque than the solid shaft of the same material, length and weight. Obviously a hollow shaft is stronger than a solid shaft

7.27.2 Comparison by Weight

Ratio of weights when both the hollow and solid shafts are of same material, length and transmit the same torque.

From the torsion equation,

$$\frac{T_h}{\frac{\pi}{2}(R_0^4 - R_1^4)} = \frac{\tau_h}{R_0} \text{ and } \frac{T_s}{\frac{\pi}{2}R^4} = \frac{\tau_s}{R}$$

Since $T_h = T_s$ and $\tau_h = \tau_S$ (same material), the above identities give

$$\frac{R_0^4 - R_i^4}{R_0} = R^3 \quad \text{or} \quad R_0^{\ 3}\left[1 - \frac{R_i^{\ 4}}{R_0^{\ 4}}\right] = R^3$$

or $$R = R_0\left(1 - \frac{1}{n^4}\right)^{1/3} \qquad \text{where } n = \frac{R_0}{R_i}$$

$$\frac{\text{Weight hollow shaft}}{\text{Weight of solid shaft}} = \frac{W_h}{W_s} = \frac{\rho\pi(R_0^4 - R_i^2)l}{\rho\pi R^2 l}$$

$$= \frac{R_o^2 - R_i^2}{R^2} = \frac{R_o^2}{R^2}\left(1 - \frac{1}{n^2}\right)$$

Upon substitution of the value of

$$R_o/R = \frac{1}{\left(1 - \frac{1}{n^4}\right)^{1/3}}, \text{ we get}$$

$$\frac{W_h}{W_s} = \frac{1-\frac{1}{n^2}}{\left(1-\frac{1}{n^4}\right)^{2/3}} = \frac{(n^2-1)}{n} \times \frac{(n^4)^{2/3}}{(n^4-1)^{2/3}} = \frac{(n^2-1)n^{2/3}}{(n^4-1)^{2/3}}$$

If the ratio $\frac{R_o}{R_i} = n = 2$

$$\frac{W_h}{W_s} = \frac{(2^2-1)\times 2^{2/3}}{(2^4-1)^{2/3}} = \frac{3\times 2^{2/3}}{(15)^{2/3}} = 0.782$$

This shows that for the same material, length and given torque, weight of hollow shaft will be less. Hollow shafts are thus economical compared to solid shafts as regards torsion.

7.28 COMPOSITE SHAFTS

7.28.1 Shafts in Series

Shafts are said to be connected in series when the driving torque is applied at one end and equivalent resistant torque at the other. Fig. 7.47 shows two shafts connected in series and forming a continuous length. The shafts may be of same or different dimensions and also of same or different materials. For such a composite shaft :

(*i*) The torque is same at both the ends.

$T_1 = T_2 = T$

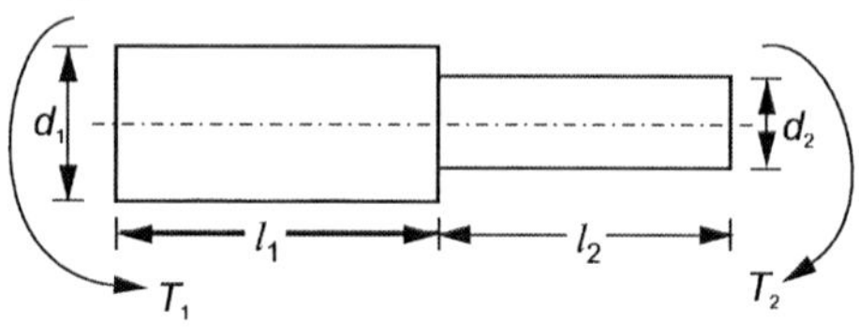

Fig 7.47 Two shafts connected in series

(*ii*) The total angle of twist between the two ends is the sum of the angles of twist of the different shafts.

$$\theta = \theta_1 + \theta_2 = \frac{Tl_1}{C_1 J_1} + \frac{Tl_2}{C_1 J_2}$$

and

$$\frac{\theta_1}{\theta_2} = \frac{l_1}{l_2} \times \frac{C_2}{C_1} \times \frac{J_2}{J_1} = \frac{l_1}{l_2} \times \frac{C_2}{C_1}\left(\frac{d_2}{d_1}\right)^4$$

If the shaft segments are made of the same material $C_1 = C_2$ and then the above expressions may be rewritten as

$$\theta = \frac{T}{C}\left(\frac{l_1}{l_1} + \frac{l_2}{l_2}\right)$$

and
$$\frac{\theta_1}{\theta_2} = \frac{l_1}{l_2}\left(\frac{d_2}{d_1}\right)^4$$

7.28.2 Shafts in Parallel

Two shafts are said to be connected in parallel when one shaft surrounds the other as shown in Fig. 7.48.

For such a shaft combination :

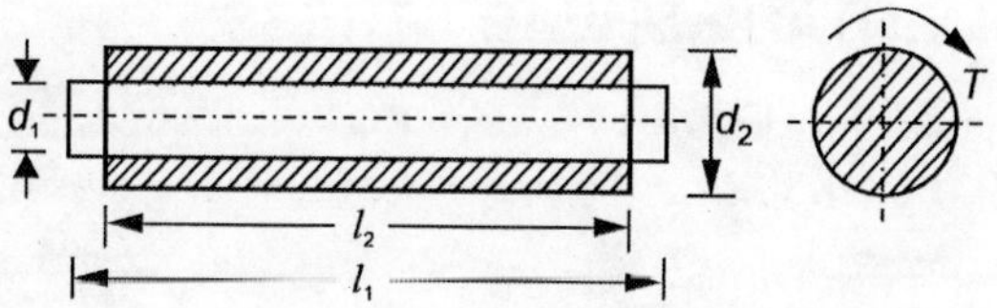

Fig. 7.48 Shafts in parallel

(*i*) Twist is same for both the shafts

$$\theta = \theta_1 = \theta_2$$

or
$$\frac{T_1 l_1}{C_1 l_1} = \frac{T_2 l_2}{C_2 l_2}$$

$$\frac{T_1}{T_2} = \frac{J_1}{J_2} \times \frac{C_1}{C_2} \times \frac{l_2}{l_1}$$

(*ii*) The applied torque is divided between the two shafts depending upon their dimensions and elastic properties

$$T = T_1 + T_2 = \frac{C_1 \theta_1 J_1}{l_1} = \frac{C_2 \theta_2 J_2}{l_2}$$

and
$$= \theta\left(\frac{C_1 J_1}{l_1} = \frac{C_2 J_2}{l_2}\right)$$

If both the shafts are of same material ($C_1 = C_2 = C$) and equal lengths ($l_1 = l_2 = l$), the equations can be rewritten as

$$T_1 l_1 = T_2 l_2; \ \frac{T_1}{T_2} = \frac{l_2}{l_1}$$

and $$T = \frac{\theta C}{l}(J_1 + J_2);\ \theta = \frac{Tl}{C(J_1 + J_2)}$$

Such a situtation ($T = T_1 + T_2$ and $\theta = \theta_1 = \theta_2$) also arises when the shaft ends are fixed and are subjected to a torque at the common junction (Fig. 7.49).

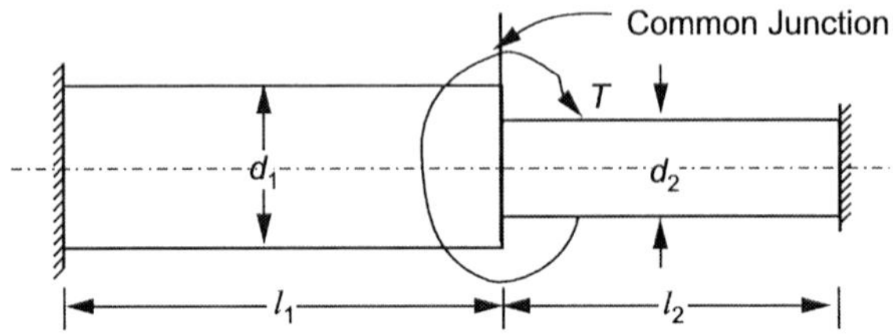

Fig. 7.49 Shafts ends are fixed

7.29 COMPLIMENTARY SHEAR STRESS

Consider a rectangular block *ABCD* subjected to shear stress of intensity τ at the forces *AB* and *CD* (Fig. 7.50 (*a*).

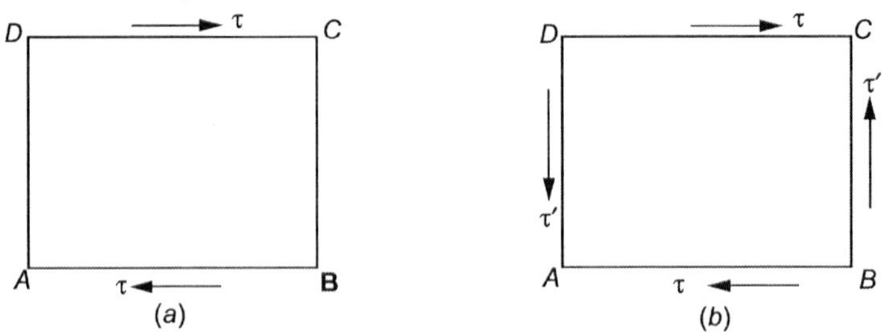

Fig. 7.50 ABCD subjected to shear stress

If the block has unit thickness perpendicular to the plane of the paper, then shear forces acting on these faces will be $\tau \times AB$ or $\tau \times DC$. These forces constitute a couple of moment

$$\text{Moment} = \text{Force} \times \text{Perpendicular distance}$$
$$= (\tau \times AB) \times BC \text{ or } (\tau \times DC) \times BC$$

This couple tends to rotate the block in clockwise direction. To keep the block in equilibrium, shear stress t' must act on the other two faces of the block. The direction of these stresses has to be such that these provide an equal but opposite couple (Fig. 7.50(*a*). Thus, under equilibrium conditions:

$$(\tau \times AB) \times BC = (\tau' \times BC) \times AB; \qquad \tau' = \tau$$

The stress τ' is called *complimentary shear stress.* Thus, if a rectangular or square block is in equilibrium, shear stress in one plane is accompanied by an equal complimentary shear stress in the other plane. The state of shear depicted in fig. 7.50(*b*) is called *simple* or *pure shear* as no other type of stress is acting.

Let attention be focused on a cubical element *ABCD* of each side l (Fig. 7.50). The block is in simple shear, *i.e.*, the faces *AB* and *CD* are subjected to shear stress of intensity τ, and complimentary shear stress τ´ of equal magnitude on the faces *DA* and *BC*.

Considering equilibrium of the free body diagram of the triangular portion *ABD*,

Shear forces on face $DA = \tau' \times (l \times l) = \tau\, l^2$ $(\because \tau' = \tau)$

Shear force on face $AB = \tau \times (l \times l) = \tau\, l^2$

Resultant of these forces is $\sqrt{2}\ \tau l$, and this is balanced by force σ_n acting normal to diagonal *BD*.

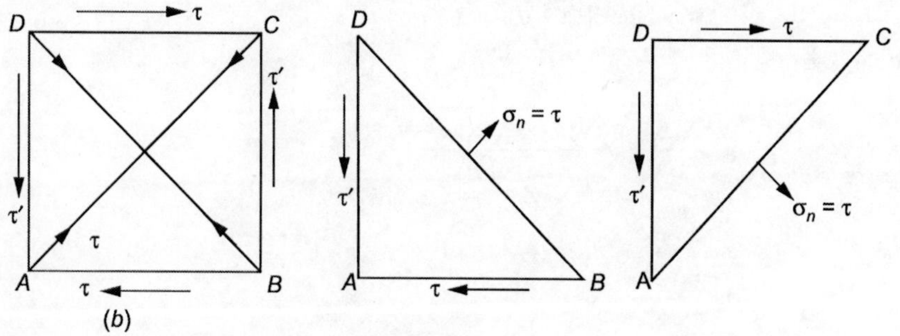

Fig. 7.51 Resultant of forces

Sectional area of face $BD = \sqrt{l^2 + l^2} = \sqrt{2}l$

∴ Normal stress on diagonal BD,

$$\sigma_n = \frac{\sqrt{2}\tau l}{\sqrt{2}l} = \tau$$

Likewise, equilibrium of free body diagram of triangular portion *DCA* would give normal stress on diagonal *AC*

$$\sigma_n = \tau$$

The above analysis shows that when a element is in a state of simple shear, tensile (along diagonal *AC*) and compressive (along diagonal *BD*) stresses equal in magnitude to the shear stress act on planes inclined at 45° to the planes of simple shear.

7.29.1 Hydrostatic Stress, Volumetric Strain and Bulk Modulus

When a body is immersed in a fluid to a large depth, the body gets subjected to equal external pressure at all points on the body. This external pressure is compressive in nature, and is called *hydrostatic stress.*

The hydrostatic stress causes change in volume of the body, and this change of volume per unit volume is called volumetric strain, ε_υ.

$$\varepsilon_\upsilon = \frac{\Delta V}{V}$$

Consider a cuboid of sides x, y and z subjected to hydrostatic stress and let its sides change by dx, dy and dz respectively. Then change in volume,

$$\Delta V = (x + dx)\,(y + dy)\,(z + dz) = xyz$$

Neglecting product of small changes,

$$\Delta V = (xyz + xydz + yzdx = xzdy) - xyz$$

$$= y\,z\,dx + x\,z\,dy + x\,y\,dz$$

∴ Volumetric strain,

$$\varepsilon_\upsilon = \frac{\Delta V}{V} = \frac{yzdx + xzdy + xydz}{xyz}$$

$$= \frac{dx}{x} + \frac{dy}{y} + \frac{dz}{z}$$

$$= \varepsilon_x + \varepsilon_y + \varepsilon_z$$

Apparently volumetric strain equals the sum of the linear normal strains in x, y and z directions.

Within elastic limits, the hydrostatic stress is directly proportional to volumetric strain. Mathematically,

hydrostatic stress ∝ volumetric strain

= constant of proportionality × volumetric strain

or $\quad \sigma_\upsilon = K\,\varepsilon_\upsilon; \qquad K = \dfrac{\sigma_v}{\varepsilon_v}$

The constant of proportionality K is called the bulk modules or *modulus of volume expansion* of the material. Physically, the bulk modulus K is a measure of the resistance of a material to change of volume without change of shape or form.

7.30 EXTENSION OF A TAPERED BAR

Consider a circular bar that tapers uniformly from diameter d_1 at the bigger end and diameter d_2 at the small end, which is subjected to axial tensile load P (Fig. 7.52).

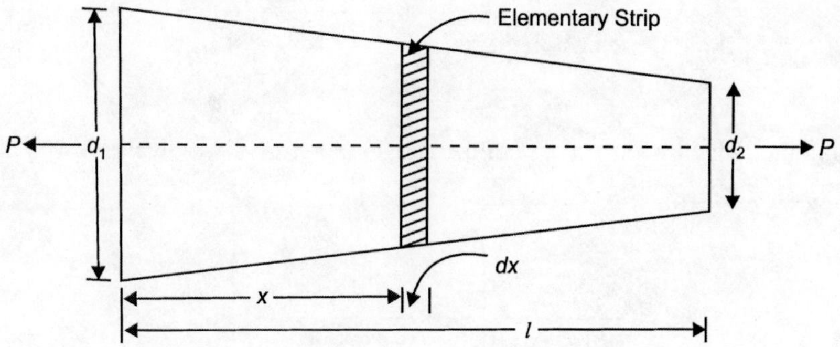

Fig. 7.52 A tapered bar

Let us take an elementary strip of length d_x at distance x from the bigger end. Diameter of the elementary strip,

$$d_x = d_1 - (d_1 - d_2)\,\frac{x}{l}$$

$$= d_1 - k\,x \text{ where } k = (d_1 - d_2)/l$$

Cross-sectional area of the strip,

$$A_x = \frac{\pi}{4} d_x^2 = \frac{\pi}{4}(d_1 - kx)^2$$

Stress in the strip,

$$\sigma_x = \frac{P}{A_x} = \frac{P}{\frac{\pi}{4}(d_1 - kx)^2} = \frac{4P}{\pi(d_1 - kx)^2}$$

Strain in the strip,

$$\varepsilon_x = \frac{\sigma_x}{E} = \frac{4P}{\pi(d_1 - kx)^2 E}$$

Elongation of the strip

$$\delta l_x = \varepsilon_x dx = \frac{4P}{\pi(d_1 - kx)^2 E}$$

The total elongation of this tapering bar can be worked out by integrating the above expression between the limits $x = 0$ to $x = l$

$$\delta l = \int_0^1 \frac{4P\,dx}{\pi(d_1 - kx)^2 E} = \frac{4P}{\pi E}\int_0^1 \frac{dx}{(d_1 - kx)^2}$$

$$= \frac{4P}{\pi E}\left[\frac{(d_1 - kx)^{-1}}{(-1)\times(-k)}\right]_0^1 = \frac{4P}{\pi EK}\left[\frac{1}{d_1 - kx}\right]_0^1$$

Putting the value of $k \dfrac{(d_1 - d_2)}{l}$ in the above expression, we obtain

$$\delta l = \frac{4Pl}{\pi E(d_1 - d_2)}\left[\frac{1}{d_1 - \dfrac{(d_1 - d_2)l}{l}} - \frac{1}{d_1}\right]$$

$$= \frac{4Pl}{\pi E(d_1 - d_2)}\left[\frac{1}{d_2} - \frac{1}{d_1}\right]$$

$$= \frac{4Pl}{\pi E(d_1 - d_2)} \times \frac{d_1 - d_2}{d_1 d_2}$$

$$= \frac{4Pl}{\pi E d_1 d_2}$$

If the bar is of uniform diameter d throughout its length, then

$$\delta l = \frac{4Pl}{\pi E d^2} = \frac{Pl}{\frac{\pi}{4} d^2 E} = \frac{Pl}{AE}$$

7.31 EXTENSION OF BAR DUE TO SELF-WEIGHT

7.31.1 Bar of Uniform Section

Consider a bar of cross-section. area A and length l hanging freely under its own weight (Fig.7.53). Take a small element of length dy at distance y from the lower end. If w is the specific weight (weight per unit volume) of the bar material, then total tension at section m-n equals weight of the bar for length y and is given by

$$P = w\,A\,y$$

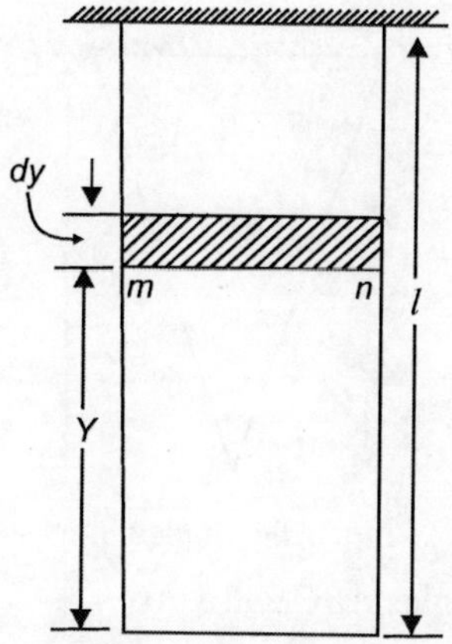

Fig. 7.53 A bar of uniform section

As a result of this load, the elemental length dx elongates by a small amount Δx, and

$$\Delta x = \frac{P\,dy}{AE} = \frac{w\,A\,y}{AE}\,dy = \frac{w}{E}\,y\,dy \qquad ...(i)$$

The total change in length of the bar due to self weight is worked out by integrating the above expression between the limits $y = 0$ and $y = l$.

$$\therefore \qquad \delta l = \int_0^1 \frac{w}{E}\,y\,dy = \frac{w}{E}\left[\frac{y^2}{2}\right]_0^2 = \frac{w}{E}\frac{l^2}{2} \qquad ...(ii)$$

If W is the total weight of the bar ($W = w\,A\,l$), then $w = W/Al$. In that case, total extension of the bar

$$\delta l = \left(\frac{W}{Al}\right)\frac{l^2}{2E} = \frac{Wl}{2\,AE} \qquad ...(iii)$$

Thus, total extension of the bar due to self-weight is equal to the extension that would be produced if one-half of the weight of the bar is applied at its end.

7.31.2 Conical Bar (Elongation Due to Self-Weight)

Consider a conical bar ABC of length l with its diameter d fixed rigidly at AB. Take a small element of length dy at distance y from the lower end point C.

Total tension P at section DE equals weight of the bar for length y and is given by

$$P = w \times \frac{1}{3}\left(\frac{\pi}{4}d_s^2\,y\right) \qquad ...(iv)$$

where w is the specific weight of the bar material and $d_s = DE$ is the diameter of the elementary strip.

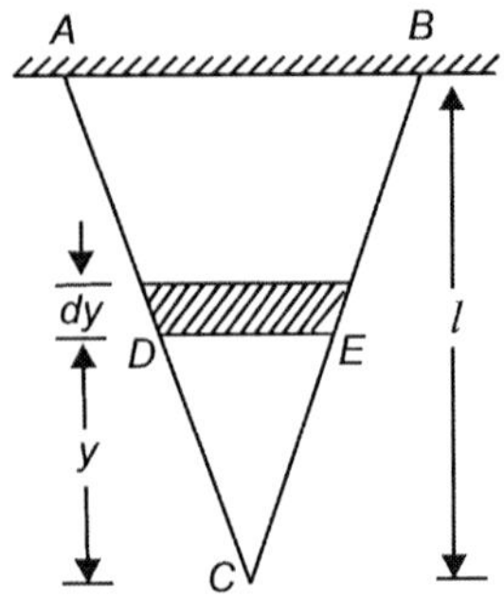

Fig. 7.54 Conical bar

From the similarity of triangles *ABC* and *DEC*,

$$\frac{AB}{DE} = \frac{l}{y} \quad \text{or} \quad DE = AB\frac{y}{l} = d\frac{y}{l}$$

$$P = w \times \frac{1}{3}\left[\frac{\pi}{4}\frac{d^2 y^2}{l^2} y\right]$$

$$= w\frac{\pi}{12}\frac{d^2}{l^2} y^3 \qquad \ldots(v)$$

As a result of this load, the elemental length *dy* elongates by a small amount Δy, and

$$\Delta y = \frac{P\,dy}{AE} = w\frac{\pi}{12}\frac{d^2}{l^2} y^3 dy + \frac{\pi}{4}\frac{d^2 y^2}{l^2} E = \frac{w}{3E} y\,dy \qquad \ldots(vi)$$

The total change in length of the conical bar due to self-weight is worked out by integrating the above expression between the limits $y = 0$ to $y = 1$

$$\delta l = \frac{w}{3E}\int_0^1 y\,dy = \frac{w}{3E}\left[\frac{y^2}{2}\right]_0^1 = \frac{wl^2}{6E} = \frac{\rho g}{6E} l^2 \qquad \ldots(vii)$$

where ρ is the mass density of the bar material.

Tapering Rod (Elongation due to self-weight)

Consider a tapering rod hung vertically and firmly fixed at the top position. The rod is of length l and it tapers uniformly *A* from diameter d_1 to d_2. Let the sides *AC* and *BD* meet at point *E* when produced. Extension δl in the length l of the rod *ABDE* due to self-weight is

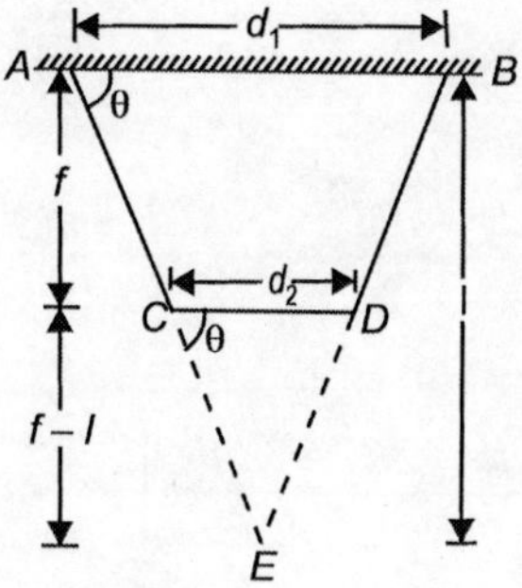

Fig. 7.55 Tapering rod

δl = extension of conical rod *AEB* due to self weight – extension due to conical segment *CED* due to self weight - extension in rod length 1 due to weight of segment *CED*

$$= \frac{wl'^2}{6E} = \frac{w(l'-l)^2}{6E} - \frac{4Pl}{\pi E d_1 d_2}$$

where *P* is the weight of segment *CED* and w is the specific weight of bar material

$$P = w\frac{\pi}{4} \times \left[\frac{1}{3} d_2^2 \times (l' - l)\right] \quad ...(viii)$$

Substituting this value of tensile load *P* in the above expression, we get

$$\sigma l = \frac{wl'^2}{6E} = \frac{w(l'-l)^2}{6E} = \frac{w(l'-l)l}{3E}\frac{d_2}{d_1} \quad ...(ix)$$

From the geometrical configuration

$$\cot\theta = \frac{d_2/2}{(l'-l)} = \frac{d_1/2}{l'}$$

or $$\frac{d_2}{l'-l} = \frac{d_1 l^2}{l'}$$

This gives: $$l' = \frac{d_1 l}{d_1 - d_2} \text{ and } l' - l = \frac{d_2 l}{d_1 - d_2}$$

$\therefore$ $$\delta l = \frac{w}{6E}\left(\frac{d_1 l}{d_1 - d_2}\right)^2 - \frac{w}{6E}\left(\frac{d_2 l}{d_1 - d_2}\right)^2 - \frac{w}{6E}\left(\frac{d_2 l}{d_1 - d_2}\right) l \times \frac{d_2}{d_1}$$

$$= \frac{wl^2}{3E}\left[\frac{d_1^2}{2(d_1-d_2)^2} - \frac{d_2^2}{2(d_1-d_2)^2} - \frac{d_2^2}{(d_1-d_2)d_1}\right]$$

$$= \frac{wl^2}{3E}\left[\frac{d_1^2 - d_1 d_2^2 - 2d_2^2(d_1-d_2)}{2d_1(d_1-d_2)^2}\right]$$

$$= \frac{wl^2}{6E}\left[\frac{d_1^2 - 2d_2^2 - 3d_1 d_2^2}{d_1(d_1-d_2)^2}\right] \quad \ldots(x)$$

If the rod is conical, *i.e.*,

$$d_2 = 0$$

$$\delta l = \frac{wl^2}{6E}$$

which is the same expression as derived earlier.

7.32 BAR OF UNIFORM STRENGTH

When a bar is designed in such a way that it has uniform stress maintained at all its sections, it, is called a bar of uniform strength. This is achieved by having a variation in area from one end to the other end.

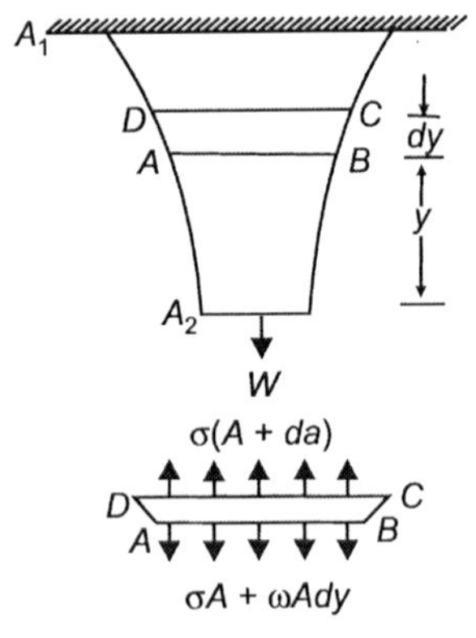

Fig. 7.56 Bar of uniform strength

Fig. 7.56 shows that one such bar subjected to an axial tensile load W. The bar is of length l and its area increases from A_2 at lower end to A_1 at the upper end. Let attention be focused at strip $ABCD$ which has area A at distance y from the lower and area $(A + dA)$ at distance $(y + dy)$ from the lower end.

For this strip to be in equilibrium state,

Total force acting upward = Total force acting downwards

$$\sigma(A + dA) = \sigma A + wA\, dy$$

where σ is the uniform stress intensity and w is the weight per unit volume of the strip.

Simplification gives :

$$\frac{dA}{A} = \frac{w}{\sigma} dy$$

Upon integration :

$$\log, A = \frac{w}{\sigma} y + c$$

The value of the constant of integration c can be evaluated by noting that at $y = 0$, $A = A_2$. That gives

$$c = \log_e A_2$$

Accordingly, we can write:

$$\log_e A = \frac{w}{\sigma} y + \log_e A_2$$

$$\log_e \frac{A}{A_2} = \frac{w}{\sigma} y \text{ and } \log_e \frac{A_1}{A_2} = \frac{w}{\sigma} l$$

The usual practice is to write the above expression as

$$\frac{A_1}{A_2} = (e)^{wl/\sigma}$$

SOLVED EXAMPLES

Example 7.1: The following data was recorded during tensile test made on a standard tensile test specimen:

Original diameter and gauge length = 10 mm and 50 mm; minimum diameter at fracture = 50 mm; distance between gauge points at fracture = 60 mm; load at yield point and at fracture = 40 kN and 45 kN; maximum load that specimen could take = 80 kN.

Make calculations for (*a*) yield strength, ultimate tensile strength and breaking strength (*b*) percentage elongation and percentage reduction in area after fracture (*c*) nominal and true stress and fracture.

Solution:

$$\text{Original area } A_o = \frac{\pi}{4} (10)^2 = 78.5 \text{ mm}^2$$

$$\text{Final area } A_f = \frac{\pi}{4} (5)^2 = 19.620 \text{ mm}^2 \text{ (at fracture)}$$

(*a*) $$\text{Yield strength} = \frac{40\times10^3}{78.5} = 509.55 \text{ N/mm}^2$$
$$= 78.5 \times 10^6 \text{ N/m}^2 = 509.55 \text{ MPa}$$

$$\text{Ultimate tensile strength} = \frac{80\times10^3}{78.8} = 1015.22 \text{ N/mm}^2$$
$$= 1015.21 \times 10^6 \text{ N/m}^2 = 1015.21 \text{ MPa}$$

$$\text{Breaking strength} = \frac{45\times10^3}{78.5} = 573.24 \text{ N/mm}^2$$
$$= 572.26 \times 10^6 \text{ N/m}^2 = 572.24 \text{ MPa}$$

It may be noted that calculations for all the strengths are made on the basis of original area at the test section

(*b*) $$\text{Percentage elongation} = \frac{60-50}{50} \times 100 = 20\%$$

$$\text{Percentage reduction in area} = \frac{78.5-19.62}{78.5} \times 100 = 75.00\%$$

(*c*) $$\text{Nominal stress} = \frac{\text{load at fracture}}{\text{original area}} = \frac{45\times10^3}{78.5} = 573.24 \text{ N/mm}^2$$

The nominal strength and the breaking strength are synonymous

$$\text{True stress} = \frac{\text{load at fracture}}{\text{original area}} = \frac{45\times10^3}{78.5} = 573.25$$

Example 7.2: A rectangular block 250 mm × 100 mm is subjected to axial load as follows:

600 kN tensile in the direction of its length; 900 kN compressive on the 250 mm × 150 mm faces, and 750 kN tensile on 250 mm × 100 mm faces.

Assuming Young's modulus $E = 2 \times 10^5$ and Poisson's ratio $\mu = 0.25$ determine:

(*a*) strains in the direction of each force.
(*b*) change in volume of the block due to the loads specified, and
(*c*) modulus of rigidity and bulk modulus for the material of block.

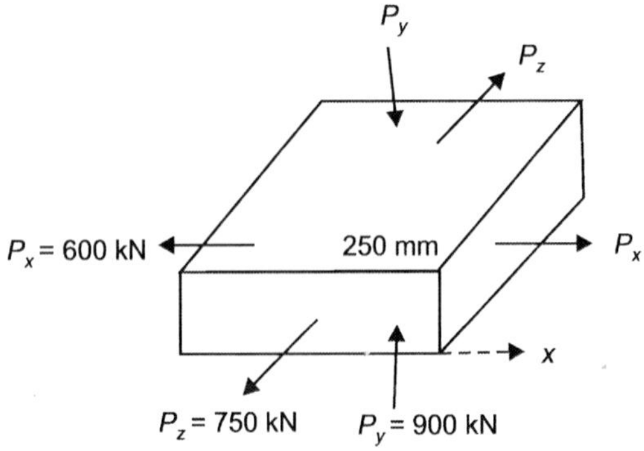

Fig. 7.57 A rectangular block

Solution:

Volume of block, $V = 250 \times 150 \times 100 = 375 = 10^4$ mm^3

Stress in the three-directions are

$$s_x = \frac{600 \times 10^3}{150 \times 100} = 40\text{N/mm}^3 \text{ (tensile)}$$

$$s_y = \frac{900 \times 10^3}{250 \times 150} = 24\text{N/mm}^3 \text{ (compressive)}$$

$$s_z = \frac{750 \times 10^3}{250 \times 100} = 30\text{N/mm}^3 \text{ (tensile)}$$

The resultant strains along the three directions can be worked out by taking the effect of individual stress.

Stain in the x-direction

$$E_x = \text{Strain due to } \sigma_x - \text{Strain due to } \sigma_y - \text{Strain due to } \sigma_z$$

$$= \frac{\sigma_x}{E} - \mu\frac{\sigma_y}{E} - \mu\frac{\sigma_z}{E}$$

$$= \frac{40}{2 \times 10^5} - \frac{0.25(-24)}{2 \times 10^5} - \frac{0.25 \times 30}{2 \times 10^5}$$

$$= \frac{40 + 6 - 7.5}{2 \times 10^{-5}} = 19.25 \times 10^{-5}$$

Likewise: $$\varepsilon_y = \frac{\sigma_x}{E} - \mu\frac{\sigma_y}{E} - \mu\frac{\sigma_z}{E}$$

$$= \frac{24}{2 \times 10^5} - \frac{0.25 \times 40}{2 \times 10^5} - \frac{0.25 \times 30}{2 \times 10^5}$$

$$= \frac{-24 - 10 - 7.5}{2 \times 10^5} = -20.75 \times 10^{-5}$$

And $$\varepsilon_x = \frac{\sigma_x}{E} - \mu\frac{\sigma_y}{E} - \mu\frac{\sigma_z}{E}$$

$$= \frac{30}{2 \times 10^5} - \frac{0.25 \times 40}{2 \times 10^5} - \frac{0.25(-24)}{2 \times 10^5}$$

$$= \frac{30 - 10 + 6}{2 \times 10^5} = 13 \times 10^{-5}$$

For a rectangular body subjected to three mutually perpendicular forces, volumetric strain is given by

$$\varepsilon_v = \varepsilon_x + \varepsilon_y + \varepsilon_z$$
$$= (19.25 - 20.75 + 13) \times 10^{-5} = 11.5 \times 10^{-5}$$

∴ Change in volume,

$$\delta V = \varepsilon_v \times V$$
$$= 11.5 \times 10^{-5} \times (375 \times 10^4) = 431.25 \text{ cm}^3$$

From the relations

$$E = 2C\,(1 + \mu) = 3K\,(1 - 2\mu), \text{ we have}$$

Modulus of rigidity

$$C = \frac{E}{2(1+\mu)} = \frac{2\times10^5}{2(1+0.25)} = 0.8\times10^5 \text{ N/mm}^2$$

$$\text{Bulk modulus } K = \frac{E}{3(1-24)} = \frac{2\times10^5}{3(1.2\times0.25)} = 1.33\times10^5 \text{ N/mm}^2$$

Example 7.3: A rod 25 mm in diameter and 2 metre long is subjected to an axial pull of 35 kN. If the modulus of elasticity of the material of the rod is 210 GN/m^2. Determine: (*i*) the stress (*ii*) the strain, and (*iii*) the elongation of the rod.

Solution: Given:

Length of rod, $l = 2$ m or 2000 mm

Diameter of the rod, $d = 25$ mm

Axial Pull, $P = 35$ kN $= 35 \times 10^3$N

Modulus of Elasticy, $E = 210$ GN/m^2 $= 210 \times 10^9$ N/m^2 $= 210\times10^9$ N/mm^2

Area of cross-section of the rod,

$$A = \frac{\pi}{4}d^2$$

$$= \frac{\pi}{4}(25)^2 = 490.6 \text{ mm}^2$$

$$= 210 \times 10^9 \text{ N/m}^2$$
$$= 210 \times 10^3 \text{ N/mm}^2$$

$$(i) \text{ Stress} \quad (\sigma) = \frac{P}{A} = \frac{35\times10^3}{490.62} = 71.34 \text{ N/mm}^2$$

$$(ii) \text{ Strain} \quad (c) = \frac{\sigma}{E} = \frac{71.34}{210\times10^3} = 0.00034$$

(*iii*) Elongation, $\delta l = l \times$ strain

$$= 2000 \times 0.00034 = 0.68 \text{ mm}$$

Example 7.4: The ultimate stress, for a hollow steel column which carries an axial load of 2.5 MN is 500 N/mm^2. If the external diameter of the column is 250 mm, determine the internal diameter. Take the factor of safety as 4.

Solution: Given:

Axial load, $P = 2.5 \text{ MN} = 2.5 \times 10^6$ N

External diameter, $D = 250$ mm

Ultimate stress = 500 N/mm^2

Let d = Internal diameter of the column

$\therefore$ Cross-sectional area of the column,

$$A = \frac{\pi}{4}(D^2 - d^2)$$

$$= \frac{\pi}{4}(250^2 - d^2)\text{mm}^2 \qquad \ldots(i)$$

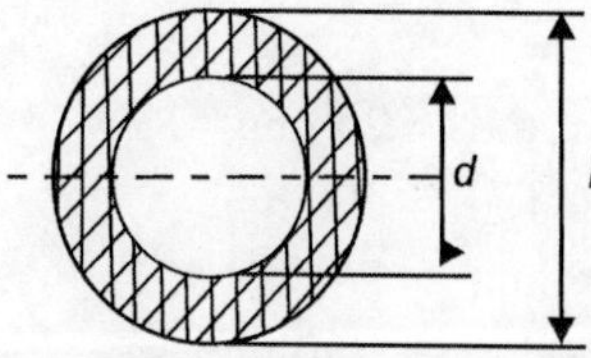

Fig. 7.58 A hollow steel column.

Working stress, $\sigma = \dfrac{\text{Ultimate stress}}{\text{Factor of safety}}$

$$= \frac{500}{4} = 125 \text{ N/mm}^2$$

Now, $\sigma = \dfrac{P}{A}$

or $A = \dfrac{P}{\sigma} = \dfrac{2.5 \times 10^6}{125} = 20{,}000 \text{ mm}^2 \qquad \ldots(ii)$

From equations (*i*) and (*ii*), we get

$$\frac{\pi}{4}(250^2 - d^2) = 20{,}000$$

or $62500 - d^2 = \dfrac{20{,}000 \times 4}{\pi} = 25477.70$

or $d^2 = 62500 - 25477.70 = 37022.29$

$\therefore$ $d = 192.41$ mm

Example 7.5: A tensile test was conducted on a mild steel rod. The following data was obtained from the test:

(*i*) Diameter of the steel rod = 20 mm
(*ii*) Gauge length of the rod = 200 mm
(*iii*) Load at elastic limit = 260 kN
(*iv*) Extension at a load of 150 kN = 0.20 mm
(*v*) Maximum load = 500 kN
(*vi*) Total extension = 60 mm
(*vii*) Diameter of the rod at failure = 15 mm

Determine:

(*a*) The Young's modulus
(*b*) The stress at elastic limit
(*c*) The percentage elongation
(*d*) The percentage reduction in area.

Solution: Given:

$$\begin{aligned} \text{Diameter of the steel rod} &= 20 \text{ mm} \\ \text{Gauge length of the rod} &= 200 \text{ mm} \\ \text{Load at elastic limit} &= 260 \text{ kN} \\ \text{Extension at a load of 150 kN} &= 0.20 \text{ mm} \\ \text{Maximum load} &= 500 \text{ kN} \\ \text{Total extension} &= 60 \text{ mm} \\ \text{Diameter of the rod at failure} &= 15 \text{ mm} \end{aligned}$$

Original cross-sectional area of the rod

$$= \frac{\pi}{4} D^2$$

$$= \frac{\pi}{4} (20)^2 = 314.16 \text{ mm}^2$$

(*a*) To find Young's modulus, first calculate the value of stress and strain within elastic limit. A load of 150 kN (which is within elastic limit) and corresponding extension of 0.21 mm is given

$$\therefore \quad \text{Stress} = \frac{\text{Lead}}{\text{Area}} = \frac{150 \times 103}{314.16} = 477.46 \text{ N/mm}^2$$

$$\therefore \quad \text{Strain} = \frac{\text{Increase in length (or extension)}}{\text{Original length (or gauge length)}}$$

$$= \frac{0.20}{200} = 0.001$$

∴ Young's Modulus,

$$E = \frac{\text{Stress}}{\text{Strain}} = \frac{477.46}{0.001}$$

$$= 477463 \text{ N/mm}^2 = 477.46 \text{ kN/mm}^2$$

(*b*) Stress at elastic limit

$$= \frac{\text{Load at elastic limit}}{\text{Original cross-sectional area}}$$

$$= \frac{260 \times 10^3}{314.66} = 827.60 \text{ N/mm}^2$$

(*c*) Percentage elongation

$$= \frac{\text{Total increase in length}}{\text{Original length}} \times 100$$

$$= \frac{60 \times 100}{200} = 30\%$$

(*d*) Area of the rod at the failure

$$= \frac{\pi}{4}(15)^2 = 176.62 \text{ mm}^2$$

Percentage reduction in area

$$= \frac{\text{Original area} - \text{Area at the failure}}{\text{Original area}} \times 100$$

$$= \frac{314.16 - 176.62}{314.16} \times 100 = 43.78\%$$

Example7.6: Fig. 7.59 shows a steel bar consisting of three lengths. Find the stresses in the three parts and the total elongation of the bar for an axial pull of 40 kN. Take $E = 2.1 \times 10^6$ N/mm^2.

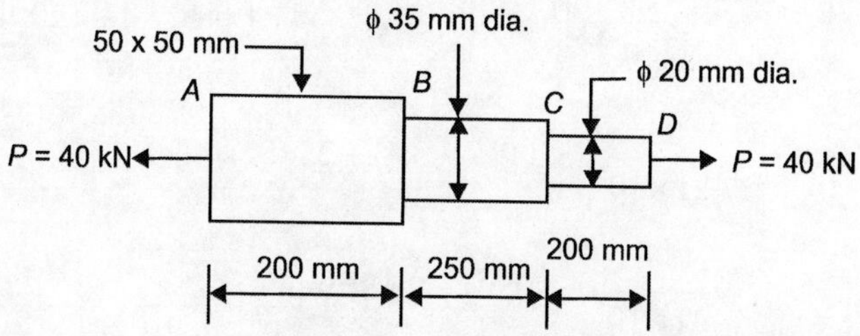

Fig. 7.59 A steel bar

Solution:

Given pull, $P = 40$ kN = 4000 N

Part AB:

$$\text{Area, } A_1 = 50 \times 50 = 2500 \text{ mm}^2$$

$$\text{Length, } l_1 = 200 \text{ mm}$$

$$\text{Stress, } \sigma_1 = \frac{P}{A_1} = \frac{40000}{2500} = 16 \text{ N/mm}^2$$

Part BC:

$$\text{Area, } A_2 = \frac{\pi}{4}(35)^2 = 962.11 \text{ mm}^2$$

$$\text{Length, } l_2 = 250 \text{ mm}$$

$$\text{Stress, } \sigma_2 = \frac{P}{A_2} = \frac{40000}{962.11} = 41.58 \text{ N/mm}^2$$

Part CD:

$$\text{Area, } A_3 = \frac{\pi}{4}(20)^2 = 100\pi \text{ mm}^2$$

$$\text{Length, } l_2 = 200 \text{ mm}$$

$$\text{Stress, } \sigma_3 = \frac{P}{A_3} = \frac{40000}{100\pi} = 127.32 \text{ N/mm}^2$$

Total elongation,

$$\delta l = \delta l_1 + \delta l_2 + \delta l_3$$

$$= \frac{\sigma_1 l_1}{E_1} + \frac{\sigma_2 l_2}{E_2} + \frac{\sigma_3 l_3}{E_3}$$

$$= \frac{1}{E}\left[\sigma_1 l_1 + \sigma_2 l_2 + \sigma_3 l_3\right]$$

$$= \frac{1}{21 \times 10^5}(15 \times 200 + 41.58 \times 250 + 127.32 \times 200)$$

or

$$\delta l = \frac{1}{21 \times 10^5}(3200 + 10395 + 25464)$$

$$= \frac{1}{21 \times 10^5}(39059) = 0.186 \text{ mm}$$

:. Total elongation, $\delta l = 0.186$ mm.

Example 7.7: A straight bar 450 mm long is 10 mm in diameter for the first 200 mm length and 20 mm in diameter for the remaining length. If the bar is subjected to an axial push of 10 kN, determine decrease in length of the bar. Take modulus of elasticity of bar material $E = 2 \times 10^5$ N/mm^2.

Solution: Refer Fig. 7.60.

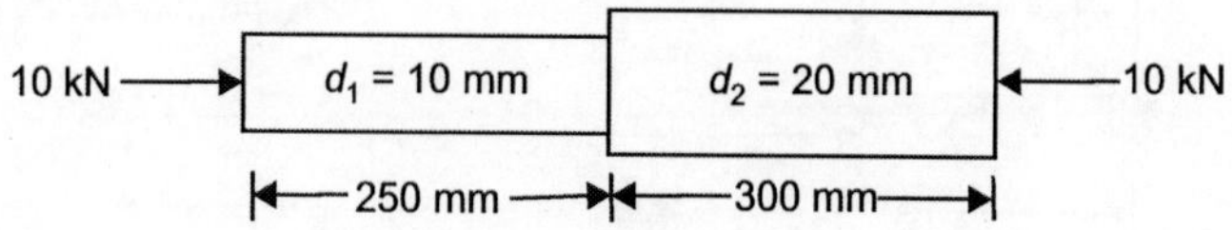

Fig. 7.60 A straight bar

$$A_1 = \frac{\pi}{4}(10)^2 = 78.5 \text{ mm}^2; A_2 = \frac{\pi}{4}(20)^2 = 314 \text{ mm}$$

$$l_1 = 250 \text{ mm and } l_2 = 300 \text{ mm}$$

Let δl_1 and δl_2 be the contractions in parts 1 and 2 of the bar respectively. It may be recalled that

$$E = \frac{\sigma}{\varepsilon} = \frac{P/A}{\delta l / l} = \frac{Pl}{A\delta l} \quad \text{or} \quad \delta l = \frac{Pl}{AE}$$

Then $$\delta l_1 = \frac{P_1 l_1}{A_1 E_1} \text{ and } \delta l_2 = \frac{P_2 l_2}{A_2 E_2}$$

$\therefore$ Total decrease in length of the bar,

$$\delta l = \delta l_1 + \delta l_2 = \frac{P_1 l_1}{A_1 E_1} + \frac{P_2 l_2}{A_2 E_2}$$

Both the segments are of same material ($E_1 = E_2$) and are subjected to the same push ($P_1 - P_2$).

$\therefore$ $$\delta l = \frac{P}{E}\left(\frac{l_1}{A_1} + \frac{l_2}{A_2}\right) = \frac{10 \times 10^3}{2 \times 10^5}\left(\frac{250}{78.5} + \frac{300}{314}\right)$$

$$= 0.05\ (3.184 + 0.955) = 0.2069 \text{ mm}$$

Example 7.8: The bar shown in Fig. 7.61 is subjected to an axial pull of 150 kN. Determine diameter of the middle portion, if stress there is limited to 125 N/mm^2. Proceed to determine the length of this middle portion if total extension of the bar is specified as 0.15 mm. Take modulus of elasticity of bar material $E = 2.5 \times 10^5$ N/mm^2.

Solution: Refer Fig.7.61.

Each of the segment of this composite bar is subjected to axial pull $P = 150$ kN.

Stress in the middle portion

$$\sigma_2 = \frac{\text{axial pull}}{\text{area}}$$

$$= \frac{150 \times 10^3}{\frac{\pi}{4} d_2^2}$$

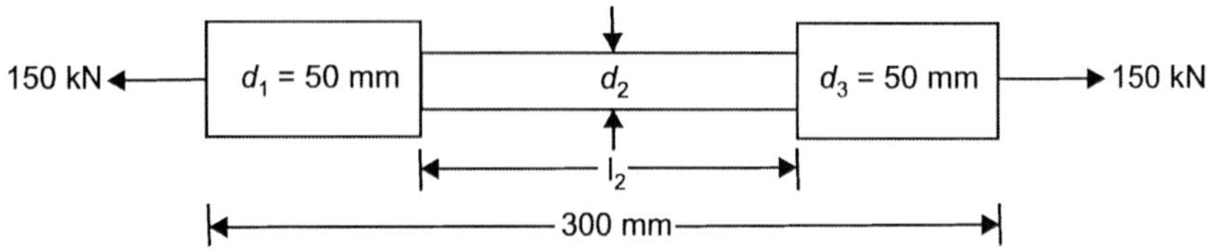

Fig. 7.61 The bar

Since stress is limited to 125 N/mm^2,

$$125 = \frac{150 \times 10^3}{\frac{\pi}{4} d_2^2}; \; d_2^2 = \frac{150 \times 10^3}{\frac{\pi}{4} \times 125} = 1528.66$$

∴ Diameter of middle portion

$$d_2 = \sqrt{1528.66} = 39.1 \text{ mm}$$

(*ii*) Stress in the end portion,

$$d_1 = d_3 = \frac{150 \times 10^3}{\frac{\pi}{4} (50)^2} = 76.43 \text{ N/m}^2$$

Total change in length of the bar,

= change in length of end portions

+ change in length of mid portion

$$\delta l = \frac{P_1 l_1}{A_1 E_1} + \frac{P_3 l_3}{A_3 E_3} + \frac{P_2 l_2}{A_2 E_2} = \frac{\sigma_1 l_1}{E_1} + \frac{\sigma_3 l_3}{E_3} + \frac{\sigma_2 l_2}{E_2}$$

Here $E_1 = E_2 = 2 \times 10^5$ N/mm^2, $s_1 = s_3 = 76.43$ N/m^2 and combined length $(l_1 + l_3) = (300 - l_2)$. Further, total extension of the bar is specified as 0.15 mm.

$$\therefore \quad 0.15 = \frac{76.43}{2 \times 10^3} \times (300 - l_2) + \frac{125 l_2}{2 \times 10^5}$$

or $\quad 0.15 \times 2 \times 10^5 = 22929 - 76.43\, l_2 + 125\, l_2$

or $\quad 3000 - 22929 = 48.57\, l_2$

∴ length of middle portion,

$$l_2 = \frac{3000 - 22929}{48.57} = 145.58 \text{ mm}$$

Example 7.9: A brass bar is subjected to axial forces as shown in Fig. 7.62. Find the total change in the length bar. Take $E = 1 \times 10^5$ N/mm^2.

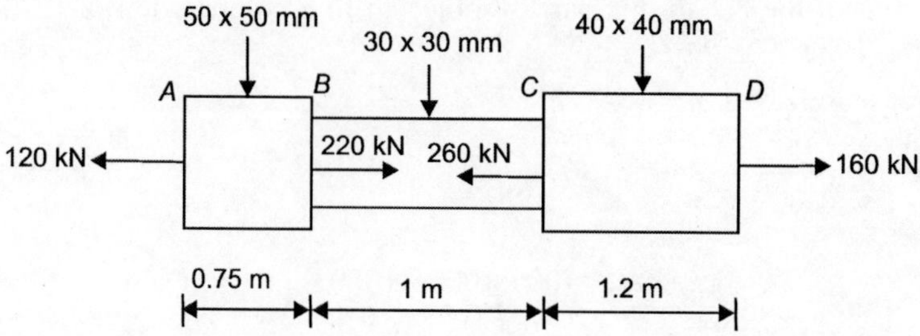

Fig. 7.62 A brass bar

Solution: First of all make the free body diagram for the load-on parts.

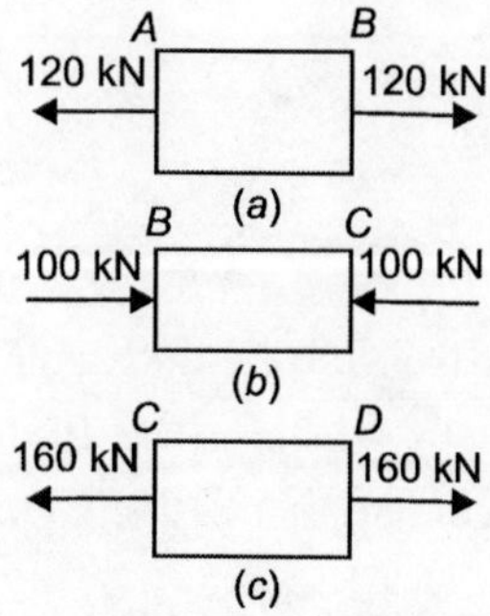

Fig. 7.63 Diagram of the load

Part *AB*:

Area, $A_1 = 50 \times 50 = 2500$ mm^2

Length, $l_1 = 0.75$m $= 750$ mm

The section of the bar in this part is subjected to a tension of 120 kN (Refer Fig. 7.63)

∴ Increase in length of *AB*,

$$\delta l_1 = \frac{\sigma_1 l_1}{E}$$

$$= \frac{P_1 \times l_1}{A_1 \times E}$$

$$= \frac{120 \times 10^3 \times 750}{2500 \times 1 \times 10^5} = 0.36 \text{ mm}$$

Part *BC*:

Area, $A_2 = 30 \times 30 = 900 \text{ mm}^2$

Length, $l_2 = 1\text{m} = 1000 \text{ mm}$

The section of the bar in this part is subjected to a compression of (220 – 120) = 100 kN (Refer Fig. 7.63)

∴ Decrease in length of *BC*,

$$\delta l_2 = \frac{\sigma_2 1_2}{E}$$

$$= \frac{P_2 \times l_2}{A_2 \times E} = \frac{100 \times 1000 \times 1000}{900 \times 1 \times 10^5} = 1.11 \text{mm}$$

Part *CD*:

Area, $A_3 = 40 \times 40 = 1600 \text{ mm}^{2:}$

Length, $l_3 = 1.2 \text{ m} = 1200 \text{ mm}$

The section of the bar in this part is subjected to a tension of 160 kN (Refer Fig. 7.63)

∴ Increase in length of *CD*,

$$\delta l_3 = \frac{\sigma_3 1_3}{E} = \frac{P_3 \times l_3}{A_3 \times E} = \frac{160 \times 1000 \times 1200}{1600 \times 1 \times 10^5} = 1.2 \text{mm}$$

∴ Net change in length of the bar, $\delta l = 8l, - 8l, + 8l_3$

$= 0.36 - 1.11 + 1.2 = +0.45$ (Extension)

(Taking +ve sign for increase in length and –ve sign for decrease in length).

Example 7.10: Refer Fig. 7.64 for the tapering bar of circular cross-section which is subjected to point loads as shown. The section *AB* has a uniform diameter of 60 mm and for the section *CD*, the uniform diameter is 300 mm. Determine load P_3 for equilibrium and net change in length of bar. Take modulus of elasticity of the bar material as $E = 200$ GPa.

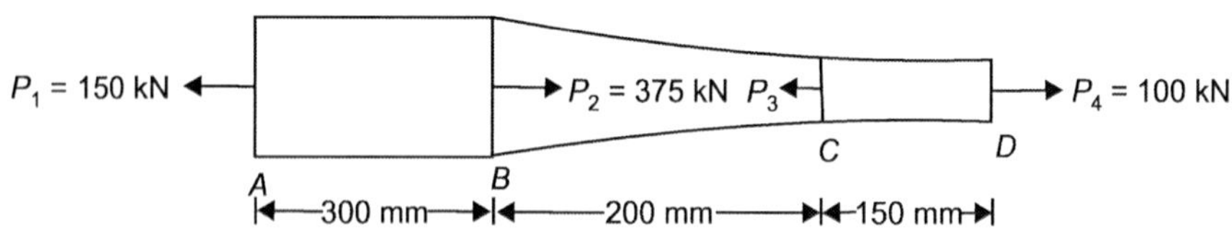

Fig. 7.64 The tapering bar

Solution:

Modulus of elasticity $E = 200 \text{ GPa} = 2 \times 10^5 \text{ N/mm}^2$

Considering equilibrium of forces along the axis of bar,

$P_1 + P_3 = P_2 + P_4$; $150 + P_3 = 375 + 100$

∴ Force $P_3 = 375 + 100 - 150 = 325$ kN

The forces acting on each segment of the bar are as shown below:

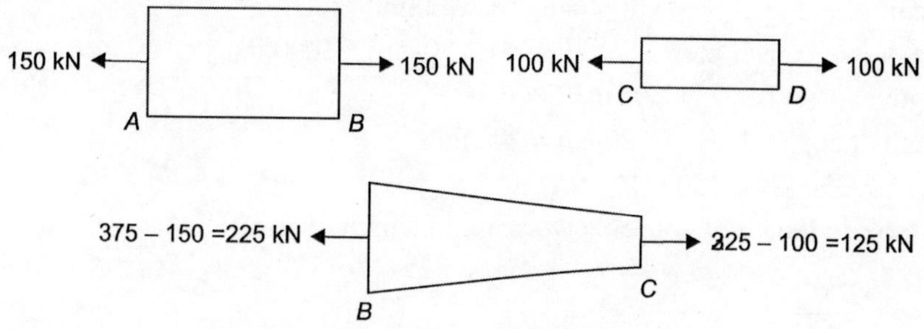

Fig. 7.65 Forces acting on each segment

Recalling that extension/contraction for a uniform bar is given by *Pl*/*AE* and that for a tapering bar is $4P/\pi E\, d_1, d_2$, we have

Extension of segment *AB*;

$$\delta l_1 = \frac{(150\times 10^3)\times 300}{\frac{\pi}{4}(60)^2\times(2\times 10^5)} = 0.079\,\text{mm}$$

Contraction of segment *BC*;

$$\delta l_2 = \frac{4\times(225\times 10^3)\times 200}{\pi\times(2\times 10^5)\times 60\times 30} = 0.1592\,\text{mm}$$

Extension of segment *CD*;

$$\delta l_3 = \frac{(100\times 10^3)\times 150}{\frac{\pi}{4}\times(30)^2\times(2\times 10^5)} = 0.1063\,\text{mm}$$

∴ Net change in length of member

$$\delta l = \delta l_1 + \delta l_2 + \delta l_3 = 0.079 - 0.1592 + 0.1063 = 0.0261\ \text{mm}$$

Example 7.11: Two copper bars and a steel bar together support a load of 350 kN as shown in Fig. 7.66. Find the stress in the rods. Young's modulus for steel is twice that of copper.

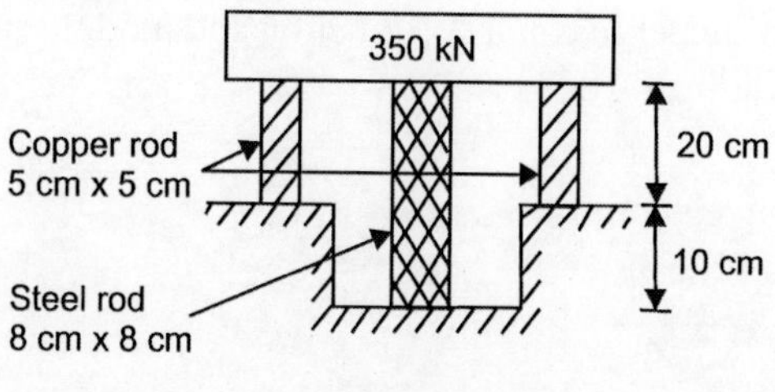

Fig. 7.66 Two copper bars and a steel bar

Solution: Given:

Area of steel bar, $A_S = 8 \times 8 = 64 cm^2 = 6400 \ mm^2$

Area of two copper bars, $A_C = 2(5 \times 5) = 50 \ cm^2 = 5000 \ mm^2$

Length of steel bar, $l_S = 30cm = 300mm$

Length of copper bar, $l_C = 20 \ cm = 200 \ mm$

load, $P = 350 \ kN = 350 \times 10^3 \ N$

Decrease in length of copper = Decrease in length of steel

$$\frac{\sigma_C}{E_C} \times l_C = \frac{\sigma_S}{E_S} \times l_S$$

$$\sigma_S = \frac{\sigma_C}{E_C} \times \frac{l_C}{l_S} \times E_S$$

$$= \frac{\sigma_C \times 200 \times 2E_C}{E_C \times 300} \qquad (\because E_S = 2E_C \text{ given})$$

$$\sigma_S = \frac{4}{3}\sigma_C$$

The load applied = Load on steel + Load on copper

$$350 \times 10^3 = \sigma_S A_s + \sigma_C A_C$$

$$= \frac{4}{3}\sigma_C \times 6400 + \sigma_C \times 5000 = 13533.33 \ \sigma_C$$

$$\sigma_C = \frac{350 \times 10^3}{13533.33} = 25.86 \ N/mm^2$$

Stress in copper = $25.86 \ N/mm^2$

Stress in steel, $\sigma_S = \frac{4}{3} \ \sigma_C = \frac{4}{3} \times 25.68 = 34.48 \ N/mm^2$.

Example 7.12: Two steel rods and one copper rod lie equidistant in a vertical plane and together support a load of 25 kN as shown in Fig. 7.67. Each rod is 10 mm in diameter and these rods share the load equally. Determine the stress to which each rod would be subjected when the system is to support an addition load of 10 kN. All the rods undergo equal deformation and modulus of elasticity of steel equals twice that of copper.

Solution:

Initial load carried by each rod

$$= \frac{25 \times 10^3}{3} = 7000 \, N$$

$$\text{Stress in each rod} = \frac{7000}{\frac{\pi}{4}(10)^2} = \text{N/mm}^2 = 89.17 \text{ N/mm}^2$$

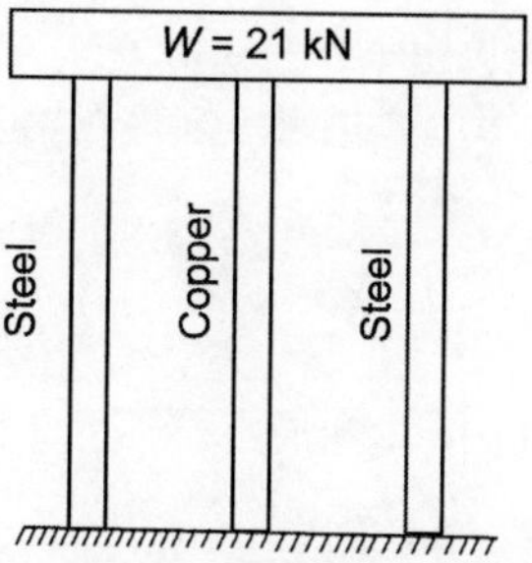

Fig. 7.67 Two steel rods and one copper rod

Division of additional load is as follows,

additional load = load carried by copper rod + load carried by two steel rods

$$10 \times 10^3 = \sigma_C A_C + 2\,\sigma_S A_S$$

The rods undergo equal deformation and are of same length; their strains are equal and accordingly

$$\frac{\sigma_S}{E_S} = \frac{\sigma_C}{E_C}$$

$$\sigma_S = \frac{E_S}{E_C}\sigma_C = 2\sigma_C$$

Expression (*i*) may then be written as

$$10 \times 10^3 = \sigma_C \times \frac{\pi}{4}(10)^2 + 2\,(2\,\sigma_C) \times \frac{\pi}{4}(10)^2 = 392.5\,\sigma_C$$

or $$\sigma_C = \frac{10 \times 10^3}{392.5} = 25.48 \text{ N/mm}^2 \text{ and } \sigma_S = 2 \times 25.48 = 50.96 \text{ N/mm}^2$$

$\therefore$ Total stress in copper = 89.17 + 25.48 = 114.65 N/mm^2

Total stress in steel = 29.17 + 50.96 = 140.13 N/mm^2

Example 7.13: A reinforced concrete column is 300 mm × 300 mm in section. The column is provided with 8-bars of 20 mm diameter. The column carries a load of 360 kN. Find the stresses in concrete and the steel bars.

(Take $E_S = 2.1 \times 10^5$ N/mm^2 and $E_C = 0.14 \times 10^5$ N/mm). (Refer Fig. 7.68)

Solution: Given:

Diameter of steel bar = 20 mm Number of steel bars = 8;

Load, $P = 360$ kN $= 360 \times 10^3$N;

Modulus of Elasticity for steel, $E_S = 2.1 \times 1.05$ N/mm^2;

Modulus of Elasticity for concrete, $E_C = 0.14 \times 10^5$ N/mm^2

Area of steel bars (A_S)

$$A_S = 8 \times \frac{\pi}{4}(20)^2 = 314.2 \times 8 = 2513.6 \text{ mm}^2 \text{ say } 2514 \text{ mm}^2$$

Actual area of concrete (A_C)

$$Ac = 400^2 - 2514 = 157486 \text{ mm}^2$$

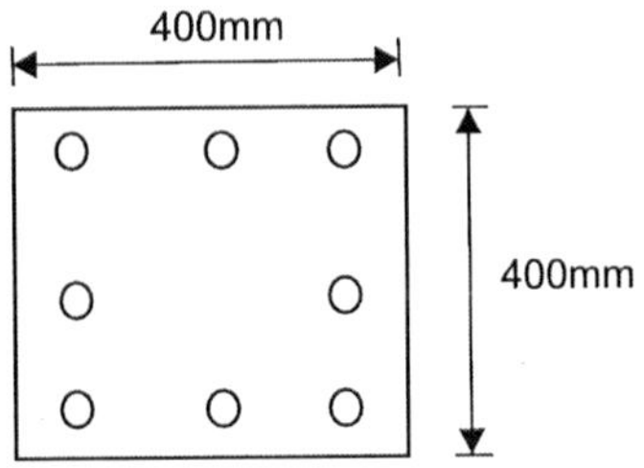

Fig. 7.68 A reinforced concrete column

Let the stresses in concrete and steel be σ_C and σ_S respectively.

Strain in concrete = Strain in steel

$$\frac{\sigma_C}{E_C} = \frac{\sigma_S}{E_S}$$

$$\sigma_S = \frac{\sigma_C \times E_S}{E_C} = \frac{2.1}{0.14}\sigma_C$$

$$\therefore \quad \sigma_S = 26P_C$$

Load on steel + Load on concrete = Total load on the column

i.e. $$\sigma_S A_S + \sigma_C A_C = P$$

$$15\sigma_C A_S + \sigma_C A_C = P$$

$$15\sigma_C \times 2514 + \sigma_C \times 157486 = 360 \times 10^3$$

$$195196\ \sigma_C = 1.84 = 360 \times 10^3 \text{ N/mm}^2$$

$$\sigma_S = 15 \times 1.84 = 27.6 \text{ N/mm}^2.$$

Example 7.14: (*a*) A short hollow cast iron column of 300 mm external diameter and 160 mm internal diameter is filled with concrete as shown in Fig.7.69. The column carries a load of 350 kN. If the modulus of elasticity of cast iron is 6 times that of concrete, find the stresses in the cast iron and the concrete. (*b*) What must be the internal diameter of the cast iron column if a load of 400 kN is to be carried, the stresses and the external diameter being unchanged.

Solution:

(*a*) Load on column,

$$P = 350 \text{ kN} = 350 \times 10^3 \text{ N}$$

Area of cross-section of cast iron,

$$A_1 = \frac{\pi}{4}(300^2 - 160^2) = 50{,}554 \text{ mm}^2$$

Area of cross-section of concrete,

$$A_2 = \frac{\pi}{4}(160)^2 = 20096 \text{ mm}^2$$

Strain in cast iron = Strain in concrete

$$\frac{\sigma_1}{E_1} = \frac{\sigma_2}{E_2}$$

or

$$\sigma_1 = \frac{\sigma_1 \times E_1}{E_2}$$

$$= \frac{6E_2 \times \sigma_2}{E_2} \qquad (\because E_1 = 6E_2)$$

$$\sigma_1 = 6\sigma_2$$

Total load on column = Load on cast iron + Load on concrete

$$350 \times 10^3 = \sigma_1 A_1 + \sigma_2 A_2$$

$$= 6\sigma_2 \times 50554 + \sigma_2 \times 20096$$

or

$$350 \times 10^3 = 70650\sigma_2$$

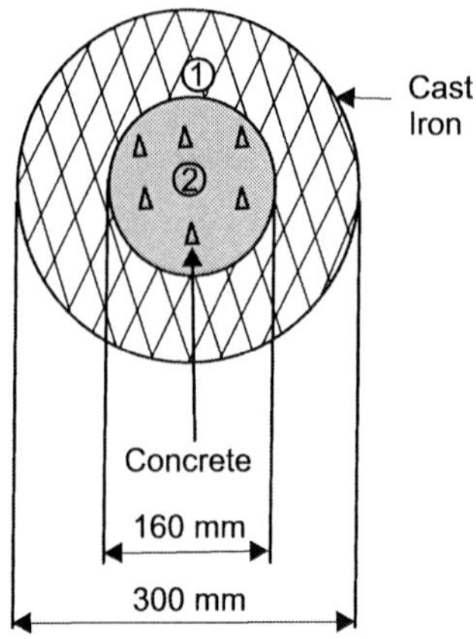

Fig. 7.69 A cast iron column

$$\therefore \qquad \sigma_2 = \frac{350 \times 10^3}{70650} = 4.95 \text{ N/mm}^2$$

$$\sigma_2 = 4.95 \text{ N/mm}^2$$

Stress in concrete,

$$\sigma_2 = 4.95 \text{ N/mm}^2$$

Stress in cast iron,

$$\sigma_1 = 6\sigma_2$$

$$= 6 \times 4.95 = 29.7 \text{ N/mm}^2$$

$$\sigma_1 = 29.7 \text{ N/mm}^2$$

(*b*) Total load on column = 400 kN

$$= 400 \times 10^3 \text{ N}$$

Stress in cast iron,

$$\sigma_1 = 29.7 \text{ N/mm}^2$$

Stress in concrete,

$$\sigma_2 = 4.95 \text{ N/mm}^2$$

External diameter of cast iron column, $D = 300$ mm

Let $\qquad d$ = internal diameter of cast iron column in mm.

Area of cross-section of cast iron,

$$A_1 = \frac{\pi}{4} (300^2 - d^2) \text{ mm}^2$$

Area of cross-section of concrete,

$$A_2 = \frac{\pi}{4} d^2 \text{ mm}^2$$

Total load on column = Load on cast iron + Load on concrete

$$400 \times 10^3 = \sigma_1 A_1 + \sigma_2 A_2$$

$$= 29.7 \times \frac{\pi}{4}(300^2 - d^2) + 4.95 \times \frac{\pi}{4} d^2$$

$$= 2098305 - 23.31\, d^2 + 3.88\, d^2$$

or $$19.43\, d^2 = 2098305$$

$$d^2 = 107993.05$$

$$d = 328.62 \text{ mm}$$

Example 7.15: Two vertical rods of steel and copper are firmly secured at their upper ends and lie at a distance of 100 cm apart. Each rod is 3 m long and 20 mm in diameter. A horizontal rigid cross bar connects the lower ends of the rods and on it is placed a load of 40 kN. Determine (*a*) location of the load on the cross bar if it remains horizontal after being loaded, (*b*) What will be the inclination of the cross bar if the load is hung at its mid-span?

Take E for steel = 2×10^5 N/mm², E for copper = 1×10^5 N/mm² and length of each rod =3m

Solution:

Refer Fig. 7.70 for the arrangement of the composite system.

Area of each rod $A = \frac{\pi}{4}(20)^2 = 314 \text{ mm}^2$

Since the cross-bar remains horizontal, each rod experiences the same amount of strain. Thus,

$$\frac{\sigma_S}{E_S} = \frac{\sigma_C}{E_C}$$

$\therefore$ $$\sigma_S = \frac{E_S}{E_C}\sigma_C = \frac{2\times10^5}{1\times10^5}\sigma_C = 2\sigma_C$$

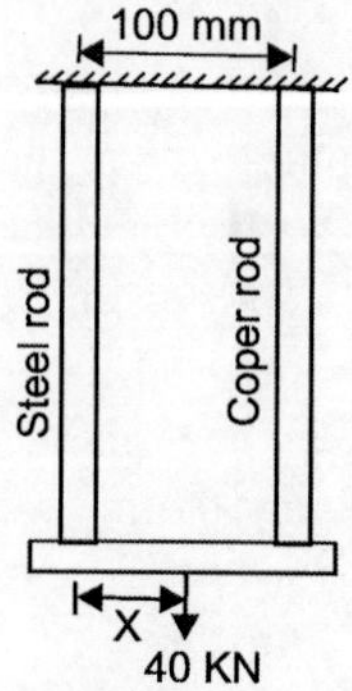

Fig. 7.70 Two vertical rods of steel and copper

Further, the total load is shared by both the rods. Therefore,

$$P = P_S + P_C = \sigma_s A_s + \sigma_C A_C$$

or $$40 \times 10^3 = 2\sigma_C \times 314 + \sigma_C \times 314 = 942 \qquad (\because \sigma_S = 2\sigma_C)$$

and $$\sigma_C = \frac{40\times10^3}{942} = 42.462 \text{ N/mm}^2$$

and $$\sigma_S = 2\sigma_C = 2 \times 42.46 = 84.92 \text{ N/mm}^2$$

Load carried by steel rod, $P_S = \sigma_S A_S = 84.92 \times 314 = 26665$ N

Load carried by copper rod, $P_C = \sigma_C A_C = 42.46 \times 314 = 13332$ N

Let the load of 40 kN be applied at x mm from the steel rod. Then taking moments about the steel rod, we have

$$P_C \times 100 = P \times x;\ 13332 \times 100 = 40 \times 10^3 \times x$$

$$\therefore \quad x = \frac{13332\times100}{40\times10^3} = 33.33 \text{ mm}$$

Thus, the load should be placed at a distance of 33.33 mm from the centre line of steel rod if the cross bar is to remain horizontal.

(*b*) When the 40 kN load is hung at mid span of the cross-bar, then load carried by each wire = 20 kN

$$\delta_C = \frac{P_C l}{A_C E_C} = \frac{(20\times10^3)\times(3\times10^3)}{314\times(1\times10^5)} = 1.910 \text{ mm}$$

$$\delta_S = \frac{P_S l}{A_S E_S} = \frac{(20\times10^3)\times(3\times10^3)}{314\times(2\times10^5)} = 0.955 \text{ mm}$$

If θ is the inclination of the rigid bar with the horizontal, then

$$\tan\theta = \frac{1.910-0.955}{100} = 0.00995$$

$$\theta = \tan^{-1}(0.00995) = 547°$$

Example 7.16: A steel rod 20 mm diameter passes centrally through a steel tube of 25 mm internal diameter and 35 mm external diameter. The tube is 500 mm long and is closed by rigid washers, of negligible thickness, which are fastened by nuts threaded on the rod. The nuts are tightened and a load of 1.5 kN is exerted on the tube. Calculate the stresses in the tube and the rod. How these stresses would be affected when one nut is tightened by one-quarter of turn relative to the other. Take $E = 2 \times 10^5$ N/mm² and 4 threads per 10 mm (*i.e.* pitch of bolt threads = 2.5 mm).

Solution:

$$\text{Area of tube } A_t = \frac{\pi}{4}(35^2 - 25^2) = 471 \text{ mm}^2$$

$$\text{Area of rod } A_t = \frac{\pi}{4}(20^2) = 314 \text{ mm}^2$$

When the nut is tignten the tube would be compressed and the rod will be elongated. Since no external forces have been applied, tensile load on the rod = compressive load on the tube.

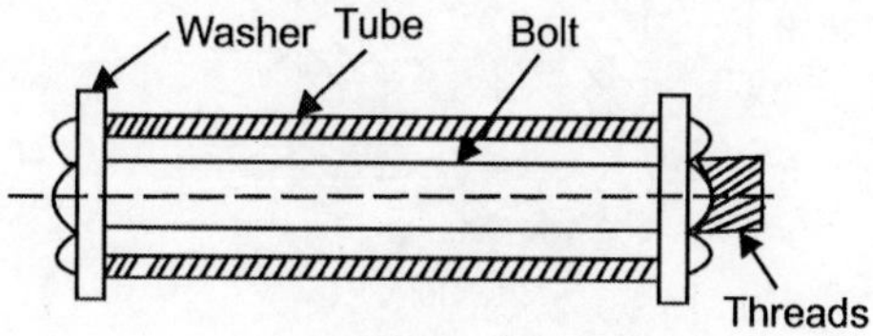

Fig. 7.71 A steel rod

(*i*) When compressive load on the tube is 15 kN:

$$\text{Stress in the tube } \sigma_t = \frac{15\times10^3}{471} = 31.85 \text{ N/mm}^2 \text{ (compressive)}$$

$$\text{Stress in the rod } \sigma_r = \frac{15\times10^3}{314} = 47.77 \text{ N/mm}^2 \text{ (tensile)}$$

(*ii*) *When nut is tighten by one - quarter of a turn:*

Let σ_t' and σ_r' be the stresses in the tube and rod respectively due to further tightening of the nut. Then

$$\sigma_t' A_r = \sigma_t'; \ \sigma_r' = \frac{At}{Ar}\sigma_t' = \frac{471}{314}\sigma_t' = 1.5\sigma_t'$$

$$\text{Reduction in length of tube} = \frac{\sigma_r'}{E}l = \frac{\sigma_t'}{2\times10^5}\times500 = 0.002\sigma_r'$$

$$\text{Increase in length of rod} = \frac{\sigma_r'}{E}l = \frac{1.5\sigma_t'}{2\times10^5}\times500 = 0.037\sigma_r'$$

Now, contraction of tube + extension of rod = axial advance of the nut

$$0.003\ \sigma_t' + 0.0045\ \sigma_t' = \frac{1}{4}\times\frac{1}{4}\ (10) = 0.625$$

$$\sigma_t' = \frac{0.625}{0.002+0.0037} = 125 \text{ N/ mm}^2 \text{ (compressive)}$$

$$\sigma_r' = 1.5\ \sigma_t' = 1.5 \times 83.33 = 187.5 \text{ N/mm}^2 \text{ (tensile)}$$

Example 7.17: Two equal vertical rods, one of steel and other of bronze each of *x*-sectional area 100 mm^2 and of length 300 mm are fastened rigidly at the upper ends to a level support and the distance between the centres of two rods being 0.80 metre. A cross bar fixed to the rods at the lower ends carries a load of 25000 N such that the cross bar remains horizontal even after loading. Find the stress in each rod and the position of the load on the bar. (Take E for steel = 2×10^5N/mm^2 and E for bronze = 1.2×10^5N/n-mm^2).

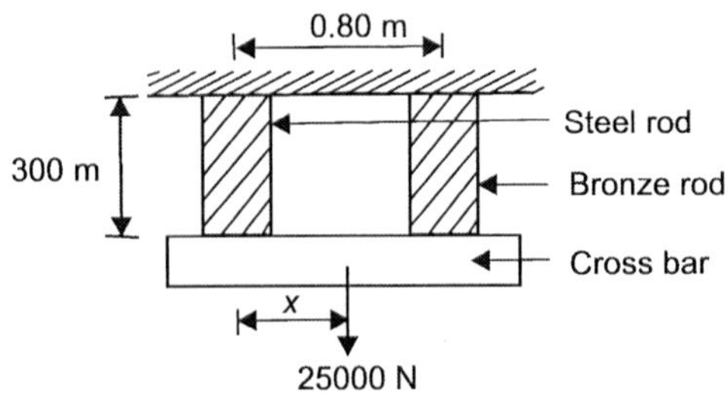

Fig. 7.72 Two equal vertical rods

Solution: Let

σ_S = Stress in steel rod
σ_b = Stress in bronze rod
E_S = Modulus of elasticity in steel
E_b = Modulus of elasticity in bronze

Since the cross bar remains horizontal after loading, the strain of these rods will be equal.

∴ Strain in steel = Strain in bronze

$$\frac{\sigma_S}{E_S} = \frac{\sigma_b}{E_b}$$

or

$$\sigma_S = \frac{\sigma_b \times E_S}{E_b}$$

$$= \frac{\sigma_b \times 2 \times 10^5}{1.2 \times 10^5} = 1.67\sigma_b$$

Total load = Load on steel rod + Load on bronze rod

$25000 = P_s + P_b$
$25000 = \sigma_s A_s + \sigma_b A_b$
$25000 = 1.67\sigma_b \times 100 + \sigma_b \times 100$
$25000 = 267\sigma_b$

∴ $\sigma_b = \dfrac{25000}{267} = 93.63 \text{ N/mm}^2$

∴ $\sigma_s = 1.67\sigma_b$
$= 1.67 \times 93.63 = 156.36 \text{ N/mm}^2$

Load carried by steel rod,

$P_s = \sigma_s \times A_s$
$= 156.36 \times 100 = 15636 \text{ N}$

Load carried by bronze rod,

$$P_b = \text{Total load} - P_S$$

$$25000 - 15636 = 9363.79 \text{ N}$$

Let the load be placed at a distance x metre from the steel rod (Refer Fig. 7.42) Taking moments about the steel rod, we get

$$P_b \times 0.80 = 25000 \times x$$

$$9363.79 \times 0.80 = 20000\ x$$

$$\therefore \quad x = \frac{9263.79 \times 0.80}{25000} = 0.299 \text{ m}$$

Example 7.18: A screw driver corresponds to a solid steel shaft of 3 mm diameter. Make calculations for the maximum torque that can be transmitted by it. Maximum permissible shear stress for steel is 250 MN/m^2.

Solution: $\tau = 250 \text{ MN/ m}^2 = 250 \text{ N/ mm}^2$

From the torsion equation,

$$\frac{T}{J} = \frac{\tau}{R}, \text{ we have}$$

$$T = \frac{\tau}{R} J = \frac{\tau}{d/2} \times \frac{\pi}{32} d^4 = \frac{\pi}{16} \pi d^3$$

$$= \frac{\pi}{16} \times 250 \times (3)^3 = 1324.7 \text{ Nmm} = 1.325 \text{ Nm}$$

Example 7.19: Determine the torque required to give a twist of 10 rounds to a steel wire which is of 0.6 mm diameter and 30 m length. For steel, modulus of rigidity $C = 80$ GPa.

Solution:

$C = 80 \text{ GPa} = 80 \times 10^9 \text{ N/m}^2 = 80 \times 10^3 \text{ N/mm}^2$

$\theta = 10 \text{ rounds} = 10 \times 2\pi = 62.8$ radians

From the torsion equation,

$$\frac{T}{J} = \frac{C\theta}{l}, \text{ we have}$$

$$T = \frac{C\theta}{l} \times J = \frac{C\theta}{l} \times \frac{\pi}{32} d^4$$

$$= \frac{80 \times 10^3 \times 62.8}{30 \times 10^3} \times \left(\frac{\pi}{32} \times 0.6^4\right) = 2.1297 \text{ Nmm}$$

Example 7.20: A circular steel shaft of 25 mm diameter of subjected to a torque of 500 Nm. Determine (*a*) the maximum shear stress developed in the shaft, (*b*) the shear stress at a distance of 10 mm from the centre of shaft, and (*c*) the angular twist over 1 m length of the shaft.

Take modulus of rigidity for the shaft material, $C = 80$ GPa.

Solution:

$$T = 500 \text{ Nm} = 500 \times 10^3 \text{ Nmm}$$
$$G = 800 \text{ GPa} = 80 \times 10^9 \text{ N/m}^2 = 80 \times 10^3 \text{ N/mm}^2$$

$$J = \frac{\pi}{32} d^4 = \frac{\pi}{32} (25)^4 = 38330 \text{ mm}^4$$

Torsion equation:

$$\frac{T}{J} = \frac{\tau}{R} = \frac{C1}{l}$$

(*a*) Maximum shear stress occurs at the surface of shaft, *i.e.*, at $R = 25/2 = 12.5$ mm

$$\tau = \frac{\tau}{R} \times R \times \frac{500 \times 10^3}{38330} \times 12.5 = 163 \text{ N/mm}^2$$

(*b*) Shear stress at $r = 10$ mm

Shear stress at any point is proportional to its distance from the axis of shaft. Therefore,

$$\tau_r = \frac{T}{J} \times r = \frac{163}{12.5} \times 10 = 130.4 \text{ N/mm}^2$$

(*c*) Length of the shaft $l = 1$ m $= 1000$ mm

$$\text{Angular twist } \theta = \frac{Tl}{CJ} = \frac{500 \times 10^3 \times 1000}{(80 \times 10^3) \times 38330} = 0.163 \text{ radian} = 9.344°$$

Example 7.21: Calculate the maximum torque that can be transmitted by a 35 cm diameter shaft if the twist in the shot is not to exceed 1 degree in 15 diameter of length and also the shear stress is limited to 50 MPa. Take modulus of rigidity for the shaft material $C = 85$ GPa.

Solution:

$$\theta = 1° = \frac{\pi}{180} \times 1 = 0.01744 \text{ radian}$$

$$C = 85\text{GPa} = 85 \times 10^9 \text{ N/m}^2;$$
$$\tau = 50 \text{ MPa} = 50 \times 10^6 \text{ N/m}^2$$
$$l = 15\text{d} = 15 \times 35 = 525 \text{ cm} = 5.25\text{m}$$

$$J = \frac{\pi}{32} d^4 = \frac{\pi}{32} (0.35)^4 = 0.001472 \text{ m}^4$$

Torsion equation:

$$\frac{T}{J} = \frac{\tau}{R} = \frac{C\theta}{l}$$

Considering stress as the basis:

$$T = \frac{C\theta}{l} = J = \frac{85 \times 10^9 \times 0.01744}{5.25} \times 0.001472 = 415637 \text{ Nm}$$

The maximum torque that the shaft can transmit under the specified conditions will be smaller of the values calculated above.

∴ Maximum permissible torque T = 415637 Nm = 415.64 k Nm

Example 7.22: A hollow shaft is to transmit a uniform torque of 30 kNm. The total angular twist in 2.5 m length of the shaft is not to exceed 2 degree and the allowable shearing stress is 85 MPa. Determine the external and internal diameters of the shaft. Take modulus of rigidity of the shaft material as 80 GPa.

Solution:

$$T = 30 \text{ kNm} = 30 \times 10^6 \text{ Nmm};\ \tau = 85 \text{ MPa} = 85 \text{ N/mm}^2$$

and $$C = 80 \text{ GPa} = 80 \times 10^3 \text{ N/mm}^2$$

From torsion equation;

$$\frac{T}{J} = \frac{C\theta}{l}, \text{ we have}$$

$$\theta = \frac{Tl}{CJ};\ \frac{\pi}{180} \times 2 = \frac{30 \times 10^6 \times 2.5 \times 10^3}{80 \times 10^4 \times \pi(d_0^4 - d_1^4)/32}$$

$$d_0^4 - d_1^4 = \frac{30 \times 10^6 \times 2.5 \times 10^3}{80 \times 10^4} \times \frac{32}{\pi} \times \frac{180}{2\pi} = 273.84 \times 10^6 \qquad \ldots(i)$$

Again from torsion equation,

$$\frac{T}{J} = \frac{\tau}{R}, \text{ we have}$$

$$\tau = \frac{TR}{J};\ 85 = \frac{30 \times 10^6 \times d_0/2}{\pi(d_0^4 - d_1^4)/32}$$

$$\frac{d_0^4 - d_1^4}{d_0} = \frac{30 \times 10^6}{85 \times \pi} \times \frac{32}{2} = 1.798 \times 10^6 \qquad \ldots(ii)$$

From equations (*i*) and (*ii*)

$$d_0 = \frac{273.84 \times 10^6}{1.798 \times 10^6} = 152.3 \text{ mm}$$

Substituting d_0 = 152.3 mm in expression (*i*), we get

$$(152.3)^4 - d_i^4 = 273.84 \times 10^6$$

$$d_i^4 = (152.3)^4 - 273.84 \times 10^6 = 538 \times 10^6 - 273.84 \times 10^6$$

$$d_i = (264.16 \times 10^6)^{1/4} = 127.49 \text{ mm}$$

Thus, external and internal diameters for the hollow shaft are 152.3 mm and 127.49 mm respectively.

Example 7.23: A solid circular transmits 80 kW of power while turning 200 revolutions per minute. Workout suitable diameter of the shaft if the shear is limited to 60 MN/m^2 and twist in the shaft is not to exceed 1 degree in 2 metres of its length. Assume uniform turning moment and take modulus of rigidity for the shaft material C =100 GN/m^2.

Solution:

$$P = \frac{2\pi NT}{60}\,;\ 80 \times 10^3 = \frac{2\pi \times 200 \times T}{60}$$

$$\therefore \qquad T = \frac{60 \times 70 \times 10^3}{2\pi \times 200} = 3822 \text{ Nm}$$

Considering stress as *the basis* :

$$\tau_s = 60 \text{ MN/ m}^2 = 60 \times 10^6 \text{ N/mm}^2$$

$$\frac{T}{J} = \frac{\tau}{R};\ J = \frac{\pi}{32}d^4 \text{ and } R = \frac{d}{2}$$

Then $$T = \frac{\pi}{16}d^3\tau;\ 3822 = \frac{\pi}{16}d^3 \times 60 \times 10^6$$

$$d^3 = \frac{3822 \times 16}{\pi \times 60 \times 10^6} = 0.0003246;\ d = 0.0689 \text{ m} = 68.9 \text{ mm}$$

Considering angle of twist as the basis :

$$\theta = 1° = 0.01744 \text{ rad}$$

$$\frac{T}{J} = \frac{C\theta}{l};\ \frac{3822}{\frac{\pi}{32}d^4} = \frac{(100 \times 10^9) \times 0.01744}{2}$$

$$d^4 = \frac{3822 \times 32 \times 2}{\pi \times (100 \times 10^9) \times 0.01744} = 0.00004466$$

$$d = 0.0817\text{m} = 81.7 \text{ mm}$$

The required shaft diameter will be the larger of the two values determined.

∴ Suitable shaft diameter d = 81.7 mm

Example 7.24: A shaft to transmit 50 kW of power will be running at 100 rev/min. The maximum twist in 3m length of the shaft is limited to 1.5 degree. If maximum torque is 30 per cent more than mean or average torque, calculate: (*a*) diameter of the shaft, and (*b*) maximum shear stress developed in the shaft.

Assume entirely elastic action and take modulus of rigidity for the shaft material C =135 GN/m^2.

Solution:

$$\theta = 1.5 \text{ degree} = \frac{1.5 \times \pi}{180} = 0.02617 \text{ rad}$$

$$P = \frac{2\pi NT}{60};\ 50 \times 10^3 = \frac{2\pi \times 100 \times T}{60};\ T = 4777 \text{ Nm}$$

∴ Maximum torque transmitted $T_{max} = 1.3 \times 4777 = 6210$ Nm

(*a*) $$\frac{T_{max}}{J} = \frac{C\theta}{l};\ \frac{6210}{\frac{\pi}{32}d^4} = \frac{(135 \times 10^9) \times 0.02617}{3}$$

$$d^4 = \frac{6210 \times 32 \times 3}{\pi \times (135 \times 10^9) \times 0.02617} = 0.00005374$$

$$d = 0.0856 \text{ m} = 85.6 \text{ mm}$$

(*b*) $$\frac{T_{max}}{J} = \frac{\tau}{R};\ J = \frac{\pi}{32}d^4 \text{ and } R = \frac{d}{2}$$

Then $$\tau = \frac{16T_{max}}{\pi d^3} = \frac{16 \times 6210}{\pi \times (0.0856)^3} = 50.45 \times 10^6 \text{ N/m}^2$$

Check: $$\tau = \frac{C\theta}{l} \times R = \frac{(135 \times 10^9) \times 0.02617}{3} \times \left(\frac{0.0856}{2}\right)$$

$$= 50.4 \times 10^6 \text{ N/m}^2$$

Example 7.25: During tensile test on a 25 mm diameter test piece, an extension of 0.1 mm was observed over a gauge length of 200 mm when a load of 50 kN was applied. Subsequently, the same rod was subjected to 120 Nm torque in tension test and a twist of 0.025 radian was noted over 200 mm length. Calculate the Poisson's ratio and elastic moduli of the material of shaft.

Solution: Tensile test : stress

$$\sigma = \frac{50 \times 10^3}{\frac{\pi}{4}(25)^2} = 101.91 \text{ N/mm}^2$$

$$\text{Strain } \varepsilon = \frac{0.1}{200} = 0.0005$$

∴ Modulus of elasticity

$$E = \frac{\sigma}{\varepsilon} = \frac{101.91}{0.0005} = 2.038 \times 10^5 \text{ N/ mm}^2 = 203.8 \text{ GPa}$$

Torsion Test : From torsion equation, $C = \dfrac{Tl}{J\theta}$

Modulus of rigidity

$$C = \frac{(120 \times 10^3) \times 200}{\frac{\pi}{32}(25)^4 \times 0.025} = 0.78643 \times 10^5 \text{ N/m}^2 = 78.64 \text{ GPa}$$

From the correlation

$$E = 2\,C\,(1 + \mu),$$

$$\text{Poisson's ratio } \mu = \frac{E}{2C} - 1 = \frac{203.8}{2 \times 78.64} - 1 = 0.2958$$

From the correlation

$$E = 3\,K\,(1 - 2\mu),$$

$$\text{Bulk modulus } K = \frac{E}{3(1-2\mu)} = \frac{203.8}{3(1-2\times 0.2958)} = 166.34 \text{ GPa}$$

Example 7.26: A hollow circular shaft with an external diameter d and internal diameter $d/2$ is replaced by a solid shaft of diameter d.

(*a*) Find the ratio of torques transmitted by hollow and solid shafts if equal stresses are induced in the two shafts.

(*b*) Find the ratio of maximum shear stress induced in the hollow and solid shafts if the two shafts transmit equal torques.

Solution:

Let suffix h and s represent the hollow and solid shafts.

From torsion equation, $T = \dfrac{\tau J}{R}$

∴ For the hollow shaft

$$T_h = \tau \times \frac{\pi[d^4 - (d/2)^4]/32}{d/2} = \frac{15}{256}\pi\tau d^3 \qquad \ldots(i)$$

For the solid shaft

$$T_s = \tau \times \frac{\pi d^4/32}{d/2} = \frac{1}{16}\pi\tau d^3 \qquad \ldots(ii)$$

From expressions (*i*) and (*ii*),

$$\frac{T_h}{T_s} = \frac{15}{256} \times \frac{16}{1} = \frac{15}{16}$$

(*b*) From torsion equation,

$$\tau = \frac{TR}{J}$$

∴ For the hollow shaft

$$\tau_h = \frac{T\,d/2}{\pi[d^4 - (d/2)^2]/32} = \frac{256T}{15\pi d^3} \quad ...(iii)$$

For the solid sheaf

$$\tau_s = \frac{T\,d/2}{\pi d^4/32} = \frac{16T}{\pi d^3} \quad ...(iv)$$

From expressions (*iii*) and (*iv*)

$$\frac{\tau_h}{\tau_s} = \frac{256}{15} \times \frac{1}{16} = \frac{16}{15}$$

Example 7.27: A solid shaft of 36 mm diameter has been replaced by a hollow shaft of same length and material and of external diameter 40 mm. Both the shafts are acted upon by the same torque and have same maximum shear stress induced in them. Compare the weights and angular twists for the two shafts.

Solution:

we know torsion equation, $\frac{T}{J} = \frac{\tau}{R}$

For the solid shaft : $\frac{T}{\pi d^4/32} = \frac{\tau}{d/2}$

For the hollow shaft : $\frac{T}{\pi(d_o^4 - d_i^4)/32} = \frac{\tau}{d_o/2}$

Since both the shafts are subjected to same torque and have same maximum shear stress induced in them, the above expressions give

$$d_i = d_o\left[1 - \left(\frac{d}{d_o}\right)^3\right]^{1/4} = 40\left[1 - \left(\frac{36}{40}\right)^3\right]^{1/4} \quad 28.86 \text{ mm}$$

Both the shafts are of the same material and length, and accordingly the ratio of the weights of the shafts would be same as the ratio of their cross-sectional areas

$$\frac{W_{\text{hollow}}}{W_{\text{solid}}} = \frac{d_o^2 - d_i^2}{d^2} = \frac{40^2 - 28.86^2}{36^2} = 0.592$$

From torsion equation, we have $\theta = \frac{T}{J}\frac{l}{C}$. For the same torque applied, material and length of shaft $\theta \propto \frac{1}{J}$. Accordingly

$$\frac{\theta_{\text{hollow}}}{\theta_{\text{solid}}} = \frac{d^4}{d_o^4 - d_i^4} = \frac{36^4}{40^4 - 28.86^4} = 0.9$$

Example 7.28: A compound shaft 1.5 m long fixed at one end is subjected to a torque of 15 kNm at the free end and of 20 kNm at the junction point as shown in Fig. 7.73 (*a*).

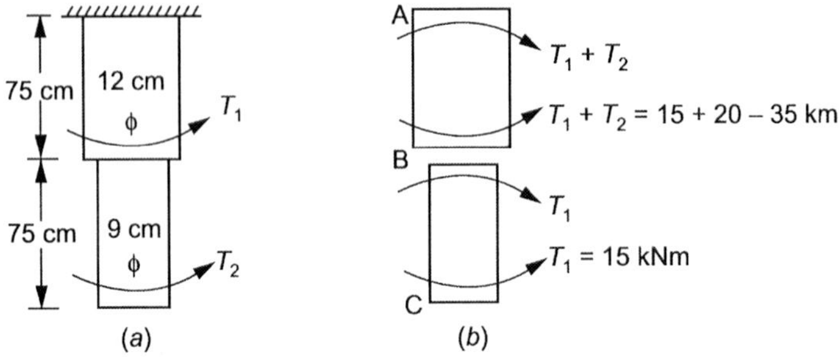

Fig. 7.73 A compound shaft

Determine

(*a*) Maximum shearing in each portion of the shaft,

(*b*) Angle of twist at the junction of the two sections and at the free end.

Take modulus of rigidity of shaft material $C = 0.82 \times 10^5$ N/mm^2.

Solution:

Torque in lower portion BC = 15 kNm = 15×10^6 Nmm

Torque in upper portion AB = 35 kNm = 35×10^6 Nmm

From torsion equation,

$$\frac{T}{J} = \frac{C\theta}{l}$$

θ_1 = angle of twist of C (free end) relative to B (junction)

$$= \frac{Tl}{CJ} = \frac{15 \times 10^6 \times 750}{0.82 \times 10^5 \times \pi \times 90^4 / 32} = 0.0213 \text{ radian}$$

θ_2 = angle of twist of B (junction) relative to A (fixed end)

$$= \frac{35 \times 10^6 \times 750}{0.82 \times 10^5 \times \pi \times 120^4 / 32} = 0.0157 \text{ radian}$$

θ_1 and θ_2 are in the same direction. Hence net angular twist of free end with respect to fixed end A is

$$\theta = \theta_1 + \theta_2 = 0.0213 + 0.0157$$
$$= 0.037 \text{ radian} = 2.12 \text{ degree}$$

(*b*) From torsion equation,

$$\frac{T}{J} = \frac{\tau}{R}, \text{ we have}$$

$$\tau = \frac{TR}{J} = \frac{T \times d/2}{\pi d^4/32} = \frac{16T}{\pi d^3}$$

$$\text{Shear stress in the lower segment} = \frac{16 \times 15 \times 10^6}{\pi \times (90)^3} = 104.85 \text{ N/mm}^2$$

$$\text{Shear stress in the upper segment} = \frac{16 \times 35 \times 10^6}{\pi \times (120)^3} = 103.21 \text{ N/mm}^2$$

Example 7.29: A solid steel cube of 1 m side is immersed in water to a depth of 600 m. Make calculations for the decrease in the length of its sides and in its volume. For steel, bulk modulus is 135 GPa and for water, mass density is 1000 kg/m^3.

Solution:

Hydrostatic pressure at 600 m depth is

$$\sigma_v = w\,h = \rho\,g\,h$$
$$= 1000 \times 9.81 \times 600 = 5.886 \times 10^6 \text{ N/ m}^2$$

Volumetric strain

$$\sigma_v = \frac{5.886 \times 10^6}{135 \times 10^9} = 43.55 \times 10^{-6}$$

Change in volume

$$dV = \text{volumetric strain} \times \text{original volume}$$
$$= (43.55 \times 10^{-6}) \times 1 = 43.55 \times 10^{-6} \text{ m}^3$$
$$= 43555 \text{ mm}^3 \text{ (decrease)}$$

The volumetric strain ε_v is given approximately by the sum of the linear normal strains in x, y and z direction. That is

$$\varepsilon_v = \varepsilon_x + \varepsilon_y + \varepsilon_Z$$

Here $\varepsilon_x + \varepsilon_y + \varepsilon_Z$

$$\therefore \quad \varepsilon_x = \frac{\varepsilon_v}{3} = \frac{43.55 \times 10^{-6}}{3} = 14.518 \times 10^{-6}$$

$\therefore$ Change in length of the side of cube,

$$\varepsilon_v = \text{linear strain} \times \text{original length}$$
$$= (14.519 \times 10^{-6}) \times 1 = 14.518 \times 10^{-6}\text{m}$$
$$= 0.014518 \text{ mm (decrease)}$$

Example 7.30: A vertical bar fixed at the upper end, and of uniform strength carries an axial load of 12 kN. The bar is 2.5 m long and having a specific weight of 0.0001 N/mm^3. If the area of the bar at the upper end is 500 mm^2, find the area at the lower end.

Solution:

For a bar of uniform strength, the stress remains same everywhere and it equals

$$\sigma = \frac{W}{A_1} = \frac{12 \times 10^3}{500} = 24 \text{ N/mm}^2$$

Further, the areas A_1 (at the upper end) and A_2 (at the lower end) are related by the expression

$$\log_e \frac{A_1}{A_2} = \frac{wl}{\sigma} = \frac{0.0001 \times (2.5 \times 1000)}{24} = 0.01042$$

or

$$\frac{A_1}{A_2} = 1.01047$$

$$A_2 = \frac{A_1}{1.01047} = 494.87 \text{ mm}^2.$$

HIGHLIGHTS OF THE CHAPTER

1. **Stress:** The resistance per unit area, offered by a body against deformation is known as stress.

$$\sigma = \frac{P}{A}$$

where P = External force or load

A = Cross-sectional area

2. Types of stress
   1. Direct stress
   2. Indirect stress
   3. Combined stress
3. **Strain:** The ratio of change of dimension of the body to the original dimension is known as strain. It is a dimensionless quantity.
4. **Hook's Law:** When a material loaded within classic limit, the stress is propotional to strain.
5. **Young's Modulus:** It is the ratio of normal stress to normal strain within elastic limit.
6. **Modulus of Rigidity:** It is the ratio of shear stress to shear strain is known as modulus of rigidity.
7. **Bulk Modulus:** It is the ratio of normal stress to the volumetric strain.
8. **Poisson's Ratio:** It is the ratio of lateral strain to linear strain.
9. Relation between elastic constants

$$\mathbf{E} = 2G(1 + v)$$

$$E = 3k(1 - 2v)$$

$$E = \frac{9kG}{3k + G}$$

10. Factor of safety = $\dfrac{\text{Ultimate stress}}{\text{Working stress}}$

11. Beam: A structural member which carries load at right angle to the Gx is of the beam.

12. Types of Beams

(*i*) Cantillever beam (*ii*) Simply supported beam
(*iii*) Overhanging beam (*iv*) Fixed beam

13. Types of Load:

(*i*) Point load (*ii*) Distributed load
(*iii*) Uniformly distributed load
(*iv*) Uniformly varying load

14. The axis at which the stress is zero called neutral axis.

15. Bending equation

$$\frac{M}{I} = \frac{\sigma}{y} = \frac{E}{R}$$

where

M = Bending moment in Nmn
I = Moment of inertia - mm^4
σ = Bending stress in N/mm^2
y = Distance of fibre form neutral axis
E = Young's modulus of elasticity in N/mm^2

16. Torque is a product of the force applied and radius of the shaft.

17. Torsion equation

$$\frac{T}{J} = \frac{G\theta}{l} = \frac{z}{R}$$

Where

T = Torque
R = Radius of shaft
l = Length of the shaft
z = Shear stress
G = Modulus of rigidity
θ = Angle of twist

18. Power transmitted by shaft $P = \dfrac{2\pi NT}{60}$ watts.

REVIEW QUESTIONS

1. What do you mean by stress?

2. Explain the types of stress.

3. Write a short note on:

(*i*) Normal strain

(*ii*) Shear strain

(*iii*) Volumetric strain

(*iv*) Longitudinal.

4. State Hook's law.

5. Define Young's Modulus of elasticity.

6. What do you understand by bulk modulus of elasticity?

7. What is Poisson's ratio?

8. Establish the relation between Young's modulus and bulk modulus.

9. Write a short note on principle of superposition.

10. Give a brief account of various types of loads.

11. What are composite shafts?

■■■

8

Manufacturing System

8.1 MANUFACTURING SYSTEM

Manufacturing system may be defined as a collection of integrated equipment and human resources, whose function is to perform one or more processing and assembly operations on a raw material part or set of parts.

The integrated equipment includes production machines and tools, material handling and work positioning devices and computer systems. Human resources are required to keep the system in running. Generally manufacturing system is used in the larger production unit as shown in Fig. 8.1.

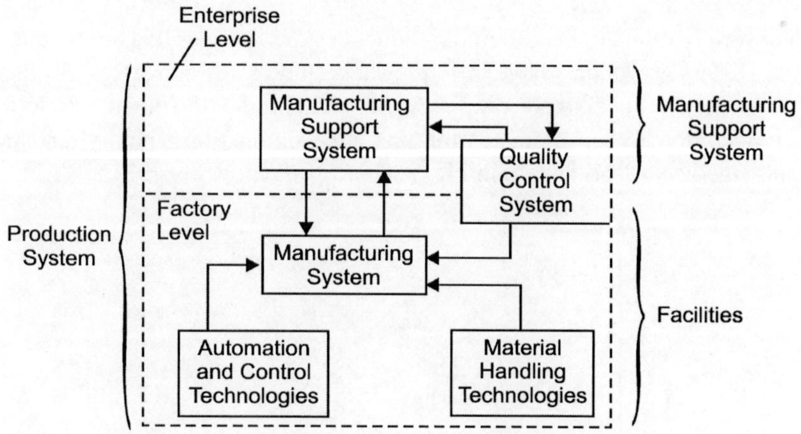

Fig. 8.1 The position of the manufacturing system in the large production unit

Production system means to manufacture products and assembled manufactured these products. Production system involves the collection of people, equipment, and procedures for the manufacturing operation of a company or organisation.

Production system can be divided into two catogories as shown in Fig. 8.2.

1. **Facilities:** The facilities of the production system consist of the factory, the equipment in the factory, and the way the equipment orgainsed.
2. **Manufacturing Support System**: It is the set of procedures used by the company to manage production and to solve the technical and logistics problems faced in ordering materials, moving work through the factory and assuring that product most quality standards.

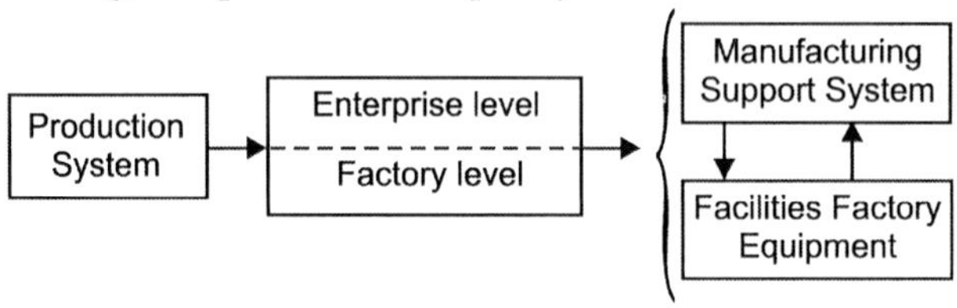

Fig. 8.2 Production system consists of facilities and manufacturing system

8.2 CLASSIFICATION OF MANUFACTURING SYSTEMS

Classification of manufacturing systems depends upon the variety of manufacturing system. Which are as follows:

1. Part or product variety
2. Types of operations performed
3. Number of workstation and system lay out
4. Types of automation

1. Part or Product Variety

Product variety refers to the different product designs or types that are produced in a plant. Different products have different shapes and sizes and styles. So there is an inverse correlation between product variety and production quantity in terms of factory operation. When product variety is high, production quantity tends to be low and *vice-versa*. It is shown in Fig. 8.3.

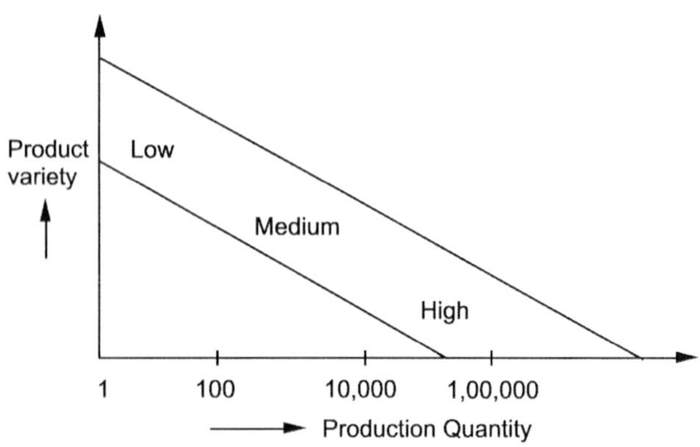

Fig. 8.3 Relationship between product variety and production quantity

Based on the quantity of product, manufacturing activity can be classified into three categories:

(*i*) Job shop production

(*ii*) Batch production

(*iii*) Mass production

(*i*) **Job Shop Production:** Low volume (quantity range of 1 to 100 units/year) of production and variety of work characterize job shop production. Job shop production is commonly used to meet specific customer orders requiring one or very few numbers of the product having a very specific design and specifications. Examples of the products made from job shop production includes a special tool, a space vehicle, and easy chair for comfort, etc.

(*ii*) **Batch Production:** Batch production involves the manufacture of medium quantity (100 –10,000 units annually) of an item or product. In which a batch of one product is made, after the facility is changed over to produce a batch of the next product, and so on. The orders for the product may be repetitive or non-repetitive. Examples of the products made in batch production includes textbooks, furniture and soon.

(*iii*) **Mass Production:** Mass production is characterized by the manufacturing of identical products in bulk quantities (10,000 to millions of units per year). In mass production environment, the entire organization is dedicated to manufacture a particular type of product. Examples of mass production are pens, refills, screws, cars and scooters. Since the machines and equipments are involved in the manufacturing of single type of product, it must be for special purpose rather than for general purpose. Hence the investment required for machine tools and equipments is high.

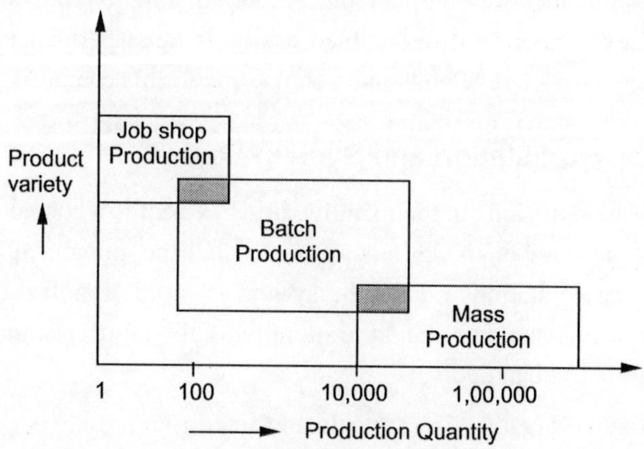

Fig. 8.4 Types of layouts used for different levels of production quantity and product variety.

2. Types of Operations Performed

Manufacturing systems are divided into basic operations *i.e.*

(*a*) Processing operations

(*b*) Assembly operations

(***a***) **Processing Operations:** As processing operation is a transformation of a work material from raw material to finished product. It adds value by changing the geometry, properties or appearance of the raw material.

A general model of a processing operation is shown in Fig. 8.5.

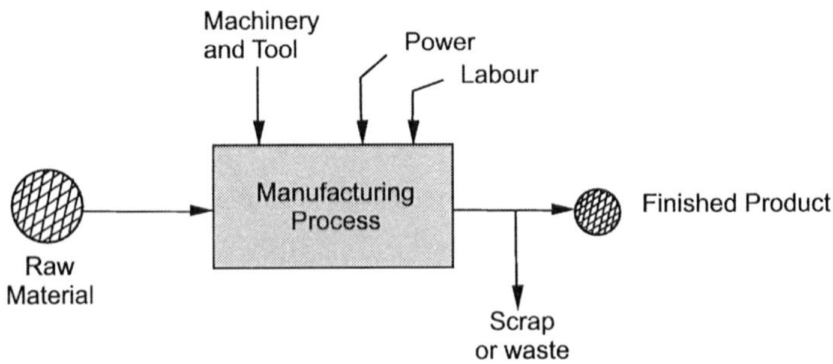

Fig.8.5 Processing operations

Material is fed into the process, energy is applied by the machinery and tooling to transform the material and completed workpart exits the process. Some scrap or waste are produced as a product of the process.

(***b***) **Assembly Operations:** In these operations, two or more separate parts are joined to form a new entity. Components of the new entity are connected together either permanently or semi-permanently. Permanently joining process includes welding, brazing, soldering, adhesive bonding and rivets etc. They cannot be disassembled easily. In semi-permanently, threaded fasteners (*e.g.*, screw, bolts, nuts etc.) come in this category.

3. Number of Workstation and System Layout

The number of workstations in the manufacturing system is a convenient measure of its size. As the number of stations are increased, the amount of work can be increased. But more stations means the system is more complex and therefore more difficult to manage and maintain, more workers, more machines and more parts which are being handled.

Workstation may be organised in variable and fixed routing.

Variable routing means the workpieces are transported through a variety of different station sequence. It includes job of shop production and batch productions. While

Fixed routing means work-pieces always flow through the same sequence of stations. It includes line production as shown in Fig. 8.6.

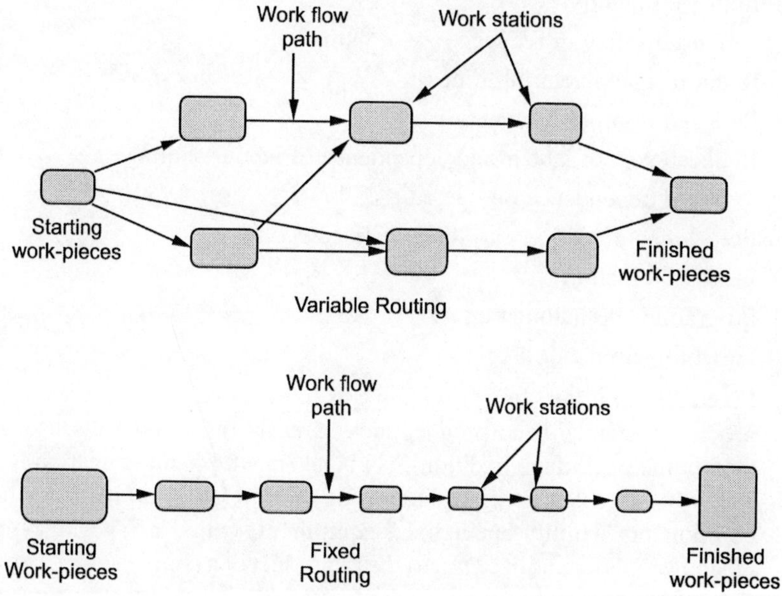

Fig. 8.6 Variable and Fixed Routing

According to the number of stations and the layout of the stations, it has three levels:

(*i*) **Single Station:** It consists only one work station usually including a production machine that may be manually operated, semi-automated or fully-automated,

(*ii*) **Multi-stations with Variable Routing:** It consists of two or more stations including one for processing or assembling of different parts etc.

(*iii*) **Multi-stations with Fixed Routing:** It consists of two or more workstations as a production line.

4. Types of Automation

Automation is the technology by which a process or procedure is accomplished without human assistance.

Do not confuse between the terms mechanization and automation. Mechanization means something is done or operated by machinery and not by hand *e.g.* the automatic feed on a lathe machine is not automation. Where automation means a system in which many or all of the processes in the production, movement and inspection of parts and material are performed under control of the self-operating devices called controllers.

There are some reasons for automation in the manufacturing systems like:

1. Increased productivity.
2. Improved quality.
3. Increased safety or reduced risk of humans.
4. Reduced manufacturing time.
5. Reduced in process inventory.
6. Reduced cost of labour and dependence of labour shortages.
7. Reduced dependence on operator skills

Automation can be classified into three main categories *i.e.*

(*i*) Fixed automation
(*ii*) Programmable automation
(*iii*) Flexible automation

(*i*) **Fixed Automation:** It is also known as hard automation, which is used to produce a standardized product such as gears, nuts and bolts, etc. Even though the operating conditions can be changed, fixed automation is used for very large quantity production of one or few marginally different components. Highly specialized equipment, called as special purpose machine tools, are utilized to produce a product or a component of a product very efficiently and at high production rates such that the unit costs are low relative to the alternative methods of production.

(*ii*) **Programmable Automation:** In programmable automation, we can change the design of the product or even change the product by changing the program. It is useful for the low quantity production of large number of different components. The equipment used for the production is designed to be adaptable or programmable. The production is normally carried out in batches.

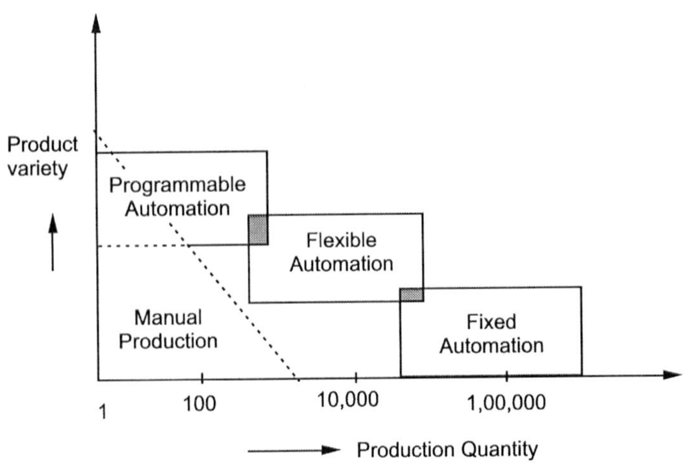

Fig. 8.7 Three types of automation relative to production quantity and product variety

(*iii*) **Flexible Automation:** Third category between the fixed automation and programmable automation is called flexible automation. This is also called flexible manufacturing system (FMS). This allows manufacturing different products on the same equipment in any order or mix.

8.3 COMPONENTS OF A MANUFACTURING SYSTEM

There are four components generally used in a manufacturing system :

1. Human workers
2. Production machine
3. Material handling system
4. Computer control system

1. Human Workers

Human workers are required, to operate the system even if all the manufacturing systems in the factory are automated, some works performed by human workers are as:

(*i*) **Equipment Maintenance:** Skilled workers will be required to maintain and repair the automated system in the factory when these systems breakdown.

(*ii*) **Programming and Computer Operation:** Part programming for NC and CNC machine is carried by human workers and to install new version of software packages and execute the programs.

(*iii*) **Engineering Project Work:** For upgradation of production machines, tool designing and continues improvement in projects, skilled engineers are required in the factory.

2. Production Machine

Most of the actual processing or assembly work is accomplished by machines in the modern manufacturing system.

The machines can be classified as:

(*i*) Manual operated machine

(*ii*) Semi-automated machine

(*iii*) Fully automated machine

(*i*) **Manual Operated Machine:** These machines are operated by human workers. The machine provides the power for the operation and the worker provides the control. Conventional machine tools (*e.g.* lathes, milling machines, drill etc.) come in this category. The worker must attend the machine continuously.

(*ii*) **Semi-automated Machine:** These machines work under some form of programme control and a human worker looks the machine for loading and unloading the product or part *e.g.* CNC lathe work by the part programme but a worker is required for unloading the finished part and loading the next workpiece at the end of the part programme. So, worker has to attend the machine but not continuously.

(*c*) **Fully Automated Machine:** It is the extension of semi-automated machine. In this machine, any work is performed under programme control and no human worker is required for workpiece. The worker may look the machine after every tenth workpiece or every hundredth workpiece *e.g.* injection moulding plants, where the moulding machine run on automatic cycle, but periodically the collection-bin, full of moulded part beside the machine, must be taken away and replaced by an empty bin.

3. Material Handling System

Material handling is the movement, storage, protection and control of materials throughout the manufacturing and distribution process.

Material handling system performs many functions for processing and assembly operations. These are:

(*a*) Loading and Unloading

(*b*) Positioning

(*c*) Transporting

(*d*) A Temporary Storage

(*a*) **Loading and Unloading:** These material handling functions occur at each workstation. Loading means moving the workpiece into the production machine from a source inside the plant. Unloading means when the production operation is completed, that workpiece is removed from the production machine and placed in a container at the workstation or transported to the next work station in the processing sequence.

(*b*) **Positioning:** Workpiece must be positioned in the production machine for accuracy and precision work. Positioning in the production machine is accomplished by a workholder *i.e.* a device that accurately locates, orients and clamps the part for the operation and resists any forces that may occur during processing *e.g.* Jigs, Fixtures and Chucks etc.

(*c*) **Transporting:** In manufacturing system workpiece moves between the workstations. The transport function can be accomplished by either manually or material handling equipments. Some material handling equipments are:

(*i*) *Industrial trucks:* Industrial trucks are divided into two types *i.e.* non-powered and powered.

Non-powered trucks are platforms or containers with wheels that are pushed or pulled by human workers to move materials.

Powered industrial trucks are steered by human workers. They are mechanized for movement.

(*ii*) *Automated guided vehicles* (*AGVs*): These are battery powered, operated, automatically steered vehicles that follow defined pathways on the floor. The path should be clean without any obstacle on the path. AGVs are used to move unit loads between the loaded and unloaded stations in the plant. Routing variations are the possible, means that different loads move between different stations.

(*iii*) *Monorails:* These are self-propelled vehicles that ride on a fixed rail system that is either on the floor or suspended from the ceiling. These vehicles are operated independently and are usually driven by electric motors.

(*iv*) *Conveyors:* Conveyors are used to move materials over fixed path. Conveyors can be either powered or non-powered.

In powered conveyor mechanical drive system is used for movement and in non-powered conveyor movement is used to move the material by gravity. Conveyor maybe of different types, *i.e*. Roller conveyor, Belt conveyor etc.

(*v*) *Cranes and hoists:* These are the devices used for lifting, lowering and transporting materials, often used for very heavy loads. Hoists accomplish *vertical* lifting both manual operated and powered types are available. Cranes provide horizontal travel and generally include one or more hoists.

(*d*) A Temporary Storage: The purpose of a temporary storage in this system is to make sure that work should be going on. Workstation should not fail or stop due to workpiece. Workpiece prior to and after manufacturing can be stored as:

(*i*) *Bulk storage:* It is simple storing material in an open floor area, fitted on pallets.

(*ii*) *Rack systems:* Rack systems are structural frame for storing material vertically.

(*iii*) *Shelving and bin:* Steel shelf is to store material. It comes in standard width, depth and heights. It can include bins, which are containers for loose items.

(*iv*) *Drawer storage:* Drawer storage system is more convenient than shelves. In shelves to find out the material is difficult if its level is too high or too deep. Drawer storage system is used with pulling out or in and it is used for tools, hardware and other small items.

4. Computer Control System

A computer is required to control the automated and semi-automated equipment for overall coordination in a manufacturing system.

Following are the functions of a computer control in a manufacturing system:

(*a*) *Material handling control:* It controls the material handling system and link with the work-stations for efficient activities.

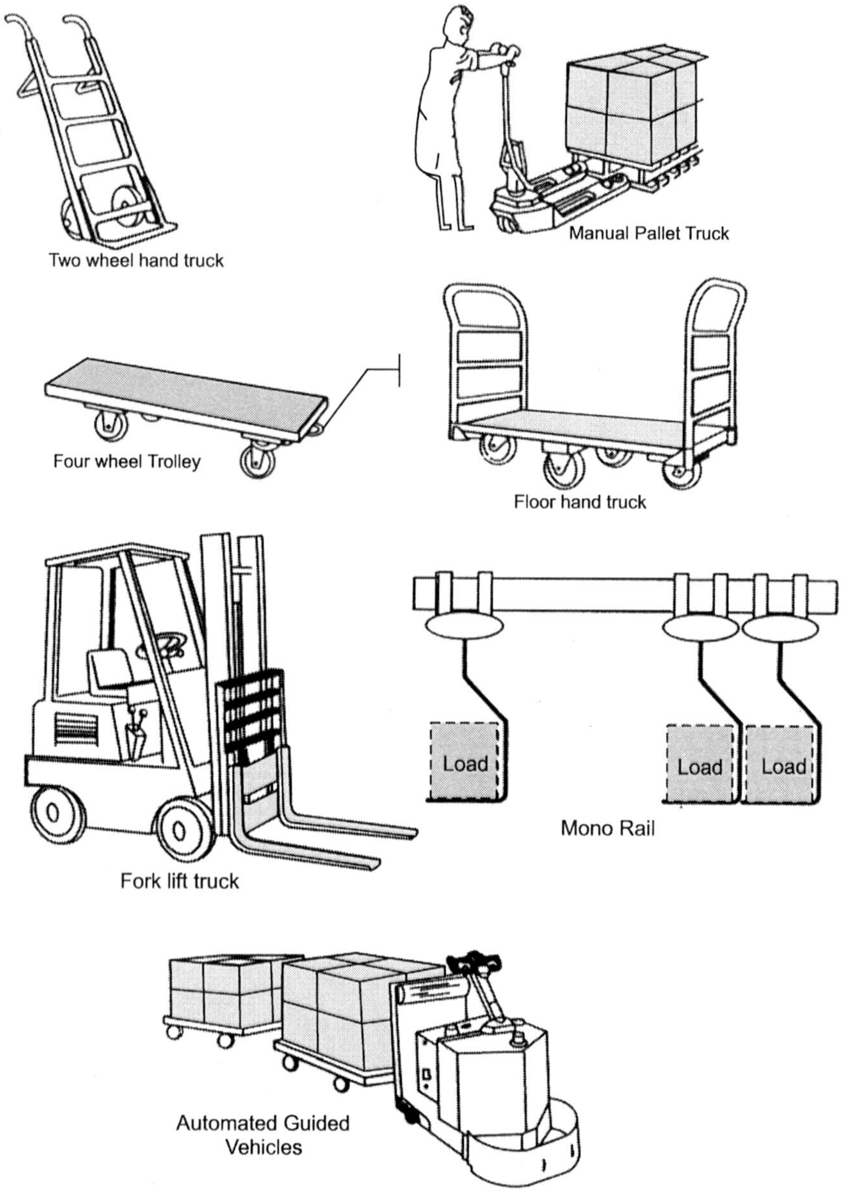

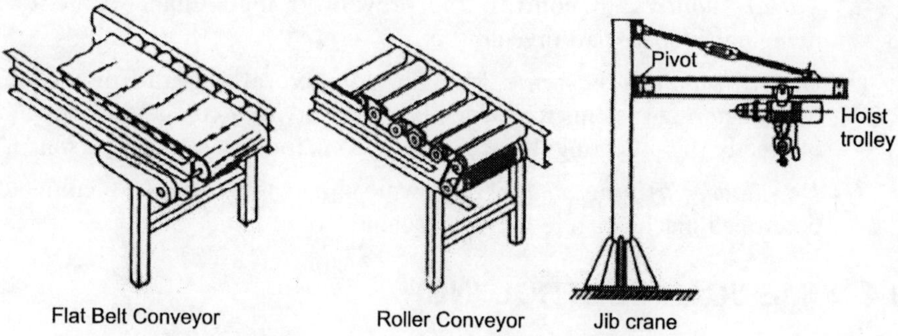

Fig. 8.8 Transporting equipments

(*b*) *Quality control:* It detects the defective work units produced in the system and rejects it.

(*c*) *Safety control:* It ensures that system should work in safe conditions to protect both the human workers and equipments of the system.

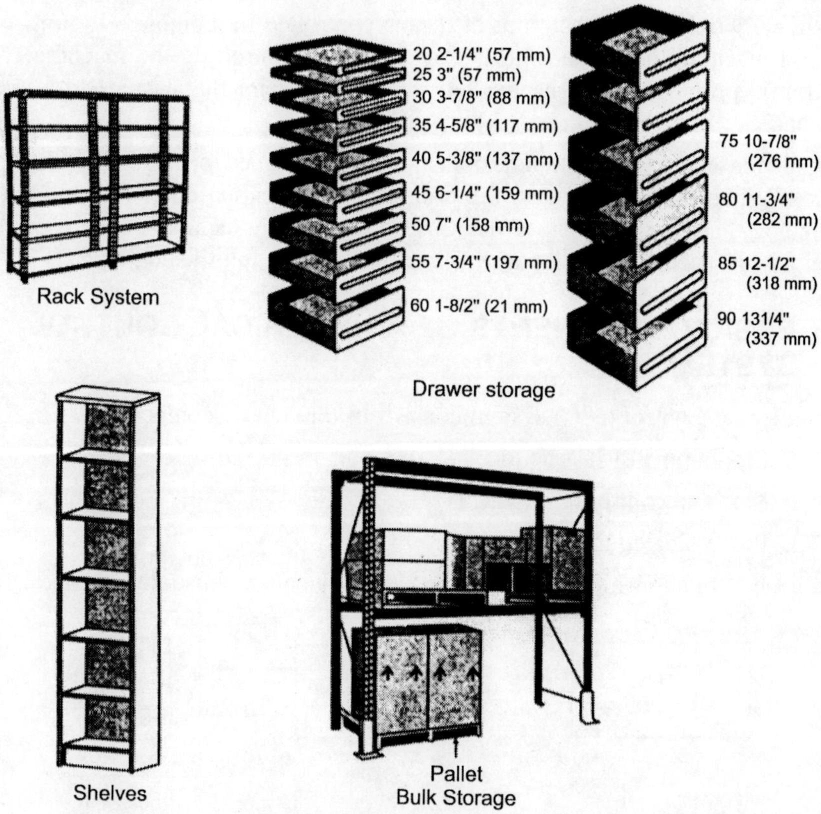

Fig. 8.9 Temporary storage

(*d*) *Failure control:* It controls the preventive maintenance schedules maintaining spare part inventory etc.

(*e*) *Operation management:* Manage the overall operations of the manufacturing system either directly (by supervisory computer control) or indirectly (by preparing the necessary reports for management personnel).

(*f*) *Download part programme:* To download part programmes to computer controlled machines (*e.g.* CNCs machine tools).

8.4 NUMERICAL CONTROL (NC)

Numerical control is defined as a form of programmable automation in which the process is controlled by numbers, letters and symbols. A machine tool is said to be numerically controlled if it operates in a semi-automatic or automatic cycle as per instructions transmitted to it in a coded form. The electronics Industries Association defines numerical control as "A system in which actions are controlled by the direct insertion of numerical data at some time. The system must automatically interpret at least some portion of these data."

In simple words, numerical control means controlled by numbers. In numerical control machine tools, the input information for controlling the machine. Tool motion is provided through punched paper tape or magnetic tapes in a coded language.

In numerical control, a programme of instruction is designed for a particular workpart or job. When the job changes, the programme of instructions is changed. This technology has been applied to a wide variety of operations including turning, milling, assembly, inspection, sheet metal working, welding and profile cutting etc.

8.5 BASIC COMPONENTS OF A NUMERICAL CONTROL SYSTEM

A Numerical Control (N.C) system consists of three basic components:

1. Programme of instruction
2. Machine control unit (M.C.U)
3. Processing equipment or Machine tool.

The general relationship among the three components is illustrated in Fig. 8.10.

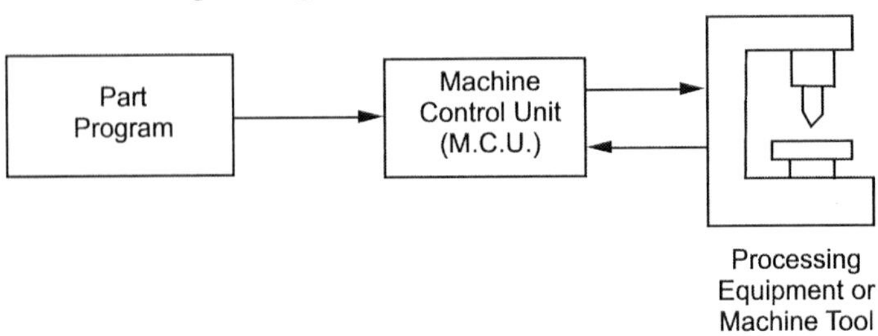

Fig. 8.10 Basic components of an NC system

Part program is supplied to MCU and MCU (Machine Control Unit) controls the operation unit of machine tool.

1. Programme of Instruction

The programme of instruction is the detailed step-by-step command that direct the machine tool what to do and in what sequence. In machine tool application, the programme of instruction is called a part programme, and the person who prepares the program is called Part programmer. The part programme is written in coded form and contains all the information needed for machining the components. The part programme is fed to the machine control unit through some input medium. Various types of input medium are:

(*a*) ***Punched Cards*:** Punched cards are used for feeding a small amount of information to NC system. Usually in such cases several dozen cards are required for the operation of the machine tool during a shift. These cards are available with 45 or 80 columns and with 12 rows. Hence, information can be recorded in any of 960 position in an 80-column card. Hence, large number of cards are required for long programme. These cards generally bend during decoding in card readers (Card reading machines). Due to this disadvantage and improper handling of these cards, they are almost obsolete now a days.

(*b*) ***Magnetic Tape***: Some NC machines use 1-inch (25.4 mm) wide magnetic tapes to store data. This tape is similar to the tape which is used in audio system but is of higher quality. But it has not much applications because of the interference from near by electrical equipment such as transformer etc. It has a tendency to erase or scramble some information on the tape.

(*c*) ***Punched Tape*:** Punched tape is widely used for feeding the programme to numerical control systems. There are various types of paper tapes used in NC system, but the standard format for tape size and configuration, issued by Electronic Industries Association of USA (EIA) and International Standard Organisation (ISO), are universally accepted.

A standard tape is 1 inch (25.4 mm) wide. The punched tape has capacity of storing 10 characters per 25 mm length. A punched tape is shown in Fig. 8.11.

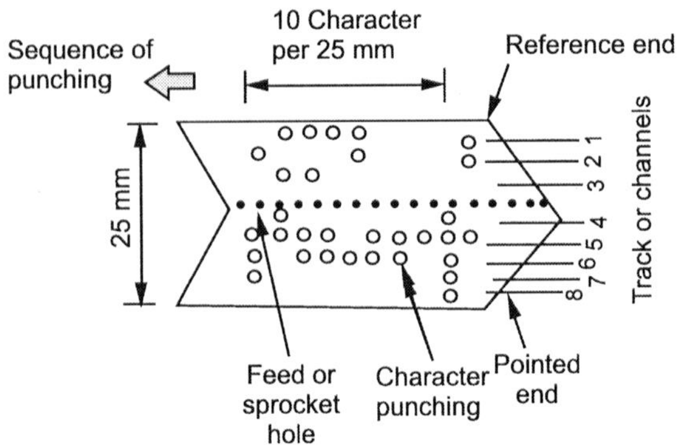

Fig. 8.11 25 mm wide punched tape

There are 8 tracks on the tape, which are used for punching the information in coded form. The edge adjacent to track 1 is called Reference edge. A row of small holes between track 3 and 4 is used for feeding the tape into the tape reader. The information required to machine the components is punched on the tape by a tape punching device.

(*d*) ***Floppy Disk*****:** It has channels or tracks to record and store magnetic bits of information. Long part programme can be stored conveniently on floppy disks. It is a compact and handy device for storing the part programme of NC machine. Information of floppy disk is decoded by a reading head on the MCU of NC system.

2. Machine Control Unit (MCU)

The second basic component of numerical control system is machine control unit. It is also called the -Controller unit. It receives the part programme through various input media and converts this information into a language that the machine tool can understand and operate according to convert a part into finished one.

Main elements of NC controller units are as follows:

(*a*) Tape Reader

(*b*) Buffer storage

(*c*) Feeding channels and drives

(*d*) Gauging device

(*e*) Feedback unit

(*f*) Sequence controls

(*g*) Control panel.

(*a*) ***Tape Reader*:** Tape reader is used to read the data registered on the punched tape. The punched tape readers commonly used are:

(*i*) Mechanical (Electro-Mechanical)

(*ii*) Optical (Photo-Electrical)

(*iii*) Pneumatic

(*i*) *Mechanical tape reader:* The principle of a simple mechanical device for reading the punched tape is shown in Fig. 8.12. If there is no hole in the tape the contacts remain open but when a hole is present in the tape, the contacts close.

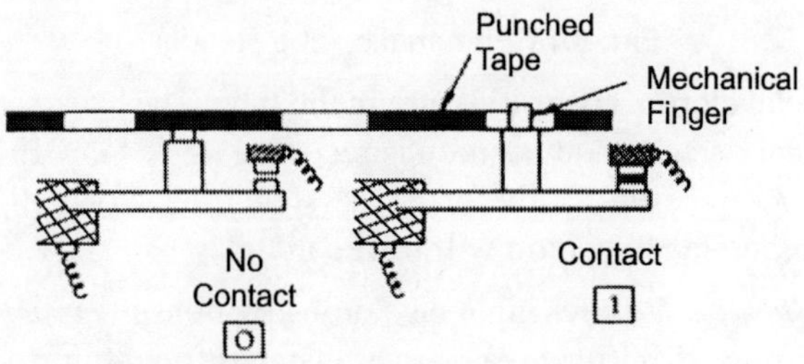

Fig. 8.12 Mechanical tape reader

(*ii*) *Optical tape reader:* It is based on the principle that if a beam of light falls on a photoelectric cell, it generates an electric signal. It is shown in Fig. 8.13.

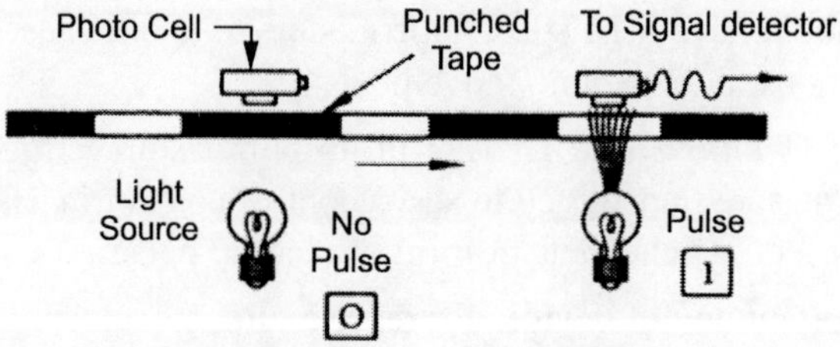

Fig. 8.13 Optical tape reader

Whenever a hole is present in the tape, light from the light source passes through the hole and energies the corresponding photocell, which converts the light energy into electrical energy to produce a pulse *i.e.* ON position. When there is no hole, the light from the light source does not reach the photocell, hence no signal is produced and the position is recorded as OFF.

(*iii*) *Pneumatic tape reader:* A pneumatic tape reader is shown in Fig. 8.14. Air is allowed to pass through the positions on the punched tape where a hole is present, and where there is no hole, the tape does not allow the air to pass.

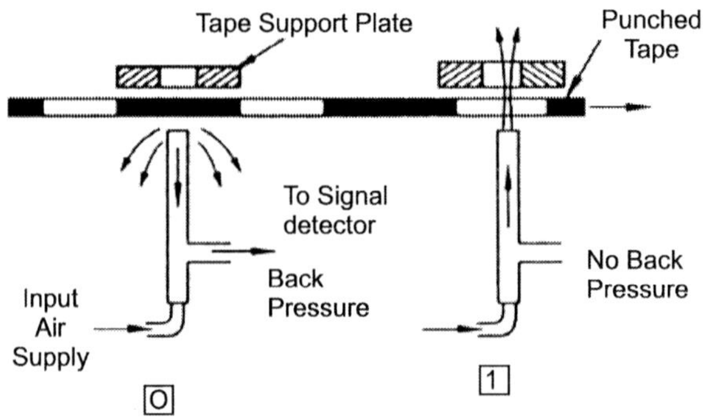

Fig. 8.14 Pneumatic Tape Reader

That air developes a back pressure in the tube. The back pressure is sensed by a signal detector and position is recorded as '0' *i.e.* OFF. If air does not create back pressure in the tube, the signal detector will not detect any back pressure and position will be recorded as 1 *i.e.* ON.

(*b*) ***Buffer Storage***: It stores input data or instructions in *proper* logical blocks. The purpose of buffer area is to transfer information to other areas of controller unit at a faster rate. If there were no buffer storage then the controller would wait until the tape reader sends next set of informations. This would cause little pause in transfer of information which would result in pause in different motion of machine tool and cause tool marks on workpiece. Hence, machine control units which do not have buffer storage areas should have high speed tape readers to avoid pauses in transferring informations and machine movements.

(*c*) ***Feeding Channels and Drives***: From buffer storage, part programming instructions are transmitted to servo control units of machine tool through different feeding channels in form of electric pulses.

An important unit of devices for reproducing movements specified in part programme is drive. In contouring NC system either a step drive or servo drive is used. The step motor can either reproduce the movements specified in the part programme or it can become a servo motor which controls a power drive.

(*d*) ***Gauging Devices:*** It measures the machine movements to determine if the servo control units have given correct command.

(*e*) ***Feedback Unit***: Feedback units fed information back from gauging device for comparison so that machine moves to correct location. These devices send back information to MCU so that machine table slide position can be compared with required dimension of input data. If there is any difference the drive motor is directed through electrical pulses to make necessary corrections.

Feedback systems are of two types:

(*i*) Analog

(*ii*) Digital

(*i*) *Analog:* Potentiometer is an example of analog transducer. They produce an electrical voltage which varies, as input shaft is rotated. This voltage is in proportion to rotation of input shaft.

(*ii*) *Digital:* Digital feedback units are attached to lead screw of machine tool. They change or convert rotary motion of machine lead screw to individual or discrete electrical pulses. This series of pulses can be counted by a counter to indicate how lead screw has turned which further indicates the amount machine table has moved.

(*f*) ***Sequence Controls*:** It is synchronizing unit whose function is to synchronize activities of other elements of MCU. Tape reader reads and decodes input information from punched tape and sends it to buffer storage from where signals are sent to machine tool by feeding channels and then movement of operative units is feedback to MCU by feedback devices and so on. All these operations must be synchronized and this is achieved by sequence controls.

(*g*) ***Control Panel*:** For efficient control of machine tool, operator should be able to control whole NC system and this is done through control panel. Which contains dials and switches with which operator runs machine. It has two parts:

1. Operator's Control panel
2. Machine Control panal

3. Procesing Equipment or Machine Tool

The third part of the numerical control system is the machine tool itself, it is the output generating component of NC system. Input is given to NC system in form of part programme by different input devices. MCU transmits this programme of instruction to machine tool to generate different relative motions between workpiece and cutting tool so that stock is removed from raw material to generate different shapes on it.

8.6 NUMERICAL CONTROL PROCEDURE

To utilise numerical control in manufacturing, the following steps must be accomplished:

1. Process Planning
2. Part Programming
3. Tape Preparation
4. Tape Verification
5. Production.

1. **Process Planning:** This involves the interpretation of engineering drawings of the work part in terms of the manufacturing processes to be used. This is accomplished by preparing a route sheet. The route sheet is a listing of the sequence of operations which must be performed on the work part along with the required machining data like speeds, feeds, depth of cut, tools used etc.

2. **Part Programming:** Part programmer plans the process for the portions of the job to be accomplished by numerical control. Part programmer is responsible for planning the sequece of machining steps to be performed by NC and to document these in a special format. There are two ways to program numerical control machines: (*a*) Computer assisted part programming (*b*) Manual part programming

   (*a*) *Computer assisted part programming:* In this programming the tedious computational work and cost required in manual part programming is transferred to the computer.

   (*b*) *Manual part programming:* In this the machining instructions are prepared on a form called part programme manuscript. The manuscript is a listing of the relative cutter/workpiece positions which must be followed to machine the part.

3. **Tape Preparation:** A punched tape is prepared from the part programmer's NC process plan. In manual part programming, the punched tape is prepared directly from the part programme manuscript on a typewriter like device equipped with tape punching capability. In computer assisted part programming, the computer interprets the list of part programming instructions, performs the necessary calculations to convert this into a detailed set of machine tool motion commands, and then controls a tape punching device to prepare the tape for the specific NC machine.

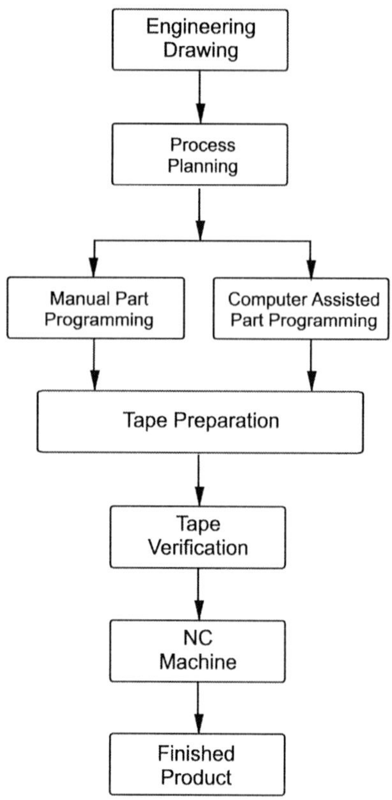

Fig. 8.15 The procedure of production through numerical control

4. **Tape Verification:** The prepared tape is checked by running it through a computer programme which plots the various tool movements on paper. In this way, major errors in the tape can be found out.
5. **Production:** The final step in the numerical control procedure is to use the part programme in production. This involves ordering the raw workparts, specifying and preparing the tooling and any special fixturing that may be required and setting up the NC machine tool for the job.

8.7 CLASSIFICATION OF NC MACHINE

NC machine tools can be classified in different ways as:

1. **Based on the Control System:** Based on the control system features, NC machines are classified as:

   (*a*) Point-to-point system

   (*b*) Straight line system or straight cut system

   (*c*) Contouring system or Continuous path system

   (*a*) Point-to-point System: In this system, the machine tool performs machining operations at selected positions and does not affect the workpiece while moving from one point to the next. This system is the simplest point-to-point system and also called a positioning system. According to Fig. 8.16, after drilling the hole at position A, the tool moves to position B, and then move to hole position C. A drilling machine is the best example of point-to-point system.

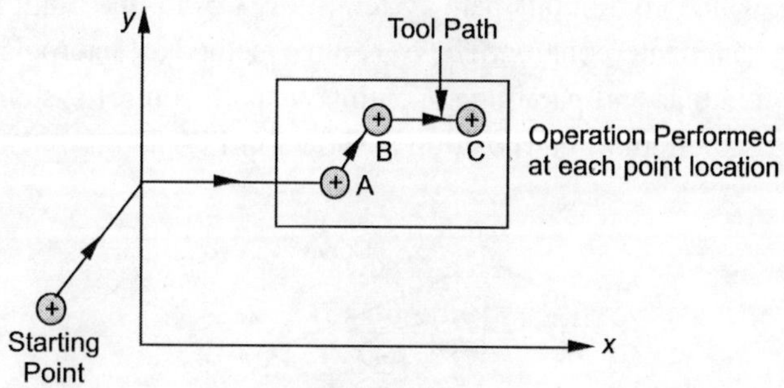

Fig. 8.16 Point-to-point (Positioning) NC System

   (*b*) Straight Line System or Straight Cut System: It is also known as straight line system. It is an extension of point to point system. In this system tool moves at a controlled feed rate in one axis direction at a time. It is there for at appropriate for performing milling operations to fabricate work pieces of rectangular configurations as shown in Fig. 8.17.

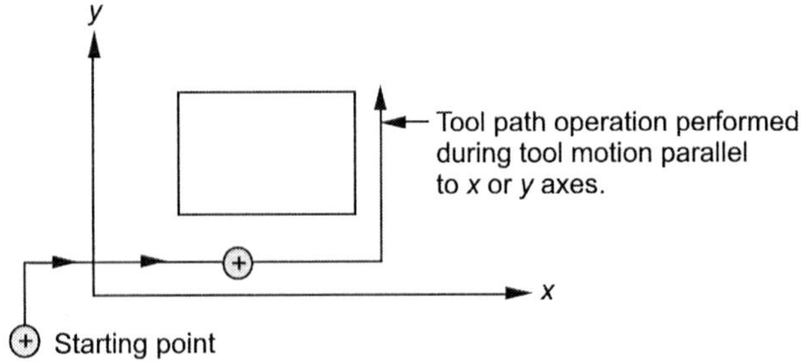

Fig. 8.17 Straight cut system

(*c*) **Continuous Path System:** Continuous or contouring path system is the most complex, flexible and expensisve type of machine tool control. It is capable of performing both point to point and straight cut operations. It is a system in which motion is controlled on move than one axis movement of machine tool continuously and simultaneously. The path of cutter is continuously controlled to generate the desired geometry of the workpiece. This enables machining of profiles, contours and curved surfaces.

The system requires control of several slides simultaneously, so that their relative positions and velocities are established at requisite points throughout the operations.

Milling and turning operations are the common example of continuous path system.

A continuous path system is shown in Fig. 8.18. The method by which continuous path system moves from one point to another is called interpolation. There are three methods of interpolation—linear, circular and parabolic. Continuous path control system with linear and circular interpolation is shown in Fig. 8.18.

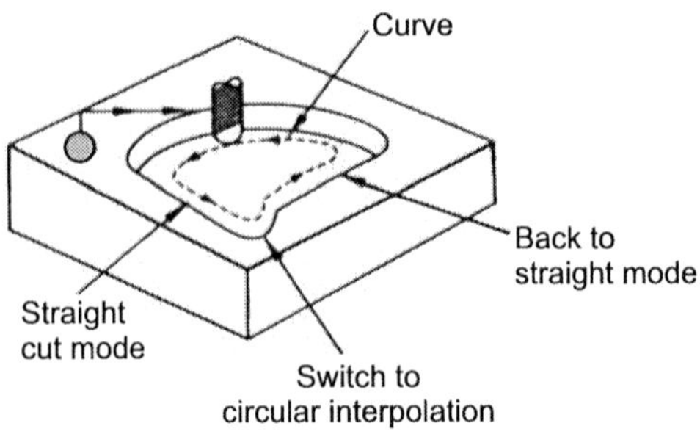

Fig. 8.18 Contouring (Continuous path) NC system.

2. **Based on Control Loops:** Based on control loops, NC machines are classified as:

   (*a*) Open loop system

   (*b*) Close loop system

(*a*) *Open loop system*: In an open *loop system,* there is no feedback or return signal to provide the information to the controller about the effect of signal that it produces. The system has no method of checking that the servomotors have accurately moved the machine slides to the programmed distance. The accuracy of the system is solely a function of motor's ability to pass through the exact number of steps which is provided at its input.

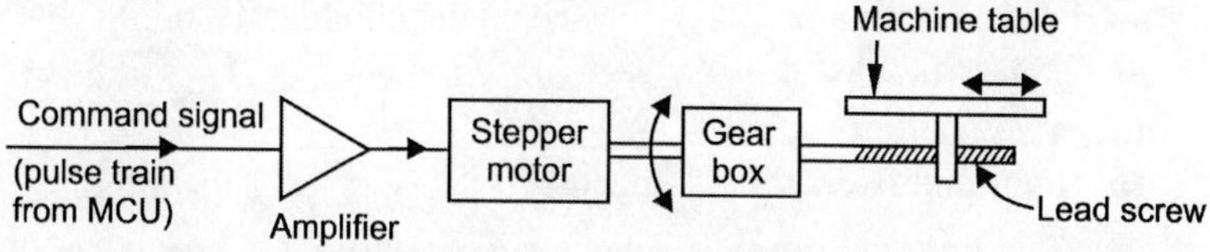

Fig. 8.19 Open loop control system

(*b*) *Close loop system*: The *closed loop system* has a feedback built into the system which automatically monitores the position of the tool. The output, *i.e.,* the distance the machine tool moves is measured and compared to the program input. The distance between the actual and the programmed value is called the error. The control system is designed in such a way that the error is eliminated. Zero error implies that the position of tool is automatically controlled until it is in the right position. Cutting tool edge relative to datum point will be measured. Measuring the slide movements with measuring device provides the positional feedback. The position measuring devices are either rotary or linear measuring transducer.

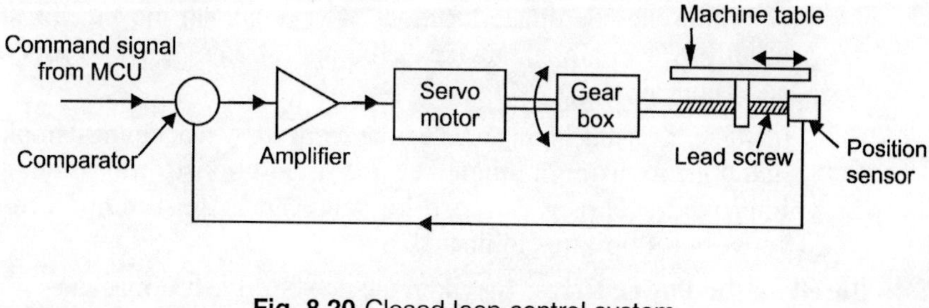

Fig. 8.20 Closed loop control system

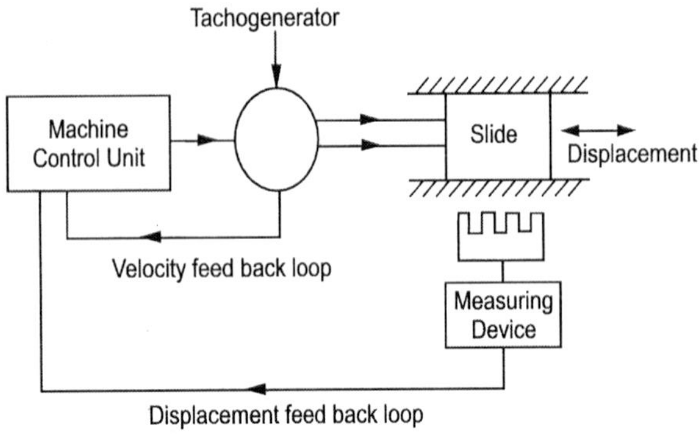

Fig. 8.21 Block diagram of close loop control system

3. **Based on the Number of Axis:** Based on the number of axis, NC machines may be classified as:
   (*a*) Two and three axes NC machine
   (*b*) Four and five axes NC machine

   (*a*) *Two and three axes NC machine:* Two axes NC machine normally consists of the *x* and y axes. The *x* dimension is written first, followed by they dimension, *i.e.* (*x*, *y*). In a two axes NC machine, the machine table moves left and right to position the cutter.

   A longitudinal movement of the machine table on this machine is an *x* axis move. A cross or saddle movement is a *y* axis move. In numerical control programming, positive sign need not to be written but minus sign must be written.

   In three axes machines, there will be one more axis, perpendicular to the above two axes. By the simultaneous control of all the 3 axes, complex surfaces can be machined.

   (*b*) *Four and five axes NC machine:* Some of the NC machines are equipped with four and five axes contouring capabilities. These capabilities are generally secondary and tertiary axes which parallels the *x*, *y* and *z* axes plus some additional rotational movements *A*, *B*, *C*, within the coordinate locations. The rotational movements can usually sweep through all axes in complex cutting tool movements and applications.

   In higher axis machining, cycle time reduced by machining complex components using a single set up. In addition to time savings, improved accuracy can also be achieved as positioning errors between set-ups are eliminated.

4. **Based on the Power Drive:** Based on the power drive NC machines may be classified as:
   (*a*) Electro-mechanical system,
   (*b*) Hydraulic system,
   (*c*) Pneumatic system

(*a*) *Electro-mechanical system:* Electro-mechanical drives are the most precise and the most commonly used power source. In this system, electric motor may be used for controlling both positioning and contouring machines. The speed of a either AC or DC motor and the torque and direction of rotation need to be controlled. The speed of a DC motor can be controlled by varying either the field or the armature supply. The clutch controlled motor can either be an AC or DC motor. They are generally used for small machine tools because of heat losses in the clutches. Split field motors are the simplest form of motors and can be controlled in a manner according to the machine tool. These are small and generally run at high maximum speeds and so require reduction gears of high ratio. Separately excited motors are used with control systems for driving the slides of large machine tools.

(*b*) *Hydraulic system:* These hydraulic systems may be used with positioning and contouring machine tools of all sizes. These systems may be either in the form of rams or motor. In these motors oil under pressure is supplied by a pump to the motor's blade. Due to the hydraulic pressure, the continuously blades rotate at high speed. Hydraulic motors are smaller than electric motors of equivalent power. There are several types of hydraulic motors. The advantage of using hydraulic motors is that they can be very small and have considerable torque. This means that they can be incorporated in servo-system which require having a rapid response.

(*c*) *Pneumatic system:* Pneumatic system have pneumatic motor which are similar in working and design to hydraulic motors.

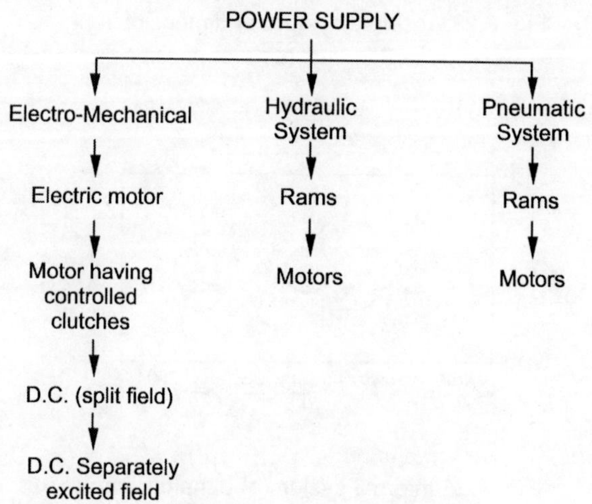

Fig. 8.22 Power supply

8.8 TOOL POSITIONING SYSTEMS

In NC programming the following two types of positioning systems are used:

1. **Absolute System:** In this system the positions are indicated from a fixed zero point or reference point. The reference point or zero point is fixed before starting the operation on the machine.

   A major advantage of using absolute system is that it is very easy to check and correct a program written using this method. If a mistake is made in the value of any dimension in a particular block, it will affect that dimension only and once the error is corrected there will be no further problems.

   An example of this type of dimensioning is shown in Fig. 8.24 where all tool positions *i.e.* the centers of the hole circles, are marked with reference to fixed zero point.

   In this case, the zero or reference point in the left hand bottom corner of the workpiece and all the distances are given from this point.

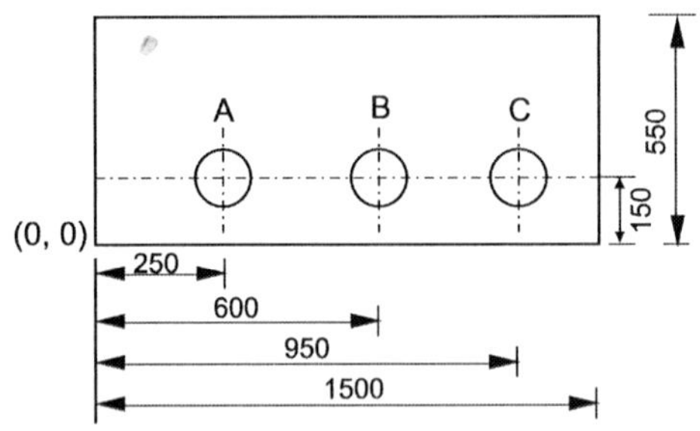

Fig. 8.23 Absolute system of dimensioning a component to indicate tool positions

2. **Incremental System**

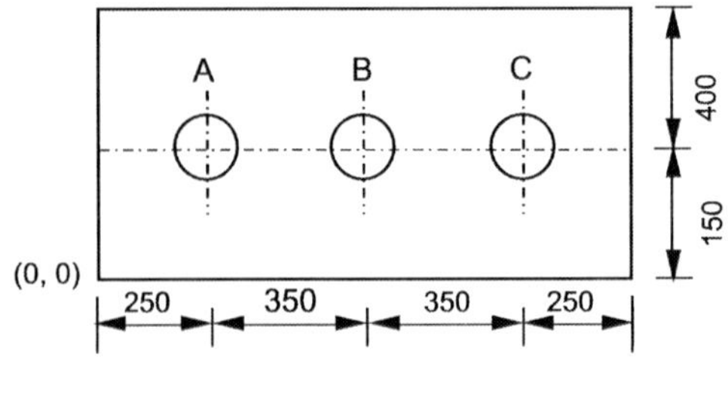

Fig. 8.24 Internal system of dimensioning a component to indicate tool positions

Table 8.1: Showing the coordinate values

Point	Absolute system		Incremental system	
	x	*y*	*x*	*y*
A	250	150	250	150
B	600	150	350	0
C	950	150	350	0

It is difficult to check a part program written in incremental dimension mode. An example of this type of dimensioning, on the same component as discussed in Fig. 8.23 is shown in Fig. 8.24 wherein all the dimensions are given *i.e.* the locations of all the tool positions are marked, with reference to a previously known location. Table *I* shows the coordinate values of both absolute and incremental mode.

8.9 AXIS OF MOTION

Axis of motion means an axis in which the cutting tool can be moved with respect to a workpiece. This movement is achieved by the motion of the machine tool slides. The main three axes of motion will be referred to as *x, y,* and *z*-axes. The *x* and *y*-axis of motion are horizontal and at right angles to each other. The *x* axis is from left to right, and the *y*-axis is from front to back. The *z*-axis of motion is perpendicular to both *x* and *y* in order to create a right hand coordinate system, as shown in Fig. 8.25.

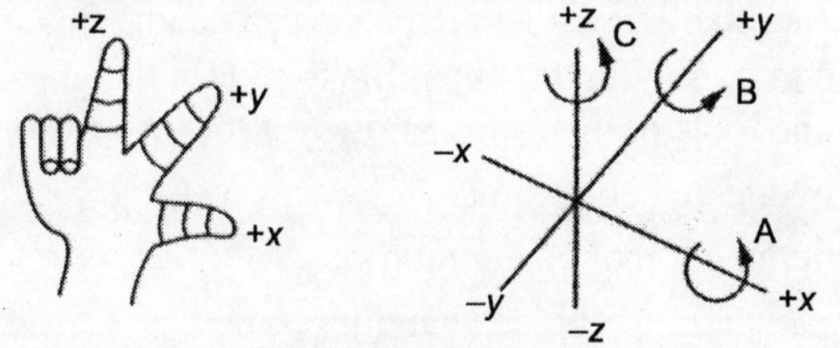

Fig. 8.25 Linear and Rotary motion

A positive command at the *z*-axis moves the cutting tool away from the workpiece. Rotary motions around the axis parallel to *x, y* and *z* are designated *A, B* and *C* respectively.

8.10 APPLICATION OF NC

The most common application of NC is for metal cutting machine tools such as milling, drilling, boring, turning and grinding etc. In addition to metal machining it can be used for other operations also *e.g.* press working, welding machines, inspection machines, assembly machines, tube bending, flame cutting, automatic riveting and so on.

The following situations favour the use of NC machine tools:

1. Complex part where human error is likely in conventional machining processes.
2. Low to medium batch quantity production of repetitive precision quality parts.
3. Components that require 100% inspection because of accuracy and repeatability.
4. Situations/Components that require reduction in lead time. (It is the time between the design stage to the finished stage of the manufacturing).
5. Complex machining operations.
6. Part subject to frequent design modifications.
7. Close tolerance must be held on the work part.
8. Investment on tooling and fixtures inventory is high.

8.11 ADVANTAGES OF NC SYSTEM

Following are the advantages of Numerical Controls system:

1. **Reduced Manufacturing Lead Time:** Jobs can be set up more quickly with NC and only fewer set-ups are generally required with NC, so the lead time to deliver a job to the customer is reduced.
2. **Reduced Non-productive Time:** Parts can be produced in less time and therefore, are likely to be less expensive. The idle time (non-cutting time) is reduced to minimum. It is possible to reduce the non-productive time in NC machine tools in the following ways:
   (*a*) by reducing set-up time
   (*b*) by reducing tool changing time
   (*c*) by reducing workpiece handling time
   (*d*) by reducing the number of set-ups
3. **Reduced Fixturing:** NC requires fixtures which are simpler and less costly to fabricate because the positioning is done by the NC tape rather than the jig or fixture. So, expensive jigs and fixtures are reduced or eliminated depending upon the part geometry.
4. **Reduced Parts Inventory:** Reduced set-up time and shorter lead time with numerical control, the amount of inventory carried by company is reduced.
5. **Reduced Scrap:** In NC machining operations, certain errors like incorrect hole locations are almost eliminated and like oversizing of holes may be reduced drastically, thus scrap is reduced.
6. **Reduced Floor Space Requirements:** Since one NC machining centre can perform a variety of machining operations that have to be carried out

on several conventional machine tools, thus reducing the number of machine tools on the shop floor. So, floor space required in an NC shop is usually less than in a conventional shop.

7. **Greater Accuracy:** Part can be produced more accurately because of automation and the absence of interrelated human factors.
8. **Greater Manufacturing Flexibility:** The modification to part design can be very easily translated into manufacturing by the simple changes in part programmes without expensive and time consuming changes in jigs, fixtures and tooling. This will add to the flexibility at manufacturing.
9. **Lower Inspection:** Inspection time is reduced, since all the parts in a batch are identical, proper care is taken about tool compensations and tool wear in part programme preparation and operation. Normally, it is necessary to inspect the first component only.
10. **Improved Quality Control:** NC is suitable for complicated workparts where the chances of human mistakes are high. Numerical control produces parts with greater accuracy, reduced scrap, and lower inspection. It improves quality control.
11. **Accurate Timing and Costing:** Machining times and costs are more accurately predictable since all the elements involved in the manufacturing have to be throughly analysed before a part programme is prepared.
12. **Longer Tool Life:** Tools can be utilized at optimum feeds and speeds that can be programmed.
13. **Elimination of Special Tools:** The certain types of form tools is completely eliminated in NC machines. This is because the profile generated can be programmed, even if it involves three dimensions.
14. **Elimination of Operator Error:** The machine is controlled by program of instructions stored in the machine control unit. The program is checked before it goes to the machine so no error will occur in the job. So fatigue, boredom or inattention by an operator will not affect the quality or duration of the machining.

8.12 DISADVANTAGES OF NC MACHINE

NC machine have a number of advantages but there are certain limitations which are given below:

1. **Higher Investment Cost:** The cost of an NC machine tool is much high compared to an equivalent conventional machine tool. The cost is often 5 to 10 times higher. The cost of tooling is also high. This requires a very high initial investment. All this makes the machine hourly rate high. As a result, it is necessary to utilize the machine tool for a large percentage of time.

2. **Skill Operators:** In NC system, skill required to operate an NC machine is generally high. This requires part programmers, tool setters, punch operators and maintenance staff who are more educated and trained than conventional machine operators.
3. **Training NC Personnel:** Special training is required for the personnel managing NC machine tools. NC manufacturing requires training of personnel both for software as well as hardware. Part Programmers are trained to write instruction in desired languages for the machine on the shop floor.
4. **Higher Maintenance Cost:** As NC is a complex and sophisticated technology, it also requires higher investments for maintenance in terms of wages of highly skilled personnel and expensive spares. The need for maintenance engineers trained in all the subsystems present such as mechanical, hydraulic, pneumatic and electronics makes the job more difficult.

8.13 COMPUTER NUMERICAL CONTROL (CNC)

CNC may be defined as the NC system whose MCU is a dedicated, stored programme computer and used to perform some or all of the basic numerical control functions in accordance with control program stored in the read-write memory of the computer.

CNC controls are also referred to as soft wired NC systems because most of their control functions are implemented by the control software programmes. In CNC, the program is entered once and then stored in the computer memory. In CNC, the computer allows for the following: storage of additional programes, program editing, running of program from memory, machine and control diagnostics, special routines, inch/metric, incremental/absolute switchability.

Term Associated with CNC

- **Computer-aided Design (CAD):** Use of computers and special software at the various stages in the design of a product or component.
- **Computer-aided Manufacturing (CAM):** Use of computers to control different phases of machining and manufacturing.
- **Computer Integrated Manufacturing (CIM):** A computer system that controls all phases of production from management, design, manufacturing, quality control, sales and marketing etc.
- **Computer Numerical Control (CNC):** A microcomputer/microprocessor with a programmed numerical control that activates the machine servos and spindle drives, and controls the machining operations.

- **Direct Numerical Control (DNC):** A manufacturing system where a number of machines are connected to a single computer which is designed to provide instructions to each machine on demand.
- **Adaptive Control System (ACS):** An extension of CNC system in which the calculations and setting of the operating like speed, feel and depth of cut are done during machining by the control system itself.

The Salient Features of a CNC

1. The computer used with CNC system is generally dedicated one *i.e.,* it is assigned specifically to a single machine tool. The dedicated computer is usually a minicomputer constituting an integral part of the CNC controller.
2. The computer receives the coded instructions (input data) in numerical form, processes this information and produces output data that causes the machine tool to function.
3. The input data may be fed directly to the computer or through punched or perforated tape, magnetic tape, floppy disk or specially designed sensors. Once the part programme is read, it is stored in memory for repetitive machining.

   The programme is a set of instructions that outline each step a machine must take to produce a component.
4. The minicomputer of the CNC controller essentially consists of two parts namely the programmable (soft wired) computer module and the hardwired logical circuitary. The controller serves to perform the following functions:
   - readily programmable tool and fixture offsets
   - storage of part programmes
   - selection of absolute/incremental modes and of positioning/ contouring control
   - conversion between system of units for dimensions
   - cutter compensation; compensation for any change in the dimensions of cutting tool

A conventional NC system is characterized by the following aspects:

(*i*) The instructions (program) are recorded and stored in coded language on a paper tape.

(*ii*) The tape reader reads and feeds the programme into the machine control unit (MCU) where the data processing unit (DPU) processes the input data and passes on the information regarding the various controls to the control loop unit (CLU).

(*iii*) There is fixed circuit interconnections of DPU and CLU that determines the entire data input and data handling sequence including control functions. These functions pertain to tool and fixture offsets, positioning or contouring.

(*iv*) The circuitry is hard wired and cannot be altered easily; the circuitry of MCU has to be rewired if some modification is to be done.

(*v*) There is no provision of storage and for every component, the tape needs to be loaded again.CNC machines belong to the following three categories:

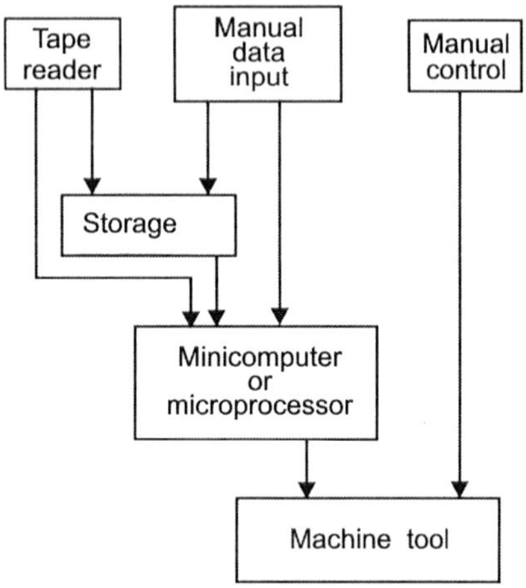

Fig. 8.26 CNC system machine tool

- **Mainframe Computer:** A large rigid machine containing a large memory capacity; the machine performs general-purpose data processing and can do more than one job at the same time
- **Minicomputer:** Small size and compact; is dedicated one, *i.e.*, assigned specifically to a single machine tool
- **Microcomputer:** A one chip microprocessor that is designed for simple applications. However, it has the capability to perform at least the arithmetic-logic and control-logic functions of the central processing unit.

Applications of CNC Machines

The technology of CNC machines is being successfully employed in

- drilling, boring, milling and grinding machines
- welding and flame cutting machines
- pipe bending and coil winding machines.

The acceptability in the diverse areas is due to the advantages of accuracy, reliability, repeatability and productivity which it offers to the manufacturing industry.

In modern CNC systems, end to end component design is highly automated using CAD/ CAM programs. The programs produce a computer file that is interpreted to extract the commands needed to operate a particular machine, and then load into the CNC motions for production.

8.14 ADVANTAGES AND DISADVANTAGES OF CNC

Advantages

1. **One Time Data Entry:** Once the program of instructions entered manually or by a punch tape into the computer, it stored and never need to enter again.
2. **Program Editing on Machine:** The program of instructions can be edited for any error on the computer without the editing in the tape.
3. **Unit Conversion:** CNC can easily changes the units of inches to international system of units or *vice-versa.*
4. **More flexibility:** The flexibility of changing the job is more as compare to NC system as only program of instruction is to be changed which is already store in computer.
5. **Demo before Actual Machining:** CNC offers flexibility to check the program of instructions by virtual machining on computer monitor rather than actual machining. If there is any error, it can be corrected on the computer.
6. **Technology Costs are Decreasing:** CNC machines are now within the financial reach of small manufacturing shops.

Disadvantages

1. **High Initial Investment:** CNC system is more costly than NC system
2. **High Running Cost:** CNC requires special tools, coolant, lubricant and atmosphere which increase the running cost.
3. **Not Suitable for Job Production:** Because of high running cost CNC only for mass production.
4. **Skilled Operator Required:** To operate the CNC machine and to with the program of instructions skilled operator is required which again increase the speed of running due to higher salary.
5. **Higher Maintenance Cost:** As CNC machine is more complex than NC the maintenance cost is also higher.
6. **Increase Unemployment:** Single CNC machine can complete the work at many NC machines and create unemployment.

8.15 COMPARISON BETWEEN NC AND CNC

	NC	CNC
1.	Every time program of instruction is to be entered.	Program of instruction entered only one time.
2.	Can not edit the data after entering in to the machine.	Data can be edit even after entered mini computer.
3.	After preparation the punch tape, units can not be changed.	Units can be changed in the computer.
4.	No computer is required.	Mini or micro computer is required.
5.	No display for virtual machining.	Monitor is used for virtual machining display.
6.	Less costly as compare to CNC.	More costly than NC.
7.	Less running cost as compare to CNC.	Running cost is higher than NC.
8.	Non-optimum feed and speed.	Optimum feed and speed.
9.	Uses less reliable punch tape again and again.	Punch tape is used only one time.
10.	No storage capacity for program of instructions.	Program of instructions can be stored in the memory of computer.

8.16 DIRECT NUMERICAL CONTROL (DNC)

Direct Numerical Control (DNC) is a manufacturing system in which a number of machines are interconnected and controlled by a computer through direct connection. In case of NC or CNC systems, one computer is used to control one machine tool. But, in case of DNC system, one computer can be used to control more than 100 machines.

In DNC, computer is designed to provide instructions to each machine tool on demand. Fig. 8.27 illustrates the general DNC configuration.

The DNC system consists of four general components:

1. Central computer
2. Bulk memory, which stores the NC program
3. Telecommunication lines
4. Machine tools.

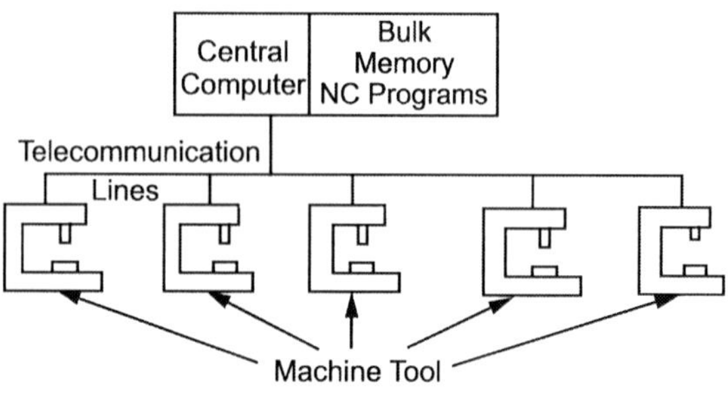

Fig. 8.27 Direct Numerical Control system

The computer calls the program instructions from bulk storage and sends them to the individual machines as the need arise. It also receives feedback data from the machines. This two way information flow occurs which means that each machine requests for instructions that must be satisfied simultaneously. Similarly, the computer must always be ready to receive information from the machines and to respond accordingly.

8.17 COMPARISON BETWEEN CONVENTIONAL MACHINES (OPERATION CONTROL) AND NC MACHINES

Figures 8.28 (*a*) and (*b*) shows procedural steps for a conventional operation controlled machines and a numerically controlled machine tool.

In a conventional machine tool, the component drawing and the initial planned/decided values of speed, feed, depth of cut etc., are given to the operator.

The operator reads the component drawing and the information in the drawing is fed into the machine tool via hand wheels, levers, switches etc. The operator monitors the cutter position during machining and could make any necessary corrections to ensure suitable output.

In a NC machine tool, the operator is replaced by the data processing part of the system and the control unit.

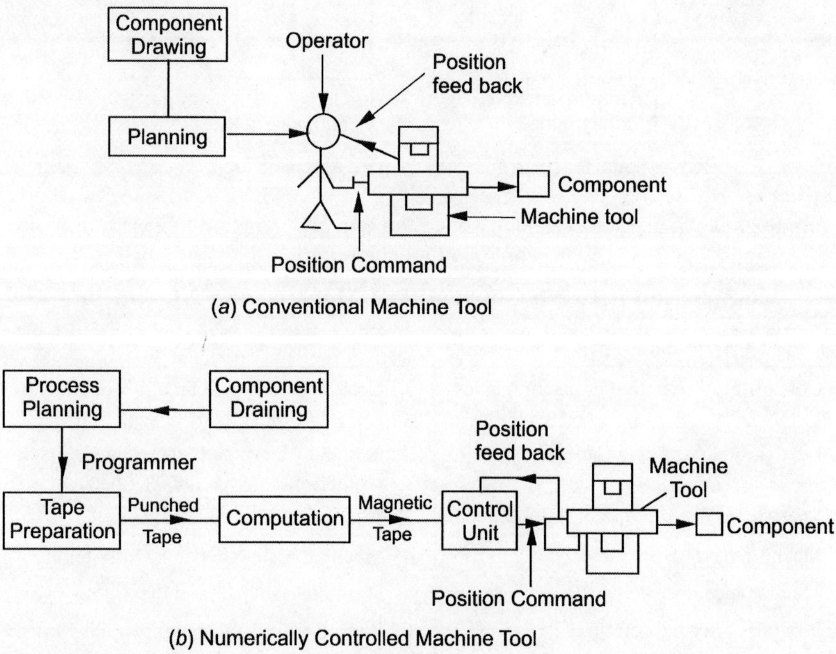

Fig. 8.28 Comparison between conventional and NC Machines

The component drawing is translated into a form acceptable to control unit. This is the data processing part of the system where the coordinate information is recorded on a punched tape or on a magnetic tape by means of a teleprinter. The tape is then fed to the control unit. The control unit is now able to send the position command signals to slide way transmission elements of machine tool, etc. The command signal is constantly compared with the actual position achieved.

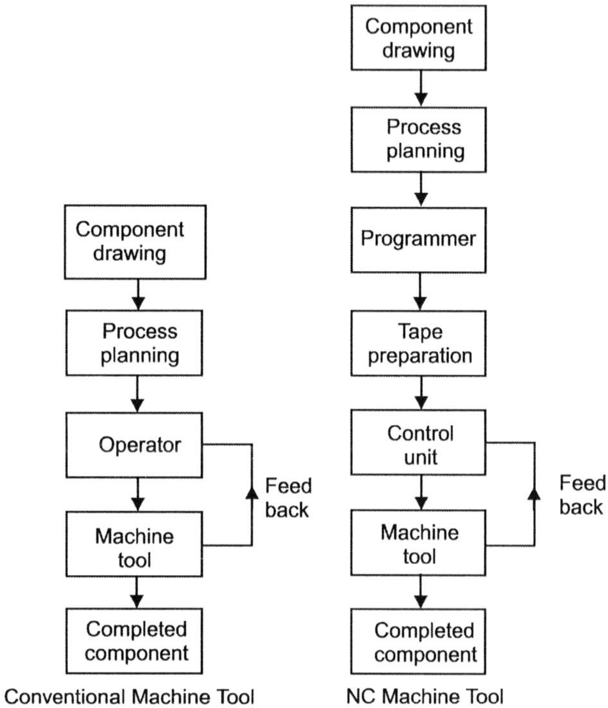

Fig. 8.29 Block diagram Working sequence of a NC machine tool vis a operator controlled machine tool

Comparison between Conventional Machine and NC Machne

Conventional Machine	NC Machine
1. Cost of the conventional machine is less.	1. Cost of the NC machine is around two to five times more than conventional machine.
2. Less skilled workers required.	2. A high order of skill and trained personnel required to operate the NC machine.
3. There is no need of air conditioning.	3. Air condition system is essential.
4. Machine have rate is normal.	4. Machine hour rate is generally higher.
5. Maintenance cost is less.	5. Maintenance cost is higher.
6. It requires normal tooling.	6. It requires special tooling which works on very high operating conditions.

Contribution Machine	NC Machine
7. Rate of production is normal.	7. Rate of production is higher and complicated profiles can be prepared with good accuracy.
8. Features are limited.	8. Automated tool changing and selection of tools according to the sequence of operations. Which reduces set up time.

HIGHLIGHTS OF THE CHAPTER

1. A collection of integrated equipment and human resources, where function is to perform one or more processing operations on raw material part called manufacturing system.

2. Types of automation.

(*i*) Human workers (*ii*) Programmable automation

(*iii*) Flexible automation

3. Components of a manufacturing system

(*i*) Human workers (*ii*) Production machine

(*iii*) Material handing system (*iv*) Computer control system

4. A programmable automation in which process is controlled by numbers, letters and sysmbols called numerical control (NC).

5. Components of NC system are :

(*i*) Programme of instruction (*ii*) A machine control unit (MCU)

(*iii*) Machine tool

6. Steps used in numerical control procedure are :

(*i*) Process planning (*ii*) Part programming

(*iii*) Tape preparation (*iv*) Tape verification

(*v*) Production

7. Ways of classification of NC machine are:

(*i*) Based on conrol system (*ii*) Based on control loop

(*iii*) Based on the number of axes (*iv*) Based on power drive or the actuation system.

8. Tool dimensioning system.

(*i*) Absolute system (*ii*) Incremental system

9. An NC system where MCU is dedicated, stored programme computers are used to perform some basic numerical control function which are stored in read - write memory of the computer called CNC.

10. A manufacturing system in which number of machines are interconnected controlled by a computer through direct connection called DNC.

11. Component of DNC.

(*i*) Control computer

(*ii*) Both memory, which stores the NC program

(*iii*) Telecommunication lines

(*iv*) Machine tools.

REVIEW QUESTIONS

1. What is a manufacturing system ?
2. What are the components of a manufacturing system?
3. Classify the manufacturing system.
4. Explain the automating and its types.
5. Write a short note on numerical control machine.
6. What are the basic components of a numerical control system?
7. What do you mean by input media?
8. What do you mean by machine control unit?
9. What is the procedure of a numerical control?
10. Classify the numerical control machine.
11. What do you mean by point to point system ?
12. What is a straight line system?
13. What do you mean by continuous from system?
14. What is the difference between open loop and closed loop system?
15. What do you mean by absolute position system?
16. What is an incremental system?
17. What do you mean by axis of option?
18. What are the applicants of NC system?
19. Explain the advantages and disadvantages of NC system.
20. Write a note on computer numerical control (CNC)?
21. Explain the components of a CNC.
22. Write a short role on Direct Numerical Control (DNC).
23. Write advantages and disadvantage of CNC.
24. Differentiate between NC and CNC.
25. Differentiatate between commercial and NC system.

■■■

Index

D

E

F

N

O

P

Q

R

■■■